AF844289

Quality Control and Production of Biological Control Agents

Theory and Testing Procedures

Quality Control and Production of Biological Control Agents

Theory and Testing Procedures

Edited by

J.C. van Lenteren

Laboratory of Entomology
Wageningen University
Wageningen
The Netherlands

CABI Publishing

CABI Publishing is a division of CAB International

CABI Publishing
CAB International
Wallingford
Oxon OX10 8DE
UK

Tel: +44 (0)1491 832111
Fax: +44 (0)1491 833508
E-mail: cabi@cabi.org
Web site: www.cabi-publishing.org

CABI Publishing
44 Brattle Street
4th Floor
Cambridge, MA 02138
USA

Tel: +1 617 395 4056
Fax: +1 617 354 6875
E-mail: cabi-nao@cabi.org

© CAB International 2003. All rights reserved. No part of this publication may be reproduced in any form or by any means, electronically, mechanically, by photocopying, recording or otherwise, without the prior permission of the copyright owners.

A catalogue record for this book is available from the British Library, London, UK.

Library of Congress Cataloging-in-Publication Data

Quality control and production of biological control agents : theory and testing procedures / edited by J.C. van Lenteren.
p. cm.
Includes bibliographical references (p.).
ISBN 0-85199-688-4
1. Biological pest control agents. 2. Biological pest control agents industry--Quality control. I. Lenteren, J. C. van.
SB975 .Q35 2003
632'.96--dc21 2002151406

ISBN 0 85199 688 4

Typeset by Columns Design Ltd, Reading, Berkshire
Printed and bound in the UK by Biddles Ltd, Guildford and King's Lynn

Contents

Contributors

D. Babendreier, Swiss Federal Research Station for Agroecology and Agriculture, Reckenholzstrasse 191, CH-8046 Zurich, Switzerland.

F. Bigler, Swiss Federal Research Station for Agroecology and Agriculture, Reckenholzstrasse 191, CH-8046 Zurich, Switzerland.

S. Bjørnson, Department of Biology, Saint Mary's University, 923 Robie Street, Halifax, Nova Scotia, Canada B3H 3C3.

K.J.F. Bolckmans, Koppert Biological Systems, PO Box 155, 2650 AD Berkel and Rodenrijs, The Netherlands. e-mail KBolckmans@koppert.nl

G. Burgio, Department of Agroenvironmental Sciences and Technologies (DISTA), University of Bologna, via F. Re 6, 40126 Bologna, Italy.

H. Cain, Bio-Bee Biological Systems Sde Eliyahu, Bet Shean Valley, 10810 Israel.

P. De Clercq, Laboratory of Agrozoology, Department of Crop Protection, Faculty of Agricultural and Applied Biological Sciences, Ghent University, Coupure Links 653, B-9000 Ghent, Belgium.

L.D. Forster, Department of Entomology, University of California, Riverside, CA 92521, USA.

C.S. Glenister, IPM Laboratories, Inc., 980 Main Street, Locke, NY 13092-0300, USA. e-mail ipmlabs@implabs.com

S. Grenier, UMR INRA/INSA de Lyon, Biologie Fonctionnelle, Insectes et Intéractions, Institut National des Sciences Appliquées, Bât. Pasteur, 20 av. A. Einstein, 69621 Villeurbanne Cedex, France. e-mail sgrenier@jouy.inra.fr

D. Grzywacz, CABI Bioscience, Silwood Park, Ascot, Berkshire SL5 7TA, UK.

A. Hale, The Bug Factory, 1636 East Island Highway, Nanoose Bay, British Columbia, Canada V9P 9A5.

R.F. Hoekstra, Laboratory of Genetics, Wageningen University, Arboretumlaan 4, 6703 BD Wageningen, The Netherlands. e-mail rolf.hoekstra@wur.nl

H.M.T. Hokkanen, Department of Applied Biology, University of Helsinki, Finland.

N.E. Jenkins, CABI Bioscience, Silwood Park, Ascot, Berkshire SL5 7TA, UK. e-mail n.jenkins@cabi.org

J.N. Klapwijk, Koppert Biological Systems, PO Box 155, 2650 AD Berkel and Rodenrijs, The Netherlands.

S. Kuske, Swiss Federal Research Station for Agroecology and Agriculture, Reckenholzstrasse 191, CH-8046 Zurich, Switzerland.

N.C. Leppla, Department of Entomology and Nematology, University of Florida, Natural Area Drive, PO Box 110630, Gainesville, FL 32611-0603, USA. e-mail ncl@gnv.ifas.ufl.edu
W.J. Lewis, Insect Biology and Population Management Research Laboratory, USDA-ARS, PO Box 748, Tifton, GA 31793, USA. e-mail wjl@tifton.cpes.peachnet.edu
A.J.M. Loomans, Laboratory of Entomology, Wageningen University, PO Box 8031, 6700 EH Wageningen, The Netherlands.
R.F. Luck, Department of Entomology, University of California, Riverside, CA 92521, USA. e-mail rluck@citrus.ucr.edu
A. Luczynski, Biobugs Consulting Ltd, 16279 30B Ave., Surrey, British Columbia, Canada V4P 2X7.
I. Menzler-Hokkanen, Department of Applied Biology, University of Helsinki, Finland.
L. Nunney, Department of Biology, University of California, Riverside, CA 92521, USA. e-mail nunney@citrus.ucr.edu
D.R. Papaj, Department of Ecology and Evolutionary Biology, University of Arizona, Tucson, AZ 85721, USA.
C. Schütte, Laboratory of Entomology, Wageningen Agricultural University, PO Box 8031, 6700 EH Wageningen, The Netherlands.
S. Steinberg, Bio-Bee Biological Systems Sde Eliyahu, Bet Shean Valley, 10810 Israel. e-mail s_stein@bio-bee.com
R. Stouthamer, Laboratory of Entomology, Wageningen University, PO Box 8031, 6700 EH Wageningen, The Netherlands. Present address: Department of Entomology, University of California at Riverside, Riverside, CA 92521, USA. e-mail richards@ucracl.ucr.edu
M.B. Thomas, CABI Bioscience, Silwood Park, Ascot, Berkshire SL5 7TA, UK.
M.G. Tommasini, CRPV (Centro Ricerche Produzioni Vegetali), Via Vicinale Monticino 1969, 47020-Diegaro di Cesena (FC), Italy. e-mail tommasini@crpv.it
J.H. Tumlinson, Insect Biology and Population Management Research Laboratory, USDA-ARS, PO Box 14565, Gainesville, FL 32604, USA.
J.C. van Lenteren, Laboratory of Entomology, Wageningen University, PO Box 8031, 6700 EH Wageningen, The Netherlands. e-mail Joop.vanLenteren@wur.nl
P.C.J. van Rijn, CABI Bioscience, Silwood Park, Ascot, Berkshire SL5 7TA, UK.
J. van Schelt, Koppert Biological Systrems, PO Box 155, 2650 AD Berkel and Rodenrijs, The Netherlands.
L.E.M. Vet, Laboratory of Entomology, Wageningen University, PO Box 8031, 6700 EH Wageningen, The Netherlands; and Netherlands Institure of Ecology, PO Box 40, 6666 ZG Heteren, The Netherlands. e-mail vet@nioo.knaw.nl
F.L. Wäckers, Laboratory of Entomology, Wageningen University, PO Box 8031, 6700 EH Wageningen, The Netherlands; and Netherlands Institute of Ecology, PO Box 40, 6666 ZG Heteren, The Netherlands. e-mail waeckers@nioo.knaw.nl
E. Wajnberg, INRA, 37 Blvd du Cap, 06600 Antibes, France. e-mail wajnberg@antibes.inra.fr

Preface

The use of biological control agents is increasing worldwide and there are now many companies mass producing and selling such organisms. However, there is a great need for quality control in the production and use of these natural enemies, because deterioration of mass-reared biological control agents leads to failures in pest management. The area of quality control is rather new for biological control workers. Therefore, the first book on this topic specifically for biological control agents contains several chapters with background information, before discussing the quality control guidelines that have recently been developed.

The first section of the book is devoted to emergence of quality control for natural enemies. In Chapter 1 the need for quality control for mass-produced biological control agents is discussed. In Chapter 2 the aspects of total quality control for the production of natural enemies are described.

The second section of the book – the basis of variability in foraging behaviour of natural enemies – comprises chapters dealing with background information on sources of variation in behaviour that are regularly encountered, but not understood and often misinterpreted in mass rearing. In Chapters 3 and 4, factors are analysed that induce the variability in searching behaviour of natural enemies, and technologies are described that illustrate how to manage this variation. Searching behaviour is influenced by the insect's genetic constitution, its physiological state and its experience. Chapter 5 presents an overview of the information on the topic of food ecology of natural enemies, and illustrates that a certain physiological state is needed before a natural enemy is able to search for hosts. These chapters make it clear that insight into behavioural variability in the foraging behaviour of natural enemies is a prerequisite for proper mass rearing and efficient application of natural enemies in pest management.

The third section focuses on how to cope with this variation. In Chapter 6 a population genetic perspective is given on how to manage captive populations. Examples of adaptation to captive rearing and of the trade-off with field performance are presented. Chapter 7 discusses the effects of a transfer of natural enemies from the field to a mass production facility, such as reduction of fitness and enhancing the possibility of fixation of deleterious mutations in the population by genetic drift. Ways to prevent these negative effects are presented. In Chapter 8 the possibilities and advantages of unisexual reproduction for biological control are discussed. Some evidence is found for two advantages of unisexual reproduction: (i) unisexuals are cheaper to produce in mass rearing than sexuals, and (ii) in classical biocontrol projects they are more easily established. In Chapter 9, mass production of natural enemies on artificial media is reviewed, particularly with regard to their quality. Chapter 10 reviews pathogens of mass-produced natural enemies and pollinators, and the effects of these pathogens on performance of the infected organisms.

The fourth section gives an overview of the species of natural enemies that are mass produced worldwide. Chapter 11 reviews the species that are commercially available. Chapter 12 discusses mass production, storage, shipment and release of natural enemies. In Chapter 13 the currently highly relevant topic of risk assessment of exotic natural enemies is addressed.

The fifth section contains chapters that decribe developments towards quality control testing of natural enemies. Chapter 14 gives an overview of developments in North America, and Chapter 15 reviews the European situation. In Chapter 16 an addition to the currently used laboratory quality control tests is described. Chapter 17 discusses quality in the context of a biological control agent's reproductive success in terms of the offsprings' characteristics that allow them to maximize their reproduction in the field on the targeted pest.

The sixth and final section deals with actual quality control tests. Chapter 18 illustrates how quality control of fungal and viral biological control agents can be standardized. Chapter 19 provides a description of the guidelines that are currently used for quality control of commercially produced natural enemies, and discusses future improvements of these guidelines. Chapter 20 presents basic statistical methods for analysis of the data obtained with the quality control tests of the previous chapter.

The quality control guidelines described in this book will certainly undergo modifications in the coming years. First, I expect that simple tests will be included to determine the flight capacity of mass-reared biocontrol agents. Next, semi-field and field performance tests will be developed. Finally, based on extensive testing by the mass production industry and comparison of results of the current tests with those of the new flight and performance tests, a new set of criteria will likely evolve.

J.C. van Lenteren, October 2002, Perugia, Italy
Laboratory of Entomology, Wageningen University, The Netherlands

Acknowledgements

First of all, I would like to thank all participants in the EC programme 'Designing and implementing quality control of beneficial insects: towards more reliable biological pest control'. It was very satisfying to see the initially difficult contacts between academia and industry develop into real collaboration, and this book is the result of that collaboration.

Next, I thank the Entomology Section, Department of Arboriculture and Plant Protection of the University of Perugia (Italy) for providing space, library facilities and an intellectually attractive atmosphere during sabbaticals in 2001 and 2002 to work on this book. Particularly I thank Prof.dr. Ferdinando Bin for his hospitality. Prof. Bin is also thanked for allowing me to use the scanning electron micrograph picture of *Trissolcus basalis* for the cover of this book.

Further, I thank Franz Bigler and Norm Leppla of the global IOBC (International Organization for Biological Control of Noxious Animals and Weeds) working group 'Quality Control of Mass Reared Arthropods' for helping to develop the initial framework for *Quality Control and Production of Biological Control Agents*. All authors are thanked for having been very cooperative in handing in their manuscript in on time. CAB International, the *Journal of Insect Behavior* (Kluwer Academic/Plenum Publishers) and the journal *Environmental Entomology* (Entomological Society of America) are thanked for granting permission to use earlier published material. Johannes Steidle and Joop van Loon are thanked for allowing me to use their excellent, and a currently unpublished review paper for updating Chapters 3 and 4. The following persons are thanked for reviewing (parts of) chapters: R. Albajes, F. Bigler, F. Bin, K. Bolckmans, E. Conti, H.M.T. Hokkanen, J. Klapwijk, N. Leppla, A.J.M. Loomans, J.J.A. van Loon, R.F. Luck, R. Romani, G. Salerno, R.F. Luck, W. Ravensberg, B. Roitberg, P.C.J. van Rijn, J.L.M. Steidle, M.B. Thomas, M.G. Tommasini, A. van Lenteren and L.E.M. Vet. Wilma Twigt assisted in the compilation of reference lists. The Foundation for Integrative Agriculture funded the writing of this book. Lastly, I thank Tim Hardwick, Claire Gwilt, Rachel Robinson and Elaine Coverdale at CAB International for efficiently taking care of all matters related to production of this book.

The studies described in several chapters of this book have been carried out with financial support from the Commission of the European Communities, Agriculture and Fisheries RTD programme CT93-1076 'Designing and implementing quality control of beneficial insects: towards more reliable biological pest control', and RTD programme CT97-3489 'Evaluating environmental risks of biological control introductions into Europe'. Ideas expressed in this book do not necessarily reflect the views of the commission and in no way anticipates the commission's future policy in this area.

1 Need for Quality Control of Mass-produced Biological Control Agents

J.C. van Lenteren
Laboratory of Entomology, Wageningen University, PO Box 8031, 6700 EH Wageningen, The Netherlands

Abstract

Mass-rearing of natural enemies often takes place in small companies with little know-how and understanding of conditions influencing performance, which may result in natural enemies of bad quality and failures with biological control. This makes robust quality control programmes a necessity. Background information is presented on the activity of mass-producing natural enemies, the emergence of the development of quality control worldwide is sketched, basic considerations for quality control are outlined and difficulties encountered when developing quality control are discussed.

Introduction

Augmentative biological control, where large numbers of natural enemies are periodically introduced, is commercially applied on a large area in various cropping systems worldwide (van Lenteren, 2000a; van Lenteren and Bueno, 2002). It is a popular control method applied by professional and progressive farmers and stimulated by the present international attitudes in policies of reducing pesticide use. Initially, augmentative biological control was used to manage pests that had become resistant to pesticides. Now it is applied because of efficacy and costs, which are comparable with conventional chemical control. Farmers are also motivated to use biological control to reduce environmental effects caused by pesticide usage.

Worldwide, more than 125 species of natural enemies are commercially available for augmentative biological control (Anon., 2000; Gurr and Wratten, 2000). This form of control is applied in the open field in crops that are attacked by only a few pest species, and it is particularly popular in greenhouse crops, where the whole spectrum of pests can be managed by different natural enemies (van Lenteren, 2000b). Its popularity can be explained by a number of important benefits when compared with chemical control: there are no phytotoxic effects on young plants, premature abortion of fruit and flowers does not occur, release of natural enemies takes less time and is more pleasant than applying pesticides, several key pests can be controlled only with natural enemies, there is no safety or re-entry period after release of natural enemies, which allows continuous harvesting without danger to the health of personnel, biological control is permanent and the general public appreciates biological control.

© CAB International 2003. *Quality Control and Production of Biological Control Agents: Theory and Testing Procedures* (ed. J.C. van Lenteren)

Two forms of periodic releases with natural enemies are generally distinguished: the inundative and the seasonal inoculative method. The inundative-release method is where beneficial organisms are collected, mass-reared and periodically released in large numbers to obtain immediate control of a pest (i.e. use as a biotic insecticide). Pest control is mainly obtained from the released natural enemies and not from their offspring. Inundative releases are applied to crops where viable breeding populations of the natural enemy are not possible, in crops where the damage threshold is very low and rapid control is required at very early stages of infestation or in crops where only one generation of the pest insects occurs. An example is the use of *Trichogramma* spp. against the cornborer in maize in Europe (Bigler, 1994). The seasonal inoculative-release method is where natural enemies are collected, mass-reared and periodically released into short-term crops (6–12 months) and where many pest generations occur. A relatively large number of natural enemies is released to obtain both immediate control and a build-up of the natural-enemy population for control throughout the same growing season. This method can be applied when the growing method of a crop prevents control extending over many years – for example, in greenhouses where the crop together with the pests and natural enemies are removed at the end of the growing season. The method is distinctly different from the inundative method and more closely resembles the inoculative or classical biocontrol method because control is obtained for a number of generations of the pest and control would be permanent if the crop were grown for a much longer period. The seasonal inoculative-release method has been developed in Europe during the last three decades and is applied with great commercial success in greenhouses. Two well-known natural enemies used in this approach are the spider-mite predator *Phytoseiulus persimilis* and the whitefly parasitoid *Encarsia formosa* (van Lenteren, 1995).

Augmentative biological control is applied worldwide. Data about the current use of augmentation are sometimes very hard to obtain (e.g. for Russia) and estimates are therefore incomplete. A worldwide review from 1977 (Ridgway and Vinson, 1977) provides data about the use of natural enemies in the USSR (on 10 million ha), China (1 million ha), West Europe (< 30,000 ha) and North America (< 15,000 ha). Since that review, many new natural enemies have become available (Anon., 2000) and activities have strongly increased in Latin America (van Lenteren and Bueno, 2002). The best-known examples of augmentative biological control are those: (i) where the egg parasitoid *Trichogramma* is used for control of Lepidoptera in various crops (Smith, 1996); and (ii) where a whole set of different natural enemies (parasitoids, pathogens and predators) is used to manage pests in greenhouses (Albajes *et al.*, 1999). The total world area under augmentative biological control was recently estimated to be about 16 million ha (van Lenteren, 2000a).

For a long time, natural enemies were produced without proper quality control procedures. Poorly performing natural enemies resulted in failures of biological control and a low profile of this pest-control method (e.g. P. DeBach, Riverside, California, 1976, and P. Koppert, Berkel and Rodenrijs, The Netherlands, 1980, personal communications). Quality control was touched upon by several biological control workers in the 20th century, but the first papers seriously addressing the problem appeared only in the 1980s (van Lenteren, 1986a).

Emergence of Quality Control

Trends in commercial mass production of natural enemies

The appearance and disappearance of natural-enemy producers have characterized commercialization of natural enemies over the past 30 years. Only a few producers active in the 1970s are still in business today. In addition to many small insectaries producing at the 'cottage-industry' level, three large facilities (i.e. having more than 50 persons employed) exist that provide material of good quality. At these three production sites, more than 5–10 million individuals per species per week are produced (van Lenteren and Woets, 1988; van Lenteren and Tommasini, 1999), and these facilities provide the full spectrum of natural enemies needed for an entire integrated pest management (IPM) programme in a specific commodity (Albajes *et al.*, 1999). As the sale of biological control agents is still an emerging market that is influenced by small competing companies, product quality and prices are continuously affected by competitive pressure. While such pressure may in the short term be profitable for growers due to lower costs of natural enemies, in the long run such price competition could lead to biological control failures. Natural enemies were properly evaluated before commercial use some 20 years ago, but nowadays some species of natural enemies are sold without tests under practical cropping situations that show that the natural enemies are effective against the target pest (van Lenteren and Manzaroli, 1999). Lack of stability at the producer's level has resulted in the sale and use of natural enemies of poor quality or with inadequate guidance. These problems have in some cases resulted in failure of biological control and have influenced the development of IPM in a very negative way.

Natural-enemy producers are a rather diverse group. Rearing of natural enemies can be a full-time business or a part-time activity of growers. But natural enemies may also be reared by companies in associated industries, such as seed companies or producers of fertilizers. In some cases, production of natural enemies has been started by a research group with governmental support and later continued as a private endeavour. The number of biological control agents that are commercially available has increased dramatically over the past 25 years (Fig. 1.1; see also Chapter 11). Today, more than 125 natural-enemy species are on the market for biological pest control, and about 30 of these are produced in commercial insectaries in very large quantities (Table 1.1). Worldwide, there are about 85 commercial producers of natural enemies for augmentative forms of biological control: 25 in Europe, 20 in North America, six in Australia and New Zealand, five in South Africa, about 15 in Asia (Japan, Korea, India, etc.) and about 15 in Latin America. The worldwide turnover of natural enemies of all producers was estimated to be US$25 million in 1997, and about US$50 million in 2000, with an annual growth of 15–20% in the coming years (K. Bolckmans, Berkel and Rodenrijs, The Netherlands, 2001,

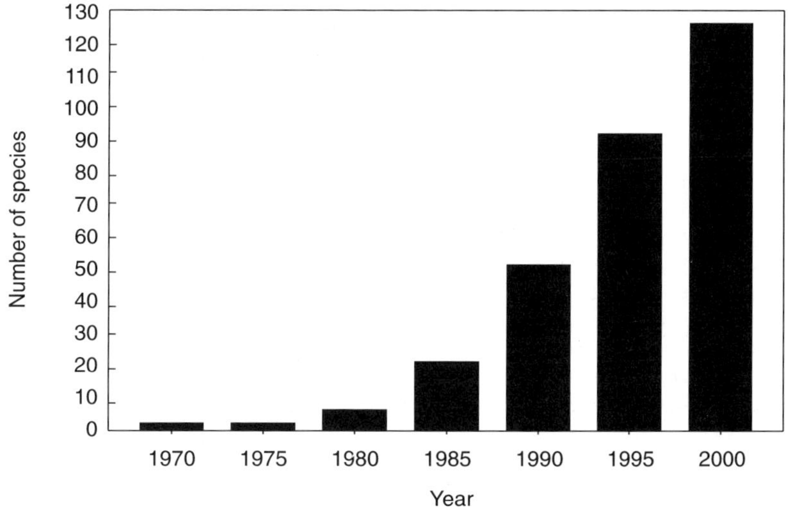

Fig. 1.1. Number of species of natural enemies commercially available for biological control.

Table 1.1. Major species of biological control agents commercially available for pest control.

Biological control agent	Pest species
Amblyseius (*Neoseiulus*) *degenerans* Berlese	*Frankliniella occidentalis* (Pergande) *Thrips tabaci* Lindeman
Aphelinus abdominalis Dalman	*Macrosiphum euphorbiae* (Thomas) *Auleurocorthum solani* Kaltenbach
Aphidius colemani Viereck	*Aphis gossypii* Glover *Myzus persicae* Sulzer
Aphidius ervi Halliday	*Macrosiphum euphorbiae*
Aphidoletes aphidimyza Rondani	Aphids
Chrysoperla carnea (Stephens)	Aphids
Cryptoleamus montrouzieri Mulsant	Pseudococcidae, Coccidae
Dacnusa sibirica Telenga	*Liriomyza bryoniae* (Kaltenbach) *Liriomyza trifolii* (Burgess) *Liriomyza huidobrensis* (Blanchard)
Delphastus pusillus (LeConte)	Whiteflies
Diglyphus isaea Walker	*Liriomyza bryoniae* *Liriomyza trifolii* *Liriomyza huidobrensis*
Encarsia formosa Gahan	*Trialeurodes vaporariorum* (Westwood) *Bemisia* spp.
Eretmocerus eremicus Rose & Zolnerowich (formerly *E. californicus*)	*Bemisia* spp.
Eretmocerus mundus Mercet	*Bemisia* spp.
Harmonia axyridis (Pallas)	Aphids
Heterorhabditis megidis Poinar	*Otiorhynchus sulcatus* (F.)
Hippodamia convergens Guerin-Meneville	Aphids
Hypoaspis aculeifer (Canestrini)	*Rhizoglyphus echinopus* Fumouzze and Robin, Sciaridae
Hypoaspis miles (Berlese)	*Rhizoglyphus echinopus*, Sciaridae
Leptomastidea abnormis Girault	Pseudococcidae
Leptomastix dactylopii (Howard)	*Planococcus citri* (Risso)
Leptomastix epona (Walker)	Pseudococcidae
Lysiphlebus testaceipes (Cresson)	*Aphis gossypii*
Macrolophus caliginosus Wagner	Whiteflies
Neoseiulus californicus (McGregor)	*Tetranychus urticae* Koch
Neoseiulus cucumeris (Oudemans)	*Frankliniella occidentalis* *Thrips tabaci*
Opius pallipes Wesmael	*Liriomyza bryoniae*
Orius insidiosus Say	Thrips
Orius laevigatus Fieber	Thrips
Orius majusculus Reuter	Thrips
Phytoseiulus persimilis Athias-Henriot	*Tetranychus urticae*
Steinernema feltiae (Filipjev)	Sciaridae and two other spp.
Trichogramma evanescens Westwood	Lepidoptera
Verticillium lecanii (A. Zimmerman) Viégas	Whiteflies/aphids

personal communication). Currently, more than 75% of all activities in commercial augmentative biocontrol (expressed in monetary value) take place in northern Europe and North America. Emerging markets are those of Latin America, South Africa, Mediterranean Europe and Japan and Korea in Asia. In addition to the commercial producers, there are many natural-enemy production units funded by the government, such as in Brazil (40 facilities), China (many, number unknown), Colombia (more than 20 facilities), Cuba (more than 200 facilities), Mexico (30 facilities) and Peru (more than 20 facilities). For prices of natural enemies in Europe and the USA, see van Lenteren *et al.* (1997) and Cranshaw *et al.* (1996), respectively.

Commercial natural-enemy producers rear mainly predators and parasitoids (see Table 1.1). Only a few companies produce microbial agents, such as nematodes, entomopathogenic fungi, bacteria or viruses. Chemical companies are the main producers of microbial agents and it is expected that all activities in this area will in the future be exclusively the domain of the pesticide industry. Mass-rearing methods for parasitoids and predators are usually developed on an ad hoc basis, an approach that may result in natural enemies of poor quality. The technology for rearing natural enemies on 'unnatural' hosts and host plants or on artificial diets is not yet well developed (see Chapter 9) and seems to be hampered not only by physiological problems but also by ethological and ecological ones (requirements for associative learning of host-habitat and host-finding cues (see Chapters 3 and 4)). Conflicts between attributes favoured in mass-rearing programmes and those needed for field performance form another obstacle for the cost-effective production of natural enemies. Artificial selection that occurs during mass rearing may lead to reduced performance of natural enemies (see below, and Chapters 6 and 7). The suggested cures for this problem are often expensive and time-consuming and are therefore very seldomly applied.

Professional natural-enemy producers may have research facilities, procedures for monitoring product quality, an international distribution network, promotional activities and an advisory service. The market for high-quality, effective natural enemies will certainly increase with the growing demand for unsprayed food and a cleaner environment. The growing pesticide-resistance problems will also move growers to adopt biological control methods.

Initial developments in the area of mass production, quality control, storage, shipment and release of natural enemies (Chapter 12) have decreased production costs and led to better product quality, but much more can be done. Innovations in long-term storage (e.g. through induction of diapause), shipment and release methods may lead to a further increase in natural-enemy quality, with a concurrent reduction in costs, thereby making biological control easier and economically more attractive to apply. Even if the natural enemies leave the insectary in good condition, shipment and handling by the producers, distributors and growers may result in deterioration of the biological control agents before they are released.

Quality control programmes that address not only natural-enemy numbers but also natural-enemy quality (field performance) are a necessity. Simple and reliable quality control programmes for natural enemies are now emerging as a result of intensive cooperation between researchers and the biological control practitioners, and it is expected that these developments will result in a rapid improvement of the biological control industry.

The International Organization for Biological Control/European Community (IOBC/EC) initiative on quality control

Although augmentative types of biological control of arthropod pests have been applied since 1926, large-scale production of natural enemies began only after the Second World War (DeBach, 1964; van Lenteren and Woets, 1988). Initial mass-rearing efforts involved the production of not more than several thousand individuals per week of three natural enemies: the spider-mite predator *P. persimilis,* the whitefly parasitoid *E. formosa* and the lepidopteran egg parasitoid *Trichogramma* sp. None of the early publications on commercial aspects of biological control mention the topic of quality control of natural enemies (e.g. Hussey and Bravenboer, 1971). Quality control is mentioned in relation to biological control only in the mid-1980s, and shortly after that the topic gained more interest (van Lenteren, 1986a,b). The Fifth Workshop of the IOBC Global Working Group, 'Quality Control of

Mass-Reared Arthropods' (Bigler, 1991), in Wageningen, The Netherlands, formed the starting-point for a heated discussion among producers of natural enemies and scientists on how to approach quality control in the commercial setting at that time.

A series of workshops, some partly and others largely funded by the EC, followed in Horsholm, Denmark, in 1992, Rimini, Italy, in 1993 (Nicoli *et al.*, 1993; van Lenteren *et al.*, 1993), Evora, Portugal, in 1994 (van Lenteren, 1994), Antibes, France, in 1996 (van Lenteren, 1996) and in Barcelona, Spain, in 1997 (van Lenteren, 1998). As a result of these meetings, quality control guidelines were written for more than 20 species of natural enemies, and these have been tested and adopted by commercial producers of biological control agents in Europe (van Lenteren, 1998; van Lenteren and Tommasini, 1999). The guidelines cover features that are relatively easy to determine in the laboratory (e.g. emergence, sex ratio, lifespan, fecundity, adult size, predation/parasitism rate). Work is now focused on the development of: (i) flight tests; and (ii) a test relating these laboratory characteristics to field efficiency.

Recently, the International Biocontrol Manufacturers Association (IBMA) has taken the initiative to update and further develop quality control guidelines and fact sheets. Their first meeting, with the participation of the most important European mass producers of natural enemies and representatives of mass producers from Canada and the USA under the umbrella of the Association of Natural Bio-control Producers (ANBP), took place in September 2000 in The Netherlands and was followed up by a meeting in North America in 2001. The quality control guidelines for more than 30 species of natural enemies developed so far are presented in Chapter 19.

State of affairs concerning application of quality control worldwide

Currently, quality control guidelines as presented in Chapter 19 are applied by several companies that mass-produce natural enemies in Europe and North America. Depending on the size of the company and the number of natural-enemy species they produce, they may apply from one to more than 20 tests. Through correspondence and literature search, the following information was obtained for other countries.

In the former Soviet Union, quite a lot of work was done during the 1980s on quality control of *Trichogramma*, a parasitoid that was used on several million hectares for control of various lepidopteran pests. References to this work, as well as examples of USSR quality control programmes, can be found in a Russian paper in the Proceedings of the First International Symposium on *Trichogramma* and other egg parasitoids (Voegele, 1982), in three papers authored by Russian researchers in the Proceedings of the Second International Symposium on *Trichogramma* and other egg parasites (Voegele *et al.*, 1988) and several papers published in later proceedings of this working group (two papers in Wajnberg and Vinson (1991), third symposium; five papers in Wajnberg (1995), fourth symposium). Most of the elements of quality control discussed in these papers are included in the current quality-control guidelines described in Chapter 19 of this book, with the exception of an interesting test to evaluate searching and dispersal ability in a maze in the laboratory, developed by Greenberg (1991). This test was later used by Silva *et al.* (2000) to measure the performance of *Trichogramma* in the laboratory and to predict its dispersal capacity in the field. Disappointingly, it appeared that the laboratory bioassay with the maze did not properly predict the dispersal capacity of *Trichogramma*.

Information on quality control of mass-produced natural enemies used in China is not easy to trace, although inundative and seasonal inoculative forms of biological control are used on about 1 million ha. Aspects of quality control are described in two Chinese papers in the Proceedings of the First International Symposium on *Trichogramma* and other egg parasitoids (Voegele, 1982), in about ten papers authored by Chinese researchers in the Proceedings of the Second International

Symposium on *Trichogramma* and other egg parasites (Voegele *et al.*, 1988), in five papers by Chinese in Wajnberg and Vinson (1991) (third symposium) and in four papers by Chinese in Wajnberg (1995) (fourth symposium). Details are not described here because very few papers specifically address quality control and most of the useful components of the Chinese quality-control studies are included in the present guidelines for *Trichogramma* and other egg parasitoids given in Chapter 19. An exception is a simple quality control method that I saw demonstrated in one of the *Trichogramma* mass-production units in the Biocontrol Station of Shun-de County, near the town of Ghuanzhou, Province of Guangdong, China. Parasitoids were reared on silkworm eggs, adult parasitoids were allowed to emerge on the dark side of the room and fresh host eggs were offered on the light side of the room near a window about 3 m away from the dark side, so the freshly emerged parasitoids had to fly several metres before they could parasitize hosts. In this way, non-flying parasitoids were prevented from reproducing (J.C. van Lenteren, Guangdong, China, November 1986, personal observation).

Australian producers are applying one full quality control guideline – the one for *Aphytis* as specified in Chapter 19 – and are using elements of the other IOBC/EC guidelines described in Chapter 19. There are no Australian publications on quality control. A set of guidelines for natural enemies that are specifically applied in Australia is in development. Genetic diversity and rejuvenation of laboratory material with field-collected natural enemies form a specific point of interest of Australian producers (all information from D. Papacek, Australia, April 2001, personal communication). In New Zealand, elements of the IOBC/EC guidelines are used for quality control of about five species of natural enemies, and critical-point standards for quality checks during the production process are in development; there are no publications from New Zealand on quality control (R. Rountree, New Zealand, April 2001, personal communication). In Japan, elements of the IOBC/EC guidelines are used for quality control of several species of natural enemies that are imported from Europe or produced in Japan; there are no Japanese publications on quality control (E. Yano, Japan, April 2001, personal communication). Elements of quality control are applied in India to evaluate the quality of mass-reared *Trichogramma* (Kaushik and Arora, 1998; Swamiappan *et al.*, 1998).

The Insectary Society of Southern Africa is actively developing a set of minimum quality control standards for insects commercially for sale as biocontrol agents and other purposes, developments are discussed in biennial insect-rearing workshops and progress is reported in the proceedings of these workshops (see, for example, Conlong, 1995) (D. Conlong, South Africa, April 2001, personal communication). In several other African countries, such as Benin, Kenya, Nigeria, Sudan and Zambia, quality control is applied (Conlong, 1995; Conlong and Mugoya, 1996; van Lenteren, Africa, 1983–2001, personal observation), but it is not easy to trace published material providing detail about the methodology, with the exception of work done at the International Institute for Tropical Agriculture (IITA) (e.g. Yaninek and Herren, 1989).

The situation concerning quality control in Latin America is even less clear than in other areas of the world. Recently, two rather detailed papers appeared on quality control of a tachinid parasitoid (Aleman *et al.*, 1998) and predatory mites (Ramos *et al.*, 1998), as performed in Cuba. Also, a book edited by Bueno (2000) provides examples of quality control for microbials, predatory mites and predatory and parasitic insects in Brazil, but few details about methodology are provided. Based on the vast areas under augmentative biological control in Latin America (van Lenteren and Bueno, 2003), I suppose that there is much more done on quality control than could be traced in the literature.

The Objectives of Quality Control

Quality control programmes are applied to mass-reared organisms to maintain the quality of the population. The overall quality of an organism can be defined as the ability to function as intended after release into the field. The aim of quality control programmes is to check whether the overall quality of a species is maintained, but that is too general a statement to be manageable. Characteristics that affect overall quality have to be identified. These characteristics must be quantifiable and relevant for the field performance of the parasitoid or predator. This is a straightforward statement, but very difficult to actually put into practice (Bigler, 1989).

Rather than discussing the development of quality control in strictly scientific terms, this discussion will outline a more pragmatic approach. The aim of releases of mass-produced natural enemies is to control a pest. In this context, the aim of quality control should be to determine whether a natural enemy is still in a condition to properly control the pest. Formulated in this way, we do not need to consider terms like maximal or optimal quality, but rather acceptable quality. Some researchers believe the aim of quality control should be to keep the quality of the mass-reared population identical to that of the original field population. Not only is this an illusion (see Chapters 6 and 7), but it is also an unnecessary and expensive goal to pursue. Another important consideration is that quality control programmes are not applied for the sake of the scientist, but as a necessity. Leppla and Fisher (1989) formulated this dilemma as: 'Information is expensive, so it is important to separate "need to know" from "nice to know".' Only if characteristics to be measured are very limited in number, but directly linked to field performance, will companies producing natural enemies ever be able to apply quality control programmes on a regular basis.

Basic Considerations for Quality Control

Genetic changes in laboratory colonies

The problem of quality control of beneficial insects can be approached from two sides:

1. Measure how well the biological control agent functions in its intended role. If it does not function well enough, trace the cause and improve the rearing method.
2. List what changes we can expect when a mass rearing is started; measure these and, if the changes are undesirable, improve the rearing method.

The disadvantage of the first method is that changes may have occurred that cannot be corrected because the material has already changed so much that the original causes of the observed effects cannot be identified. The disadvantage of the second method is that too many measurements may be needed. The second approach has the advantage that potential problems are foreseeable and corrections can be made in time. Bartlett (1984a), for example, approaches the problem from the second viewpoint. He states that many authors have suggested remedial measures for assumed genetic deterioration, but that causes for deterioration are not easily identified. Identification demands detailed genetic studies, and it is often difficult to define and measure detrimental genetic traits. Bartlett (1984a) concludes:

> I believe an unappreciated element of this problem is that the genetic changes taking place when an insect colony is started are natural ones that occur whenever any biological organism goes from one environment to another. These processes have been very well studied as evolutionary events and involve such concepts as colonisation, selection, genetic drift, effective population numbers, migration, genetic revolutions, and domestication theory.

In two other articles, Bartlett (1984b, 1985) discusses what happens to genetic variability in the process of domestication, what factors might change variability and which ones might be expected to have little or no effect. In laboratory domestication, those insects are selected that have suitable genotypes to survive in this new environment, a process called winnowing by Spurway (1955) or, less appropriately but widely used, 'forcing insects through a bottleneck' (e.g. Boller, 1979). The changes that a field population

Table 1.2. Factors influencing changes in field populations after introduction into the laboratory.

1. Laboratory populations are kept at constant environments with stable abiotic factors (light, temperature, wind, humidity) and constant biotic factors (food, no predation or parasitism). There is no selection to overcome unexpected stresses. The result is a change of the criteria that determine fitness and a modification of the whole genetic system (Lerner, 1958)
2. There is no interspecific competition in laboratory populations, resulting in a possible change in genetic variability (Lerner, 1958)
3. Laboratory conditions are made suitable for the average, sometimes even for the poorest, genotype. No choice of environment is possible as all individuals are confined to the same environment. The result is a possible decrease in genetic variability (Lerner, 1958)
4. Density-dependent behaviours (e.g. searching efficiency) may be affected in laboratory situations (Bartlett, 1984b)
5. Mate-selection processes may be changed because unmated or previously mated females will have restricted means of escape (Bartlett, 1984b)
6. Dispersal characteristics, specifically adult flight behaviour and larval dispersal, may be severely restricted by laboratory conditions (Bush *et al.*, 1976)

may undergo when introduced into the laboratory are given in Table 1.2.

Variability in performance traits is usually abundantly present in natural populations (Prakash, 1973) and can remain large even in inbred populations (Yamazaki, 1972). But differences between field and laboratory environments will result in differences in variability. When natural-enemy cultures are started, part of the 'open population' from the field, where gene migration can occur and environmental diversity is large, is brought into the laboratory and becomes a 'closed population'. Thereafter, all future genetic changes act on the limited genetic variation present in the original founders (Bartlett, 1984b, 1985; Chapters 6 and 7). The size of the founder population will directly affect how much variation will be retained from the native gene pool. Although there is no agreement on the size of founder populations needed for starting a mass production, a minimum number of 1000 individuals is suggested (Bartlett, 1985). Founder populations for commercial cultures of a number of natural enemies were, however, much smaller, sometimes fewer than 20 individuals (for examples, see van Lenteren and Woets, 1988). Fitness characteristics appropriate for the field environment will be different to those for the laboratory. These environments will place different values on the ability to diapause or to locate hosts/prey or mates. Such laboratory selection forces may produce a genetic revolution (Mayr, 1970) and new, balanced gene systems will be selected for (Lopez-Fanjul and Hill, 1973).

One of the methods often suggested to correct for genetic revolutions is the regular introduction of wild individuals from the field. But, if the rearing conditions remain the same in the laboratory, the introduced wild individuals will be subjected to the same process of genetic selection. Furthermore, if a genetic differentiation has developed between laboratory and field populations this may lead to genetic isolation (Oliver, 1972) and usually the laboratory-selected population will take over. Also, positive correlations have been found between the incompatibility of such races and the differences between the environments (laboratory, field) where the races occur (e.g. Jaenson, 1978; Jansson, 1978), and for the length of time that the two populations have been isolated. Given these processes, introduction of native individuals to mass-rearing colonies is likely to be useless if incompatibility between field and laboratory populations is complete. If one wants to introduce wild genes, it should be done regularly and from the start of a laboratory rearing onwards. It should not be delayed until problems occur. Introducing field-collected insects into mass rearing also poses risks of introduction of parasitoids, predators or pathogens into the colony (Bartlett, 1984b).

Another effect of laboratory colonization can be inbreeding – mating of relatives and production of progeny that are more genetically homozygous than when random mating occurs in large populations. Genetically homozygous individuals often expose harmful traits. The degree of inbreeding is directly related to the size of the founder population. Because artificial selection in the laboratory often results in an initial decrease in population size, the rate of inbreeding increases. The result is often a definite and rapid effect on the genetic composition of the laboratory population (Bartlett, 1984b). Inbreeding can be prevented by various methods that maintain genetic variability (Joslyn, 1984), including the following:

1. Precolonization methods: selection and pooling of founder insects from throughout the range of the species to provide a wide representation of the gene pool, resulting in a greater fitness of the laboratory material.
2. Postcolonization methods:
   a. Variable laboratory environments (variation over time and space). Although the concept of varying laboratory conditions is simple, putting it into practice is difficult. Consider for example the investments for rearing facilities with varying temperatures, humidities and light regimes, or the creation of possibilities to choose from various diets or hosts, or the provision of space for dispersal, etc.
   b. Gene infusion: the regular rejuvenation of the gene pool with wild insects.

A fundamental question concerning inbreeding is: how large must the population size be to keep genetic variation sufficiently large? Joslyn (1984) says that, to maintain sufficient heterogeneity, a colony should not decline below the number of founder insects. The larger the colony, the better. Very few data are available about effective population size; Joslyn (1984) mentions a minimum number of 500 individuals.

The above discussion suggests several criteria to be considered before a mass-rearing colony is started (Table 1.3, after Bartlett, 1984b).

A broader approach to quality control

Chambers and Ashley (1984), Leppla and Fisher (1989) and Leppla (Chapter 2) put quality control in a much wider perspective. These papers are food for thought for all engaged in mass production of beneficial arthropods. They present some refreshing and, for most entomologists, new ideas. These authors approach quality control from the industrial side and consider three elements as essential: product control, process control and production control. Product control rejects faulty products and production

Table 1.3. Criteria to be considered before starting a mass-rearing programme.

1. The effective number of parents at the start of a mass rearing is much lower than the number of founder individuals, so start with a large population
2. Compensate for density-dependent phenomena
3. Create a proper balance of competition, but avoid overcrowding
4. Set environmental conditions for the best, not the worst or average, genotype; use fluctuating abiotic conditions
5. Maintain separate laboratory strains and cross them systematically to increase F_1 variability
6. Measure frequencies of biochemical and morphological markers in founder populations and monitor changes
7. Develop morphological and biochemical genetic markers for population studies
8. Determine the standards that apply to the intended use of the insects, and then adapt rearing procedures to maximize those values in the domesticated strain

Table 1.4. Obstacles in mass rearing of natural enemies.

1. Production of good-quality natural enemies at low costs may be difficult (Beirne, 1974; Chapters 11 and 12)
2. Artificial diets are often not available for natural enemies (Beirne, 1974; Chapter 9)
3. Techniques that prevent selection pressures leading to genetic deterioration are usually lacking (Mackauer, 1972, 1976; Chapters 6 and 7)
4. Cannibalism by predators or superparasitism by parasitoids generally occurs (Chapter 9)
5. Rearing on unnatural hosts/prey or under unnatural conditions may cause behavioural changes in preimaginal and imaginal conditioning (Morrison and King, 1977; Vet *et al.*, 1990; Chapters 3, 4 and 9)
6. Reduced vigour can occur when natural enemies are reared on unnatural hosts (Morrison and King, 1977; Chapter 9)
7. Reduced vigour can also be the result when natural enemies are reared on hosts that are reared on an unnatural host diet (Morrison and King, 1977; Chapter 9)
8. Contamination of the rearing by pathogens may occur (Bartlett, 1984b; Chapter 10)

control maintains consistency of production output. Process control tells how the manufacturing processes are performing. These elements of quality control are seldom applied to arthropod mass-rearing programmes.

Mass rearing, usually done by small private companies, is developed by trial and error. Knowledge of mass-rearing techniques is often limited in such organizations and the time or money for extensive experimentation is lacking. If success is to be obtained, quality control of the end-product is essential, but producers are generally more than happy if they can meet deadlines for providing certain numbers of natural enemies. Although most experts on quality control have adopted tools and procedures needed to regulate the processes of arthropod production so that product quality can be assured (Chambers and Ashley, 1984), such tools and procedures are not yet widely used by the many small companies that compose 95% of all producers. The main reason most of the small companies do not develop and use such product, process and production controls is that they lack the extra financial resources that are required. This limitation can be a serious constraint for starting producers.

Quality control seems to be developed best when mass rearing is done in large governmentally supported units. Chambers and Ashley (1984) state that entomologists often concentrate too much on production control, while they are at best only partially controlling production processes and products. Quality control is frequently, but wrongly, seen as an alarm and inspection system that oversees and intimidates production personnel.

Difficulties Encountered When Developing Quality Control

Obstacles in mass rearing of arthropods

Artificial selection forces in mass rearing may lead to problems related to performance of natural enemies in the field if rearing conditions differ strongly from the situation in which natural enemies are to be released (Table 1.4). For example, if temperature in the mass-rearing facility differs considerably from the field situation, synchronization problems between natural enemy and pest insect can be expected. Also, rearing on non-target hosts or host plants (Chapter 9) can create problems with natural-enemy quality or recognition by natural enemies of essential semiochemicals.

Any of the obstacles mentioned in Table 1.4 may be encountered in mass-production programmes. One of the main obstacles to economic success seems to be the difficulty to produce qualitatively good natural enemies at a low price. But, with a strongly decreasing number of pesticides available, with increasing costs per unit of volume for chemical pesticides and implementation of pesticide levies, as is currently taking place in several countries, the aspect of relatively high costs of natural enemies will disappear. Also, effective techniques to mass-produce natural enemies on artificial diets are often not available. Fewer than ten species of natural enemies can be produced on artificial diets, but their field performance may be poorer than that of natural enemies reared on a host insect (Chapter 9). Although mass production on artificial diets may lead to reduction of costs, the risks of changing natural-enemy effectiveness should not be underrated (see below). Another obstacle for mass production is the lack of techniques to prevent selection pressures leading to genetic deterioration of the mass-produced organisms. Through such deterioration, the natural enemy could lose its effectiveness (Boller, 1972; Boller and Chambers, 1977).

Cannibalism among predators may make individual rearing (e.g. for *Chrysopa* spp.) or rearing at relatively high prey densities (e.g. for *Amblyseius* and *Phytoseiulus* spp.) necessary and will lead to high rearing costs. Superparasitism with parasitoids has the same effect. Rearing of parasitoids and predators under 'unnatural' conditions on 'unnatural' hosts or prey or on artificial media may change their reactions to natural-host or host-plant cues as a result of missing or improper preimaginal or imaginal conditioning (Chapters 3 and 4). Rearing parasitoids on unnatural hosts may lead to reduced vigour as a result of an inadequate supply of nutrition (quantity or quality) from the unnatural host; the same effect can occur when the host is reared on an unnatural diet, even if the host itself remains apparently unaffected (Chapter 9).

Finally, the rearings can be infected by pathogens (Chapter 10). One of the problems often encountered in insect rearing is the occurrence of pathogens and microbial contaminants, leading to high mortality, reduced fecundity, prolonged development, small adults, wide fluctuations in the quality of insects or direct pathological effects. Goodwin (1984), Shapiro (1984), Sikorowski (1984), Singh and Moore (1985), Bjørnson and Schütte (Chapter 10) and Stouthamer (Chapter 8) give information on the effects of microorganisms on insect cultures and the measures available to minimize or eliminate the pathogens or contaminations. Further, they discuss the recognition of diseases and microorganisms in insect rearing and the common sources of such microbial contaminants. The most common microbial contaminants encountered in insect rearing are fungi, followed by bacteria, viruses, protozoa and nematodes. The field-collected insects that are used to start a laboratory colony are a major source of microbial contaminants. The second main source is the various dietary ingredients. Disinfection of insects and dietary ingredients is recommended to prevent such contaminations. The causes of microbial contamination are usually rapidly found, but elimination of pathogens from insect colonies is difficult (Bartlett, 1984a; Chapter 10).

Behavioural variation in natural enemies

The variation and changes in behaviour of natural enemies that can be caused by rearing conditions are manifold. The main question is whether erratic behaviour of natural enemies can be prevented or cured. This issue, together with a thorough theoretical background, is discussed in Chapters 3 and 4. Most ecologists are aware that variability in natural-enemy behaviour occurs frequently. It is important to know how natural enemies function in agroecosystems because such understanding may help in designing systems where natural enemies can play an even more important role in inundative and seasonal inoculative releases.

The core of natural-enemy behaviour, host-habitat and host-location behaviour, shows great variability, which often leads to inconsistent results in biological control. Most studies aimed at understanding such variability have focused on extrinsic factors as causes for any inconsistencies seen in foraging behaviour. Typically, however, foraging behaviour remained irregular even when using precisely the same set of external stimuli. Two types of adaptive variation have been distinguished in the foraging behaviour of natural enemies: genetically fixed differences and phenotypic plasticity. In order to understand erratic behaviour and to be able to manipulate such variation, biological control researchers need to know the origins and width of variation (Chapters 3 and 4). Foraging behaviour can also be strongly influenced by the physiological condition of the natural enemy. Natural enemies face varying situations in meeting their food, mating, reproductive and safety requirements. The presence of strong chemical, visual or auditory cues, cues related to the presence of enemies of the natural enemy and (temporary) egg depletion can all reduce or disrupt the response to cues used to find hosts. For example, hunger may result in increased foraging for food and decreased attention to hosts. In that case, the reaction to food and host cues will be different from when the natural enemy is well fed (Chapters 3, 4 and 5).

The sources of intrinsic variation in foraging behaviour (genetic, phenotypic and those related to the physiological state) are not mutually exclusive but overlap extensively, even within a single individual. The eventual foraging effectiveness of a natural enemy is determined by how well the natural enemy's net intrinsic condition is matched with the foraging environment in which it operates.

Managing variability in behaviour of natural enemies

In order to be efficient as biological control agents, natural enemies must be able to effectively locate and attack a host and stay in a host-infested area until most hosts are attacked. Efficiency as a biological control agent is used here in the anthropocentric sense (i.e. our purposes for pest control), which does not necessarily mean efficiency from a natural-selection viewpoint. Management of natural-enemy variation is particularly important when species are mass-produced in the laboratory, especially if rearing is done in factitious hosts (Chapter 9). Such laboratory rearings remove natural enemies from the context of natural selection and expose them to artificial selection for traits that are useless in the field (van Lenteren, 1986a). In addition to the genetic component, associative learning may lead to many more changes in behavioural reactions (Chapters 3 and 4).

Managing genetic qualities

Successful predation or parasitism of a target host in a confined situation does not guarantee that released individuals will be suitable for that host under field conditions. When selecting among strains of natural enemies, we need to ensure that the traits of the natural enemies are appropriately matched with the targeted use situations in the field. Natural-enemy populations should perform well on the target crop and under the specific climate conditions.

Managing phenotypic qualities

Without care, insectary environments lead to agents with weak or distorted responses. If we understand the sources and mechanism of natural-enemy learning, we can, in theory, provide the appropriate level of experience to correct such defects before releasing the natural enemies. Also, prerelease exposure to important stimuli can help improve the responses of natural enemies through associative learning, leading to reduction in escape response and increased arrestment in target areas.

Managing physical and physiological qualities

Natural enemies should be released in a physiological state in which they are most responsive to herbivore or plant stimuli and will not be hindered in their responses by deprivations that interfere with host searching. Thus, adult parasitoids should be well fed (honey or sugar source available in mass rearing; Chapter 5), have had opportunities to mate and have had their preoviposition period before releases are made.

Laboratory rearing and field performance of natural enemies

In view of all these obstacles, one of the first conclusions is that it would be best to rear the natural enemies in as natural a situation as possible, a conclusion that is supported by a number of researchers with experience in mass production (see, for example, King and Morrison, 1984; Bigler, 1989). Another important conclusion based on recent information about learning is that the host habitat and the host should provide the same cues in mass rearing as in the field, or, if this is not possible, the natural enemies should be exposed to these cues after rearing but before being released in the field. The problems that remain, even when rearing is done as naturally as possible, are related to obstacles 3, 4, 5 and 8 in Table 1.4. Anyone starting a mass-rearing facility should be prepared not only to overcome these obstacles but also to recognize the conflicting requirements placed on natural enemies in a mass-production programme and during field performance (Table 1.5).

Development and Implementation of Quality Control

Natural enemies are often mass-produced under conditions that are very different from those found in commercial crops. Because of these differences, most of the points listed in Table 1.5 are applicable and must be considered in quality control programmes. The development of quality control programmes for natural-enemy production has been rather pragmatic. Guidelines have been developed for more than 30 species of natural enemies (Chapter 19) and descriptions of the development of various quality control tests included in these guidelines can be found in van Lenteren (1996, 1998) and van Lenteren and Tommasini (1999). The guidelines developed until now refer to product-control procedures, not to production or process control. They were designed to be as uniform as possible so that they can be used in a standardized manner by many producers, distributors, pest-management advisory personnel and farmers. These tests should preferably be carried out by the producer after all handling procedures just before shipment. It is expected that the user (farmer or grower) will only perform a few aspects of the quality test, e.g. per cent emergence or number of live adults in the package. Some tests are to be carried out frequently by the producer, i.e. on a daily, weekly or batch-wise basis. Others will be done less frequently, i.e. on an annual or seasonal basis, or when rearing procedures are changed. In the near future, large cage tests, flight tests and field-performance tests will be added to these guidelines (Chapters 16 and 19). Such tests are needed to show the relevance of the laboratory measurements. Laboratory tests are only adequate when a

Table 1.5. Conflicting requirements concerning performance of natural enemies in a mass-rearing colony and under field conditions.

Natural-enemy features that are valued in mass rearing	Natural-enemy features that are important for field performance
1. Polyphagy (makes rearing on unnatural host/prey easier)	Monophagy or oligophagy (more specific agents often have a greater pest-reduction capacity)
2. High parasitism or predation rates at high pest densities	High parasitism or predation rates at low pest densities
3. No strong migration as a result of direct or indirect interference	Strong migration as a result of direct or indirect interference
4. Migration behaviour unnecessary and unwanted, ability to disperse minimal	Migration behaviour essential
5. Associative learning not appreciated	Associative learning appreciated

good correlation has been established between the laboratory measurements, flight tests and field performance.

In addition to the quality control tests, fact-sheets for natural enemies and pests will be prepared to inform new quality control personnel and plant-protection services on biological details.

Conclusions

Companies just beginning the production of a natural enemy are often rather ignorant about the obstacles and complications entailed in mass-rearing programmes. New producers are even more ignorant about the development and application of quality control. A special point of concern is the lack of knowledge about the sources of variability of natural-enemy behaviour and methods to prevent genetic deterioration of natural enemies.

Quality control programmes should be designed to obtain acceptable quality, not necessarily the best possible quality. The number of necessary tests will be smallest if the natural enemies are reared under the same conditions as those under which they also have to function in the field in terms of the same climate, host and host plant. The more artificial the rearing conditions become, and the more the natural enemies are 'handled' before use (removed from the plant or host, counting, containerization, gluing to substrate, manipulation to induce diapause, shipment, release method, etc.), the larger the number of tests that will have to be performed. Also, under these circumstances, prerelease training of the natural enemies may be needed so that they can perceive relevant cues from the pest insect or infested plant.

Simple quality control procedures for natural enemies have been designed for about 30 natural-enemy species and are currently being developed for additional natural-enemy species. The quality control criteria now used relate to product control and are based on laboratory measurements that are easy to carry out. These criteria will be complemented with flight tests and field-performance tests.

If the biological control industry is to survive and flourish, the production of reliable natural enemies that meet basic quality standards is essential.

Acknowledgements

European producers of natural enemies are thanked for cooperation in the design of quality control guidelines. The following persons are thanked for assisting in obtaining information about quality control activities outside Europe: V.H.P. Bueno (Brazil), D. Conlong (South Africa), D. Papacek (Australia), R. Rountree (New Zealand) and E. Yano (Japan). Development of quality control guidelines was financially supported by the Commission of the European Communities, Directorate General for Agriculture, Concerted Action CT93-1076, 'Designing and Implementing Quality Control of Beneficial Insects: Towards More Reliable Biological Pest Control'. This chapter does not necessarily reflect the Commission's views and in no way anticipates the Commission's future policy in this area.

References

Albajes, R., Gullino, M.L., van Lenteren, J.C. and Elad, Y. (eds) (1999) *Integrated Pest and Disease Management in Greenhouse Crops.* Kluwer Academic Publishers, Dordrecht, The Netherlands, 545 pp.

Aleman, J., Plana L., Vidal, M., Llanes, G. and Delgado, M. (1998) Criterios para el control de la calidad en la cria masiva de *Lixophaga diatraeae*. In: Hassan, S.A. (ed.) *Proceedings of the 5th International Symposium on* Trichogramma *and Other Egg Parasitoids, 4–7 March 1998, Cali, Colombia.* Biologische Bundesanstalt für Land-und Forstwirtschaft, Darmstadt, pp. 97–104.

Anon. (2000) 2001 Directory of least-toxic pest control products. *IPM Practitioner* 22, 1–38.

Bartlett, A.C. (1984a) Establishment and maintenance of insect colonies through genetic control. In: King, E.G. and Leppla, N.C. (eds) *Advances and Challenges in Insect Rearing*. US Department of Agriculture, Agricultural Research Service, Southern Region, New Orleans, Louisiana, p. 1.

Bartlett, A.C. (1984b) Genetic changes during insect-domestication. In: King, E.G. and Leppla, N.C. (eds) *Advances and Challenges in Insect Rearing*. US Department of Agriculture, Agricultural Research Service, Southern Region, New Orleans, Louisiana, pp. 2–8.

Bartlett, A.C. (1985) Guidelines for genetic diversity in laboratory colony establishment and maintenance. In: Singh, P. and Moore, R.F. (eds) *Handbook of Insect Rearing*, Vol. 1. Elsevier, Amsterdam, The Netherlands, pp. 7–17.

Beirne, B.P. (1974) Status of biological control procedures that involve parasites and predators. In: Maxwell, F.G. and Harris, F.A. (eds) *Proceedings of the Summer Institute on Biological Control of Plant Insects and Diseases*. University Press of Mississippi, Jackson, Mississippi, pp. 69–76.

Bigler, F. (1989) Quality assessment and control in entomophagous insects used for biological control. *Journal of Applied Entomology* 108, 390–400.

Bigler, F. (ed.) (1991) *Quality Control of Mass Reared Arthropods. Proceedings 5th Workshop IOBC Global Working Group 'Quality Control of Mass Reared Arthropods', 25–29 March 1991, Wageningen, The Netherlands*. Swiss Federal Research Station for Agronomy, Zurich, 205 pp.

Bigler, F. (1994) Quality control in *Trichogramma* production. In: Wajnberg, E. and Hassan, S.A. (eds) *Biological Control with Egg Parasitoids*. CAB International, Wallingford, UK, pp. 93–111.

Boller, E.F. (1972) Behavioral aspects of mass-rearing of insects. *Entomophaga* 17, 9–25.

Boller, E.F. (1979) Behavioral aspects of quality in insectary production. In: Hoy, M.A. and McKelvey, J.J. (eds) *Genetics in Relation to Insect Management*. Rockefeller Foundation, New York, pp. 153–160.

Boller, E.F. and Chambers, D.L. (1977) Quality aspects of mass-reared insects. In: Ridgway, R.L. and Vinson, S.B. (eds) *Biological Control by Augmentation of Natural Enemies*. Plenum, New York, pp. 219–236.

Bueno, V.H.P. (ed.) (2000) *Controle biologico de pragas: producao massal e controle de qualidade*. Editora UFLA, Lavras, Brazil, 215 pp. (in Portuguese).

Bush, G.L., Neck, R.W. and Kitto, G.B. (1976) Screwworm eradication, inadvertent selection for noncompetitive ecotypes during mass rearing. *Science* 193, 491–493.

Chambers, D.L. and Ashley, T.R. (1984) Putting the control in quality control in insect rearing. In: King, E.G. and Leppla, N.C. (eds) *Advances and Challenges in Insect Rearing*. US Department of Agriculture, Agricultural Research Service, Southern Region, New Orleans, Louisiana, pp. 256–260.

Conlong, D.E. (1995) Small colony initiation, maintenance and quality control in insect rearing. In: *Proceedings 4th National Insect Rearing Workshop, Grahamstown, South Africa, 3 July 1995*, 39 pp.

Conlong, D.E. and Mugoya, C.F. (1996) Rearing beneficial insects for biological control purposes in resource poor areas of Africa. *Entomophaga* 41, 505–512.

Cranshaw, W., Sclar, D.C. and Cooper, D. (1996) A review of 1994 pricing and marketing by suppliers of organisms for biological control of arthropods in the United States. *Biological Control* 6, 291–296.

DeBach, P. (ed.) (1964) *Biological Control of Insect Pests and Weeds*. Cambridge University Press, Cambridge, 844 pp.

Goodwin, R.H. (1984) Recognition and diagnosis of diseases in insectaries and the effects of disease agents on insect biology. In: King, E.G. and Leppla, N.C. (eds) *Advances and Challenges in Insect Rearing*. US Department of Agriculture, Agricultural Research Service, Southern Region, New Orleans, Louisiana, pp. 96–129.

Greenberg, S.M. (1991) Evaluation techniques for *Trichogramma* quality. In: Bigler, F. (ed.) *Quality Control of Mass Reared Arthropods. Proceedings 5th Workshop IOBC Global Working Group 'Quality Control of Mass Reared Arthropods', 25–29 March 1991, Wageningen, The Netherlands*. Swiss Federal Research Station for Aronomy, Zurich, pp. 138–145.

Gurr, G. and Wratten, S. (eds) (2000) *Measures of Success in Biological Control*. Kluwer Academic Publishers, Dordrecht, 448 pp.

Hussey, N.W. and Bravenboer, L. (1971) Control of pests in glasshouse culture by the introduction of natural enemies. In: Huffaker, C.B. (ed.) *Biological Control*. Plenum, New York, pp. 195–216.

Jaenson, T.G.T. (1978) Mating behaviour of *Glossina pallides* Austen (Diptera, Glossinidae): genetic differences in copulation time between allopatric populations. *Entomologia Experimentalis et Applicata* 24, 100–108.

Jansson, A. (1978) Viability of progeny in experimental crosses between geographically isolated popula-

tions of *Arctocorisa carinata* (Sahlberg, C.) (Heteroptera, Corixidae). *Annales Zoologici Fennici* 15, 77–83.
Joslyn, D.J. (1984) Maintenance of genetic variability in reared insects. In: King, E.G. and Leppla, N.C. (eds) *Advances and Challenges in Insect Rearing*. US Department of Agriculture, Agricultural Research Service, Southern Region, New Orleans, Louisiana, pp. 20–29.
Kaushik, H.D. and Arora, R.K. (1998) *Trichogramma*: research and use in India. In: Hassan, S.A. (ed.) *Proceedings of the 5th International Symposium on* Trichogramma *and Other Egg Parasitoids, 4–7 March 1998, Cali, Colombia*. Biologische Bundesanstalt für Land- und Forstwirtschaft, Darmstadt, pp. 155–176.
King, E.G. and Morrison, R.K. (1984) Some systems for production of eight entomophagous arthropods. In: King, E.G. and Leppla, N.C. (eds) *Advances and Challenges in Insect Rearing*. US Department of Agriculture, Agricultural Research Service, Southern Region, New Orleans, Louisiana, pp. 206–222.
Leppla, N.C. and Fisher, W.R. (1989) Total quality control in insect mass production for insect pest management. *Journal of Applied Entomology* 108, 452–461.
Lerner, I. (1958) *Genetic Basis of Selection*. John Wiley & Sons, New York, 298 pp.
Lopez-Fanjul, C. and Hill, W.G. (1973) Genetic differences between populations of *Drosophila melanogaster* for quantitative traits. II. Wild and laboratory populations. *Genetical Research* 22, 60–78.
Mackauer, M. (1972) Genetic aspects of insect control. *Entomophaga* 17, 27–48.
Mackauer, M. (1976) Genetic problems in the production of biological control agents. *Annual Review of Entomology* 21, 369–385.
Mayr, E. (1970) *Populations, Species, and Evolution*. Harvard University Press, Cambridge, Massachusetts, 493 pp.
Morrison, R.K. and King, E.G. (1977) Mass production of natural enemies. In: King, E.G. and Leppla, N.C. (eds) *Advances and Challenges in Insect Rearing*. US Department of Agriculture, Agricultural Research Service, Southern Region, New Orleans, Louisiana, pp. 183–217.
Nicoli, G., Benuzzi, M. and Leppla, N.C. (eds) (1993) *Proceedings 7th Global IOBC Workshop Quality Control of Mass Reared Arthropods, 13–16 September 1993, Rimini, Italy*, 240 pp.
Oliver, C.G. (1972) Genetic and phenotypic differentiation and geographic distance in four species of Lepidoptera. *Evolution* 26, 221–241.
Prakash, S. (1973) Patterns of gene variation in central and marginal populations of *Drosophila robusta*. *Genetics* 75, 347–369.
Ramos, M., Aleman, J., Rodriguez, H. and Chico, R. (1998) Estimacion de parametros para el control de calidad en crias de *Phytoseiulus persimilis* (Banks) (Acari: Phytoseiidae) empleando como presa a *Panonychus citri* McGregor (Acari: Tetranychidae). In: Hassan, S.A. (ed.) *Proceedings of the 5th International Symposium on* Trichogramma *and Other Egg Parasitoids, 4–7 March 1998, Cali, Colombia*. Biologische Bundesanstalt für Land- und Forstwirtschaft, Darmstadt, pp. 109–118.
Ridgeway, R.L. and Vinson, S.B. (1977) *Biological Control by Augmentation of Natural Enemies. Insect and Mite Control with Parasites and Predators*. Plenum Press, New York, 480 pp.
Shapiro, M. (1984) Micro-organisms as contaminants and pathogens in insect rearing. In: King, E.G. and Leppla, N.C. (eds) *Advances and Challenges in Insect Rearing*. US Department of Agriculture, Agricultural Research Service, Southern Region, New Orleans, Louisiana, pp. 130–142.
Sikorowski, P.P. (1984) Microbial contamination in insectaries. In: King, E.G. and Leppla, N.C. (eds) *Advances and Challenges in Insect Rearing*. US Department of Agriculture, Agricultural Research Service, Southern Region, New Orleans, Louisiana, pp. 143–153.
Silva, I.M.M.S., van Meer, M.M.M., Roskam, M.M., Hoogenboom, A., Gort, G. and Stouthamer, R. (2000) Biological control potential of *Wolbachia*-infected versus uninfected wasps: laboratory and greenhouse evaluation of *Trichogramma cordubensis* and *T. deion* strains. *Biocontrol Science and Technology* 10, 223–238.
Singh, P. and Moore, R.F. (eds) (1985) *Handbook of Insect Rearing*. Elsevier, Amsterdam, The Netherlands, Vol. 1, 488 pp.,Vol. 2, 514 pp.
Smith, S.M. (1996) Biological control with *Trichogramma*: advances, successes, and potential of their use. *Annual Review of Entomology* 41, 375–406.
Spurway, H. (1955) The causes of domestication: an attempt to integrate some ideas of Konrad Lorenz with evolution theory. *Journal of Genetics* 53, 325–362.
Swamiappan, M., Muthuswami, M. and Sithanantham, S. (1998) Quality control of mass reared *Trichogramma* in commercial laboratories in Tamil Nadu, India. In: Hassan, S.A. (ed.) *Proceedings of the 5th International Symposium on* Trichogramma *and Other Egg Parasitoids, 4–7 March 1998, Cali, Colombia*, pp. 105–108.

van Lenteren, J.C. (1986a) Evaluation, mass production, quality control and release of entomophagous insects. In: Franz, J.M. (ed.) *Biological Plant and Health Protection.* Fischer, Stuttgart, Germany, pp. 31–56.

van Lenteren, J.C. (1986b) Parasitoids in the greenhouse: successes with seasonal inoculative release systems. In: Waage, J.K. and Greathead, D.J. (eds) *Insect Parasitoids.* Academic Press, London, pp. 341–374.

van Lenteren, J.C. (1994) Quality control guidelines for 21 natural enemies. *Sting, Newsletter on Biological Control in Greenhouses, Wageningen* 14, 3–24.

van Lenteren, J.C. (1995) Integrated pest management in protected crops. In: Dent, D. (ed.) *Integrated Pest Management.* Chapman & Hall, London, pp. 311–343.

van Lenteren, J.C. (1996) Designing and implementing quality control of beneficial insects: towards more reliable biological pest control. In: *Proceedings Quality Control Meeting, 13–18 February 1996, Antibes, France,* 22 pp.

van Lenteren, J.C. (1998) Quality control guidelines. *Sting, Newsletter on Biological Control in Greenhouses, Wageningen* 18, 32 pp.

van Lenteren, J.C. (2000a) Measures of success in biological control of arthropods by augmentation of natural enemies. In: Gurr, G. and Wratten, S. (eds) *Measures of Success in Biological Control.* Kluwer Academic Publishers, Dordrecht, pp. 77–103.

van Lenteren, J.C. (2000b) A greenhouse without pesticides: fact or fantasy? *Crop Protection* 19, 375–384.

van Lenteren, J.C. and Bueno, V.H.P. (2003) Augmentative biological control in Latin America as seen from a worldwide perspective. *BioControl* (in press).

van Lenteren, J.C. and Manzaroli, G. (1999) Evaluation and use of predators and parasitoids for biological control of pests in greenhouses. In: Albajes, R., Gullino, M.L., van Lenteren, J.C. and Elad, Y. (eds) *Integrated Pest and Disease Management in Greenhouse Crops.* Kluwer Academic Publishers, Dordrecht, The Netherlands, pp. 183–201.

van Lenteren, J.C. and Tommasini, M.G. (1999) Mass production, storage, shipment and quality control of natural enemies. In: Albajes, R., Gullino, M.L., van Lenteren, J.C. and Elad, Y. (eds) *Integrated Pest and Disease Management in Greenhouse Crops.* Kluwer Academic Publishers, Dordrecht, The Netherlands, pp. 276–294.

van Lenteren, J.C. and Woets, J. (1988) Biological and integrated control in greenhouses. *Annual Review of Entomology* 33, 239–269.

van Lenteren, J.C., Bigler, F. and Waddington, C. (1993) Quality control guidelines for natural enemies. In: Nicoli, G., Benuzzi, M. and Leppla, N.C. (eds) *Proceedings 7th Global IOBC Workshop Quality Control of Mass Reared Arthropods, 13–16 September 1993, Rimini, Italy,* pp. 222–230.

van Lenteren, J.C., Roskam, M.M. and Timmer, R. (1997) Commercial mass production and pricing of organisms for biological control of pests in Europe. *Biological Control* 10, 143–149.

Vet, L.E.M., Lewis, W.J., Papaj, D.R. and van Lenteren, J.C. (1990) A variable-response model for parasitoid foraging behavior. *Journal of Insect Behavior* 3, 471–490.

Voegele, J. (ed.) (1982) *Proceedings of the 1st International Symposium on* Trichogramma, *20–23 April 1982, Antibes, France.* Les Colloques de l'INRA 9, Paris, 307 pp.

Voegele, J., Waage, J. and van Lenteren, J.C. (eds) (1988) *Proceedings of the 2nd International Symposium on* Trichogramma *and Other Egg Parasites, 10–15 November 1986, Guangzhou, China.* Les Colloques de l'INRA 43, Paris, 644 pp.

Wajnberg, E. (ed.) (1995) *Proceedings of the 4th International Symposium on* Trichogramma *and Other Egg Parasitoids, 4–7 October 1994, Cairo, Egypt.* Les Colloques de l'INRA 73, Paris, 226 pp.

Wajnberg, E. and Vinson, S.B. (eds) (1991) *Proceedings of the 3rd International Symposium on* Trichogramma *and Other Egg Parasitoids, 23–27 September 1990, San Antonio, USA.* Les Colloques de l'INRA 56, Paris, 246 pp.

Yamazaki, T. (1972) Detection of single gene effect by inbreeding. *Nature* 240, 53–54.

Yaninek, J.S. and Herren, H.R. (1989) *Biological Control: a Sustainable Solution to Crop Pest Problems in Africa.* IITA, Ibadan, Nigeria, 194 pp.

2 Aspects of Total Quality Control for the Production of Natural Enemies

N.C. Leppla
Department of Entomology and Nematology, University of Florida, Natural Area Drive, PO Box 110630, Gainesville, FL 32611-0603, USA

Abstract

In this chapter the following questions are addressed: (i) what happens when yields of natural enemies decline and the cause is not evident? (ii) how are yields maintained while decreasing inputs and thereby improving efficiency? (iii) how are complaints resolved about the postproduction performance of natural enemies? and (iv) how are decisions made to correct apparent problems or improve the production system? A system of total quality control (TQC) is described, because it is an uncomplicated structure for organizing and addressing the major steps in producing, using and improving natural enemies. TQC can assist in carefully evaluating trade-offs in the system and judiciously investing resources, critical functions for commercial biological control. 'Of all concepts in the quality function, none is so far-reaching or vital as "fitness for use"' (Juran *et al.*, 1974). To be marketable, products and services must meet the expectations of users in terms of price, reliability and performance. In this sense, 'fitness for use' is the definition of quality for producers of natural enemies and their customers.

Introduction: Why Practise Quality Control in the Production and Use of Natural Enemies?

Quality control is practised in the production of natural enemies, at least intuitively, at some level as a measure of the success or failure of the production system. Adequate yields indicate that rearing operations have been performed efficiently. In a small, hands-on organization, there can be a sense about whether each step in the rearing process has been accomplished adequately. However, what happens when yields decline and the cause is not evident? How are yields maintained while decreasing inputs and thereby improving efficiency? How are complaints resolved about the postproduction performance of natural enemies? How are decisions made to correct apparent problems or improve the production system?

Information is required to determine the status of each rearing operation in the system and the quality of the final insect product (Leppla and Ashley, 1989). The cost of obtaining this information should be recovered in reduced incidence of problems and increased efficiencies. It is not an added expense but an integral function in natural-enemy production (Leppla and King, 1996). Typically, data are derived from a representative sample of rearing units, i.e.

© CAB International 2003. *Quality Control and Production of Biological Control Agents: Theory and Testing Procedures* (ed. J.C. van Lenteren)

oviposition cages and containers for holding parasitized hosts. Measurements may include the number and condition of hosts, the sex ratio and fecundity of the natural enemy, the yield of the final product and the emergence, condition and longevity of adult parasitoids (Williams and Leppla, 1992; van Lenteren and Tommasini, 1999). Predators are monitored similarly, except that the host may be replaced by artificial diet. The number of units sampled is usually small and the data are acquired in a way that makes them easy to analyse (Chambers and Ashley, 1984).

Optimization or troubleshooting of the production system is accomplished by analysing production units, not batches. Batches are generally worker shifts, days or weeks that combine the products of individual units. For example, a shift may produce a certain number of parasitized eggs, regardless of the number of oviposition cages. This measurement of yield combines the variability from all of the cages and obscures the rate of parasitization in individual cages. How can we know the number and identity of cages that are producing well versus those that are having some difficulty? Cages with problems could be at the end of the process line, set up by an inexperienced worker, positioned in an unfavourable environment or associated with some other cause. The source of the problem can be corrected only if the affected production units can be identified. Otherwise, we just know that yields have declined and there is a problem somewhere in the system. Changes made intentionally to improve the system must be monitored similarly by sampling individual production units. It can be very costly to attempt to manage an entire natural-enemy production system without knowing the condition of its individual units.

What is Total Quality Control for the Production of Natural Enemies?

Total quality control (TQC) is an uncomplicated structure for organizing and addressing the major steps in producing, using and improving natural enemies (Leppla and Fisher, 1989; Leppla, 1994). More generically, it is:

> An effective system for integrating the quality-development, quality-maintenance, and quality-improvement efforts of the various groups in an organization so as to enable marketing, engineering, production, and service at the most economical levels which allow for full customer satisfaction.
>
> (Feigenbaum, 1983)

TQC is composed of eight generic subdivisions: management, research, methods development, material, production, utilization, personnel and quality control (Fig. 2.1). Although often not individually identified, all of these elements are present in pest-management systems based on mass-reared natural enemies and each has internal control functions. Coordination across these interdependent subdivisions and feedback to management provide a means of ensuring production of the most efficacious natural-enemy products and eliminating unnecessary costs (Fig. 2.2). These products must be monitored and evaluated during and after production, and while being used, to assure that they meet expectations.

A TQC system begins with the ability to raise a natural enemy that is effective in controlling populations of a specific pest (Leppla, 1989). This ability entails methods to accurately identify and effectively collect, handle, house, feed, cycle and harvest an adequate number of natural enemies. Standard operating procedures (SOPs) are described for all rearing operations, ranging from acquisition and storage of supplies to maintenance and preparation of reports. Workers who actually perform the operations should participate in writing or at least reviewing the SOP steps. In practice, the details and potential pitfalls of SOPs often exist only in the experience of senior workers. Detailed procedures must be documented along with associated standards of performance. Check sheets may be developed to keep track of their completion.

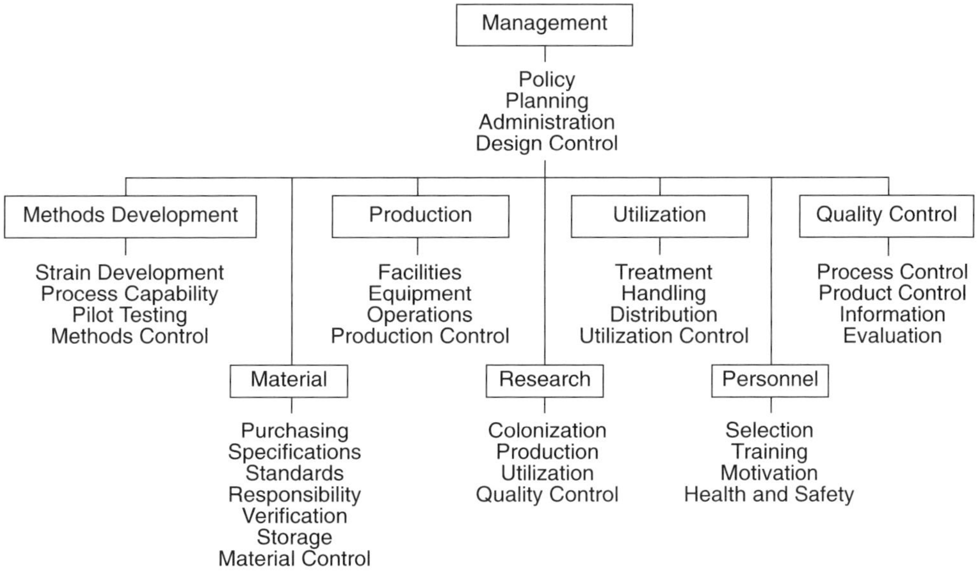

Fig. 2.1. Production system for natural enemies composed of generic subdivisions and associated functions.

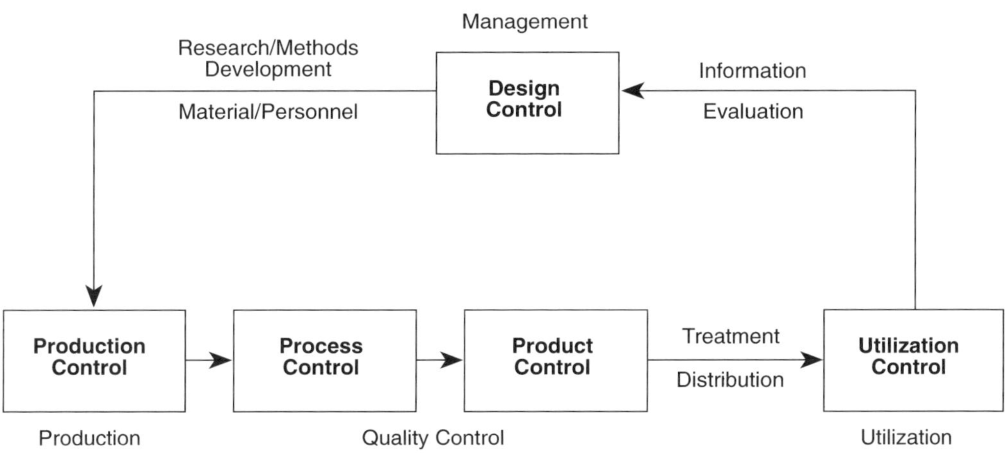

Fig. 2.2. Total quality-control system emphasizing the control functions (boxes) within generic subdivisions of the production system for natural enemies. The feedback loop provides a means of optimizing the production and use of the natural enemies.

What is the Relationship Between Specifications and Standards?

There has been some confusion about the differences between specifications and standards as they apply to insect rearing. They are closely related terms and both can be measures of quality. However, specification is defined as: 'The document that prescribes the requirements with which the product or service has to conform' (ANSI/ASQC, 1987). Standard equals grade, 'An indicator of category or rank related to features or characteristics that cover different sets of needs or products or services intended for the same functional use' (ANSI/ASQC, 1987). In other words, a standard is the level of quality at which a specification is written. A specification for a natural enemy can be written at a required standard of quality based on biological, production and market options. The natural enemy is typically characterized according to its species description (identity and purity), life history and behaviour. For example, a particular species of parasitoid may oviposit a maximum of 500 eggs per female but we specify that females from our colony must produce an average (± SD) of 250 ± 25 eggs. This specification becomes the standard by which we measure our success in producing the natural enemy. If this standard is too difficult or expensive to achieve in practice, it can be lowered to 200 ± 50 eggs. Standards are relative to requirements or expectations, which should be realistic (Boller and Chambers, 1977).

What are Production, Process and Product Control?

The functions of production, process and product control are performed within the production and quality control subdivisions of the production system for natural enemies (Fig. 2.1). Since they are applied and interpreted differently in various industrial manufacturing fields (Besterfield, 1986), they must be defined specifically for use in mass-rearing arthropods. Production is responsible for most inputs and therefore performs production control. Process and product control are performed by the quality control subdivision as functions that support production. It is important to distinguish between the functions of the production and quality control subdivisions because, rather than inputs, the quality control subdivision generally monitors outputs. The quality of processes and products is determined by sampling insects, measuring key characteristics and comparing the results to established specifications and standards, usually by means of process-control charts.

Production control is the monitoring and maintenance of all rearing inputs in terms of personnel, materials, equipment, schedules, environments, SOPs and so forth. Most production failures can be traced to deficiencies in production control and are due to errors caused by workers, unpredictable changes in materials or loss of environmental control. Consequently, problems are prevented and troubleshooting initiated by first focusing on the performance of production SOPs, abnormal appearance of the materials and arthropod stages and environmental deviations.

Process control is the evaluation of key components of the manufacturing processes as they are employed along the production line (Feigenbaum, 1983). In rearing systems, process control is accomplished by determining the constancy of immature arthropod stages as a means of predicting quality and identifying sources of increased efficiency. It is particularly important when unforeseen, detrimental changes occur or inputs are modified intentionally. The process-control information is used by the production subdivision to make any necessary adjustments. A common example is the addition of more females to an oviposition cage as the number of fertile eggs per female declines, before determining and correcting the cause of the decline.

Product control is the same in arthropod rearing as it is in other industrial processes:

> the control of products at the source of production and through field service, so that departures from the quality specification can be corrected before defective or non-conforming products are manufactured and the proper service can be maintained in the field.
>
> (Feigenbaum, 1983)

Thus, the performance of natural enemies is measured and evaluated at the production facility and critical points during their transportation, application and impact on the target pest. Feedback is provided to optimize production, field performance and customer satisfaction.

What is the Management Subdivision of Total Quality Control?

TQC provides a very powerful and flexible system for producing natural enemies because it encompasses all of the necessary elements that must be considered (Fig. 2.2). The manager's role is to establish policies, plan the production effort, provide administration and exercise design control. A production system is designed by applying TQC to determine priorities and assign resources. This framework helps the manager avoid common errors of omission, such as adequate storage of materials or the health and safety of employees. Moreover, it indicates subdivisions that can be combined or split. Examples of amalgamation include research and methods development, production and material, and management and personnel. Utilization and quality control are often stand-alone activities. In addition to visualizing and defining the entire system, the design-control function of TQC enables the manager to make informed decisions. The feedback loop from planning to evaluation also provides for the involvement and education of the customer (Penn *et al.*, 1998). TQC can assist in carefully evaluating trade-offs in the system and judiciously investing resources, critical functions for commercial biological control.

A typical example of decision making by the managers of an arthropod production system is whether or not to change a colonized strain, formerly based on time, intuition or weight of opinion. This question is central to the production of natural enemies but has probably been deliberated in the greatest detail during nearly 50 years of mass-rearing the screw-worm fly, *Cochliomyia hominivorax* (Coquerel). Initially, a strain from Texas was used to conduct tests of the sterile-insect technique on the Caribbean island of Curaçao and in Florida. Eradication was achieved on Curaçao with the Texas strain in 1954 but, for Florida, a new strain was established by collecting from 12 locations in Florida and one in Georgia. This Florida strain was used to eradicate the screw-worm from both the south-east and the south-west by 1966. Between 1966 and 1975, five strains from Texas and Mexico were mass-reared in succession at the Mission, Texas, facility (Meyer, 1987). It became apparent during these years that strains differed greatly in their ability to adapt to production and that their establishment and maintenance depended on the quality of the rearing system. Screw-worm production was moved to a new facility at Tuxtla Gutierrez, Mexico, in 1976 and seven Mexican strains were reared during the next 8 years (Marroquin, 1985). Unpredictably, some strains performed well in the rearing facility and field while others failed, so a strain-development programme was initiated to collect, rear and test new strains in advance of their use for eradication. Managers obtained feedback on the production and field performance of the current strain to compare with rearing and field-testing data on potential replacement strains. Generally, as strain development and mass-rearing capabilities improved, strains were retained and remained effective in the field for longer periods of time. Annual strain replacement is no longer automatic.

Conclusion

TQC for the production of natural enemies accounts for the major variables in planning, implementing, managing and improving the system. It helps to increase production efficiency and cost-effectiveness, rapidly identify and correct the causes of rearing problems, ensure the effectiveness of natural enemies in the field and have the information necessary to optimize their use in pest management. SOPs are established with reasonable specifications (what result is expected) and standards (what quality is expected) for arthropod products. These standards can be achieved by carefully controlling production (SOPs, facilities and equipment), processes (indicated by insect stages) and products (stage that is delivered and used). TQC is a means of planning, organizing and managing the subdivisions and functions of natural-enemy production systems.

References

ANSI/ASQC (1987) *Quality Systems Terminology*. American National Standards Institute/American Society for Quality Control, ANSI/ASQC A3–1987. Milwaukee, Wisconsin, 10 pp.

Besterfield, D.H. (1986) *Quality Control*, 2nd edn. Prentice-Hall Publishers, Englewood Cliffs, New Jersey, 368 pp.

Boller, E.F. and Chambers, D.L. (1977) Concepts and approaches. In: Boller, E.F. and Chambers, D.L. (eds) *Quality Control, An Idea Book for Fruit Fly Workers*. IOBC/WPRS Bulletin, pp. 4–13.

Chambers, D.L. and Ashley, T.R. (1984) Putting the control in quality control in insect rearing. In: King, E.G. and Leppla, N.C. (eds) *Advances and Challenges in Insect Rearing*. Agricultural Research Service, USDA, US Government Printing Office, Washington, DC, pp. 256–260.

Feigenbaum, A.V. (1983) *Total Quality Control*, 3rd edn. McGraw-Hill Publishers, New York, 851 pp.

Juran, J.M., Gryna, F.M., Jr and Bingham, R.S., Jr (1974) *Quality Control Handbook*, 3rd edn. McGraw-Hill Publishers, New York, 1780 pp.

Leppla, N.C. (1989) Laboratory colonization of fruit flies. In: Robinson, A.S. and Hooper, G. (eds) *World Crop Pests* 3B, *Fruit Flies, Their Biology, Natural Enemies and Control*. Elsevier Publishers, Amsterdam, pp. 91–103.

Leppla, N.C. (1994) Principles of quality control in mass-reared arthropods. In: Nicoli, G., Benuzzi, M. and Leppla, N.C. (eds) *Proceedings of Seventh Workshop of the IOBC Working Group on Quality Control of Mass-reared Arthropods, Rimini, 13–16 September 1993.* IOBC, Montpellier, France, Bulletin 1–11.

Leppla, N.C. and Ashley, T.R. (1989) Quality control in insect mass production: a review and model. *Bulletin Entomological Society of America*, Winter, 33–44.

Leppla, N.C. and Fisher, W.R. (1989) Total quality control in insect mass production for insect pest management. *Journal of Applied Entomology* 108, 452–461.

Leppla, N.C. and King, E.G. (1996) The role of parasitoid and predator production in technology transfer of field crop biological control. *Entomophaga* 41, 343–360.

Marroquin, R. (1985) Mass production of screwworm in Mexico. In: Graham, O.H. (ed.) *Symposium on Eradication* of *the Screwworm from the United States and Mexico*. Miscellaneous Publications of the Entomological Society of America 62, Lanham, Maryland, pp. 31–36.

Meyer, N.L. (1987) *History of the Mexico–United States Screwworm Eradication Program*. Vantage Press, New York, 367 pp.

Penn, S.L., Ridgway, R.L., Scriven, G.T. and Inscoe, M.N. (1998) Quality assurance by the commercial producer of arthropod natural enemies. In: Ridgway, R.L., Hoffman, M.P., Inscoe, M.N. and Glenister, C.S. (eds) *Mass-reared Natural Enemies: Application, Regulation and Needs.* Proceedings of Thomas Say Publications in Entomology, Entomological Society of America, Lanham, Maryland, pp. 202–230.

van Lenteren, J.C. and Tommasini, M.G. (1999) Mass production, storage, shipment and quality control of natural enemies. In: Albajes, R., Gullino, M.L. and van Lenteren, J.C. (eds) *Integrated Pest and Disease Management in Greenhouse Crops.* Kluwer Publishers, Dordrecht, pp. 276–294.

Williams, D.W. and Leppla, N.C. (1992) The future of augmentation of beneficial arthropods. In: Kauffman, W.C. and Nechols, J.R. (eds) *Selection Criteria and Ecological Consequences of Importing Natural Enemies.* Proceedings of Thomas Say Publications in Entomology, Entomological Society of America, Lanham, Maryland, pp. 87–102.

3 A Variable-response Model for Parasitoid Foraging Behaviour

L.E.M. Vet,[1,2] W.J. Lewis,[3] D.R. Papaj[4] and J.C. van Lenteren[1]

[1]*Laboratory of Entomology, Wageningen University, PO Box 8031, 6700 EH Wageningen, The Netherlands;* [2]*Netherlands Institute of Ecology, PO Box 40, 6666 ZG Heteren, The Netherlands;* [3]*Insect Biology and Population Management Research Laboratory, USDA-ARS, PO Box 748, Tifton, GA 31793, USA;* [4]*Department of Ecology and Evolutionary Biology, University of Arizona, Tucson, AZ 85721, USA*

Abstract

An important factor inducing variability in foraging behaviour in parasitic wasps is experience gained by the insect. Together with the insect's genetic constitution and physiological state, experience ultimately defines the behavioural repertoire under specified environmental circumstances. A conceptual variable-response model based on several major observations of a foraging parasitoid's responses to stimuli involved in the host-finding process is presented in this chapter. These major observations are that: (i) different stimuli evoke different responses or levels of response; (ii) strong responses are less variable than weak ones; (iii) learning can change response levels; (iv) learning increases originally low responses more than originally high responses; and (v) host-derived stimuli serve as rewards in associative learning of other stimuli. The model specifies how the intrinsic variability of a response will depend on the magnitude of the response and predicts when and how learning will modify the insect's behaviour. Additional hypotheses related to the model concern how experience with a stimulus modifies behavioural responses to other stimuli, how animals respond in multi-stimulus situations, which stimuli act to reinforce behavioural responses to other stimuli in the learning process and, finally, how generalist and specialist species differ in their behavioural plasticity. It is postulated that insight into behavioural variability in the foraging behaviour of natural enemies may be a help, if not a prerequisite, for the efficient application of natural enemies in pest management and for developing quality control tests of biocontrol agents.

Introduction

Much information has become available on cues utilized by parasitoids during foraging for hosts or food over the past decade (Chapters 4 and 5; Vet and Dicke, 1992; Godfray, 1994; Wäckers, 1994; Vet *et al.*, 1995; Lewis *et al.*, 1998; Powell, 1999). At the same time, it became clear that foraging behaviour could no longer be considered to be fixed and predictable, but rather it varies in response to the insect's physiological condition and genetic composition as well as to environmental factors. However, the quest for factors inducing variability in parasitoid foraging behaviour has largely centred on the influence of learning. Experience in either preadult or adult stages modifies adult

© CAB International 2003. *Quality Control and Production of Biological Control Agents: Theory and Testing Procedures* (ed. J.C. van Lenteren)

behaviour (see below). Learning may be loosely defined as 'any change in behaviour with experience' (for a discussion of the definition of learning, see Papaj and Prokopy, 1989; Vet *et al.*, 1995). It can impinge on every phase of parasitoid foraging from habitat location to host acceptance. Associative learning (defined as 'the establishment through experience of an association between two stimuli or between a stimulus and a response') has now been demonstrated in a number of parasitoid species (Lewis and Tumlinson, 1988; Vet, 1988; Turlings *et al.*, 1989; Vet and Groenewold, 1990; Vet *et al.*, 1995), and it appears to be a general phenomenon in the Hymenoptera.

Studies on sources of variability in parasitoid behaviour other than learning (including both genetic and non-genetic sources) are still rare, even more than 10 years after we published essential parts of the current chapter (e.g. Lewis *et al.*, 1990; Steidle and van Loon, 2002). Prévost and Lewis (1990) demonstrated genetic variability in responses to host-plant odours and studies by Mollema (1988) point to genetic variability in host-selection behaviour. The animal's physiological state will specify its responsiveness to stimuli, especially to those related to essential resources (Chapter 5; Tinbergen, 1951; Nishida, 1956; Herrebout, 1969; Herrebout and van der Veer, 1969; Gould and Marler, 1984; Dicke *et al.*, 1986; Lewis and Takasu, 1990; Wäckers, 1994).

Apart from the interest in behavioural variation from a theoretical standpoint (where we ask whether plasticity in behaviour is adaptive or if such plasticity affects the evolution of other behaviours (Papaj and Prokopy, 1989)), there is an applied side to understanding the mechanisms that generate behavioural variation. Ultimately, the effectiveness of natural enemies in controlling populations of insect pests is in part associated with this variability. Understanding its nature may result in its manipulation to our benefit (see, for example, Gross *et al.*, 1975; Wardle and Borden, 1986; van Lenteren, 1999) and thus insight into behavioural variability is a help, if not a prerequisite, for the efficient application of biological control agents (Lewis *et al.*, 1990, 1997). Also, the acquired knowledge about the basis of behavioural variability is expected to assist in the development of quality control tests.

In this chapter, we argue that certain key stimuli evoke absolute responses that are conservative to change in both an ontogenetic and an evolutionary sense. As such they act as an 'anchor' by which responses to other stimuli are altered freely in a reliable manner. Other key stimuli arise through association with the original key stimuli and act to accelerate learning of new stimuli. Even for insects of a given genetic constitution, physiological state and degree of experience, a behavioural response to a given stimulus varies both among individuals and over repeated observations of the same individual. Variability in a response will depend on the magnitude of the response. The impact of learning will relate to the magnitude and variability of behavioural responses. These ideas are presented in a conceptual variable-response model based on several major observations of a foraging parasitoid's responses to assorted host or host-microhabitat stimuli.

Observations Underpinning the Model

Five observations made in our collective studies of parasitoid foraging behaviour inspired the model: (i) different stimuli evoke different responses or levels of response; (ii) strong responses are less variable than weak ones; (iii) learning can change response levels; (iv) learning increases originally low responses more than originally high responses; and (v) for naïve females, host-derived stimuli serve as key stimuli (rewards) in associative learning of other stimuli.

Different stimuli evoke different responses or levels of response

A naïve female parasitoid searching for hosts in which to lay eggs encounters a variety of environmental stimuli. Consequently, foraging typically involves a sequence of responses to some of these stimuli, first to long-range cues (usually for locating and selecting proper habitats) and then to close-range cues (usually for detecting and selecting hosts). The stimuli and motor patterns evoked by foraging cues are diverse and include a variety of plant and host chemicals, such as volatiles, towards which the parasitoid walks or flies, and non-volatiles, which the parasitoid is arrested, antennates or probes with her ovipositor (e.g. Godfray, 1994; Dicke and Vet, 1999). Stimuli may also be physical in nature, including light, which induces migratory flight behaviour, and sound or mechanical vibrations, which elicit orientation responses to hosts (see Lewis *et al.*, 1976; Vinson, 1976, 1981, 1984; van Alphen and Vet, 1986; Steidle and van Loon, 2002).

Let it be assumed that natural selection has set the strength of the response to each of the stimuli involved. The outcome of this selection will not be without some developmental constraint, but naïve animals would nevertheless be expected to show the highest responses to those stimuli that, in evolutionary time, are predictably correlated with high reproductive success. Support for this functional argument is found in work with parasitoids of *Drosophila* larvae, where differential responses to odours from different host-food substrates or from substrates in different stages of decay is adaptive (Vet, 1983; Vet and Jansen, 1984; Vet *et al.*, 1984). Differential responses (with or without plausible adaptive functions) are reported not only for several species attacking Drosophilidae (Vet *et al.*, 1984; Vet, 1985), but also for parasitoids of other host types (e.g. Drost *et al.*, 1988; Sheehan and Shelton, 1989; see also references given by Lewis *et al.*, 1976; Vinson, 1976, 1981, 1984; van Alphen and Vet, 1986; Steidle and van Loon, 2002). A rank order in foraging stimuli, supporting our idea, has frequently been found during the past 10 years (e.g. Potting *et al.*, 1995; Du *et al.*, 1996; Steidle and Schöller, 1997; Steidle, 2000).

It is not surprising to observe that parasitoids do not respond to each possible stimulus with the same response intensity, as it is this mechanism that incites the expression of preferences, a phenomenon with an obvious function for each animal living in a complex environment where it has to make choice among the 'bad, good, better or best'.

Strong responses are less variable than weak ones

Many investigators of parasitoid behaviour have undoubtedly made the observation that the more strongly parasitoids respond to a stimulus, the less sensitive they are to all manner of disturbance. It is a common phenomenon that parasitoids are less likely to be distracted in experiments (e.g. by the observer) from strong stimuli than from weak ones. The stronger the response, the more predictable its occurrence, as can be quantified by calculating its coefficient of variation (CV) (Sokal and Rohlf, 1981). We tested this assumption with three *Leptopilina* species, parasitoids of *Drosophila* spp. (Vet *et al.*, 1990, Table I). Females were allowed to search on a standard patch of host-food substrate until they decided to leave. Animals with the same type of foraging experience can be compared in their response to two different substrate types. The response to the substrate stimulus is expressed as the search duration. Within each pair, the longest search time corresponds to the lowest CV or, in other words, strong responses are less variable than weak ones. In addition to these data, Steidle and van Loon (2002) collected data from several other studies examining the response of parasitoids to chemical stimuli. The data they collected were also analysed for a correlation between response potential, expressed as mean response, and variability, expressed as CV (Sokal and Rohlf, 1981). In agreement with the prediction, the Spearman's rank correlation was negative, although not always significant, in all nine studies (Steidle and van Loon, 2002, Fig. 4–4).

There may be good physiological reasons for expecting this pattern in variability. When a response to a given stimulus is strong, it is less likely to be deflected by responses to other stimuli, as the insect is more liable to filter out and thus ignore sensory inputs from other stimuli that may evoke motor patterns different from that of the stimulus under investigation. This explanation appeals to the importance of allothetic mechanisms in the control of behavioural output (i.e. control by external information (see Visser, 1988)). Additionally, motor patterns under strong allothetic control may be less susceptible to alteration by idiothetic mechanisms (i.e. under control by internal information).

Learning can change response levels

Studies on the influence of experience on foraging behaviour in parasitoids focus on changes in what the animals respond to and/or changes in the strength of these responses, rather than the (probably less likely) modifications of the form of the motor patterns involved.

Preadult experience

Parasitoids develop in and emerge from hosts. Upon emergence, parasitoids have no foraging experience yet and therefore have to rely exclusively on cues that are innately attractive or cues that are acquired during the development, most probably by imprinting before or shortly after emergence of the adult (e.g. van Emden *et al.*, 1996). The specific host environment in which a parasitoid develops can influence behavioural responses by the adult (e.g. Thorpe and Jones, 1937; Vinson *et al.*, 1977; Smith and Cornell, 1978; Vet, 1983; Sheehan and Shelton, 1989; van Emden *et al.*, 1996). The adult parasitoid's response is most probably modified prior to or during eclosion through a chemical legacy from previous developmental stages (Vet, 1983, 1985; Corbet, 1985). Elegant experiments by Hérard *et al.* (1988) with *Microplitis demolitor* females suggested that the cocoon is a potential source of information learned by the parasitoid during or just after emergence. A clear distinction between preadult and adult effects of experience on adult behaviour is difficult to make (Vet and Groenewold, 1990). 'Naïve' insects have had the least possible experience with the stimuli to which they will respond. We define a naïve insect not as an insect without any experience, but as one that has had no experience beyond that which occurred during development within and eclosion from the host.

Adult experience

Experience during the adult stage has more impact on subsequent behavioural responses than experience during development (Vinson *et al.*, 1977; Jaenike, 1983; Vet, 1983; Drost *et al.*, 1988; Sheehan and Shelton, 1989). In the case of parasitoids, hosts or host products serve as key stimuli (rewards), in association with which insects either: (i) learn to respond to stimuli that previously evoked no overt response (e.g. Vinson *et al.*, 1977; Lewis and Tumlinson, 1988; Vet and Groenewold, 1990); or (ii) increase a pre-existing but weak overt response to a stimulus (i.e. so-called 'alpha conditioning' (see Carew *et al.*, 1984; Gould and Marler, 1984)).

In *Drosophila* parasitoids, responses to microhabitat odours are strongly influenced by alpha conditioning (Vet, 1983, 1985, 1988; Vet and van Opzeeland, 1984; Papaj and Vet, 1990). *Leptopilina heterotoma* females dramatically increase their responses to stimuli after having encountered them in association with oviposition in host larvae. Females experienced with apple-yeast substrate respond significantly more strongly to the odour of an apple-yeast substrate than naïve females or females experienced with another substrate (in olfactometer (e.g. Vet, 1988) and in mark–recapture experiments (e.g. Papaj and Vet, 1990)). Similar response increases to stimuli associated with hosts and, in some cases, with host by-products only, have been demonstrated in several other parasitoid species, including other eucoilids (Vet, 1983), braconids (Vinson *et al.*, 1977; Drost *et al.*, 1986, 1988; Turlings *et al.*, 1989), tachinids (Monteith, 1963), ichneumonids (Arthur, 1966, 1971), aphidiidids (Sheehan and Shelton, 1989) and trichogrammatids (Kaiser *et al.*, 1989). During the last decade, learning related to host-habitat and host searching has been found in many species of parasitoids (e.g. Bjorksten and Hoffmann 1998; Geervliet *et al.*, 1998; Steidle *et al.*, 2001).

Various types of stimuli can be involved in these learning processes. There are reports in the literature of parasitoid species learning odours, colours and shapes. Learning can increase the response to an experienced stimulus by positive associative learning, but it can also decrease the response by negative associative learning (Eisenstein and Reep, 1989).

Learning increases originally low responses more than originally high responses

With flies and parasitic insects, it has been observed that responses to less preferred stimuli are influenced more by learning than responses to more preferred stimuli (Jaenike, 1982, 1983, 1988; Prokopy *et al.*, 1982; Vet and van Opzeeland, 1984; Kaiser *et al.*, 1989; Sheehan and Shelton, 1989; Vet *et al.*, 1995). These observations may possibly account for the remarks of some colleagues working only with highly preferred stimuli that 'their' species do not seem to learn. In some studies, it is partly the method by which the behavioural response is measured that limits how much a response changes with experience. When responses are measured in terms of choice situations or percentages – and so the behavioural measure has an upper bound of 100% – there may be little scope for learning. For example, in *Asobara* species, the preference for odours of originally less preferred host substrates is increased markedly by an oviposition experience on these substrates. No such measurable effect occurs with substrate odours that are originally more preferred, as an increase in preference for these odours is barely possible (Vet and van Opzeeland, 1984; see also Drost *et al.*, 1986, 1988).

In conclusion, experimental data with parasitoids and flies suggest that this lower effect of learning on responses that are initially high is (although sometimes a methodological feature) a true behavioural phenomenon. It may reflect the existence of a maximum response to a stimulus as set by physiological constraints.

For naïve females, host-derived stimuli serve as key stimuli (rewards) in associative learning of other stimuli

As stated earlier, associative learning seems to be a major source of behavioural plasticity in parasitoids and other insects. Responses to stimuli can be acquired or enhanced by linking these stimuli to a key stimulus (reward). However, what is the nature of these reinforcing stimuli for parasitoids? Naïve insects foraging for food use stimuli unambiguously associated with feeding (e.g. sugars) as key stimuli (Chapter 5; Papaj and Prokopy, 1989; Wäckers, 1994). It is no accident that these stimuli are most frequently used in conditioning paradigms. They elicit responses that are strong and consistent. By analogy, we expect naïve parasitoids foraging for hosts to use stimuli unambiguously associated with oviposition as key stimuli, and not, for example, stimuli associated with finding the host habitat. This is in fact what is observed, for example, in *L. heterotoma,* which does not link a novel odour to the presence of a substrate, but to the presence of hosts (Vet and Groenewold, 1990).

Current knowledge indicates that the key stimuli used by naïve parasitoids in associative learning are always host-derived. These stimuli themselves generally elicit strong and predictable responses in naïve animals.

The Model

A simple conceptual model embraces these initial observations. It encompasses the following. First, parasitoids do not respond to each possible stimulus in the same way or to the same extent. Secondly, strong responses are less variable than weak ones. Thirdly, learning can change response levels. Fourthly, the extent to which experience alters a response depends on its original level and the fact that learning increases weak responses more than strong ones. Fifthly, in naïve individuals, stimuli that evoke high and predictable responses, such as those derived from the host, are most likely to function as a key stimulus to condition other stimuli.

Response potential

We first postulate a unique response potential for each stimulus perceived by a parasitoid. Note that we are speaking of potential and not realized behaviour. The response potential is a way of assigning all incoming stimuli a relative value in common units, regardless of whether those stimuli evoke fundamentally different responses. If, for example, the odour of a host substrate stimulates upwind anemotaxis in a parasitoid, any differences in the insect's walking speed in different odour plumes reflect differences in response potentials among the odours. However, for stimuli that evoke different behaviours (e.g. ovipositor probing vs. flying in the presence of odour), response potentials cannot be compared readily by external observation alone. We assume that a maximum response potential exists for a naïve individual of a given physiological state, developmental history and genetic composition (see Chapter 4). This maximum is set by constraints on the motor patterns elicited by stimuli, e.g. a maximal walking speed or a maximum ovipositor-probing frequency.

Ranking of stimuli

In Fig. 3.1, all stimuli perceived by the insect are ranked according to the strength of their response potential in the naïve insect. Each stimulus occupies a unique 'slot' along the response-potential continuum. Stimulus S_1 has the highest response potential and the response potentials to the different stimuli decrease along the abscissa. The sigmoidal shape of the distribution is based on the assumption that the distribution is actually composed of two types of stimuli: those with responses maintained by natural selection and those with responses maintained by constraint. The first group of stimuli (i.e. those maintained by natural selection) involves some stimuli that are essential in the host-location process of the parasitoid and that evoke very high, adaptive responses in the naïve insect. We can think of indispensable host-derived stimuli that evoke very weak, behaviourally neutral responses that are maintained by some constraint. Some of the latter stimuli may be components of or may overlap with the more important stimuli. It may not be cost-effective or even possible to reduce these responses to zero. It may be that these behaviourally neutral responses act as a reference library, which the animal employs as needed during associative learning. The part of the curve in between the stimuli with high and those with low response potentials are stimuli of intermediate value. 'Stimuli' $S > j$ are beyond the range of sensory perception of the animal. As these stimuli cannot be perceived, they can never be learned. This distinguishes them from other behaviourally neutral stimuli to which responses can be induced through learning.

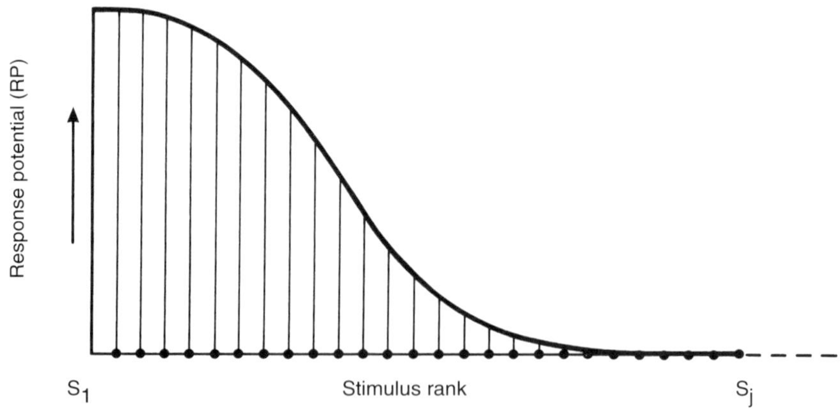

Fig. 3.1. Diagram of a female parasitoid's potential behavioural response to a variety of environmental stimuli. All stimuli perceived by the insect are ranked according to their response potential in the naïve insect. Stimuli beyond S_j are outside the range of sensory perception of the animal.

Experience

We next assume that experience can change the response potential of a stimulus and, when it does so, it moves this stimulus from one slot to another. Since a given slot can hold one and only one stimulus, this change always causes some other stimuli along the continuum to be displaced as well.

Variability

We further assume that there is always some variability when we actually measure the overt behavioural response to a certain stimulus, even if the response potential remains constant. So, given an insect of a particular genetic composition, physiological state and level of experience, the overt response can be predicted only with a certain error, as shown by the shaded area in Fig. 3.2. The magnitude of variability is assumed to depend on the strength of the response potential. When response potentials are high, they show low variability within the individual. Although this assumption is mainly empirically derived, it may be based on a plausible physiological reason (see 'Strong responses are less variable than weak ones', above). Furthermore, between individuals we expect little variation when responses to stimuli on the left-hand side of the abscissa are measured. Natural selection not only has led to these response potentials being inherently high but also has probably reduced differences in the maximal level of these response potentials between individuals of a population, which again reduces the variability in responses measured. So, when response potentials are high, actual responses appear constant and predictable and, when response potentials are lower, actual responses are assumed to show more variability within and between individuals. These responses vary over successive measurements in an unpredictable manner. When response potentials are very low, there is in reality less and less room for variability in the actual response, simply because responses cannot be lower than zero. Thus, extremely low mean responses may actually be associated with reduced variability than occurs with slightly higher mean responses. The resulting pattern of variability in actual responses over the range of response potentials is portrayed by the curve of realized variability ($V_{realized}$) positioned vertically on the left-hand side in Fig. 3.2.

Key stimuli

Finally, we assume that whether a stimulus can serve as a key stimulus (reward) for another stimulus in associative learning depends on the position of the two stimuli

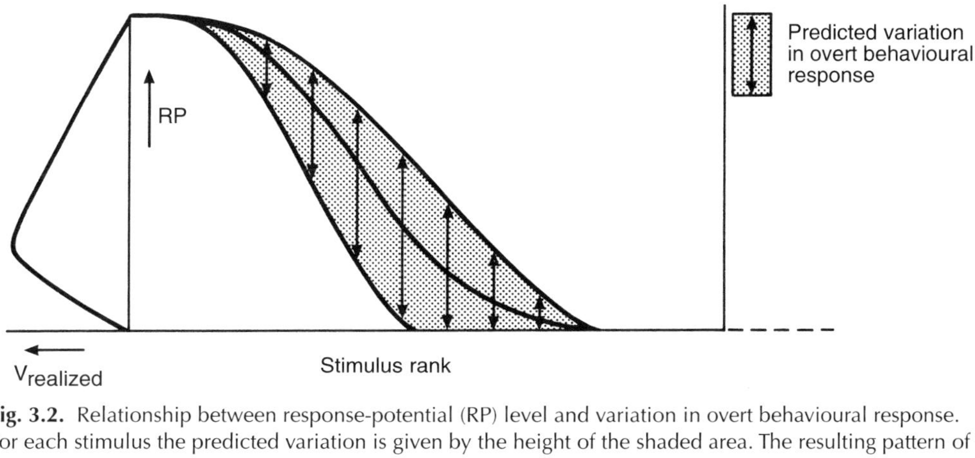

Fig. 3.2. Relationship between response-potential (RP) level and variation in overt behavioural response. For each stimulus the predicted variation is given by the height of the shaded area. The resulting pattern of variability in actual responses over the range of response potentials is given by the $V_{realized}$ curve. See text for additional explanation.

on the response potential continuum. Specifically, stimuli with the higher response potentials will be most likely to condition responses to stimuli with lower response potentials in associative learning. Moreover, we assume that the higher the response potential of the key stimulus, the greater the behavioural change it induces.

Synopsis

Figure 3.3 presents a flow diagram of the major concepts of the model. The central idea is the response potential. The lines indicate an influence or determination of one factor upon another, the arrows specifying the direction in which this occurs. This system is couched within the internal and external environment of the parasitoid.

Hypothesis Related to the Model

We can easily see that the model embraces each of our initial observations; furthermore, it enables us to formulate various testable hypotheses.

When learning changes the response to a stimulus, it should change the variability of that response accordingly

Since learning usually increases the response to a stimulus, it should also reduce the variability of that response. Thus, in general, the responses of naïve individuals should be more variable than those of experienced individuals. Several examples suggest that this is the case. Naïve and experienced *L. heterotoma* females differ in their variability (CV) in the time spent searching on two substrates (Vet *et al.*, 1990, Table II). After oviposition on a substrate, the time spent searching on that substrate increases and becomes less variable. Similarly, *Trichogramma evanescens* responds to a sex pheromone of its host, *Mamestra brassicae*, in a wind-tunnel, i.e. uses it as a kairomone (Noldus, 1988; Noldus *et al.*, 1988, 1990). The response is expressed as the length of time spent on a platform in the odour plume. Different experiences influence the mean duration, which correlates with a significant decrease in variability (Vet *et al.*, 1990, Fig. 4). Finally, *Exeristes roborator*, an ichneumonid parasitoid, was exposed to one of three conditioning treatments: (i) a natural host and

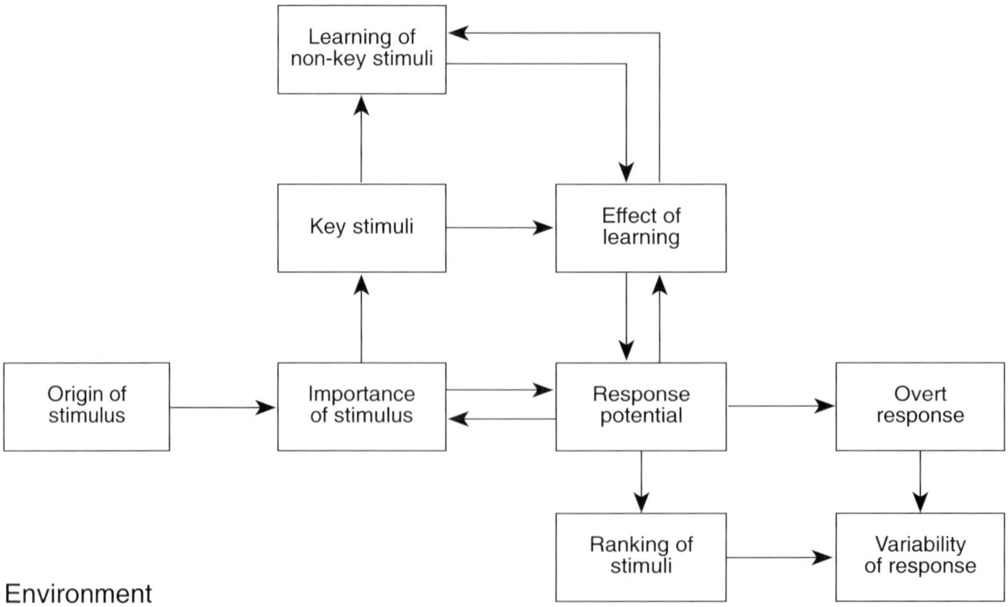

Fig. 3.3. Flow diagram of the major concepts of the variable-response model. For explanation see text.

habitat; (ii) no exposure (naïve); or (iii) conditioning to a factitious host in an artificial habitat (Vet *et al.*, 1990, Table III; based on Wardle and Borden (1986)). Its response to a natural host was then measured. The responses of the naïve parasitoids varied more than those of females experienced on the natural host. Experience with the 'wrong' host and habitat significantly reduces its response to natural hosts and simultaneously increases the response's variability.

The magnitude of the change in response for a given stimulus with a given experience depends on the level of the original response

If a given experience increases a stimulus's rank order with a certain number of steps, the change in its response potential will depend on its original position in the rank order. If it was ranked either low or high, then its response potential will change relatively little. If it was of intermediate rank, then the change will be larger (Fig. 3.2). This may explain the observations by Lewis and Tumlinson (1988), in which *Microplitis croceipes* rapidly learned some plant odours but exhibited more limited learning of other odours, e.g. vanilla (which is originally behaviourally neutral and so situated on the far right of the stimulus-rank axis). A number of stimuli with high response potentials will even evoke responses that are not variable and not subject to modification by experience. These stimuli include those that trigger motor responses known as 'fixed-action patterns' (Manning, 1972; Alcock, 1984).

A change in response to a stimulus exerted by experience can change responses to other stimuli

When experience increases the response potential of a stimulus, i.e. increases its rank order, other stimuli will be displaced and their rank order (response potential) will decrease. Furthermore, the response potential of several stimuli may increase in concert due to experience. This phenomenon has been shown for parasitoids (Vet and van Opzeeland, 1984; Drost *et al.*, 1988; Turlings *et al.*, 1989). Note that, by increasing the rank order of one stimulus, the response potentials of some stimuli will change while those of others remain unaffected. This pattern, in which experience with a given stimulus affects the response to other stimuli to differing degrees (= cross-induction of Papaj and Prokopy (1986)), has been shown for saprophagous and frugivorous insects (Jaenike, 1983; Papaj and Prokopy, 1986; Papaj *et al.*, 1989). Such cross-induction may be a selectively neutral but physiologically unavoidable side-effect of other response modifications that are adaptive. It remains to be examined in parasitoids.

The response pattern exhibited in a choice situation will be dictated by the rank order of the stimuli involved

If animals are faced with comparable stimuli (such as odours from different host plants), they should prefer the stimulus with the highest response potential. If the response potential is modified sufficiently through experience, learning may reverse the preference. For example, if *Leptopilina clavipes,* a parasitoid of mushroom-feeding *Drosophila,* is reared on a yeast substrate, its response to yeast odour increases but the increase is insufficient to displace the response to mushroom odour. However, if it oviposits in hosts on yeast, it prefers yeast odours to those of mushroom (Vet, 1983).

Key stimuli are expected most often to be those that evoke strong responses in naïve individuals, but any stimulus, whether or not it evokes a strong response in a naïve individual, can potentially act as a key stimulus for other stimuli

If we look at animals other than parasitoids, the stimuli (sugar, shock, poisons, etc.) most frequently used in conditioning paradigms are exactly those that elicit strong and consistent responses. In parasitoid foraging, oviposition-related stimuli and, generally speaking, host-derived products elicit the least variable of all responses in the naïve female. And, to the best of our knowledge, it is these stimuli that function as key stimuli in associative learning. The model assumes that the key stimulus with the highest response potential will give the strongest reinforcement. Such high-response stimuli are likely to be closely and reliably linked with the material presence of a host and its suitability for larval survival. By using such stimuli as the predominant reinforcers in associative learning processes, the insect can freely increase its responses to stimuli that are not reliable predictors of host presence and suitability in the long term (i.e. over evolutionary time) but which happen to be predictors of host presence and suitability in the short term (i.e. over the lifetime of the insect). The idea that any stimulus can potentially act as a key stimulus may account for the phenomenon of second-order conditioning. Second-order conditioning occurs when a stimulus that has been conditioned by a key stimulus becomes itself a key stimulus (Sahley, 1984). As the response potential of this conditioned stimulus increases in our paradigm, it displaces increasingly more other stimuli and is increasingly likely to be effective as a key stimulus for other stimuli. Second-order conditioning has been found in a variety of vertebrates and invertebrates (Sahley, 1984), including bees (Menzel, 1983), but has never been investigated in parasitoids.

Therefore one of the major insights of our model is perhaps this implication that many more types of stimuli can act as key stimuli than has been previously assumed, including stimuli that originally elicit little or no overt behavioural response in the naïve insect. Our model suggests that the number of key stimuli used by a parasitoid in learning will increase as increasingly more stimuli are 'confirmed' to be reliable predictors of host presence and suitability. Through this second-order conditioning, the insect effectively constructs a hierarchy of biologically meaningful causal relationships over the course of its foraging life. Acquiring a large and reliable set of key stimuli may increase the rate at which the insect learns. If this faster learning confers some reproductive benefit upon the individual, the accumulation of key stimuli should have some selective advantage.

The shape of the response-potential curve will differ among species and will reflect the ecological circumstances within which the species operates

Much attention has been devoted to the differences in foraging strategies between generalist and specialist species (e.g. Waage, 1979; Vet and Dicke, 1992; Vet *et al.*, 1995; Steidle and van Loon, 2002) and, in particular, to the possible correlation between niche breadth and learning ability (Arthur, 1971; Cornell, 1976; Daly *et al.*, 1980; Gould and Marler, 1984; Vet and van Opzeeland, 1984; van Alphen and Vet, 1986; Papaj and Prokopy, 1989; Vet and Dicke, 1992; Vet *et al.*, 1995). It is usually postulated that generalist species (because of their more variable environment) will learn more ably than specialist species. For insects the evidence for this is conflicting (Papaj and Prokopy, 1989). Perhaps we can add some 'food for thought' from the viewpoint of this variable-response model. The shape of the response curve itself can be expected to differ among species and to reflect the ecological circumstances within which each species operates. If the area under the response curve is constrained and remains relatively constant across related species, we might expect that generalist species have a flatter distribution of response potentials than specialists (Fig. 3.4). As a general rule based on our model, we expect specialists to show less variability in their responses than generalists with regard to the stimuli which they are specialist or generalist for. In addition, we can argue that, as the fraction of intermediate response potentials is greater in generalists than in specialists, the breadth of what can be learned is expected to be greater in generalist species. For parasitoids, there is some evidence that both generalists (e.g. Arthur, 1966; Vet and Schoonman, 1988; Turlings *et al.*, 1989) and specialists (e.g. Arthur, 1971; Vet, 1983; Vet and van Opzeeland, 1984; Sheehan and Shelton, 1989) can learn.

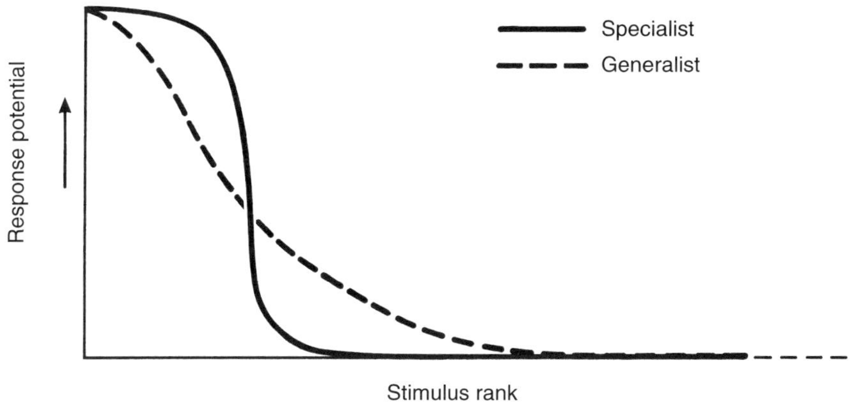

Fig. 3.4. Differences in response-potential curves between specialist and generalist parasitoid species. See text for additional explanation.

Recent studies revealed that parasitoids with a broader host range seem to use more general cues than more specialized parasitoids (Vet *et al.*, 1993; Hedlund *et al.*, 1996; Röse *et al.*, 1998; Bruni *et al.*, 2000). However, in other studies, no or only minor differences were found (Geervliet *et al.*, 1996; Cortesero *et al.*, 1997). As yet, not enough data exist to make a meaningful comparison of the relative learning ability of specialist and generalist parasitoids. In addition, most present data, with the exception of results from *Cotesia* species, include work on only distantly related species. Any comparison would risk erroneously attributing differences in learning to differences in diet breadth when in fact they are due to other factors, for example, differences in phylogeny (Papaj and Prokopy, 1989). We suggest that the lack of consensus with regard to the learning abilities of specialists and generalists may be due in part to the failure to test enough stimuli over the possible range of response levels.

Concluding Remarks

The effect of experience on the mean and variability (and thus predictability) of behavioural responses has interesting implications for the use of parasitoids in biological control. An improved predictability of natural-enemy behaviour will stimulate application of biological control (Lewis *et al.*, 1990). Unpredictable behaviour can hamper mass rearing and the development of reliable introduction schemes, lead to disinterest in the biological-control method and result in the release of exorbitantly high numbers of animals of poor quality, leading to high control costs. The postrelease migration behaviour of parasitoids away from the target area is considered to be a special problem (e.g. Ridgway *et al.*, 1981; Keller *et al.*, 1985). Increasing the mean and reducing the variability of the response to target stimuli through experience could considerably alleviate this problem.

Although our model implies that all learning in parasitoids can be reduced to simple associative processes, where reinforcement increases the response to some other stimulus, it also includes other effects of experience where an obvious reinforcement is lacking (e.g. sensitization and habituation).

A behavioural repertoire is a complex process, influenced by genes, environment, physiology and experience. Being aware of this complexity, we merely present a tool to simplify and clarify the effect of experience on behavioural responses and variability in those responses. The simplicity of the model enables us to formulate clear and testable hypotheses bearing on the desired or unavoidable manipulation of natural enemies, interspecific differences in behavioural plasticity and learning mechanisms.

The information presented in this chapter

is of particular interest to those developing quality control methods. Insight into the variability and predictability of natural-enemy behaviour will lead to adequate and realistic tests to evaluate foraging behaviour.

Acknowledgements

This chapter is the result of a cooperative programme between the Insect Biology and Population Management Research Laboratory (US Department of Agriculture (USDA), Tifton, USA), the Insect Attractants, Basic Biology and Behaviour Research Laboratory (USDA, Gainesville, USA), and the Laboratory of Entomology, Wageningen University (Wageningen, The Netherlands). The *Journal of Insect Behavior* (Kluwer Academic/Plenum Publishers) granted permission to reprint an edited version of the original Vet *et al.* (1990) paper with the same title and authors. Editing was made particularly easy with the recent extensive critical review of the 1990 paper by Steidle and Van Loon (2002).

References

Alcock, J. (1984) *Animal Behavior: an Evolutionary Approach*, 3rd edn. Sinauer, Sunderland, Massachusetts, 596 pp.

Arthur, A.P. (1966) Associative learning in *Itoplectis conquisitor* (Say) (Hymenoptera: Ichneumonidae). *Canadian Entomologist* 98, 213–223.

Arthur, A.P. (1971) Associative learning by *Nemeritis canescens* (Hymenoptera: Ichneumonidae). *Canadian Entomologist* 103, 1137–1141.

Bjorksten, T.A. and Hoffmann, A.A. (1998) Persistence of experience effects in the parasitoid *Trichogramma* nr. *brassicae*. *Ecological Entomology* 23, 110–117.

Bruni, R., Sant'-Ana, J., Aldrich, J.R. and Bin, F. (2000) Influence of host pheromone on egg parasitism by scelionid wasps: comparison of phoretic and nonphoretic parasitoids. *Journal of Insect Behavior* 13, 165–174.

Carew, T.J., Abrams, T.W., Hawkins, R.D. and Kandel, E.R. (1984) The use of simple invertebrate systems to explore psychological issues related to associative learning. In: Alkon, D.L. and Farley, J. (eds) *Primary Neural Substrates of Learning and Behavioural Change.* Cambridge University Press, Cambridge, pp. 169–184.

Corbet, S.A. (1985) Insect chemosensory responses: a chemical legacy hypothesis. *Ecological Entomology* 10, 143–153.

Cornell, H. (1976) Search strategies and the adaptive significance of switching in some general predators. *American Naturalist* 110, 317–320.

Cortesero, A.M., DeMoraes, C.M., Stapel, J.O., Tumlinson, J.H. and Lewis, W.J. (1997) Comparisons and contrasts in host-foraging strategies of two larval parasitoids with different degrees of host specificity. *Journal of Chemical Ecology* 23, 1589–1606.

Daly, M., Rauschenberger, J. and Behrends, P. (1980) Food-aversion learning in kangaroo rats: a specialist–generalist comparison. *Animal Learning and Behaviour* 10, 314–320.

Dicke, M. and Vet, L.E.M. (1999) Plant–carnivore interactions: evolutionary and ecological consequences for plant, herbivore and carnivore. In: Olff, H., Brown, V.K. and Drent, R.H. (eds) *Herbivores: Between Plants and Predators.* Blackwell Science, Oxford, pp. 483–520.

Dicke M., Sabelis, M.W. and Groeneveld, A. (1986) Vitamin A deficiency modifies response of predatory mite *Amblyseius potentillae* to volatile kairomone of two-spotted spider mite. *Journal of Chemical Ecology* 12, 1389–1396.

Drost, Y.C., Lewis, W.J., Zanen, P.O. and Keller, M.A. (1986) Beneficial arthropod behavior mediated by airborne semiochemicals. I. Flight behavior and influence of preflight handling of *Microplitis croceipes* (Cresson). *Journal of Chemical Ecology* 12, 1247–1262.

Drost, Y.C., Lewis, W.J. and Tumlinson, J.H. (1988) Beneficial arthropod behavior mediated by airborne semiochemicals. V. Influence of rearing method, host plant and adult experience on host-searching behavior of *Microplitis croceipes* (Cresson), a larval parasitoid of *Heliothis. Journal of Chemical Ecology* 14, 1607–1616.

Du, Y.J., Poppy, G.M. and Powell, W. (1996) Relative importance of semiochemicals from first and second trophic levels in host foraging behavior of *Aphidius ervi*. *Journal of Chemical Ecology* 22, 1591–1605.

Eisenstein, E.M. and Reep, R.L. (1989) Behavioral and cellular studies of learning and memory in insects. In: Kerkut, G.A. and Gilbert, L.I. (eds) *Comprehensive Insect Physiology, Biochemistry and Pharmacology*, Vol. 9: *Behaviour*. Pergamon Press, Oxford, pp. 513–547.

Geervliet, J.B.F., Vet, L.E.M. and Dicke, M. (1996) Innate responses of the parasitoids *Cotesia glomerata* and *C. rubecula* (Hymenoptera: Braconidae) to volatiles from different plant–herbivore complexes. *Journal of Insect Behaviour* 9, 525–538.

Geervliet, J.B.F., Vreugdenhil, A.I., Dicke, M. and Vet, L.E.M. (1998) Learning to discriminate between infochemicals from different plant–host complexes by the parasitoids *Cotesia glomerata* and *C. rubecula*. *Entomologia Experimentalis et Applicata* 86, 241–252.

Godfray, H.C.J. (1994) *Parasitoids – Behavioral and Evolutionary Ecology*. Princeton University Press, Princeton, New Jersey, 473 pp.

Gould, J.L. and Marler, P. (1984) Ethology and the natural history of learning. In: Marler, P. and Terrace, H.S. (eds) *The Biology of Learning*. Springer-Verlag, New York, pp. 47–74.

Gross, H.R., Lewis, W.J., Jones, R.L. and Nordlund, D.A. (1975) Kairomones and their use for the management of entomophagous insects. III. Stimulation of *Trichogramma achaeae, T. pretiosum* and *Microplitis croceipes* with host seeking stimuli at the time of release to improve their efficiency. *Journal of Chemical Ecology* 1, 431–438.

Hedlund, K., Vet, L.E.M. and Dicke, M. (1996) Generalist and specialist parasitoid strategies of using odours of adult drosophilid flies when searching for larval hosts. *Oikos* 77, 390–398.

Hérard, F., Keller, M.A., Lewis, W.J. and Tumlinson, J.H. (1988) Beneficial arthropod behavior mediated by airborne semiochemicals. IV. Influence of host diet on host-oriented flight chamber responses of *Microplitis demolitor* Wilkinson. *Journal of Chemical Ecology* 14, 1597–1606.

Herrebout, W.M. (1969) Habitat selection in *Eucarcelia rutilla Vill.* (Diptera: Tachinidae). II. Experiments with females of known age. *Zeitschrift für angewandte Entomologie* 63, 336–349.

Herrebout, W.M. and van der Veer, J. (1969) Habitat selection in *Eucarcelia rutilla Vill.* (Diptera: Tachinidae). III. Preliminary results of olfactometer experiments with females of known age. *Zeitschrift für angewandte Entomologie* 64, 55–61.

Jaenike, J. (1982) Environmental modification of oviposition behavior in *Drosophila*. *American Naturalist* 119, 784–802.

Jaenike, J. (1983) Induction of host preference in *Drosophila melanogaster*. *Oecologia* 58, 320–325.

Jaenike, J. (1988) Effects of early adult experience on host selection in insects: some experimental and theoretical results. *Journal of Insect Behavior* 1, 3–15.

Kaiser, L., Pham-Delegue, M.H., Bakchine, E. and Masson, C. (1989) Olfactory responses of *Trichogramma maidis* Pintureau and Voegele: effect of chemical cues and behavioral plasticity. *Journal of Insect Behaviour* 2, 701–712.

Keller, M.A., Lewis, W.J. and Stinner, R.E. (1985) Biological and practical significance of movement by *Trichogramma* species: a review. *Southwestern Entomologist Supplement* 8, 138–155.

Lewis, W.J. and Takasu, K. (1990) Use of learned odours by a parasitic wasp in accordance with host and food needs. *Nature* 348, 635–636.

Lewis, W.J. and Tumlinson, J.H. (1988) Host detection by chemically mediated associative learning in a parasitic wasp. *Nature* 331, 257–259.

Lewis, W.J., Jones, R.L., Gross, H.R. and Nordlund, D.A. (1976) The role of kairomones and other behavioral chemicals in host finding by parasitic insects. *Behavioral Biology* 16, 267–289.

Lewis, W.J., Vet, L.E.M., Tumlinson, J.H., van Lenteren, J.C. and Papaj, D.R. (1990) Variations in parasitoid foraging behavior: essential element of a sound biological control theory. *Environmental Entomology* 19, 1183–1193.

Lewis, W.J., van Lenteren, J.C., Phatak, S.C. and Tumlinson, J.H. (1997) A total systems approach to sustainable pest management. *Proceedings of the National Academy of Sciences, USA* 94, 12243–12248.

Lewis, W.J., Stapel, J.O., Cortesero, A.M. and Takasu, K. (1998) Understanding how parasitoids balance food and host needs: importance to biological control. *Biological Control* 11, 175–183.

Manning, A. (1972) *An Introduction to Animal Behavior*. Edward Arnold, London, 294 pp.

Menzel, R. (1983) Neurobiology of learning and memory: the honeybee as a model system. *Naturwissenschaften* 70, 504–511.

Mollema, C. (1988) Heritability of host selection behaviour of *Asobara tabida*. In: Genetical aspects of resistance in a host–parasitoid interaction. PhD thesis, University of Leiden, Leiden, The Netherlands, pp. 99–107.

Monteith, L.G. (1963) Habituation and associative learning in *Drina bohemica* Mesn. (Diptera: Tachinidae). *Canadian Entomologist* 95, 418–426.

Nishida, T. (1956) An experimental study of the ovipositional behavior of *Opius fletcheri* Silvestri (Hymenoptera: Braconidae), a parasite of the melon fly. *Proceedings Hawaiian Entomological Society* 16, 126–134.

Noldus, L.P.J.J. (1988) Response of the egg parasitoid *Trichogramma pretiosum* to the sex pheromone of its host *Heliothis zea. Entomologia Experimentalis et Applicata* 48, 293–300.

Noldus, L.P.J.J., Lewis, W.J., Tumlinson, J.H. and van Lenteren, J.C. (1988) Olfactometer and windtunnel experiments on the role of sex pheromones of noctuid moths in the foraging behaviour of *Trichogramma* spp. In: Voegele, J., Waage, J. and van Lenteren, J.C. (eds) *Proceedings of the 2nd International Symposium on* Trichogramma *and Other Egg Parasites, 10–15 November 1986, Guangzhou, China.* Les Colloques de l'INRA 43, Paris, pp. 223–238.

Noldus, L.P.J.J., van Lenteren, J.C. and Lewis, W.J. (1990) How *Trichogramma* parasitoids use moth sex pheromones as kairomones: orientation behaviour in a wind tunnel. *Physiological Entomology* 16, 313–327.

Papaj, D.R. and Prokopy, R.J. (1986) Phytochemical basis of learning in *Rhagoletis pomonella* and other herbivorous insects. *Journal of Chemical Ecology* 12, 1125–1143.

Papaj, D.R. and Prokopy, R.J. (1989) Ecological and evolutionary aspects of learning in phytophagous insects. *Annual Review of Entomology* 34, 315–350.

Papaj, D.R. and Vet, L.E.M. (1990) Odor learning and foraging success in the parasitoid, *Leptopilina heterotoma. Journal of Chemical Ecology* 16, 3137–3150.

Papaj, D.R., Opp, S.B., Prokopy, R.J. and Wong, T.T.Y. (1989) Cross-induction of host fruit acceptance in medfly: the role of fruit size and chemistry. *Journal of Insect Behavior* 2, 241–254.

Potting, R.P.J., Vet, L.E.M. and Dicke, M. (1995) Host microhabitat location by stem-borer parasitoid *Cotesia flavipes*: the role of herbivore volatiles and locally and systemically induced plant volatiles. *Journal of Chemical Ecology* 21, 525–539.

Powell, W. (1999) Parasitoid hosts. In: Hardie, J. and Minks, A.K. (eds) *Pheromones of Non-lepidopteran Insects Associated with Agricultural Plants.* CAB International, Wallingford, pp. 405–427.

Prévost, G. and Lewis, W.J. (1990) Genetic differences in the response of *Microplitis croceipes* to volatile semiochemicals. *Journal of Insect Behavior* 3, 277–287.

Prokopy, R.J., Averill, A.L., Cooley, S.S. and Roitberg, C.A. (1982) Associative learning in egglaying site selection by apple maggot flies. *Science* 218, 76–77.

Ridgway, R.L., Ables, J.R., Goodpasture, C. and Hartstack, A.W. (1981) *Trichogramma* and its utilization for crop protection in the USA In: *Proceedings Joint American–Soviet Conference 'Use of Beneficial Organisms in the Control of Crop Pests'.* USDA, Washington, DC, pp. 41–48.

Röse, U.S.R., Lewis, W.J. and Tumlinson, J.H. (1998) Specificity of systemically released cotton volatiles as attractants for specialist and generalist parasitic wasps. *Journal of Chemical Ecology* 24, 303–319.

Sahley, C.L. (1984) Behaviour theory and invertebrate learning. In: Marler, P. and Terrace, H.S. (eds) *The Biology of Learning.* Springer-Verlag, New York, pp. 181–196.

Sheehan, W. and Shelton, A.M. (1989) The role of experience in plant foraging by the aphid parasitoid *Diaretiella rapae* (Hymenoptera: Aphidiidae). *Journal of Insect Behavior* 2, 743–759.

Smith, M.A. and Cornell, H.V. (1978) Hopkins host-selection in *Nasonia vitripennis* and its implications for sympatric speciation. *Animal Behavior* 27, 365–370.

Sokal, R.R. and Rohlf, F.J. (1981) *Biometry.* Freeman, New York, 859 pp.

Steidle, J.L.M. (2000) Host recognition cues of the granary weevil parasitoid *Lariophagus distinguendus* (Hymenoptera: Pteromalidae). *Entomologia Experimentalis et Applicata* 95, 185–192.

Steidle, J.L.M. and Schöller, M. (1997) Olfactory host location and learning in the granary weevil parasitoid *Lariophagus distinguendus* (Hymenoptera: Pteromalidae). *Journal of Insect Behavior* 10, 331–342.

Steidle, J.L.M. and van Loon, J.J.A. (2002) Chemoecology of parasitoid and predator oviposition behaviour. In: Hilker, M. and Meiners, T. (eds) *Chemoecology of Insect Eggs and Egg Deposition.* Blackwell, Oxford, pp. 291–317.

Steidle, J.L.M., Lanka, J., Müller, C. and Ruther, J. (2001) The use of general infochemicals in a generalist parasitoid. *Oikos* 95, 78–86.

Thorpe, W.H. and Jones, F.G.W. (1937) Olfactory conditioning in a parasitic insect and its relation to the problem of host selection. *Proceedings Royal Society Series B, Biological Sciences* 124, 56–81.

Tinbergen, N. (1951) *The Study of Instinct.* Oxford University Press, London, 228 pp.

Turlings, T.C.J., Tumlinson, J.H., Lewis, W.J. and Vet, L.E.M. (1989) Beneficial arthropod behavior mediated by airborne semiochemicals. VIII. Learning of host-related odors induced by a brief contact experience with host by-products in *Cotesia marginiventris* (Cresson), a generalist larval parasitoid. *Journal of Insect Behavior* 2, 217–225.

van Alphen, J.J.M. and Vet, L.E.M. (1986) An evolutionary approach to host finding and selection. In: Waage, J.K. and Greathead, D.J. (eds) *Insect Parasitoids*. Academic Press, London, pp. 23–61.

van Emden, H.F., Sponagl, B., Wagner, E., Baker, T., Ganguly, S. and Douloumpaka, S. (1996) Hopkins' 'host selection principle', another nail in its coffin. *Physiological Entomology* 21, 325–328.

van Lenteren, J.C. (1999) Fundamental knowledge about insect reproduction: essential to develop sustainable pest management. *Invertebrate Reproduction and Development* 36, 1–15.

Vet, L.E.M. (1983) Host-habitat location through olfactory cues by *Leptopilina clavipes* (Hartig) (Hym.: Eucoilidae), a parasitoid of fungivorous *Drosophila*: the influence of conditioning. *Netherlands Journal of Zoology* 33, 225–248.

Vet, L.E.M. (1985) Olfactory microhabitat location in some eucoilid and alysiine species (Hymenoptera), larval parasitoids of Diptera. *Netherlands Journal of Zoology* 35, 720–730.

Vet, L.E.M. (1988) The influence of learning on habitat location and acceptance by parasitoids. In: *Parasitoid Insects*. Les Colloques de l'INRA 48, INRA, Paris, pp. 29–34.

Vet, L.E.M. and Dicke, M. (1992) Ecology of infochemical use by natural enemies in a tritrophic context. *Annual Review of Entomology* 37, 141–172.

Vet, L.E.M. and Groenewold, A.W. (1990) Semiochemicals and learning in parasitoids. *Journal of Chemical Ecology* 16, 3119–3135.

Vet, L.E.M. and Jansen, C.J. (1984) Fitness of two sibling species of *Asobara* (Braconidae: Alysiinae), larval parasitoids of Drosophilidae in different microhabitats. *Ecological Entomology* 9, 345–354.

Vet, L.E.M. and Schoonman, G. (1988) The influence of previous foraging experience on microhabitat acceptance in *Leptopilina heterotoma*. *Journal of Insect Behavior* 1, 387–392.

Vet, L.E.M. and van Opzeeland, K. (1984) The influence of conditioning on olfactory microhabitat and host location in *Asobara tabida* (Nees) and *A. rufescens* (Foerster) (Braconidae: Alysiinae), larval parasitoids of Drosophilidae. *Oecologia* 63, 171–177.

Vet, L.E.M., Janse, C., van Achterberg, C. and van Alphen, J.J.M. (1984) Microhabitat location and niche segregation in two sibling species of drosophilid parasitoids: *Asobara tabida* (Nees) and *A. rufescens* (Foerster) (Braconidae: Alysiinae). *Oecologia* 61, 182–188.

Vet, L.E.M., Lewis, W.J., Papaj, D.R. and van Lenteren, J.C. (1990) A variable-response model for parasitoid foraging behaviour. *Journal of Insect Behavior* 3, 471–490.

Vet, L.E.M., Sokolowski, M.B., MacDonald, D.E. and Snellen, H. (1993) Responses of a generalist and a specialist parasitoid (Hymenoptera: Eucoilidae) to drosophilid larval kairomones. *Journal of Insect Behaviour* 6, 615–624.

Vet, L.E.M., Lewis, W.J. and Cardé, R.T. (1995) Parasitoid foraging and learning. In: Cardé, R.T. and Bell, W.J. (eds) *Chemical Ecology of Insects 2*. Chapman & Hall, New York, pp. 65–101.

Vinson, S.B. (1976) Host selection by insect parasitoids. *Annual Review of Entomology* 21, 109–134.

Vinson, S.B. (1981) Habitat location. In: Nordlund, D.A., Jones, R.L. and Lewis, W.J. (eds) *Semiochemicals: Their Role in Pest Control*. John Wiley & Sons, New York, pp. 51–77.

Vinson, S.B. (1984) How parasitoids locate their hosts: a case of insect espionage. In: Lewis, T. (ed.) *Insect Communication*. Royal Entomological Society, London, pp. 325–348.

Vinson, S.B., Barfield, C.S. and Henson, R.D. (1977) Oviposition behavior of *Bracon mellitor*, a parasitoid of the boll weevil *(Anthonomus grandis)*. II. Associative learning. *Physiological Entomology* 2, 157–164.

Visser, J.H. (1988) Host-plant finding by insects: orientation, sensory input and search patterns. *Journal of Insect Physiology* 34, 259–268.

Waage, J.K. (1979) Foraging for patchily-distributed hosts by the parasitoid, *Nemeritis canescens*. *Journal of Animal Ecology* 48, 353–371.

Wäckers, F.L. (1994) The effect of food deprivation on the innate visual and olfactory preferences in the parasitoid *Cotesia rubecula*. *Journal of Insect Physiology* 40, 641–649.

Wardle, A.R. and Borden, J.H. (1986) Detrimental effect of prior conditioning on host habitat location by *Exeristes roborator*. *Naturwissenschaften* 73, 559–560.

4 Variations in Natural-enemy Foraging Behaviour: Essential Element of a Sound Biological Control Theory

W.J. Lewis,[1] L.E.M. Vet,[2,3] J.H. Tumlinson,[4] J.C. van Lenteren[2] and D.R. Papaj[5]

[1]Insect Biology and Population Management Research Laboratory, USDA-ARS, PO Box 748, Tifton, GA 31793, USA; [2]Laboratory of Entomology, Wageningen University, PO Box 8031, 6700 EH Wageningen, The Netherlands; [3]Netherlands Institute of Ecology, PO Box 40, 6666 ZG Heteren, The Netherlands; [4]Insect Biology and Population Management Research Laboratory, USDA-ARS, PO Box 14565, Gainesville, FL 32604, USA; [5]Department of Ecology and Evolutionary Biology, University of Arizona, Tucson, AZ 85721, USA

Abstract

Intraspecific intrinsic variation in foraging behaviour is a common but often overlooked feature of natural enemies. These variations result from adaptations to the variety of foraging circumstances encountered by individuals of the species. We discuss the importance of understanding the mechanisms governing these intrinsic variations and the development of technologies to manage them. Three major sources of variation in foraging behaviour are identified. One source for variation is genotypically fixed differences among individuals that are adapted for different foraging environments. Another source of foraging variation is the phenotypic plasticity that allows individuals to make ongoing modifications of behaviour through learning, which suits them for different host-habitat situations. A third factor in determining variation in foraging behaviour is the natural enemy's physiological state relative to other needs, such as food and mating. A conceptual model is presented for comprehensively examining the respective roles of these variables and their interactive net effect on foraging behaviour. We also discuss proposed avenues for managing these variations in applied biological control programmes.

Introduction

The often erratic performances of natural enemies limits their use as pest-control agents. In parasitoids, the ability of females to locate and attack hosts is a key determinant of how well a given parasitoid population performs. Thus, the variation in this host-location ability could be a major source of inconsistent results in biological control. The causes for variation in natural-enemy foraging behaviour are currently poorly understood, despite a substantial body of theoretical and empirical literature dealing with the subject. Most earlier investigations focused on extrinsic factors, such

© CAB International 2003. *Quality Control and Production of Biological Control Agents: Theory and Testing Procedures* (ed. J.C. van Lenteren)

as foraging environments, as the source of variation in natural-enemy searching behaviour (Waage and Hassell, 1982; Vet and Dicke, 1992; Godfray, 1994; Vinson, 1998). Very limited consideration has been given to intraspecific variation in the natural enemy's genetic composition or behavioural state.

More recent studies show that foraging responses among individuals of a parasitoid population, even to the precise same set of stimuli, can be quite variable. Further, the behaviour of a given female parasitoid is often plastic and can vary considerably, depending on the history of that individual (e.g. Wardle and Borden, 1986; Lewis and Tumlinson, 1988; Vet *et al.*, 1990, 1995; Vet and Dicke, 1992; Steidle and van Loon, 2002). Therefore, researchers hoping to use natural enemies for biological control of pests must appreciate that an effective end result will be a product of the diversity and plasticity of the naural enemy's population interacting with environmental parameters of the foraging arena.

In this chapter, we explore sources of variation in the responsiveness of parasitoids to various foraging cues, with emphasis on the intrinsic causes for this variation, and the importance to biological control programmes of a proper matching of the parasitoids' genotypic and phenotypic behavioural traits with the type environment in which they must forage. We include considerations of genotypic diversity, the influence of different physiological states on the responses by individuals and the plasticity of individual parasitoids caused by preadult and adult experiences (see Chapter 3 for elaboration of the latter subject of parasitoid learning). A model is proposed for collectively assessing these sources of variation and their sum effect on parasitoid foraging behaviour. This model can, with adaptations, also be used for predators. Information about foraging behaviour of predators is, however, much more limited that that of parasitoids, and that is the reason why this chapter is mainly focused on parasitoids. Finally, we discuss ways that this information might be used to improve biological control.

Need for Understanding Variations in Parasitoid Foraging Behaviour

Animal behaviourists often emphasize interspecific diversity, particularly when illustrating how animals adapt to the variety of problems that they encounter (Alcock, 1984). Intraspecific diversity is also recognized as a common and important feature of animal behaviour, including foraging behaviour (Roughgarden, 1979; Hoy, 1988). Intraspecific differences in foraging behaviour typically involve differences among individuals and differences in the behaviour of a given individual from one foraging occasion to the next (Papaj and Prokopy, 1989). Behaviourists generally agree that these differences are caused by the selection for mechanisms that enable individuals to cope effectively with varying circumstances under which food resources must be obtained (Matthews and Matthews, 1978; Roughgarden, 1979; Alcock, 1984; Vet *et al.*, 1995).

Interspecific variation of parasitoids has been the subject of considerable discussion relative to biological control (e.g. Waage and Hassell, 1982; Bellows and Fisher, 1999), whereas intraspecific variation has received little attention in the design and implementation of biological control programmes (Caltagirone, 1985; Hoy, 1988). The regimented production process used with conventional pesticides has perhaps dulled our appreciation of biological knowledge needed for the production and use of biological organisms versus chemical formulations (Lewis, 1981; Lewis *et al.*, 1997). We must remember that evolution by natural selection does not stop at the species level but operates at the individual level. Thus, unlike chemical compounds or other products, the definition of a species or even a strain of a parasitoid does not mean that the individuals within the species are a product of one 'blueprint' or single set of performance characteristics. Furthermore, individual organisms are quite plastic, and their behavioural traits can be altered substantially by the conditions to which they are exposed. The challenge for biological control specialists is to recognize and respond effectively to this diversity as a resource rather than an obstacle to pest-management science.

Breeders of domesticated plants and animals have long recognized and exploited genetic diversity for useful purposes. However, most augmentative biological control programmes with parasitoids differ from conventional animal-breeding and production programmes in that the parasitoids are cultured and maintained in laboratory insectaries apart from the natural environment where they must eventually perform (see Chapters 1, 11 and 12). Therefore, it will be necessary for us to use techniques to ensure that the genotypic and phenotypic traits important to their performance in the natural environment are maintained intact and even enhanced during insectary production. The development and incorporation of such technology into biological control will necessitate understanding the sources and functional mechanisms of variations in parasitoid foraging behaviour. The result will be to enhance the quality of natural enemies and to improve their performance in the field.

Waage and Hassell (1982), citing van Lenteren (1980) and various case-history reports, stated:

> perhaps the outstanding question in biological control today is whether the use of parasitoids is to remain such an art, aided largely by the knowledge of what worked last time or whether it has the potential to become a fully predictive science, aided by fundamental research and theory.

Sources of Intraspecific Variations in Foraging Behaviour

Numerous extrinsic factors, such as climatic conditions and host density, can affect foraging behaviour (Chapter 1). However, in this chapter we are concerned primarily with intrinsic sources of variation. Adaptive variations in foraging traits are necessary for a parasitoid species to deal with different foraging environments. As reported for other organisms (Bradshaw, 1965; Papaj and Rausher, 1987; Papaj and Prokopy, 1989), there are two alternative types of adaptive variation in the foraging behaviour of a parasitoid species. One type of intraspecific variation is caused by genetically fixed differences among individuals. In this case, a species may have various genotypes that have a fixed or 'hard-wired' behaviour that inherently adapts them for operating effectively under the respective conditions for which they have been selected. For example, if the host occupies several habitats, the parasitoid species may consist of strains with different capabilities for searching in each of the habitats (Boulétreau, 1986; Pak, 1988; Wajnberg and Hassan, 1994).

This genotypic diversity among individuals of a species has generally been recognized by scientists and to some extent has been incorporated into considerations for biological control (Caltagirone, 1985; Luck and Uygun, 1986; Wajnberg and Hassan, 1994).The fact that strains of parasitoids that occupy different regions with different climatic conditions are inherently more suited for their respective ecological conditions has been well documented and appreciated (e.g. Pak, 1988). Also, populations of a parasitoid species with long-standing associations with different hosts and habitats are known to differ in their affinity and behaviour relative to those host-habitat situations (e.g. Mollema, 1988; Pak, 1988). In addition to these discrete genetic differences that occur among populations, we have more recently documented subtle, but distinct and measurable, heritable differences in the behaviour patterns of individuals within a single interbreeding population (Prévost and Lewis, 1990). These within-population differences are perhaps preserved as a result of the continuing flux of circumstances that the population encounters.

A second type of intraspecific variation is within individual plasticity (phenotypic plasticity, *sensu* Roughgarden (1979)). In this type the individuals have a partly open or unfixed behaviour (plastic within certain limits). These individuals are capable of adapting by experience for foraging more effectively in any one of a variety of circumstances that may be encountered. For example, a parasitoid of hosts that occurs in several different habitats may learn (as will be discussed later) to prefer the habitat in which it encounters suitable hosts (Vet, 1983; Vet *et al.*, 1995).

Only recently have we begun to appreciate the extent to which parasitoids can learn and the importance of this plasticity to biological control considerations (van Alphen and Vet, 1986; Vet *et al.*, 1995). Several studies have shown that many species of parasitoids are able to acquire by experience an increased preference for and ability to forage in a particular environmental situation. There is evidence that a parasitoid may acquire some modifications in its foraging traits during the immature stage (Thorpe and Jones, 1937; Vinson *et al.*, 1977; Vet, 1983; Luck and Uygun, 1986; van Emden *et al.*, 1996). However, the clearest cases and those with the greatest effects have thus far been shown to be from the experience of the adult parasitoid (Vinson *et al.*, 1977; Vet, 1983; Wardle and Borden, 1986; Drost *et al.*, 1988; Sheehan and Shelton, 1989; Vet *et al.*, 1995; Steidle and van Loon, 2002). The learning process is often associative learning, where the parasitoid learns to effectively use a previously weak or neutral cue for host foraging by associating it with the host or a product of the host (Lewis and Tumlinson, 1988; Turlings *et al.*, 1989; Vet *et al.*, 1990, 1995). In this case, close-range, reliable and unconditional stimuli can serve as reinforcers for the longer range and more variable conditional stimuli (Lewis and Tumlinson, 1988; Vet *et al.*, 1990, 1995). This learning process can begin at or just before eclosion, based on the host products associated with the parasitoid's cocoon (Vet, 1983; Hérard *et al.*, 1988b). Thereafter, the parasitoid's foraging responses are modified continually according to the foraging circumstances encountered (Vet *et al.*, 1990, 1995).

The conditions under which these two alternative adaptive variances of a species, fixed and unfixed, are most likely to occur have been discussed for other organisms (Bradshaw, 1965; Papaj and Rausher, 1987). We hypothesize that, in general, the occurrence of fixed versus unfixed foraging behaviour of parasitoids is determined by the combination of two basic features of the foraging environment. These features are: (i) the extent of differences between various host, habitat and other foraging situations encountered by the individuals; and (ii) the consistency with which each different foraging situation is available to the parasitoids within and among generations. Large differences among the characteristics of foraging situations should favour parasitoids with the genetically fixed alternative, because unfixed individuals must adjust to the widely different situations by learning. On the other hand, genotypes that are fixed for a particular foraging situation would need a dependable availability of that circumstance over generations. Thus, inconsistencies in the foraging situations favour individuals with a plastic behaviour (unfixed) that can be modified for the various circumstances encountered. A chart of the expected occurrence of these behavioural types relative to various foraging situations is presented in Table 4.1.

As an apparent result of the interacting effect of these selection forces, parasitoid individuals often and perhaps most commonly show a combination of the fixed and unfixed types of behavioural traits (Vet, 1983; Drost *et al.*, 1986, 1988; Vet *et al.*, 1995). This combination is accomplished by having an inherent rank order of preferences for the various cues used to locate hosts (e.g. the preference for different host-plant odours to which the parasitoid responds). However, the rank order can be modified within generations by learning based on the circumstances encountered by the individuals (Chapter 3; Vet *et al.*, 1990). We propose that this initial inherent rank order can vary substantially among individuals of the species. Further, the frequency of occurrence of a given rank order would be determined by its profitability over generations.

Another major factor that contributes to variations in the foraging behaviour of parasitoids is their general physiological state. A number of authors have shown that the foraging behaviour of female parasitoids can be altered by their physiological state relative to other needs and conditions (Chapter 5; Nishida, 1956; Hérard *et al.*, 1988a; Nordlund *et al.*, 1988; Wäckers, 1994). Naturally, a parasitoid faces varying situations in meeting its

Table 4.1. Genetically fixed versus unfixed (plastic) foraging behaviour in parasitoids and the foraging situations that affect the expected occurrence of each type.

	Extent of differences between host and habitat situations and the consistency in availability of each situation			
Type of selection	Differences small; availability consistent	Differences large; availability consistent	Differences small; availability inconsistent	Differences large; availability inconsistent
Selection for diverse fixed adaptations	Low	High	Low	High
Selection for individual plasticity	Low	Low	High	High
Resultant type of foraging behaviour	Less need for genetic differences or plasticity	Genetically different and fixed for each type situation	Adjustable through learning as situation necessitates	Genetically diversified with overlay of plasticity

food and mating requirements and its general health can vary because of diseases and climatic conditions. The resulting physiological state of the parasitoid interacts with the genotypic and phenotypic foraging traits discussed earlier in determining how a parasitoid will respond to a foraging environment (Chapter 5; Shahjahan, 1974; Hagen and Bishop, 1979; Hamm *et al.*, 1988; Wäckers, 1994).

Model of the Factors Determining Eventual Foraging Behaviour

The sources of variation discussed above are not mutually exclusive; rather, they overlap extensively, even within a single individual. Therefore, it is important that we have a means of clearly delineating the sources, roles and interacting effects of the variations. Our conceptual model for collectively describing the various foregoing factors and the sum of their effect on the foraging behaviour of parasitoids is presented in Fig. 4.1. The three major sources of intrinsic variability in the behaviour of foraging female parasitoids are represented: (i) genetic diversity among individuals; (ii) phenotypic plasticity within individuals because of experience; and (iii) the parasitoid's physiological state relative to other needs. The behaviour manifested is also dependent on the foraging environment, so the final foraging effectiveness of a parasitoid is determined by how well the parasitoid's net intrinsic condition as a result of these three components is matched with the foraging environment in which it operates.

Genetic diversity

In Fig. 4.1, we present a hypothetical parasitoid species and three foraging environments: EA, EB and EC. Under the 'genotypic diversity' heading, we show the response potential for two representative individual genotypes, G_1 and G_2. This response potential consists of the genetically fixed maximum range of usable foraging stimuli and the ultimate level with which the parasitoid could respond to the stimuli (the total darkened area plus the shaded area). This maximum level of response to the array of stimuli is shown as a curve, which indicates that the maximum response level varies with different stimuli in its range (Vet, 1983; Drost *et al.*, 1988; Vet *et al.*, 1995). As reflected by the different range and curve configurations for G_1 and G_2, the response potential may vary substantially among individuals within a population (Hoy, 1988; Prévost and Lewis, 1990).

The activated response potential of G_1 and G_2 (darkened area) that could be realized at any given time is somewhat less than their overall potential and depends on the experience of the individual, as documented earlier and as will be further discussed below for the model. Thus, only the response potential activated at the time an individual encounters stimuli can be manifested. The balance of the response potential that is not currently activated due to the experience of the individual is the latent response potential (shaded area). In the case of naïve individuals, the active response potential is that portion that is inherently activated and thus does not require experience before it can be manifested.

Obviously, the response potential of the individuals determines the response potential of a population that they make up, and the populations in turn determine the response potential of the species. However, only the response-range parameter is shown for the populations and species in Fig. 4.1, because the response level would depend on the density as well as the genotypes of individuals making up the population and species at any given time. Horizontal alignment of the response-range lines in Fig. 4.1 with the representative environments reflects the capacity to respond to the stimuli from that environment. As shown in Fig. 4.1, the stimuli of the three representative foraging environments, EA, EB and EC, are all within the range of population P_1; furthermore, the response ranges of individuals with the representative genotype, G_1, are best aligned with these environments. However, the inherent preference of the genotype G_1, as indicated, is for environment EB. Information on how well response

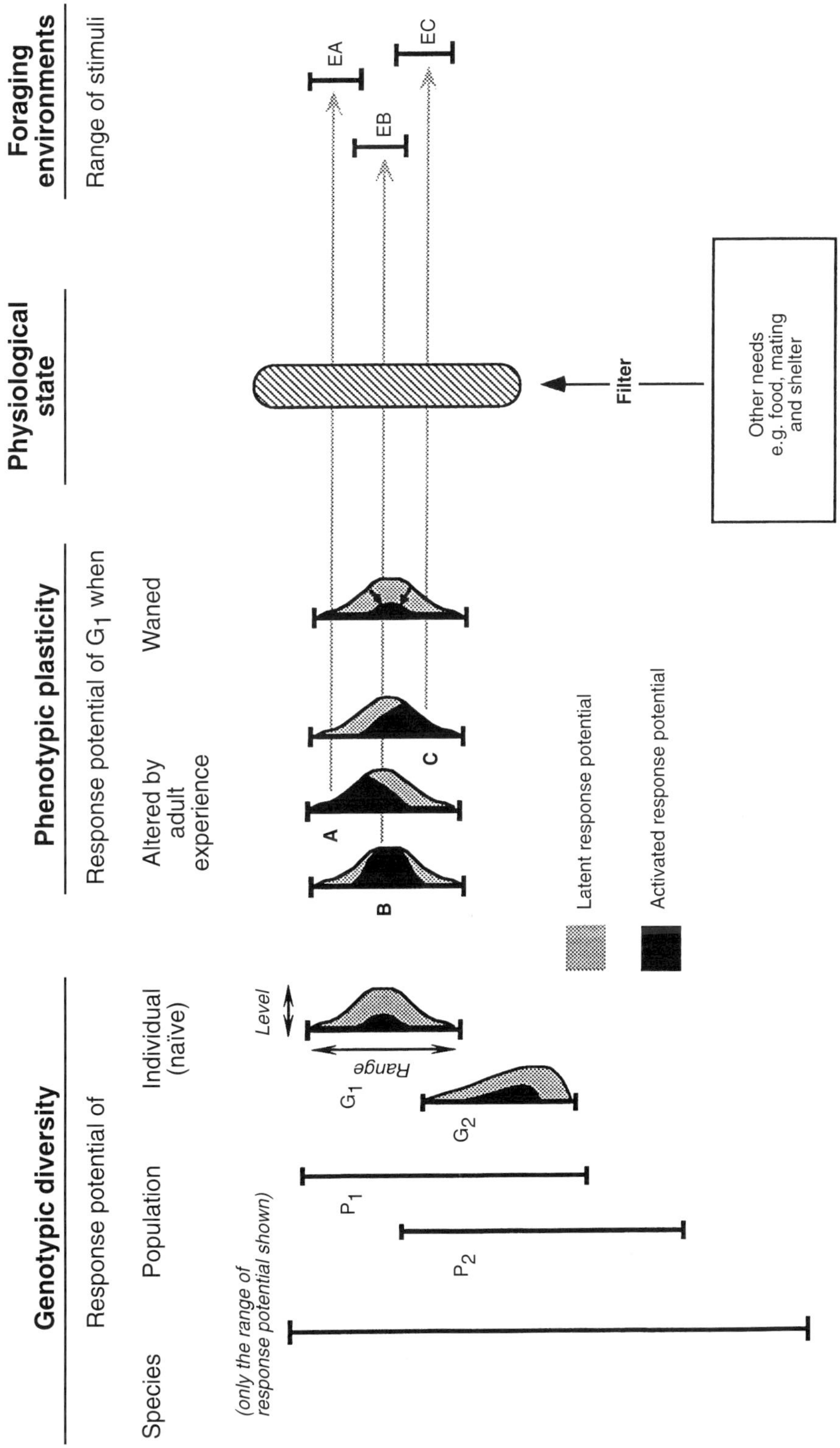

Fig. 4.1. Factors determining eventual foraging behaviour of a parasitoid (see text for explanation).

ranges match with stimuli of foraging environments is of vital importance in choosing a strain for use as biological control agent in a given environment.

For a population of parasitoids to be predictable and consistent in biological control, it must have a proper blend of genetic traits appropriate to the target environment and those traits must occur sufficiently uniformly in the population (Hoy, 1988). The need for a proper match of the parasitoid population with the target biological control environment has generally been recognized, but has been dealt with only on a gross level in applied programmes. For example, parasitoids colonized for a particular release situation have often been chosen from habitat and host circumstances similar to that expected in the targeted area (Caltagirone, 1985; Pak, 1988; Wajnberg and Hassan, 1994). We expect that in the near future the colonized parasitoids can at least be monitored with DNA-fingerprinting techniques to determine if their genetic make-up still incorporates necessary behavioural and other traits and if the important traits are occurring uniformly in the colony (e.g. Silva *et al.*, 2000).

Phenotypic plasticity

The activated response potential (Fig. 4.1, darkened area) of a foraging female is quite plastic and can be modified within the bounds of its genetic potential (Chapter 3; Vet *et al.*, 1990, 1995). The activated response potential of a parasitoid at any given time is dependent on the experience history of the individual at that moment. As discussed earlier, these modifications in response behaviour can begin during development as a result of the parasitoid's interaction with its environment. Thus, the activated response potential of the naïve adult will necessarily be altered as a routine consequence of rearing. The direction and level of the alteration as a result of rearing will depend on, among other things, the host species and host diet; these alterations have seldom, if ever, been quantified, although it has often been speculated that such changes occur (Chapters 1 and 12). Subsequently, the activated response potential of the adult parasitoid continues to change as a result of the experiences during foraging activities (see the earlier discussion on within-individual plasticity).

A hypothetical example of the changes in the activated response potential as a result of experience is shown for genotype G_1 in Fig. 4.1. As stated earlier, the genotypic response range of G_1 embraces the various stimuli from the foraging environments EA, EB and EC, as indicated by the length and alignment of its range. This hypothetical individual could develop a peak activated response potential for any of the three environments by successful experience with stimuli of that environmental situation (Vet, 1983; Wardle and Borden, 1985; Lewis and Tumlinson, 1988; Vet *et al.*, 1995). The highest activated response potential can be developed for stimuli of its more preferred environment, EB. Also, data suggest that the activated response levels for EB stimuli can be increased more quickly. Absence of reinforcement will result in a waning of the level of the activated response potential and a reversion to its naïve preference for the cues of EB (see Chapter 3 for a detailed discussion of modifications of parasitoid response potential).

Physiological state

A parasitoid's physiological state relative to other needs, such as food, mating and general health, can strongly influence the quality of its foraging behaviour. For example, if a female parasitoid is hungry, the appetitive drive for food cues may take precedence and, as a result, responses to host-related cues may be reduced (Chapter 5; Hagen and Bishop, 1979; Wäckers, 1994; Lewis *et al.*, 1998). Also, a lack of mature eggs in the ovaries can reduce the response to olfactory cues (Shahjahan, 1974). Further, the presence of other strong chemical, visual or auditory cues would probably disrupt the response to host-foraging cues by dilution. In other words, the physiological state of the parasitoid can greatly affect its propensity and ability to respond to the host-selection cues.

As shown in Fig. 4.1, the physiological state of the parasitoid relative to other needs can be considered as a gateway that filters the detection and responses to host-foraging stimuli based on priority of the needs. We feel that, until recently (Chapter 5), the influence of physiological state on the foraging behaviour of parasitoids is an area that has generally been recognized as important but one that has seldom been studied in a systematic way as needed to develop the technology important to ensuring effective and consistent host-foraging behaviour.

Variations in Responses at Different Points in the Host-selection Sequence

It is well recognized that foraging for hosts by parasitoids involves a series of steps that draw them progressively closer to their host (Salt, 1935; Flanders, 1953; Doutt, 1964). These steps were reviewed and amended by Vinson (1976, 1984a, 1998). Lewis *et al.* (1975b) proposed a basic sequence in which various behavioural acts are identified with the respective stimuli that elicit the responses. Much about the full repertoire of behaviours and specific stimuli is yet to be explained (Steidle and van Loon, 2002). Visual, tactile and chemical stimuli are all involved to some extent, but chemical stimuli appear to play a major and often dominant role for many parasitoids (e.g. Vet and Dicke, 1992).

Much more information has accumulated about mediating stimuli and close-range foraging responses than about long-range responses of parasitoids (Vinson, 1984b; Lewis *et al.*, 1985; Vet and Dicke, 1992). This was especially true before more recent advances in the knowledge of parasitoid responses to airborne odours were published (e.g. Drost *et al.*, 1986; Hérard *et al.*, 1988a; Lewis and Tumlinson, 1988; Vet and Dicke, 1992; Geervliet *et al.*, 1998). The lack of knowledge about the long-range foraging behaviour of parasitoids has been due to the greater ease with which the close-range behaviour could readily be studied in small laboratory containers. We offer arguments that behaviour-mediating stimuli and intrinsic parasitoid conditions that influence longer-range parasitoid foraging responses are more variable than close-range responses. This greater variation in the mediating stimuli and conditions required for response adds complexities to the methods and procedures required to study long-range behaviours effectively. Consequently, we seem to have avoided studies of parasitoid foraging at long range, although the information is certainly needed in designing more effective biological control.

As parasitoids negotiate a sequence of cues and approach increasingly closer to the host, there is a greater availability of direct host cues upon which to rely. Thus, there is less need for exploring, sorting and assessing indirect cues. As a hypothetical example, let us consider that a parasitoid may choose a habitat to search based on recognizable odour characteristics from a specific type of plant. Subsequently, specific parts of the plants, such as the buds or young fruit, may be scanned because of their general attraction. Upon detecting indications of an infested plant, such as damaged tissue, the parasitoid may hover close for more careful examination. If the smell of a potential host is perceived, the parasitoid may land and carefully antennate the surrounding area, particularly by-products, such as faeces and silk, indicating the presence of a candidate host. If antennation results in contact with a fresh-recognition kairomone, probing with the ovipositor may occur, followed by oviposition if a host is actually encountered. This hypothetical foraging sequence is only a general example for a parasitoid of a phytophagous host, and the number and exact types of cues in the sequence would vary among parasitoid species and types of hosts (Vinson, 1984b, 1998). We present this example to aid in illustrating some general conclusions that we shall propose.

We shall contend here that there is greater phenotypic plasticity (learning) at the long-range phase of the foraging sequence. In support of this argument, let us first consider the adaptive value of various behaviours involved in the foraging sequence of a parasitoid. The ultimate measure of success of the foraging responses – and thus the reference point from which natural selection operates

– is the actual encounter and oviposition in or upon a suitable host (e.g. Dicke and Vet, 1999). In other words, all the other responses in the foraging repertoire are of value only to the extent that they contribute to such encounters and oviposition.

A parasitoid can better perceive and use stimuli emanating directly from a host or a direct by-product of the host at close range than at longer range (Vinson, 1988). It stands to reason that the direct cues are more reliable indicators of host presence than indirect cues, such as the odour of a particular plant or other habitat odour. Of particular significance is the fact that these direct cues would be linked more consistently with hosts and ovipositional success over parasitoid generations, as would be required by natural selection for them to become genetically fixed ('hard-wired') responses. It is also apparent that at close range the parasitoids can afford to confine their responses to a more limited scope of stimuli and thus respond in a more homogeneous manner than at longer range in response to indirect cues.

Further, we have often observed that parasitoids encountering closer-range cues are less disrupted by other stimuli, such as light and movement (even touch by the observer), which reduces further the variations in behaviour at close range. This more focused behaviour at close range is similar to the tendency of other organisms to be less easily disturbed as they reach the immediate proximity of food, mates or other targets of a searching sequence (e.g. Steidle and van Loon, 2002).

At long range, the parasitoid, because of limitations in its ability to detect direct cues, must depend on indirect indicators, such as certain types, ages and parts of plants or other habitat cues typically indicating the potential presence of suitable hosts (Vet and Dicke, 1992). These indirect indicators may be cues generally associated with host presence, but the reliability of such associations within and among parasitoid generations would not be as great as that of direct cues. Moreover, at long range the parasitoids can less afford to confine their emphasis to a limited group of cues and are obliged to explore and sort a greater array of stimuli. Consequently, the overall variety of cues to be evaluated and the factors affecting the magnitude of parasitoid response are greater at longer range. Thus, there is a greater need for parasitoids to adapt their longer-range responses through learning as they encounter different situations among and within generations. Although both genetic and learned responses are probably present on both ends of the foraging sequence of parasitoids, we contend that the need for learning is greater at longer range. It is important, however, to note possible limits to our argument of more learning in the case of less reliable, longer-range cues. In discussions for other organisms, authors have suggested that, in situations of extreme unpredictability, learning may be of reduced value in tracking changes, in which case a fixed, mediocre alternative may be as suitable (Papaj and Prokopy, 1989).

As stated above, the greater variability in long-range foraging behaviour of parasitoids – and consequently the greater difficulty of its study – has resulted in a very limited knowledge of longer-range foraging behaviour, especially mechanisms governing the responses. Variability in long-range foraging behaviour may also seriously limit the use of parasitoids as dependable pest-control agents unless we can understand and manage this variability. In other words, perhaps we have the least information where the need is greatest (see preface of Drost *et al.* (1986) for further discussion of need for studies of foraging behaviour of parasitoids in response to airborne odours (Steidle and van Loon, 2002)).

Applied Considerations

A primary determinant of the effectiveness of parasitoids as biological control agents is the behaviour of the ovipositing females. Therefore, we must be able to ensure two features of their behaviour: (i) efficient location and attack of their hosts; and (ii) retention of females in the target area. To meet these requirements we must understand and manage the factors that influence the foraging behaviour of the parasitoids. Predictably effective performance of the parasitoids is a product of the proper matching of the intrinsic conditions of the searching female with the target environment. Thus, we shall discuss our need and approaches for managing both sides of this interaction.

The value and potential of managing the environmental component of the interaction would be important in all approaches for using natural enemies, including enhancement of wild populations, as well as maximizing the performance of laboratory-reared and released natural enemies. On the other hand, management of intrinsic variations in the natural enemy's response behaviour is more applicable in the situations where natural enemies are laboratory reared and released.

Managing the parasitoid component

We have discussed various aspects of genotypic and phenotypic diversity between and within parasitoid individuals that contribute to substantial variability in their foraging behaviour. In the case of natural parasitoid populations, natural selection is operating continuously to select and shape the features most effective for that environment, as depicted in Fig. 4.1. However, by laboratory colonization we remove the parasitoids from the context of natural selection and place them into an artificial environment, which may change genotypic frequencies and phenotypic consequences (Chapters 1, 6 and 12; Wardle and Borden, 1986; Hérard *et al.*, 1988b). These consequences are a particular danger in the case of inundative and seasonal inoculative programmes (van Lenteren, 2000), where propagation and release are continuously artificial and the genotypic and phenotypic traits of the field populations are dependent upon their prior laboratory colonization conditions (Chapters 1 and 6; Lewis *et al.*, 1981). In the case of inoculative-type releases, there is still an important need to manage the quality of propagated and released material, although perhaps less critical than in inundative and seasonal inoculative releases. In these inoculative cases, natural selection 'screens' the released material for the effective components for establishment. However, proper management of the colonized and released insects could greatly increase the success and speed of establishment.

We still know little of the specific features critical for parasitoid foraging behaviour and how to monitor those features. Thus, we cannot now provide a prescription for managing the variables during the production and release of parasitoid populations. Rather, our intent here is to argue for the importance of and to provide a conceptual framework for developing greater knowledge of this area.

The basic intent of this chapter is to expand our appreciation of the need for quality control procedures in the establishment, maintenance and use of colonized parasitoids and to develop methods for implementing the procedures. The quality control considerations will have to include both the genotypic and phenotypic aspects of subtle but important behavioural traits and the significant but not readily apparent ways in which various rearing and release methods might affect these traits.

Genetic qualities

When selecting a sample of a parasitoid species for establishing a laboratory colony, we need to screen the diversity of genotypic traits and ensure that the traits of the colonized population are appropriately matched with targeted use situations. To do this, we must develop bioassays that can be used to evaluate diversity, behaviour and other traits. Successful parasitism of a target host in a confined situation does not guarantee that released individuals will be suitable for that host under field conditions. The sequence of host-selection behaviours may be circumvented in laboratory confinement (Chapter 1). Various techniques and apparatuses such as olfactometers (e.g. Vet *et al.*, 1983; Vet and van Opzeeland, 1985), flight tunnels (e.g. Drost *et al.*, 1986; Elzen *et al.*, 1987; Zanen *et al.*, 1989; Noldus *et al.*, 1990; Geervliet *et al.*, 1998), ranked behavioural assays (e.g. Vinson, 1968; Lewis and Jones, 1971; Wilson *et al.*, 1974) and small field assays (e.g. Keller and Lewis, 1989; Silva *et al.*, 2000), can be used in screening and selecting material for establishing the colony.

The tests should be representative of the climatic, habitat (e.g. host-plant species, age and parts) and host-insect (e.g. age, density and distribution) situations that the parasitoid will encounter under field situations and be adequate for evaluating their ability to perform the full sequence of host-selection behaviours in those situations (Chapters 16 and 17). A good balance is needed between enough genetic diversity to cope with the fluctuations they will encounter and uniformity in the amount of diversity for consistency (Chapters 6 and 7). However, the amount of diversity desired may vary among the different traits. Similar testing techniques to monitor the colony systematically are needed for preservation of the diversity and uniformity of these traits (Chapters 1, 2, 15 and 16). Prévost and Lewis (1990) provide an example of how a flight chamber can be used to assess various genetic variations in host-finding response traits and how they can be measured and compared over generations for different colony lines.

Phenotypic qualities – learning

The response potential of a parasitoid is often a result of experiences in the preadult and adult stages: without care, insectary environments can create either weak or distorted response profiles (Chapters 1, 2 and 19). However, by understanding the sources and mechanisms of learning, we can provide the appropriate level of experience. As discussed earlier, the lack of important semiochemicals in the host-insect diet and use of factitious host insects have been shown to cause poorly responsive parasitoids. These semiochemicals can be incorporated artificially into the diet and on the hosts as synthetics or as materials such as plant extracts. This approach may be particularly important in some cases where important learning experiences occur in the immature or early adult stages (Wardle and Borden, 1985; Hérard *et al.*, 1988b; van Emden *et al.*, 1996).

Another approach that has been the subject of some experimentation is the prerelease exposure of the adult to important stimuli. Gross *et al.* (1975) showed that, with *Microplitis croceipes* (Cresson) and *Trichogramma pretiosum* Riley, exposure to host frass or host-moth scales increased the proportion of parasitism in the release area and attributed the benefit primarily to a reduction in the escape response upon emergence from the release container. Subsequent studies discussed earlier in this and the previous chapter have shown that prerelease exposure of the parasitoid to the kind of host and associated stimuli situations that they will encounter in the field can result in associative learning that enhances the parasitoids' subsequent ability to perform at the time of release.

There are genetic variations among individuals of a species as to what can be learned and to what degree. These variations are an important consideration for both the genotypic and phenotypic qualities of parasitoids.

Physical and physiological qualities

In addition to the constraints placed on learning experiences, the unnatural insectary environment can be stressful to the physical and physiological well-being of parasitoids (Chapter 1). For example, insect movements are restricted and the unnatural lighting, temperature and humidity may affect the females in a way that subsequently alters foraging behaviour. P.O. Zanen and W.J. Lewis (unpublished data) found that chilling the cocoons of *M. croceipes* severely reduced the ability of females to make flight responses to volatile host odours. Sublethal diseases (Chapter 10) that are not readily apparent may spread in the colony and affect behaviour. Hamm *et al.* (1988) reported a viral infection in an *M. croceipes* colony that accounted for reduced host-finding responses in flight chambers. Further, mating and nutritional conditions can strongly affect foraging responses (Chapter 5).

As is apparent from these points, the physical and physiological needs are very important to effective foraging behaviour. Because neither weaknesses in the learning conditions nor those of the physical or physiological state may be readily apparent from general observation, various response-evaluation techniques, such as those discussed for the genotypic traits, should be used to monitor the quality of the phenotypic traits of the colony.

Managing the environmental component

Given the release of parasitoids of adequate quality into the habitat or the presence of wild parasitoids of proper quality, we can then manage the environment side so as to maximize performance. Two basic objectives must be achieved: retention of the parasitoids in the target area and efficient host-search and attack behaviour. As a requirement for this to occur, the host plant or other substrate must be suitable and there must be a sufficient host density, otherwise the parasitoids will not remain in the area. The economic damaging threshold of the host pest can be at a density below that needed to sustain effective parasitoid foraging behaviour. Semiochemicals offer good prospects as a tool for managing the parasitoid behaviour independent of these variables.

Hagen *et al.* (1970) first manipulated a bollworm natural enemy with a semiochemical by using an artificial honeydew to attract adults of *Chrysoperla carnea* (Stephens). These chemicals provided a kairomone and food supplement, both of which served to increase predator density. Indole-acetaldehyde, a breakdown product of tryptophan found in the yeast hydrolysate of the artificial honeydew, operated as a kairomone by attracting adult lacewings into target fields (van Emden and Hagen, 1976). Other components of the artificial honeydew (sugar, water, whey–yeast hydrolysate) arrested movement of the lacewings and served as a nutritional supplement, thereby promoting oviposition. One week following an application of such a food spray to cotton, *Chrysoperla* egg density increased from one to three per plant, and the density of bollworm eggs and the number of damaged bolls declined (Hagen *et al.*, 1970).

The use of semiochemicals from plants and the host, *Heliothis zea*, has been shown to increase rates of egg parasitism by *Trichogramma* in the field. For example, parasitism of eggs of *H. zea* by *Trichogramma* spp. increased from 13% in control plots to 22% in soybeans treated with an extract of scales collected from *H. zea* (Lewis *et al.*, 1975a). Similarly, the release of a synthetic blend of the sex pheromone of *H. zea* in cotton increased parasitism of eggs from 21% in control to 36% in treated plots (Lewis *et al.*, 1982). Altieri *et al.* (1981) demonstrated that spraying various plant extracts on crops can stimulate increased rates of parasitism. For example, parasitism of eggs of *H. zea* by *Trichogramma* spp. was 21% on soybeans treated with an extract of *Amaranthus* compared with 13% on plants sprayed with water. The behaviours that lead to increased parasitism by *Trichogramma* in the presence of either plant extracts or sex pheromones are not yet fully understood, but are supposed to be based on arrestment responses of the parasitoid (e.g. Noldus *et al.*, 1988, 1990).

Application of semiochemicals to crops has not been universally successful in stimulating increased rates of mortality. For example, the use of a uniform spray of moth-scale extracts may reduce egg parasitism by *Trichogramma* spp. at low host densities, apparently by stimulating females to search too intensively where no hosts are present, thereby lowering their efficiency (Lewis *et al.*, 1979). This problem can be partly overcome by impregnating particles of diatomaceous earth with the moth-scale extract to mimic natural scales (Lewis *et al.*, 1979). When dispersed through a field, the treated particles intermittently stimulate searching by *Trichogramma*, rather than doing so continuously.

From these and other studies, it is obviously important to understand the respective roles of the various cues in the host-selection sequence and to apply the long- and close-range cues in proper proportions and distribution so as to retain the parasitoids effectively in the desired habitat without interfering with efficiency. The vacuum in understanding long-range foraging behaviour is a particular obstacle in achieving these needs.

A discovery with the larval parasitoid *M. croceipes* opened a new avenue for potential application. Lewis and Tumlinson (1988) found that, when contacting and antennating faeces of their *Heliothis* larval host, the female links plant and other volatile odours to a host-recognition kairomone in the faeces and associatively learns to fly to those volatile odours in search of hosts. Actual contact with hosts and oviposition are not a necessary part of this process. We visualize from the findings that artificial faecal pellets containing the reinforcing host-recognition kairomone and desired volatile odours can be applied to a crop at a sufficient density to reinforce search that is focused towards those volatiles. By using volatiles more prevalent in certain parts of the plant, the elicited search behaviour can be concentrated on certain portions of the plants or in other ways directed as desired. It is expected that similar phenomena and thus manipulation prospects exist for other parasitoid species.

Conclusions

There are three major sources of intrinsic variations in the foraging behaviour of individuals of a parasitoid species. One source is genotypically fixed differences among individuals that are adapted for different foraging environments. Another source is the phenotypic plasticity of individuals that allows them to modify their behaviour through learning to suit them for different host-habitat situations. A third source is the parasitoids' physiological state relative to other needs, such as food and mating.

The parasitoids' effectiveness at locating and attacking hosts is determined by the net combination of these factors, together with the conditions of their foraging environment. Therefore, our ability to obtain consistent and effective biological control with parasitoids can be strongly affected by our understanding of the mechanisms governing these sources of variation and the development of quality control techniques to manage them.

With an appropriate knowledge of these aspects of foraging behaviour, we can establish and ultimately engineer parasitoid colonies with the best genotypic qualities for their intended application. Further, we can rear, handle and release the colonies in ways that mould their phenotypic traits for optimum results, and we can manage the target environment to maximize parasitism by the released and naturally occurring parasitoids. Without such information, we are operating in a black box, in which the proper design of biological control programmes and the interpretations of their outcomes are a matter of speculation.

Many of the ideas expressed in this chapter can also be applied to the management of populations of predatory insects and mites, but, as yet, insight in foraging behaviour of predators is more limited than that of parasitoids (Steidle and van Loon, 2002).

Acknowledgements

This chapter is the result of a cooperative programme among the Insect Biology and Population Management Research Laboratory (US Department of Agriculture (USDA), Tifton, USA), the Insect Attractants, Basic Biology and Behaviour Research Laboratory (USDA, Gainesville, USA) and the Laboratory of Entomology, Wageningen University (Wageningen, The Netherlands). The journal *Environmental Entomology* (Entomological Society of America) granted permission to reprint an edited version of the original Lewis *et al.* (1990) paper with the same title and authors. Editing was made particularly easy with the recent extensive critical review of the 1990 paper by Steidle and Van Loon (2002).

References

Alcock, J. (1984) *Animal Behavior: an Evolutionary Approach*, 3rd edn. Sinauer, Sunderland, Massachusetts, 596 pp.

Altieri, M., Lewis, W.J., Nordlund, D.A., Gueldner, B.C. and Todd, J.W. (1981) Chemical interactions between plants and *Trichogramma* wasp in Georgia soybean fields. *Protection Ecology* 3, 259–263.

Bellows, T.S. and Fisher, T.W. (eds) (1999) *Handbook of Biological Control*. Academic Press, San Diego, 1046 pp.

Boulétreau, M. (1986) The genetic and coevolutionary interaction between parasitoids and their hosts. In: Waage, J. and Greathead, D. (eds) *Insect Parasitoids*. Academic Press, London, pp. 169–201.

Bradshaw, A.D. (1965) Evolutionary significance of phenotypic plasticity in plants. *Advances in Genetics* 13, 115–155.

Caltagirone, L.E. (1985) Identifying and discriminating among biotypes of parasites and predators. In: Hoy, M.A. and Herzog, D.C. (eds) *Biological Control in Agricultural IPM Systems*. Academic Press, Orlando, Florida, pp. 189–200.

Dicke, M. and Vet, L.E.M. (1999) Plant–carnivore interactions: evolutionary and ecological consequences for plant, herbivore and carnivore. In: Olff, H., Brown, V.K. and Drent, R.H. (eds) *Herbivores: Between Plants and Predators*. Blackwell Science, Oxford, pp. 483–520.

Doutt, R.L. (1964) Biological characteristics of entomophagous adults. In: DeBach, P. (ed.) *Biological Control of Insect Pests and Weeds*. Rheinhold, New York, pp. 145–167.

Drost, Y.C., Lewis, W.J., Zanen, P.O. and Keller, M.A. (1986) Beneficial arthropod behavior mediated by airborne semiochemicals. I. Flight behavior and influence of preflight handling of *Microplitis croceipes* (Cresson). *Journal of Chemical Ecology* 12, 1247–1262.

Drost, Y.C., Lewis, W.J. and Tumlinson, J.H. (1988) Beneficial arthropod behavior mediated by airborne semiochemicals. V. Influence of rearing method, host plant and adult experience on host-searching behavior of *Microplitis croceipes* (Cresson), a larval parasitoid of *Heliothis*. *Journal of Chemical Ecology* 14, 1607–1616.

Elzen, G.W., Williams, H.J.,Vinson, S.B. and Powell, J.E. (1987) Comparative flight behavior of parasitoids *Campoletis sonorensis* and *Microplitis croceipes*. *Entomologia Experimentalis et Applicata* 45, 175–180.

Flanders, S.E. (1953) Variation in susceptibility of citrus-infesting coccids to parasitization. *Journal of Economic Entomology* 46, 226–269.

Geervliet, J.B.F., Vreugdenhil, A.I., Dicke, M. and Vet, L.E.M. (1998) Learning to discriminate between infochemicals from different plant–host complexes by the parasitoids *Cotesia glomerata* and *C. rubecula*. *Entomologia Experimentalis et Applicata* 86, 241–252.

Godfray, H.C.J. (1994) *Parasitoids – Behavioral and Evolutionary Ecology*. Princeton University Press, Princeton, New Jersey, 473 pp.

Gross, H.R., Lewis, W.J., Jones, R.L. and Nordlund, D.A. (1975) Kairomones and their use for management of entomophagous insects: III. Stimulation of *Trichogramma achaeae*, *T. pretiosum*, and *Microplitis croceipes* with host-seeking stimuli at time of release to improve their efficiency. *Journal of Chemical Ecology* 1, 431–438.

Hagen, K.S. and Bishop, G.W. (1979) Use of supplemental foods and behavioral chemicals to increase the effectiveness of natural enemies. In: Davis, D.W., McMurtry, J.A. and Hoyt, S.C. (eds) *Biological Control and Insect Management*. California Agricultural Experiment Station Publication 4096, California Experiment Station, Berkeley, pp. 49–60.

Hagen, K.S., Sawall, E.F., Jr and Tassan, R.L. (1970) The use of food sprays to increase effectiveness of entomophagous insects. In: *Proceedings of the Tall Timbers Conference on Ecological Animal Control by Habitat Management*, Number 2. Tall Timbers Research Station, Tallahassee, Florida, pp. 59–81.

Hamm, J.J., Styer, E.L. and Lewis, W.J. (1988) A baculovirus pathogenic to the parasitoid *Microplitis croceipes* (Hymenoptera: Braconidae). *Journal of Invertebrate Pathology* 52, 189–191.

Hérard, F., Keller, M.A., Lewis, W.J. and Tumlinson, J.H. (1988a) Beneficial arthropod behavior mediated by airborne semiochemicals. III. Influence of age and experience on flight chamber responses of *Microplitis demolitor* Wilkinson. *Journal of Chemical Ecology* 14, 1583–1596.

Hérard, F., Keller, M.A., Lewis, W.J. and Tumlinson, J.H. (1988b) Beneficial arthropod behavior mediated by airborne semiochemicals. IV. Influence of host diet on host-oriented flight chamber responses of *Microplitis demolitor* Wilkinson. *Journal of Chemical Ecology* 14, 1597–1606.

Hoy, M.J. (1988) Biological control of arthropod pests: traditional and emerging technologies. *American Journal of Alternative Agriculture* 3, 63–68.

Keller, M.A. and Lewis, W.J. (1989) Behavior-modifying chemicals to increase the efficacy of predators and parasitoids of *Heliothis* spp. In: King, E.G. and Jackson, R.D. (eds) *Proceedings of the Workshop on Biological Control of* Heliothis*: Increasing the Effectiveness of Natural Enemies, 11–15 November 1985, New Delhi, India*. Far Eastern Regional Research Office, USDA, New Delhi, India, pp. 449–467.

Lewis, W.J. (1981) Semiochemicals: their role with changing approaches to pest control. In: Nordlund, D.A., Jones, R.L. and Lewis, W.J. (eds) *Semiochemicals: Their Role in Pest Control*. John Wiley & Sons, New York, pp. 3–12.

Lewis, W.J. and Jones, R.L. (1971) Substance that stimulates host-seeking by *Microplitis croceipes* (Hymenoptera: Braconidae), a parasite of *Heliothis* species. *Annals of the Entomological Society of America* 64, 471–473.

Lewis, W.J. and Tumlinson, J.H. (1988) Host detection by chemically mediated associative teaming in a parasitic wasp. *Nature* 331, 257–259.

Lewis, W.J., Jones, R.L., Nordlund, D.A. and Sparks, A.N. (1975a) Kairomones and their use for management of entomophagous insects. I. Evaluation for increasing rates of parasitization by *Trichogramma* spp. in the field. *Journal of Chemical Ecology* 1, 343–347.

Lewis, W.J., Jones, R.L., Nordlund, D.A. and Gross, H.R. (1975b) Kairomones and their use for management of entomophagous insects. II. Mechanism causing increase in rates of parasitization by *Trichogramma* spp. in the field. *Journal of Chemical Ecology* 1, 349–360.

Lewis, W.J., Beevers, M.D., Nordlund, D.A., Gross, H.R. and Hagen, K.S. (1979) Kairomones and their use for management of entomophagous insects. IX. Investigations of various kairomone-treatment patterns for *Trichogramma* spp. *Journal of Chemical Ecology* 5, 673–680.

Lewis, W.J., Nordlund, D.A. and Gueldner, R.C. (1981) Semiochemicals influencing behavior of entomophages: roles and strategies for their employment in pest control. In: *Les Médiateurs Chimiques.* Les Colloques de l'INRA, Paris, pp. 225–242.

Lewis, W.J., Nordlund, D.A., Gueldner, R.C., Teal, P.E.A. and Tumlinson, J.H. (1982) Kairomones and their use for management of entomophagous insects. XIII. Kairomonal activity for *Trichogramma* spp. of abdominal tips, excretion and a synthetic sex pheromone blend of *Heliothis zea* (Boddie) moths. *Journal of Chemical Ecology* 8, 695–701.

Lewis, W.J., Gross, H.R. and Nordlund, D.A. (1985) Behavior manipulation of *Trichogramma* (Hymenoptera: Trichogrammatidae). *Southwestern Entomologist* 8, 138–155.

Lewis, W.J., van Lenteren, J.C., Phatak, S.C. and Tumlinson, J.H. (1997) A total systems approach to sustainable pest management. *Proceedings of the National Academy of Sciences, USA* 94, 12243–12248.

Lewis, W.J., Stapel, J.O., Cortesero, A.M. and Takasu, K. (1998) Understanding how parasitoids balance food and host needs: importance to biological control. *Biological Control* 11, 175–183.

Luck, R.F. and Uygun, N. (1986) Host recognition and selection by *Aphytis* species: response to California red, oleander, and cactus scale cover extracts. *Entomologia Experimentalis et Applicata* 40, 129–136.

Matthews, R.W. and Matthews, J.R. (1978) *Insect Behavior.* John Wiley & Sons, New York, 507 pp.

Mollema, C. (1988) Heritability of host selection behaviour of *Asobara tabida.* In: Genetical aspects of resistance in a host–parasitoid interaction. PhD thesis, University of Leiden, Leiden, The Netherlands, pp. 99–107.

Nishida, T. (1956) An experimental study of the ovipositional behavior of *Opius fletcheri* Silvestri (Hymenoptera: Braconidae), a parasite of the melon fly. *Proceedings Hawaiian Entomological Society* 16, 126–134.

Noldus, L.P.J.J., Lewis, W.J., Tumlinson, J.H. and van Lenteren, J.C. (1988) Olfactometer and windtunnel experiments on the role of sex pheromones of noctuid moths in the foraging behaviour of *Trichogramma* spp. In: Voegele, J., Waage, J. and van Lenteren, J.C. (eds) *Proceedings of the 2nd International Symposium on* Trichogramma *and Other Egg Parasites, 10–15 November 1986, Guangzhou, China.* Les Colloques de l'INRA 43, Paris, pp. 223–238.

Noldus, L.P.J.J., van Lenteren, J.C. and Lewis, W.J. (1990) How *Trichogramma* parasitoids use moth sex pheromones as kairomones: orientation behaviour in a wind tunnel. *Physiological Entomology* 16, 313–327.

Nordlund, D.A., Lewis, W.J. and Altieri, M.A. (1988) Influence of plant-derived semiochemicals on host/prey selection behavior of entomophagous insects. In: Barbosa, P. and Letourneau, L.D. (eds) *Novel Aspects of Insect–Plant Interactions.* John Wiley & Sons, New York, pp. 65–90.

Pak, G.A. (1988) Selection of *Trichogramma* for inundative biological control. PhD thesis, Agricultural University, Wageningen, The Netherlands.

Papaj, D.R. and Prokopy, R.J. (1989) Ecological and evolutionary aspects of learning in phytophagous insects. *Annual Review of Entomology* 34, 315–350.

Papaj, D.R. and Rausher, M.D. (1987) Genetic differences and phenotypic plasticity as causes of variation in oviposition preference in *Battua philenor. Oecologia* 74, 24–30.

Prévost, G. and Lewis, W.J. (1990) Genetic differences in the response of *Microplitis croceipes* to volatile semiochemicals. *Journal of Insect Behavior* 3, 277–287.

Roughgarden, J. (1979) *Theory of Population Genetics and Evolutionary Ecology: an Introduction.* Macmillan, New York, 612 pp.

Salt, G. (1935) Experimental studies in insect parasitism. III. Host selection. *Proceedings Royal Society, London, Series B, Biological Sciences* 117, 413–435.

Shahjahan, M. (1974) *Erigeron* flowers as a food and attractive odor source for *Peristenus pseudopallipes*, a braconid parasitoid of the tarnished plant bug. *Environmental Entomology* 3, 69–72.

Sheehan, W. and Shelton, A.M. (1989) The role of experience in plant foraging by the aphid parasitoid *Diaretiella rapae* (Hymenoptera: Aphidiidae). *Journal of Insect Behavior* 2, 743–759.

Silva, I.M.M.S., van Meer, M.M.M., Roskam, M.M., Hoogenboom, A., Gort, G. and Stouthamer, R. (2000) Biological control potential of *Wolbachia*-infected versus uninfected wasps: laboratory and greenhouse evaluation of *Trichogramma cordubensis* and *T. deion* strains. *Biocontrol Science and Technology* 10, 223–238.

Steidle, J.L.M. and van Loon, J.J.A. (2002) Chemoecology of parasitoid and predator oviposition behaviour. In: Hilker, M. and Meiners, T. (eds) *Chemoecology of Insect Eggs and Egg Deposition.* Blackwell Science, Oxford, pp. 291–317.

Thorpe, W.H. and Jones, F.G.W. (1937) Olfactory conditioning in a parasitic insect and its relation to the problem of host selection. *Proceedings Royal Society Series B, Biological Sciences* 124, 56–81.

Turlings, T.C.J., Tumlinson, J.H., Lewis, W.J. and Vet, L.E.M. (1989) Beneficial arthropod behavior mediated by airborne semiochemicals. VIII. Learning of host-related odors induced by a brief contact experience with host by-products in *Cotesia marginiventris* (Cresson), a generalist larval parasitoid. *Journal of Insect Behavior* 2, 217–225.

van Alphen, J.J.M. and Vet, L.E.M. (1986) An evolutionary approach to host finding and selection. In: Waage, J.K. and Greathead, D.J. (eds) *Insect Parasitoids*. Academic Press, London, pp. 23–61.

van Emden, H.F. and Hagen, K.S. (1976) Olfactory reactions of the green lacewing, *Chrysopa carnea,* to tryptophan and certain breakdown products. *Environmental Entomology* 5, 469–473.

van Emden, H.F., Sponagl, B., Wagner, E., Baker, T., Ganguly, S. and Douloumpaka, S. (1996) Hopkins' 'host selection principle', another nail in its coffin. *Physiological Entomology* 21, 325–328.

van Lenteren, J.C. (1980) Evaluation of control capabilities of natural enemies: does art have to become science? *Netherlands Journal of Zooloogy* 30, 369–381.

van Lenteren, J.C. (2000) Measures of success in biological control of arthropods by augmentation of natural enemies. In: Gurr, G. and Wratten, S. (eds) *Measures of Success in Biological Control.* Kluwer Academic Publishers, Dordrecht, pp. 77–103.

Vet, L.E.M. (1983) Host-habitat location through olfactory cues by *Leptopilina clavipes* (Hartig) (Hym.: Eucoilidae), a parasitoid of fungivorous *Drosophila*: the influence of conditioning. *Netherlands Journal of Zoology* 33, 225–248.

Vet, L.E.M. and Dicke, M. (1992) Ecology of infochemical use by natural enemies in a tritrophic context. *Annual Review of Entomology* 37, 141–172.

Vet, L.E.M. and van Opzeeland, K. (1985) Olfactory microhabitat selection in *Leptopilina heterotoma* (Thomson) (Hym.: Eucoilidae), a parasitoid of Drosophilidae. *Netherlands Journal of Zoology* 35, 497–504.

Vet, L.E.M., van Lenteren, J.C., Heymans, M. and Meelis, E. (1983) An airflow olfactometer for measuring olfactory responses of hymenopterous parasitoids and other small insects. *Physiological Entomology* 8, 97–106.

Vet, L.E.M., Lewis, W.J., Papaj, D.R. and van Lenteren, J.C. (1990) A variable-response model for parasitoid foraging behaviour. *Journal of Insect Behavior* 3, 471–490.

Vet, L.E.M., Lewis, W.J. and Cardé, R.T. (1995) Parasitoid foraging and learning. In: Cardé, R.T. and Bell, W.J. (eds) *Chemical Ecology of Insects 2.* Chapman & Hall, New York, pp. 65–101.

Vinson, S.B. (1968) Source of a substance in *Heliothis virescens* (Lepidoptera: Noctuidae) that elicits a searching response in its habitual parasite *Cardiochiles nigriceps* (Hymenoptera: Braconidae). *Annals of the Entomological Society of America* 61, 8–10.

Vinson, S.B. (1976) Host selection by insect parasitoids. *Annual Review of Entomology* 21, 109–134.

Vinson, S.B. (1984a) How parasitoids locate their hosts: a case of insect espionage. In: Lewis, T. (ed.) *Insect Communication.* Royal Entomological Society, London, pp. 325–348.

Vinson, S.B. (1984b) The behavior of parasitoids. In: Kerkut, G.A. and Gilbert, L.I. (eds) *Comprehensive Insect Physiology, Biochemistry and Pharmacology.* Pergamon, New York, pp. 417–469.

Vinson, S.B. (1988) Comparison of host characteristics that elicit host recognition behavior of parasitoid Hymenoptera. In: Gupta, G.K. (ed.) *Advances in Parasitic Hymenoptera Research.* E.J. Brill, Kinderhook, New York, pp. 285–291.

Vinson, S.B. (1998) The general host selection behavior of parasitoid Hymenoptera and a comparison of initial strategies utilized by larvaphagous and oophagous species. *Biological Control* 11, 79–96.

Vinson, S.B., Barfield, C.S. and Henson, R.D. (1977) Oviposition behavior of *Bracon mellitor,* a parasitoid of the boll weevil *(Anthonomus grandis).* II. Associative learning. *Physiological Entomology* 2, 157–164.

Waage, J.K. and Hassell, M.P. (1982) Parasitoids as biological control agents – a fundamental approach. *Parasitology* 84, 241–268.

Wäckers, F.L. (1994) The effect of food deprivation on the innate visual and olfactory preferences in the parasitoid *Cotesia rubecula. Journal of Insect Physiology* 40, 641–649.

Wajnberg, E. and Hassan, S.A. (1994) *Biological Control with Egg Parasitoids.* CAB International, Wallingford, UK.

Wardle, A.R. and Borden, J.H. (1985) Age-dependent associative learning by *Exeristes roborator* (F.) (Hymenoptera: Ichneumonidae). *Canadian Entomologist* 117, 605–616.

Wardle, A.R. and Borden, J.H. (1986) Detrimental effect of prior conditioning on host habitat location by *Exeristes roborator. Naturwissenschaften* 73, 559–560.

Wilson, D.D., Ridway, R.L. and Vinson, S.B. (1974) Host acceptance and oviposition behavior of the parasitoid *Campoletis sonorensis* (Hymenoptera: Ichneumonidae). *Annals of the Entomological Society of America* 67, 271–274.

Zanen, P.O., Lewis, W.J., Cardé, R.T. and Mullinix, B.G. (1989) Beneficial arthropod behavior mediated by airborne semiochemicals. VI. Flight responses of female *Microplitis croceipes* (Cresson), a braconid endoparasitoid of *Heliothis* spp., to varying olfactory stimulus conditions created with a turbulent jet. *Journal of Chemical Ecology* 15, 141–168.

5 The Parasitoids' Need for Sweets: Sugars in Mass Rearing and Biological Control

F.L. Wäckers*

Laboratory of Entomology, PO Box 8031, 6700 EH Wageningen, The Netherlands

Abstract

It is generally accepted that most parasitoids and many predators require sugar sources, such as nectar or honeydew, to cover their energetic needs. Protocols for the mass rearing and release of these natural enemies often take these sugar requirements into account. Nevertheless, the choice of food sources and the methods of application are usually based on trial and error, due to the fact that basic information on food ecology of beneficial insects is scarce. In this chapter, an overview is presented of the field of parasitoid food ecology. After discussing the various ways in which parasitoid fitness can benefit from sugar feeding, various natural sugar sources are compared in respect of their function in nature and their suitability as parasitoid nutrition. Given the fact that the choice of the optimal food supplement depends on characteristics of both the food source and its consumer, either side of the equation is addressed. Sugar sources are compared in respect of their composition and the volume produced. Parasitoid characteristics addressed include taste perception, digestive efficiency and food-foraging behaviour. It is argued that the field of food ecology can help in selecting food supplements for use in parasitoid rearing as well as application in biological control.

Introduction

Due to their ability to regulate herbivore populations, parasitoids and predators play an important role both as biological control agents and as keystone species in natural ecosystems. Given this fact, it is not surprising that research interest has largely focused on how predators and parasitoids find and interact with their herbivorous prey/host (Godfray, 1994; Dicke and Vet, 1998). However, the majority of these principally carnivorous arthropods also use plant-derived foods as a source of nutrients. This vegetarian side of the menu may include various plant substrates, such as nectar, food bodies, pollen and fruits, as well as foods indirectly derived from plants (e.g. honeydew, or pycnial fluid of fungi). In some cases, predators may also feed on plant productive tissue, in which case they have to be classified as potential herbivores (Coll, 1996). The level at which predators or parasitoids depend on primary consumption varies.

Many predator species are facultative consumers of plant-derived food. This category includes predatory mites (Bakker and Klein, 1992), spiders (Ruhren and Handel, 1999),

*Present address: Netherlands Institute of Ecology, PO Box 40, 6666 ZG Heteren, The Netherlands.

© CAB International 2003. *Quality Control and Production of Biological Control Agents: Theory and Testing Procedures* (ed. J.C. van Lenteren)

predatory hemipterans (Bugg *et al.*, 1991), predacious beetles (Larochell, 1990), lacewings (Limburg and Rosenheim, 2001) and predacious wasps (Beggs and Rees, 1999). Feeding on pollen or nectar can enable these species to bridge periods of low prey availability (Limburg and Rosenheim, 2001). When combined with prey feeding, plant-derived foods can increase predator fitness over prey feeding alone (van Baalen *et al.*, 2001). A second category of natural enemies are obligatory consumers of plant-derived foods, at least during part of their life cycle. This category includes many ant species (Porter, 1989; Tobin, 1994), syrphid flies (Lunau and Wacht, 1994) and parasitoids (Jervis *et al.*, 1996).

As the nutritional ecology of predators has been extensively covered elsewhere, the focus of this chapter will be on issues concerning feeding by adult parasitoids. I shall stress that sugar feeding represents an integral part of parasitoid biology and that insight in this topic is essential to our understanding of parasitoid ecology, as well as their efficacy as biological control agents.

Nutritional Requirements of Parasitoids

During their development from parasitic larvae to free-living adults, the dietary requirements of parasitoids take an equally marked turn. While parasitoid larvae are strictly carnivorous, virtually all adult parasitoids require carbohydrates as a source of energy (Jervis *et al.*, 1996), especially for flight (Hoferer *et al.*, 2000).

While predators can often utilize both liquid and solid plant substrates (pollen, food bodies), by far the majority of parasitoids are restricted to feeding on sugar-rich solutions, such as nectar and honeydew. This group includes those species that emerge with a full complement of mature eggs (so-called preovigenic species), as well as species that continue to mature eggs during their adult life (synovigenic).

Some (usually synovigenic) parasitoid species retain a level of carnivory during their adult life, as they may feed on host haemolymph in addition to sugar feeding (Jervis and Kidd, 1986). Due to their different nutritional composition, haemolymph and nectar or honeydew are only partly interchangeable and they are believed to cover separate requirements. Sugar-rich nectar or honeydew primarily provides for the parasitoid's energetic needs. While these food sources usually contain low levels of amino acids, proteins and lipids, they might nevertheless contribute to physiological processes, such as egg maturation. Host haemolymph, on the other hand, is usually a relatively poor source of energy. In part, this can be explained by the fact that haemolymph in general contains relatively low levels of carbohydrates (Kimura *et al.*, 1992). An additional limitation lies in the fact that trehalose as the main haemolymph sugar is rather poorly metabolized by parasitoids (Wäckers, 2001). Instead, haemolymph constitutes a primary source of protein for physiological processes, such as egg maturation (Rivero and Casas, 1999). Those synovigenic species that do not engage in host feeding draw upon the protein and fat reserves transferred over from the larval stage.

Effects of Sugar Feeding on Parasitoid Fitness Parameters

Parasitoids emerge with a limited supply of energy. The nutrients transferred from the larval stage often cover no more than 48 h of the parasitoid's energetic requirements. This period is extremely brief, considering the fact that these species usually have the potential to live for weeks when suitable food is available. Part of this brief period covered by larval food reserves cannot be used to search for hosts, as parasitoids often require a preoviposition period for the maturation of their eggs. The reproductive success in the remaining narrow time-window is further limited by lack of experience, resulting in an initially slow and inefficient search (Turlings *et al.*, 1993; Vet *et al.*, 1995).

Sugar feeding can considerably increase the parasitoid's lifespan. Taking the preoviposition period and experience into account, the effective impact will be even more significant. This means that parasitoids that fail to replenish their energy reserves through sugar feeding will suffer severe fitness consequences. Carbohydrates can have a strong impact on several key fitness parameters. Sugar feeding is indispensable to parasitoid survival, a factor applying to both females (Zoebelein, 1956; Syme, 1975; van Lenteren *et al.*, 1987; Idoine and Ferro, 1988; Wäckers, 2001) and males (Zoebelein, 1956; Wäckers and Swaans, 1993). In the ideal world of the laboratory, sugar can increase parasitoid longevity up to 20-fold (Zoebelein, 1956; Syme, 1975; Idoine and Ferro, 1988; Wäckers and Swaans, 1993; Dyer and Landis, 1996).

Sugar feeding can also benefit a parasitoid's fecundity, either through a positive effect on the rate of egg maturation, through an increase in reproductive lifespan or both (Zoebelein, 1956; Hocking, 1966; Syme, 1975; Baggen and Gurr, 1998; van Lenteren, 1999; Schmale *et al.*, 2001). Finally, the feeding status can affect the parasitoid's propensity to seek out their herbivorous hosts. Telenga (1958) and van Emden (1962) found that parasitoids are more active in habitats in which flowers are in bloom than in nearby habitats without flowers. Wäckers (1994) and Takasu and Lewis (1995) demonstrated that sugar deprivation reduces host-searching efficiency, partly due to a general reduction in activity and partly due to a shift from host searching to food searching.

Each of the listed fitness parameters translates directly into the number of herbivores that can be attacked. The availability of suitable plant-provided food sources consequently has a strong impact on parasitoid mass rearing, as well as on their efficacy as biological control agents.

Choosing Food Supplements for Mass Rearing

The basic concept that parasitoid fitness can be dramatically enhanced through the simple provision of food supplements has been long engrained in parasitoid rearing. It is standard practice to provide adult insects with honey, honeydew, sugar water, fruits or other sugar sources. While the importance of food supplements is thus widely acknowledged, often little attention is given to the actual choice of the food source. This choice may be based on issues like convenience (honey can be easily obtained, does not require preparation and does not spoil), methodology (parasitoids can get stuck in liquid food media, while they have trouble imbibing solid food sources) and economy (cost). Hardly ever is the choice actually based on what should be the central issue: the question of which substrate is the optimal food for a given parasitoid.

This omission is hardly surprising in light of the fact that few comparative data exist in respect of the relative suitability of various food sources. The low priority given to this issue reflects the generally held conception that any sugar-rich liquid makes a suitable food supplement for parasitoids. In an attempt to correct this notion, I shall compare the main natural sugar sources in respect of their composition, and discuss the consequences for feeding parasitoids.

Potential Sugar Sources and Their Ecological Function

Most hymenopteran pollinators, ants and parasitoids share a dependency on sugars as their main source of energy. The ecological importance of these species, in combination with their dependency on carbohydrates, explains the fact that sugars play a central role in numerous types of mutualisms involving Hymenoptera. Sugar-feeding insects usually have a wide range of carbohydrate sources available, the most important being floral nectar, extrafloral nectar (EFN) and honeydew.

Floral nectar serves as a food reward in the mutualism between plants and their pollinators. Even though parasitoids in general are ineffective pollinators, they can freeload on this mutualism as they seek out flowers and collect floral nectar (Kevan, 1973; Jervis *et al.*, 1993). Due to the fact that many plants are pollinated by Hymenoptera (e.g. bee species), their nectar can be expected to cater to the taste and nutritional requirements of these species. As nectar requirements of hymenopteran parasitoids appear to be similar to those of honey-bees (Wäckers, 1999, 2001), the composition of floral nectar is probably suitable for hymenopteran parasitoids. Parasitoid species have been reported to feed on various types of floral nectar (Kevan, 1973; Jervis *et al.*, 1993; Idris and Grafius, 1995; Baggen and Gurr, 1998).

Extrafloral nectaries include a wide range of nectar-excreting structures (Zimmerman, 1932). Extrafloral nectaries have been described in approximately 1000 species from 93 plant families (Koptur, 1992; Whitman, 1996). They occur on a range of plant parts, including stems, leaves, fruits and flowers. Extrafloral nectaries are distinguished from their floral counterparts by the fact that they are not involved in pollination. Instead, they are thought to serve a role in an entirely different type of mutualism, in which plants use nectar to recruit predators or parasitoids. The latter return the favour by safeguarding plants against herbivory.

In a number of plant systems, it has been demonstrated that the presence of extrafloral nectar can translate into both reduced plant damage (O'Dowd and Catchpole, 1983; Wagner, 1997) and increased plant reproductive fitness (Rico-Gray and Thien, 1989; Oliveira, 1997). The above-mentioned studies have all focused on the role of EFN in plant–ant mutualisms. However, extrafloral nectaries are also frequented by a range of other carnivorous arthropods (Bugg *et al.*, 1989; Koptur, 1994; Whitman, 1996). The provision of these food supplements may serve to enhance the effectiveness of plant–spider (Ruhren and Handel, 1999), plant–predatory wasp (Torres-Hernández *et al.*, 2000) or plant–parasitoid interactions (Lingren and Lukefahr, 1977; Stapel *et al.*, 1997).

Honeydew is a generic term for sugar-rich excretions of phloem-feeding Sternorrhyncha. It is generally accepted that sap-feeding insects have to excrete carbohydrates to bring the high carbohydrate/amino acid ratio of the ingested phloem sap in balance with their nutritional requirements. Honeydew is an exception to the above-mentioned sugar sources, as it is a waste product, rather than having a primary function in mutualistic interactions. However, depending on its composition, honeydew can be eagerly collected by ants (Stadler and Dixon, 1999; Völkl *et al.*, 1999). The general tendency of ants to defend and protect sugar sources has resulted in mutualistic interaction between some honeydew producers and ants. In these instances, honeydew production has to some extent become an analogue to EFN.

Sugar-source Characteristics

Nectar and honeydew contain various sugars, amino acids, lipids and other organic compounds in more or less aqueous solutions (Baker and Baker, 1982b; Kloft *et al.*, 1985). The nutritional and energetic value of a particular nectar or honeydew is determined by its volume, its composition and the component concentrations.

Volume

The volume of floral nectar excreted and its composition are primarily a plant characteristic. In addition, however, they may be affected by other factors, such as the age of the nectary, irradiance, temperature, soil conditions and water balance (Búrquez and Corbet, 1991) and state of pollination (Gori, 1983). The duration of nectar secretion is limited by the – often brief – flowering time.

The often copious nectar volume secreted by extrafloral nectaries can exceed floral nectar production. This is in part due to high production levels, as well as to extended periods of production. As in floral nectar, the production of EFN is affected by abiotic factors (Bentley, 1977). In addition, plants can raise the secretion of nectar in response to two biotic mechanisms. Nectar production can be induced both by ant attendance (i.e. nectar removal) (Koptur, 1992; Heil *et al.*, 2000) and herbivore feeding (Koptur, 1989; Wäckers and Wunderlin, 1999; Heil *et al.*, 2001; Wäckers *et al.*, 2001). This sophisticated two-pronged mechanism allows plants to actively distribute their investments in a way that optimizes their defence.

In the case of honeydew, the volume produced and its composition depend on the sap-feeding species, as well as on plant parameters and environmental factors (Kloft *et al.*, 1985). Sap feeders can actively increase the quantity of excreted honeydew when tended by ants (Takeda *et al.*, 1982; Yao and Akimoto, 2001).

Sugar composition

Floral nectar, EFN and honeydew are principally sugar solutions. However, the sugar composition can vary in respect of both the types of saccharides and their relative proportions.

Floral nectar is generally dominated by the monosaccharides fructose and glucose and the disaccharide sucrose (Baker and Baker, 1982b). The proportions of the three main sugars are rather constant within a species, but can show wide differences between flowering species. Percival (1961) and Baker and Baker (1982b), for instance, showed that the sucrose/hexose ratios of flowering plants can vary from less than 0.1 to more than 0.999. In addition to these main nectar sugars, nectar may contain low concentrations of other carbohydrates (Table 5.1).

EFN is typically dominated by sucrose and its hexose components glucose and fructose. These are also the three most common sugars in EFN. Unlike the species-characteristic sugar ratio in floral nectar, however, the ratio of sucrose to hexose in the EFN of a given species can be much

Table 5.1. Sugars reported to occur in floral nectar, extrafloral nectar and honeydew.

	Reported to occur in	References
Monosaccharides		
Glucose	Various (extra)floral nectars	Bentley, 1977
	Honeydew	Baker and Baker, 1983; Kloft *et al.*, 1985
Fructose	Various (extra)floral nectars	Bentley, 1977; Baker and Baker, 1983
	Honeydew	Kloft *et al.*, 1985
Galactose	Extrafloral nectar	Bory and Clair Maczulajtys, 1986; Olson and Nechols, 1995
	Floral nectar	Gottsberger *et al.*, 1973
	Honeydew	Byrne and Miller, 1990
Mannose	Traces in floral nectar and fruits	Barnavon *et al.*, 2000
Rhamnose	Extrafloral nectar	Bory and Clair Maczulajtys, 1986
Disaccharides		
Sucrose	Various (extra)floral nectars	Bentley, 1977; Baker and Baker, 1983
	Honeydew	Kloft *et al.*, 1985
Maltose	Floral nectar	Baker and Baker, 1983; Belmonte *et al.*, 1994
	Coccid honeydew	Ewart and Metcalf, 1956
Melibiose	Floral nectar	Baker and Baker, 1983
	Eucalyptus manna (plant exudate)	Steinbauer, 1996
Trehalose	Honeydew	Kloft *et al.*, 1985
Trehalulose	Honeydew (whitefly)	Hendrix *et al.*, 1992
Trisaccharides		
Raffinose	Floral nectar	Baker and Baker, 1983
	Honeydew	Byrne and Miller, 1990
Melezitose	Primarily in honeydew	Kloft *et al.*, 1985; Hendrix *et al.*, 1992
	Some floral nectars	Baker and Baker, 1983
Erlose	Honeydew	Kloft *et al.*, 1985
Tetrasaccharide		
Stachyose	Floral nectar (orchids)	Baker and Baker, 1983
	Honeydew	Byrne and Miller, 1990; Davis *et al.*, 1993

more variable. In general, EFN composition shows relatively high levels of fructose and glucose (Tanowitz and Koehler, 1986; Koptur, 1994). This can be explained by the exposed nature of most extrafloral nectaries, resulting in increased microbial breakdown of sucrose. In addition to the three main sugars, several other sugars may be present (Table 5.1). Besides carbohydrates, EFN may contain variable amounts of proteins, amino acids and lipids (Baker *et al.*, 1978; Smith *et al.*, 1990). The particular amino acid composition can increase the attractiveness of EFN as a food source (Lanza *et al.*, 1993).

Honeydew differs from floral nectar and EFN as it often contains substantial amounts of oligosaccharides (Kloft *et al.*, 1985; Hendrix *et al.*, 1992). Even though the sugar composition of honeydew reflects the original composition of the phloem sap of the host plant, the sugar components and their relative quantities can be altered during the passage through the gut of the phloem feeder. On the one hand, phloem sugars such as sucrose and maltose are broken down by digestive enzymes, while, on the other hand, the sap feeders may also synthesize more complex sugars. The trisaccharides melezitose and erlose (fructomaltose), as well as the disaccharides trehalose and trehalulose, are examples of sugars that are synthesized through the action of gut enzymes on plant-derived sucrose (Mittler and Meikle, 1991; Hendrix *et al.*, 1992). The resulting sugar spectrum may range from honeydews that are almost entirely composed of the phloem sugar sucrose and its hexose components fructose and glucose to those honeydews that completely lack hexoses and are dominated by insect-synthesized oligosaccharides (Kloft *et al.*, 1985; Hendrix *et al.*, 1992; Völkl *et al.*, 1999).

Sugar concentrations

Sugar concentration is an important factor determining the uptake of a sugar source. At low concentrations, gustatory perception might be impeded (Wäckers, 1999), whereas high sugar concentrations interfere with sugar uptake (Wäckers, 2000). In floral nectar, sugar concentrations can already range from 5 to 75% at the time of nectar secretion (Dafni, 1992). Environmental conditions may further affect nectar concentrations, both indirectly, through their effects on the nectar-producing plant, and directly, through evaporation, hygroscopy or rain dilution.

Sugar concentrations of undiluted EFN range from 5 to more than 80% (Koptur, 1992; Wäckers *et al.*, 2001). In general, EFN shows much more variation in respect of sugar concentrations than floral nectar from the same plant. When protected from rain, EFN tends to be more concentrated, probably due to the fact that its exposed nature increases evaporation.

The fact that honeydew is typically available as little droplets or as a thin film on the substrate means that it is even more subjected to evaporation. As a result, sugar concentrations are often at saturation. This is likely to be a limiting factor in honeydew uptake. This problem is accentuated by the specific tendency of the honeydew sugars raffinose and melezitose to crystallize rapidly (Wäckers, 2000).

Parasitoid Characteristics

Insects often show a tendency to visit sugar sources of a certain composition (Baker and Baker, 1982a). The sugar components are an important factor determining patterns of food utilization (Inouye and Waller, 1984; Alm *et al.*, 1990; Lanza *et al.*, 1993; Josens *et al.*, 1998; Völkl *et al.*, 1999). We have seen that nectar and honeydew often vary widely in respect of their sugar composition. As a result, one frequently investigates an insect's response to individual nectar or honeydew components at well-defined concentrations, rather than studying a few arbitrary examples out of the broad range of natural nectar or honeydew compositions. In previous work, I have studied a range of sugars occurring in nectar and/or honeydew (listed in Table 5.1), as well as lactose. These 14 sugars were compared in respect of their effect on parasitoid gustatory response and longevity.

Taste perception

To test the gustatory response of *Cotesia glomerata*, food-deprived parasitoids were presented with highly concentrated (2M) solutions of individual sugars. Of the 14 sugars tested, only eight elicited feeding (Wäckers, 1999). Six sugars, including the honeydew sugar raffinose, did not elicit any feeding response in the food-deprived parasitoids. Both raffinose and mannose showed a deterrent effect when mixed with low molar solutions of sucrose. Parasitoids showed highest gustatory sensitivity (lowest acceptance threshold) to the common nectar sugars sucrose, glucose and fructose, as well as the honeydew sugar erlose.

Based on the extensive work on honeybees and ants, we know that these social Hymenoptera show distinct preferences for particular sugars (von Frisch, 1934; Vander Meer *et al.*, 1995; Tinti and Nofre, 2001), as well as certain sugar concentrations (Wykes, 1952; Waller, 1972; Baker and Baker, 1982a). In sharp contrast to this body of research, we know little or nothing about sugar preferences in parasitoids. This omission is in part due to methodological problems in assessing sugar preferences in parasitoids, as the establishment of preference requires that the test organism shows an inclination to sample, and feeds in repeated bouts. While these conditions are met in social Hymenoptera, whose foragers continuously collect food for the entire colony, solitary parasitoids feed infrequently, as their food foraging is restricted to their individuals needs. The number of parasitoid feeding events is further restricted by the fact that they can ingest and store sugar meals of up to a third of their body weight. Upon encountering a food source of sufficient quantity and quality, a hungry parasitoid will typically feed until saturation, rather than sample the food site and continue foraging for alternative sugar sources. The level of food consumption may differ depending on the sugar offered (Wäckers, 2001). However, this is at best an indirect measure of preference, as parasitoids are not making a choice based on complete knowledge of the alternatives.

Effects on longevity

To obtain a more comprehensive overview of the metabolic utilization of sugars by Hymenopteran parasitoids, the same 14 sugars were subsequently tested in respect of their effect on parasitoid longevity (Wäckers, 2001). Here again, considerable differences among sugars were found. Those sugars to which parasitoids were most sensitive in the gustatory experiment increased the parasitoid's lifespan by a factor of 15–16. A range of other sugars had a less distinct or only marginal effect. Lactose and raffinose did not significantly raise parasitoid longevity, while rhamnose actually reduced the parasitoid's lifespan significantly.

The information obtained from these studies can be of relevance to our understanding of the (un)suitability of the broad range of naturally occurring sugar sources. For instance, the poor performance of *C. glomerata* on honeydew-specific sugars might explain previous reports showing that honeydew can be an inferior food source compared with honey or sucrose (Leius, 1961; Avidov *et al.*, 1970; Wäckers, 2000).

Trade-offs Between Feeding and Reproduction

Even though oviposition and feeding represent separate behavioural categories, they can be interdependent. Obviously, feeding extends a parasitoid's reproductive lifespan, while the energetic costs of host search, oviposition and egg maturation can take toll of parasitoid longevity. Various other types of interactions between reproduction and longevity can occur as well, representing distinct trade-offs between these two fitness parameters (Fig. 5.1). Parasitoids have evolved a range of strategies to optimize these fitness conflicts. Two basic trade-offs between oviposition and feeding will be discussed below.

Host-feeding versus oviposition

In many parasitoid species, host-feeding and reproduction are mutually exclusive, as host-feeding leaves the host unsuitable for larval development. For host-feeding species, this

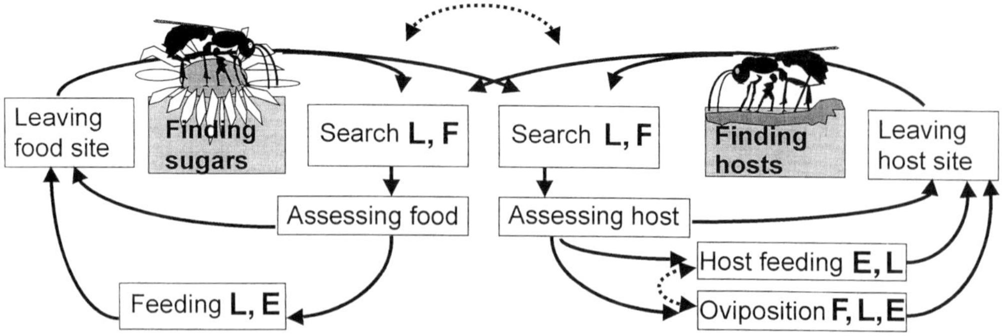

Fig. 5.1. Feeding–reproduction trade-offs in hymenopteran parasitoids. Dotted arrows indicate trade-offs between sugar foraging and host search (all parasitoid species), or between host-feeding and oviposition (host-feeding species only). Letters indicate costs and benefits to longevity (L), egg supply (E) and achieved fecundity (F).

may create the conflict of whether to use a host for current (oviposition) or future reproductive success (host-feeding) (Heimpel and Rosenheim, 1995). The question of how parasitoids balance this dual exploitation of their host resources has been the topic of optimization models (Jervis and Kidd, 1986; Kidd and Jervis, 1991), as well as empirical studies (Rosenheim and Rosen, 1992; Ueno, 1999). While earlier models assumed equal host suitability to address the effect of varying host density, later work incorporated the effect of varying host quality (Jervis and Kidd, 1986; Rosenheim and Rosen, 1992; Ueno, 1999). In the latter (more realistic) scenario, models predict that parasitoids should selectively use low-quality hosts for feeding and restrict oviposition to high-quality hosts (Jervis and Kidd, 1986; Kidd and Jervis, 1991).

Empirical studies have demonstrated that parasitoids do indeed selectively exploit their hosts according to various quality parameters. When given a choice between different host species, parasitoids tend to feed on the species that is the poorer host for parasitoid development. Parasitoids can discriminate by size, using the smaller hosts for host-feeding (Rosenheim and Rosen, 1992). Parasitoids can also use information on host developmental stage (Kidd and Jervis, 1991) or previous parasitization. In the latter case, parasitoids preferentially feed on hosts that contain offspring by conspecifics (Ueno, 1999) or heterospecifics, killing the resident parasitoid larvae.

Host search versus food foraging

A second conflict may arise from the spatial distribution of host and carbohydrate sources. This problem need not occur for those parasitoid species whose hosts are closely linked to carbohydrate-rich food sources. Examples of this category are species whose hosts excrete suitable sugars, e.g. honeydew (England and Evans, 1997; but see Wäckers, 2000), or whose hosts occur on sugar-rich substrates, such as fruits or sugar-excreting plant structures (Illingworth, 1921; Morales-Ramos *et al.*, 1996). For these parasitoids the task of locating hosts and carbohydrates is interlinked. Parasitoids from this group may show specific adaptations to the exploitation of additional carbohydrate sources (F.L. Wäckers, L. Obrist and W. Völkl, unpublished) and little or no task differentiation between food foraging and host search.

The conflict of spatial dissociation between host and carbohydrate sources is mainly acute for those parasitoids whose hosts are not reliably associated with a suitable carbohydrate source. These parasitoids have to alternate their search for hosts (reproduction) with bouts of food foraging, which requires a clear task differentiation. Consequently, parasitoids face the issue of whether to stay in a host patch, thereby optimizing short-term reproductive success, or to leave the host patch in search of food sources, a strategy that could optimize reproduction in the long term.

Parasitoids are equipped with a number of mechanisms that enable them to deal with the dichotomy between searching for hosts (reproduction) and foraging for sugar sources (energy). They possess separate categories of innate responses, which are expressed relative to their physiological needs (Wäckers and Lewis, 1994). Associative learning is also organized along separate physiological pathways. Lewis and Takasu (1990) demonstrated that host- and food-associated learning are separate entities linked to the parasitoid's physiological state. Food-deprived parasitoids typically respond by reducing their activity level, which can be a direct consequence of energetic constraints or a strategy to preserve the remaining energy. Furthermore, they start to respond to stimuli that are associated with food, such as floral odours or colours (Wäckers, 1994; Takasu and Lewis, 1995). Following feeding, parasitoids switch back to searching for hosts, choosing host-associated cues over stimuli linked to food.

Foraging for food and searching for hosts also interact on the spatial scale. Takasu and Lewis (1995) showed that parasitoids tend to concentrate their host search in the vicinity of a successful feeding experience.

Conclusion

Biological control workers have long been aware that the effectiveness of parasitoids can be enhanced through the provision of food sources (e.g. Illingworth, 1921; Wolcott, 1942; Hocking, 1966). Hocking (1966) stressed the importance of food-source availability to the success of classical biological control programmes. Others have advocated the use of food supplements to support native predators or parasitoids (Hagen, 1986; Jacob and Evans, 1998). This has resulted in several (partly successful) attempts to increase the effectiveness of biological control agents through either the use of flowering non-crop plants (Bugg *et al.*, 1987; Landis *et al.*, 2000) or the provision of artificial food sources (Hagen, 1986; Jacob and Evans, 1998). However, the choice of food sources is often not based on adequate data, due to the fact that basic information on the food ecology of beneficial insects is scarce.

Comparative studies provide a promising alternative, as they allow us to rate natural or artificial food sources in respect of their suitability as parasitoid food supplements. Based on this information we can select the optimal food supplements for use in parasitoid rearing, as well as for the enhancement of parasitoid performance in the field.

The use of suitable food supplements in mass rearing entails some specific benefits. Due to the absence of many natural mortality factors under protected mass-rearing conditions, the addition of food can enhance parasitoid longevity and fecundity to levels that exceed those resulting from food in the field. Furthermore, in mass rearing, food can be provided in the direct vicinity of the hosts, which has the advantage that the trade-off between search for hosts and food foraging does not apply. The provision of food supplements can also help reduce host-feeding (undesirable in mass rearing), especially if suitable proteins are added to the food supplement.

References

Alm, J., Ohnmeiss, T.E., Lanza, J. and Vriesenga, L. (1990) Preference of cabbage white butterflies and honey bees for nectar that contains amino acids. *Oecologia* 84, 53–57.

Avidov, Z., Balshin, M. and Gerson, U. (1970) Studies on *Aphytis coheni*, a parasite of the california red scale, *Aonidiella aurantii*, in Israel. *Entomophaga* 15, 191–207.

Baggen, L.R. and Gurr, G.M. (1998) The influence of food on *Copidosoma koehleri* (Hymenoptera: Encyrtidae), and the use of flowering plants as a habitat management tool to enhance biological control of potato moth, *Phthorimaea operculella* (Lepidoptera: Gelechiidae). *Biological Control* 11, 9–17.

Baker, D.A., Hall, J.L. and Thorpe, J.R. (1978) A study of the extrafloral nectaries of *Ricinus communis*. *New Phytologist* 81, 129–137.

Baker, H.G. and Baker, I. (1982a) Chemical constituents of nectar in relation to pollination mechanisms and phylogeny. In: Nitecki, M.H. (ed.) *Biochemical Aspects of Evolutionary Biology*. University of Chicago Press, Chicago, pp. 131–171.

Baker, H.G. and Baker, I. (1982b) A brief historical review of the chemistry of floral nectar. In: Bentley, B. and Elias, T. (eds) *The Biology of Nectaries*. Columbia University Press, New York, pp. 126–152.

Baker, H.G. and Baker, I. (1983) Floral nectar sugar constituents in relation to pollinator type. In: Jones, C.E. and Little, R.J. (eds) *Handbook of Experimental Pollination Biology*. Van Nostrand Reinhold, New York, pp. 117–141.

Bakker, F.M. and Klein, M.E. (1992) Transtrophic interactions in cassava. *Experimental and Applied Acarology* 14, 293–311.

Barnavon, L., Doco, T., Terrier, N., Ageorges, A., Romieu, C. and Pellerin, P. (2000) Analysis of cell wall neutral sugar composition, beta-galactosidase activity and a related cDNA clone throughout the development of *Vitis vinifera* grape berries. *Plant Physiology and Biochemistry* 38, 289–300.

Beggs, J.R. and Rees, J.S. (1999) Restructuring of Lepidoptera communities by introduced *Vespula* wasps in a New Zealand beech forest. *Oecologia* 119, 565–571.

Belmonte, E., Cardemil, L. and Arroyo, M.T.K. (1994) Floral nectary structure, and nectar composition in *Eccremocarpus scaber* (Bignoniaceae), a hummingbird-pollinated plant of central Chile. *American Journal of Botany* 81, 493–503.

Bentley, B.L. (1977) Extrafloral nectaries and protection by pugnacious bodyguards. *Annual Review of Ecology and Systematics* 8, 407–427.

Bory, G. and Clair Maczulajtys, D. (1986) Nectar composition and role of the extrafloral nectar in *Ailanthus glandulosa*. Composition du nectar et role des nectaires extrafloraux chez *l'Alianthus glandulosa*. *Canadian Journal of Botany* 64, 247–253.

Bugg, R.L., Ehler, L.E. and Wilson, L.T. (1987) Effect of common knotweed (*Polygonum aviculare*) on abundance and efficiency of insect predators of crop pests. *Hilgardia* 55, 1–52.

Bugg, R.L., Ellis, R.T. and Carlson, R.W. (1989) Ichneumonidae (Hymenoptera) using extrafloral nectar of faba bean (*Vicia faba* L., Fabaceae) in Massachusetts. *Biological Agriculture and Horticulture* 6, 107–114.

Bugg, R.L., Wäckers, F.L., Brunson, K.E., Dutcher, J.D. and Phatak, S.C. (1991) Cool-season cover crops relay intercropped with cantaloupe: influence on a generalist predator *Geocoris punctipes* (Hemiptera: Lygaeidae). *Journal of Economical Entomology* 84, 408–416.

Búrquez, A. and Corbet, S.A. (1991) Do flowers reabsorb nectar? *Functional Ecology* 5, 369–379.

Byrne, D.N. and Miller, W.B. (1990) Carbohydrate and amino acid composition of phloem sap and honeydew produced by *Bemisia tabaci*. *Journal of Insect Physiology* 36, 433–439.

Coll, M. (1996) Feeding and ovipositing on plants by an omnivorous insect predator. *Oecologia* 105, 214–220.

Dafni, A. (1992) *Pollination Ecology*. The Practical Approach Series (eds D. Rickwood and B.D. Hames), Oxford University Press, Oxford.

Davis, D.W., McDougall, E.M., Hendrix, D.L., Steele, D.L., Adaskaveg, J.E. and Butler, E.E. (1993) Air particulates associated with the ash whitefly. *Air and Waste* 43, 1116–1121.

Dicke, M. and Vet, L.E.M. (1998) Plant–carnivore interactions: evolutionary and ecological consequences for plant, herbivore and carnivore. In: Ollf, H., Brown, V.K. and Drent, R.H. (eds) *Herbivores between Plants and Predators*. Blackwell Science, Malden, pp. 483–520.

Dyer, L.E. and Landis, D.A. (1996) Effects of habitat, temperature, and sugar availability on longevity of *Eriborus terebrans* (Hymenoptera: Ichneumonidae). *Environmental Entomology* 25, 1192–1201.

England, S. and Evans, E.W. (1997) Effects of pea aphid (Homoptera: Aphididae) honeydew on longevity and fecundity of the alfalfa weevil (Coleoptera: Curculionidae) parasitoid *Bathyplectes curculionis* (Hymenoptera: Ichneumonidae). *Environmental Entomology* 26, 1437–1441.

Ewart, W.H. and Metcalf, R.L. (1956) Preliminary studies of sugars and amino acids in the honeydews of five species of coccids feeding on citrus in California. *Annals of the Entomological Society of America* 49, 441–447.

Godfray, H.C.J. (1994) *Parasitoids: Behavior and Evolutionary Ecology*. Princeton University Press, Princeton, New Jersey.

Gori, D.F. (1983) Post-pollination phenomena and adaptive floral changes. In: Jones, C.E. and Little, R.J. (eds) *Handbook of Experimental Pollination Biology*. Van Nostrand Reinhold, New York, pp. 31–49.

Gottsberger, G., Schrauwen, J. and Linskens, H.F. (1973) Die Zuckerbestandteile des Nektars einiger tropischen Blueten. *Portagaliae Acta Biologica* 13, 1–8.

Hagen, K.S. (1986) Ecosystem analysis: plant cultivars (HPR), entomophagous species and food supplements. In: Boethel, D.J. and Eikenbary, R.D. (eds) *Interactions of Plant Resistance and Parasitoids and Predators of Insects*. John Wiley & Sons, New York, pp. 153–197.

Heil, M., Fiala, B., Baumann, B. and Linsenmair, K.E. (2000) Temporal, spatial and biotic variations in extrafloral nectar secretion by *Macaranga tanarius. Functional Ecology* 14, 749–757.

Heil, M., Koch, T., Hilpert, A., Fiala, B., Boland, W. and Linsenmair, K.E. (2001) Extrafloral nectar production of the ant-associated plant, *Macaranga tanarius,* is an induced, indirect, defensive response elicited by jasmonic acid. *Proceedings of the National Academy of Sciences USA* 98, 1083–1088.

Heimpel, G.E. and Rosenheim, J.A. (1995) Dynamic host-feeding by the parasitoid *Aphytis melinus*: the balance between current and future reproduction. *Journal of Animal Ecology* 64, 153–167.

Hendrix, D.L., Wei, Y. and Leggett, J.E. (1992) Homopteran honeydew sugar composition is determined by both the insect and plant species. *Comparative Biochemistry and Physiology* 101B (1/2), 23–27.

Hocking, H. (1966) The influence of food on longevity and oviposition in *Rhyssa persuasoria* (L) (Hymenoptera: Ichneumonidae). *Journal of the Australian Entomological Society* 6, 83–88.

Hoferer, S., Wäckers, F.L. and Dorn, S. (2000) Measuring CO_2 respiration rates in the parasitoid *Cotesia glomerata. Mitteilungen der Deutschen Gesellschaft für allgemeine und angewandte Entomologie* 12, 555–558.

Idoine, K. and Ferro, D.N. (1988) Aphid honeydew as a carbohydrate source for *Edovum puttleri* (Hymenoptera: Eulophidae). *Environmental Entomology* 17, 941–944.

Idris, A.B. and Grafius, E. (1995) Wildflowers as nectar sources for *Diadegma insulare* (Hymenoptera: Ichneumonidae), a parasitoid of diamondback moth (Lepidoptera: Yponomeutidae). *Environmental Entomology* 24, 1726–1735.

Illingworth, J.F. (1921) Natural enemies of sugar-cane beetles in Queensland. *Queensland Bureau Sugar Experimental Station Division of Entomology Bulletin* 13, 1–47.

Inouye, D.W. and Waller, G.D. (1984) Responses of honey bees (*Apis mellifera*) to amino acid solutions mimicking floral nectars. *Ecology* 65, 618–625.

Jacob, H.S. and Evans, E.W. (1998) Effects of sugar spray and aphid honeydew on field populations of the parasitoid *Bathyplectes curculionis* (Hymenoptera: Ichneumonidae). *Environmental Entomology* 27, 1563–1568.

Jervis, M.A. and Kidd, N.A.C. (1986) Host-feeding strategies in hymenopteran parasitoids. *Biological Review* 61, 395–434.

Jervis, M.A., Kidd, N.A.C., Fitton, M.G., Huddleston, T. and Dawah, H.A. (1993) Flower-visiting by hymenopteran parasitoids. *Journal of Natural History* 27, 67–105.

Jervis, M.A., Kidd, N.A.C. and Heimpel, G.E. (1996) Parasitoid adult feeding behaviour and biocontrol – a review. *Biocontrol News and Information* 17, 11N–26N.

Josens, R.B., Farina, W.M. and Roces, F. (1998) Nectar feeding by the ant *Camponotus mus*: intake rate and crop filling as a function of sucrose concentration. *Journal of Insect Physiology* 44, 579–585.

Kevan, P.G. (1973) Parasitoid wasps as flower visitors in the Canadian high Arctic. *Anzeiger für Schädlingskunde, Pflanzenschutz und Umweltschutz* 46, 3–7.

Kidd, N.A.C. and Jervis, M.A. (1991) Host-feeding and oviposition strategies of parasitoids in relation to host stage. *Researches on Population Ecology* 33, 87–99.

Kimura, M.T., Awasaki, T., Ohtsu, T. and Shimada, K. (1992) Seasonal changes in glycogen and trehalose content in relation to winter survival of 4 temporate species of *Drosophila. Journal of Insect Physiology* 38, 871–875.

Kloft, W.J., Maurizio, A. and Kaeser, W. (1985) *Waldtracht und Waldhonig in der Imkerei.* Ehrenwirth Verlag, Munich, 328 pp.

Koptur, S. (1989) Is extrafloral nectar production an inducible defence? In: Bock, J. and Linhart, Y. (eds) *Evolutionary Ecology of Plants.* Westview Press, Boulder, Colorado, pp. 323–339.

Koptur, S. (1992) Extrafloral nectary-mediated interactions between insects and plants. In: Bernays, E. (ed.) *Insect–Plant Interactions,* Vol. IV. CRC Press, Boca Raton, Florida, pp. 81–129.

Koptur, S. (1994) Floral and extrafloral nectars of Costa Rican *Inga* trees: a comparison of their constituents and composition. *Biotropica* 26, 276–284.

Landis, D.A., Gurr, G.M. and Wratten, S.D. (2000) Habitat management to conserve natural enemies of arthropod pests in agriculture. *Annual Review of Entomology* 45, 175–201.

Lanza, J., Vargo, E.L., Pulim, S. and Chang, Y.Z. (1993) Preferences of the fire ants *Solenopsis invicta* and *S. geminata* (Hymenoptera: Formicidae) for amino acid and sugar components of extrafloral nectars. *Environmental Entomology* 22, 411–417.

Larochelle, A. (1990) The food of carabid beetles (Coleoptera: Carabidae, including Cicindelinae). *Fabreries,* Suppl. 5, 1–132.

Leius, K. (1961) Influence of food on fecundity and longevity of adults of *Itoplectis conquisitor* (Say) (Hymenoptera: Ichneumonidae). *Canadian Entomologist* 93, 771–780.

Lewis, W.J. and Takasu, K. (1990) Use of learned odours by a parasitic wasp in accordance with host and food needs. *Nature* 348, 635–636.
Limburg, D.D. and Rosenheim, J.A. (2001) Extrafloral nectar consumption and its influence on survival and development of an omnivorous predator, larval *Chrysoperla plorabunda* (Neuroptera: Chrysopidae). *Environmental Entomology* 30, 595–604.
Lingren, P.D. and Lukefahr, M.J. (1977) Effects of nectariless cotton on caged populations of *Campoletis sonorensis*. *Environmental Entomology* 6, 586–588.
Lunau, K. and Wacht, S. (1994) Optical releasers of the innate proboscis extension in the hoverfly *Eristalis tenax* L. (Syrphidae, Diptera). *Journal of Comparative Physiology A* 174, 575–579.
Mittler, T.E. and Meikle, T. (1991) Effects of dietary sucrose concentration on aphid honeydew carbohydrate levels and rates of excretion. *Entomologia Experimentalis et Applicata* 59, 1–7.
Morales-Ramos, J.A., Rojas, M.G. and King, E.G. (1996) Significance of adult nutrition and oviposition experience on longevity and attainment of full fecundity of *Catolaccus grandis* (Hymenoptera: Pteromalidae). *Annals of the Entomological Society of America* 89, 555–563.
O'Dowd, D.J. and Catchpole, E.A. (1983) Ants and extrafloral nectaries: no evidence for plant protection in *Helichrysum* spp.–ant interactions. *Oecologia* 59, 191–200.
Oliveira, P.S. (1997) The ecological function of extrafloral nectaries: herbivore deterrence by visiting ants and reproductive output in *Caryocar brasiliense* (Caryocaraceae). *Functional Ecology* 11, 323–330.
Olson, D.L. and Nechols, J.R. (1995) Effects of squash leaf trichome exudates and honey on adult feeding, survival, and fecundity of the squash bug (Heteroptera: Coreidae) egg parasitoid *Gryon pennsylvanicum* (Hymenoptera: Scelionidae). *Environmental Entomology* 24, 454–458.
Percival, M.S. (1961) Types of nectar in angiosperms. *New Phytologist* 60, 235–281.
Porter, S.D. (1989) Effects of diet on the growth of laboratory fire ant colonies (Hymenoptera: Formicidae). *Journal of the Kansas Entomological Society* 62, 288–291.
Rico-Gray, V. and Thien, L.B. (1989) Effect of different ant species on reproductive fitness of *Schomburgkia tibicinis* (Orchidaceae). *Oecologia* 81, 487–489.
Rivero, A. and Casas, J. (1999) Incorporating physiology into parasitoid behavioral ecology: the allocation of nutritional resources. *Researches in Population Ecology* 41, 39–45.
Rosenheim, J.A. and Rosen, D. (1992) Influence of egg load and host size on host feeding behavior of the parasitoid *Aphytis lingnanensis*. *Ecological Entomology* 17, 263–272.
Ruhren, S. and Handel, S. (1999) Jumping spiders (Salticidae) enhance the seed production of a plant with extrafloral nectaries. *Oecologia* 119, 227–230.
Schmale, I., Wäckers, F.L., Cardona, C. and Dorn, S. (2001) Control potential of three hymenopteran parasitoid species against the bean weevil in stored beans: the effect of adult parasitoid nutrition on longevity and progeny production. *Biological Control* 21, 134–139.
Smith, L.L., Lanza, J. and Smith, G.C. (1990) Amino acid concentrations in extrafloral nectar of *Impatiens sultani* increase after simulated herbivory. *Ecology* 71, 107–115.
Stadler, B. and Dixon, A.F.G. (1999) Ant attendance in aphids: why different degrees of myrmecophily? *Ecological Entomology* 24, 363–369.
Stapel, J.O., Cortesero, A.M., De Moraes, C.M., Tumlinson, J.H. and Lewis, W.J. (1997) Extrafloral nectar, honeydew and sucrose effects on searching behavior and efficiency of *Microplitis croceipes* (Hymenoptera: Braconidae) in cotton. *Environmental Entomology* 26, 617–623.
Steinbauer, M.J. (1996) A note on manna feeding by ants (Hymenoptera: Formicidae). *Journal of Natural History* 30, 1185–1192.
Syme, P.D. (1975) The effect of flowers on the longevity and fecundity of two native parasites of the European pine shoot moth in Ontario. *Environmental Entomologist* 4, 337–346.
Takasu, K. and Lewis, W.J. (1995) Importance of adult food sources to host searching of the larval parasitoid *Microplitis croceipes*. *Biological Control* 5, 25–30.
Takeda, S., Kinomura, K. and Sakurai, H. (1982) Short communications. Effects of ant-attendance on the honeydew excretion and larviposition of the cowpea aphid, *Aphis craccivora* Koch. *Applied Entomology and Zoology* 17, 133–135.
Tanowitz, B.D. and Koehler, D.L. (1986) Carbohydrate analysis of floral and extrafloral nectars in selected taxa of *Sansivieria* (Agavaceae). *Annals of Botany* 58, 541.
Telenga, N.A. (1958) Biological methods of pest control in crops and forest plants in the USSR. *Proceedings of the Ninth International Conference on Quarantine and Plant Protection*. Moscow, pp. 1–15.
Tinti, J.M. and Nofre, C. (2001) Response of the ant *Lasius niger* to various compounds perceived as sweet in humans: a structure–activity analysis. *Chemical Senses* 26, 231–237.

Tobin, J.E. (1994) Ants as primary consumers: diet and abundance in the Formicidae. In: Hunt, J.H. and Nalepa, C.A. (eds) *Nourishment and Evolution in Insect Societies*. Westview Press, Boulder, Colorado, pp. 279–307.

Torres-Hernández, L., Rico-Gray, V., Castillo-Guevara, C. and Vergara, J.A. (2000) Effect of nectar-foraging ants and wasps on the reproductive fitness of *Turnera ulmifolia* (Turneraceae) in a coastal sand dune in Mexico. *Acta Zoologica Mexicana* 81, 13–21.

Turlings, T.C.J., Wäckers, F.L., Vet, L.E.M., Lewis, W.J. and Tumlinson, J.H. (1993) Learning of host-finding cues by hymenopterous parasitoids. In: Papaj, D.R. and Lewis, A.C. (eds) *Insect Learning*. Chapman & Hall, New York, pp. 51–78.

Ueno, T. (1999) Multiparasitism and host feeding by solitary parasitoid wasps (Hymenoptera: Ichneumonidae) based on the pay-off from parasitized hosts. *Annals of the Entomological Society of America* 92, 601–608.

van Baalen, M., Krivan, V., van Rijn, P.C.J. and Sabelis, M.W. (2001) Alternative food, switching predators, and the persistence of predator–prey sytems. *American Naturalist* 157, 512–524.

Vander Meer, R.K., Lofgren, C.S. and Seawright, J.A. (1995) Specificity of the red imported fire ant (Hymenoptera: Formicidae) phagostimulant response to carbohydrates. *Florida Entomologist* 78 (1), 144–154.

van Emden, H.F. (1962) Observations on the effect of flowers on the activity of parasitic Hymenoptera. *Entomologist's Monthly Magazine* 98, 265–270.

van Lenteren, J.C. (1999) Fundamental knowledge about insect reproduction: essential to develop sustainable pest management. *Invertebrate Reproduction and Development* 36, 1–15.

van Lenteren, J.C., van Vianen, A., Gast, H.F. and Kortenhoff, A. (1987) The parasite–host relationship between *Encarsia formosa* Gahan (Hymenoptera: Aphelinidae) and *Trialeurodes vaporarorium* (Westwood) (Homoptera: Aleyrodidae). XVI. Food effects on oogenesis, oviposition, life-span, and fecundity of *Encarsia formosa* and other hymenopterous parasites. *Journal of Applied Entomology* 103, 69–84.

Vet, L.E.M., Lewis, W.J. and Cardé, R.T. (1995) Parasitoid foraging and learning. In: Bell, W.J. and Carde, R.T. (eds) *Chemical Ecology of Insects*, Vol. 2. Chapman & Hall, London, pp. 65–101.

Völkl, W., Woodring, J., Fischer, M., Lorenz, M.W. and Hoffmann, K.H. (1999) Ant–aphid mutualisms: the impact of honeydew production and honeydew sugar composition on ant preferences. *Oecologia* 118, 483–491.

von Frisch, K. (1934) Über den Geschmackssinn der Biene. Ein Beitrag zur vergleichenden Physiologie des Geschmacks. *Zeitschrift für Vergleichende Physiologie* 21, 1–45.

Wäckers, F.L. (1994) The effect of food deprivation on the innate visual and olfactory preferences in the parasitoid *Cotesia rubecula*. *Journal of Insect Physiology* 40, 641–649.

Wäckers, F.L. (1999) Gustatory response by the hymenopteran parasitoid *Cotesia glomerata* to a range of nectar and honeydew sugars. *Journal of Chemical Ecology* 25, 2863–2877.

Wäckers, F.L. (2000) Do oligosaccharides reduce the suitability of honeydew for predators and parasitoids? A further facet to the function of insect-synthesized honeydew sugars. *Oikos* 90, 197–201.

Wäckers, F.L. (2001) A comparison of nectar- and honeydew sugars in respect of their utilization by the hymenopteran parasitoid *Cotesia glomerata*. *Journal of Insect Physiology* 47, 1077–1084.

Wäckers, F.L. and Lewis, W.J. (1994) Olfactory and visual learning and their combined influence on host site location by the parasitoid *Microplitis croceipes*. *Biological Control* 4, 105–112.

Wäckers, F.L. and Swaans, C.P.M. (1993) Finding floral nectar and honeydew in *Cotesia rubecula*: random or directed? *Proceedings of Experimental and Applied Entomology* 4, 67–72.

Wäckers, F.L. and Wunderlin, R. (1999) Induction of cotton extrafloral nectar production in response to herbivory does not require a herbivore-specific elicitor. *Entomologia Experimentalis et Applicata* 91, 149–154.

Wäckers, F.L., Zuber, D. and Wunderlin, R. (2001) The effect of herbivory on temporal and spatial dynamics of extrafloral nectar production. *Annals of Botany* 87, 365–370.

Wagner, D. (1997) The influence of ant nests on *Acacia* seed production, herbivory and soil nutrients. *Journal of Ecology* 85, 83–93.

Waller, G.D. (1972) Evaluating responses of honey bees to sugar solutions using an artificial-flower feeder. *Annals of the Entomological Society of America* 65, 857–862.

Whitman, D.W. (1996) Plant bodyguards: mutualistic interactions between plants and the third trophic level. In: Ananthakrishnan, T.N. (ed.) *Functional Dynamics of Phytophagous Insects*. Oxford and IBH Publishing, New Delhi, pp. 207–248.

Wolcott, G.N. (1942) The requirements of parasites for more than hosts. *Science* 96, 317–318.
Wykes, G.R. (1952) The preferences of honeybees for solutions of various sugars which occur in nectar. *Journal of Experimental Biology* 29, 511–519.
Yao, I. and Akimoto, S. (2001) Ant attendance changes the sugar composition of the honeydew of the drepanosiphid aphid *Tuberculatus quercicola*. *Oecologia* 128, 36–43.
Zimmerman, M. (1932) Über die extra-floralen Nectarien der Angiospermen. *Botanisches Zentralblatt Beihefte* 49, 99–196.
Zoebelein, G. (1956) Der Honigtau als Nahrung der Insekten. *Zeitschrift für angewandte Entomologie* 38, 369–416.

6 Managing Captive Populations for Release: a Population-genetic Perspective

L. Nunney
Department of Biology, University of California, Riverside, CA 92521, USA

Abstract

The success of biological control, particularly augmentative biological control, depends upon the effective mass rearing of natural enemies. However, developing the best rearing strategy is complicated by the 'paradox of captive breeding': increasing quantity generally decreases quality. Quantity is the number of individuals produced per unit time, and it is easily measured. Quality is the ability of captive-reared individuals to function as intended in the field, and can be measured as field success relative to the success of individuals from a natural population. Such field measurements are almost always difficult and expensive. Unfortunately, quantity and quality are usually negatively correlated, since genetic adaptation to the rearing environment often adversely affects adaptation to the field environment. Here I review examples of adaptation to captive rearing and of the trade-off with field performance. Given this trade-off, the optimum management strategy is a compromise that can only be defined after extensive field tests. However, we can identify some general factors that are likely to influence the outcome of a breeding programme. A large, genetically variable founding population from a geographical region climatically similar to the release site maximizes the chance of adaptation of the control agent to the release site. While a large founding population minimizes the immediate risk of inbreeding, it may include a few genotypes preadapted to the captive rearing conditions. The success of these few genotypes can result in a genetic bottleneck. The problem can be minimized by temporarily maintaining many small breeding units, so that the reproductive success of a lot of individuals is ensured during the initial phase of domestication. However, any genetically variable population will adapt to its new artificial conditions, and the breeding facility should be designed to minimize selection for characteristics known to reduce field performance. Even so, field-adapted genotypes must be incorporated on a regular basis, either by monitored addition to the captive population or by establishing a completely new population. Alternatively, adaptation to the captive environment can be avoided by maintaining a large number of inbred (isofemale) lines. This approach, combined with prerelease crosses, can be very effective at maintaining quality. These considerations highlight an important problem associated with using genetically manipulated stocks. We must be careful that the potential benefits of genetic engineering are not squandered by incorporating beneficial genetic changes into laboratory-adapted stocks that are ill suited to field release.

© CAB International 2003. *Quality Control and Production of Biological Control Agents: Theory and Testing Procedures* (ed. J.C. van Lenteren)

Introduction

Optimizing the mass rearing of arthropods for release into the field is an extremely complex problem. Much of this complexity arises from a fundamental evolutionary conflict implicit in the mass-rearing process: the ideal captive population is maximally adapted to both the rearing conditions and the field conditions into which it will be released. In most cases, this ideal cannot be achieved and the optimal solution is a compromise between efficient rearing and good field performance. However, the details of the compromise will depend upon the genetic structure of the population. By adopting a population-genetic approach, we can avoid many of the pitfalls of establishing and managing captive populations and hopefully achieve close to the best solution.

There is an extensive literature on the evolutionary problem of how organisms in nature maintain simultaneous adaptation in two (or more) environments. This is the problem of adaptive plasticity (see Via *et al.*, 1995) and the evolutionary solution depends upon the degree of genetic correlation between high fitness in the alternative environments. A high positive correlation means that genotypes with high fitness in one environment also have high fitness in the second. However, a high negative correlation means that genotypes with high fitness in one environment have a low fitness in the other. It is the existence of negative genetic correlations that creates the evolutionary problem of adaptive plasticity and what I shall refer to as the 'paradox of captive breeding' – improving performance in the rearing facility can result in decreased performance in the field.

The paradox of captive breeding means that the optimal solution is generally a compromise between quantity and quality. Quantity is the productivity of the rearing programme, measured by numbers reared per unit time. Quality is the ability of captive-reared individuals to function as intended after field release (see Chapter 1), and can be measured relative to the field performance of individuals from the natural population. The optimal solution aims to maximize the product (quantity) × (quality), and I shall argue that, in general, this most effective strategy does not maximize either quantity or quality.

The Problem: the Trade-off Between Quantity and Quality

Measuring the trade-off

The measurement of quantity, i.e. the success of captive rearing, is a straightforward count of the numbers of individuals produced under the rearing protocol. Quantity is expected to improve with domestication, i.e. adaptation to the captive environment. This improvement can be evaluated experimentally by comparing the productivity of the established captive population to a recently wild-caught control population. The measurement of quality, i.e. field performance, is more complex and must be determined by the specifics of the release programme. While the goal is to control the numbers of some specific pest, the specific agent used may be a parasitoid, predator or, in the case of the sterile-insect technique (SIT), a conspecific male. However, at a minimum, the measure must integrate three components: first, the ability of individuals to disperse from the release site and find the target (hosts or prey in the case of natural enemies, females in the case of SIT); secondly, their ability to successfully interact with the target (parasitize the hosts or eat the prey in the case of natural enemies, or mate in the case of SIT); and, thirdly, the ability of the released individuals to survive in the field and continue to find their targets. Finally, if the project goal is to establish a self-perpetuating population of a natural enemy (classical biological control), then the ability of individuals to reproduce and to survive through unfavourable seasons defines a fourth essential component of quality.

It is important that the quality of the captive-reared population is measured relative to a recently wild-caught control population. However, in practice, it is unlikely that the field performance of a captive strain could be compared directly with that of a wild-caught population. Instead, the captive strain could be tested alone and its performance compared with some previously established 'wild' standard. To establish the standard, replicated trials (preferably run at different times and at different sites) must be conducted using a wild-caught population. These trials would establish the standard in terms of some appropriate measure of quality. The measure should reflect the success of an individual over the whole of its useful life, and include the necessity for active searching beyond the immediate release site. For example, for a parasitoid, this standard could be per cent parasitism per parasitoid per unit area measured a set number of days after release and given some typical host density.

I have been unable to find any experiments that measure both quantity and quality in strains with different degrees of domestication. Indeed, there are relatively few studies that document the consequences for biological control of a decline in quality associated with mass rearing (although see Ito, 1988; Calkins and Ashley, 1989). This is probably due to an understandable reluctance of those responsible for captive rearing to document any such decline, since the effort is likely to provoke criticism of their captive-rearing strategy. However, this attitude is misguided. The important question is not whether quality has declined, but whether the product of quality and quantity has been maximized. The theoretical expectation is that maximizing this product will almost inevitably involve some decline in quality.

Theoretical expectations

In general, quantity and quality are interchangeable, i.e. a loss of quality can be compensated for by increased quantity (Nunney, 2002). This assumption leads to the conclusion that the optimum strategy maximizes the product of these two parameters. Only if there is an interaction between quantity and quality is it necessary to maximize a more complex function. For example, if density-dependent interference among individuals caused individual field performance (quality) to decline at high density, then field density would need to be factored into the maximization. Except under these conditions, the effectiveness of mass-rearing programmes can be evaluated along the two dimensions of the quantity produced and the quality of the individuals released.

Mathematically, the effectiveness (E) of a captive-rearing programme is defined by:

$$E = Pw \quad (1)$$

where P and w are quantity (productivity) and quality (individual field performance), respectively. Furthermore, we expect that adaptation to the rearing environment (increasing P) will change (and generally decrease) w according to some function f:

$$w = f(P) \quad (2)$$

It follows from (1) and (2) that maximizing E requires:

$$\frac{d \ln f(P)}{d \ln P} = -1 \quad (3)$$

The interpretation of equations (2) and (3) is shown in Fig. 6.1. In both the upper and lower graphs, the solid curve defines $f(P)$, the relationship defining how field quality (w) changes as an initially wild-caught population adapts to captive rearing. The population is expected to gradually shift from its initial state ('wild population') to a relatively stable 'domesticated stock'. This transition is marked by some decrease in field performance. Since effectiveness is the product Pw (equation 1), points of equal effectiveness are linked on a log scale by a line of slope −1 (see Fig. 6.1, dashed lines). The maximum effectiveness is usually defined by equation (3), i.e. where one of these lines is tangential to $f(P)$. This can be seen in the upper graph of Fig. 6.1. The 'optimum strategy' shown on the graph maintains a population that is partially domesticated – at this point the gain in quantity far outweighs the loss in quality.

The lower graph differs in the shape of $f(P)$, a difference that alters the optimum strategy. In this graph, there is a local maximum near to the point of full domestication; however, the overall optimum strategy is to minimize domestication, because of the large loss in quality occurring during the initial stages of domestication. Such a pattern suggests investment in improving the rearing conditions so that the selection that results in an excessive loss of quality is reduced.

In a captive environment, we can expect the 'domesticated' stock (Fig. 6.1) to be relatively stable. However, over time, we would expect the accumulated effects of inbreeding to result in a slow general decline that reduces both quantity and quality (the lower part of the curves shown in Fig. 6.1).

Genetic Change due to Captive Rearing

Most of the studies of genetic change occurring in captive-reared populations have been on the tephritid fruit flies used in SIT. These studies are useful because they illustrate the dramatic changes that occur when insects are intensively cultured to produce very large numbers. There is no reason to believe that the data from fruit flies are unusual. Genetic changes are inevitable whenever a genetically variable population is reared in a novel environment. Natural selection will act and adaptation to the new environment will occur.

Mating behaviour

One very common adaptation to captive rearing is earlier mating and oviposition (e.g. melon fly: Miyatake, 1998; medfly: Rössler, 1975; Wong and Nakahara, 1978; Vargas and Carey, 1989; oriental fruit fly: Foote and Carey, 1987; tobacco budworm: Raulston, 1975). The time scale of this adaptation is quite rapid. Using a wild-caught population of tobacco budworm, Raulston (1975) found that, after seven generations of captive rearing, the shift to the domesticated pattern of early mating was almost complete.

More complex changes in mating behaviour may also occur. Haeger and O'Meara (1970) showed that the captive rearing of the mosquito *Culex nigripalpus* resulted from a shift in female behaviour. The mating success of colony females was about 70%, regardless of whether the males were from the colony or from the wild; however, under similar condi-

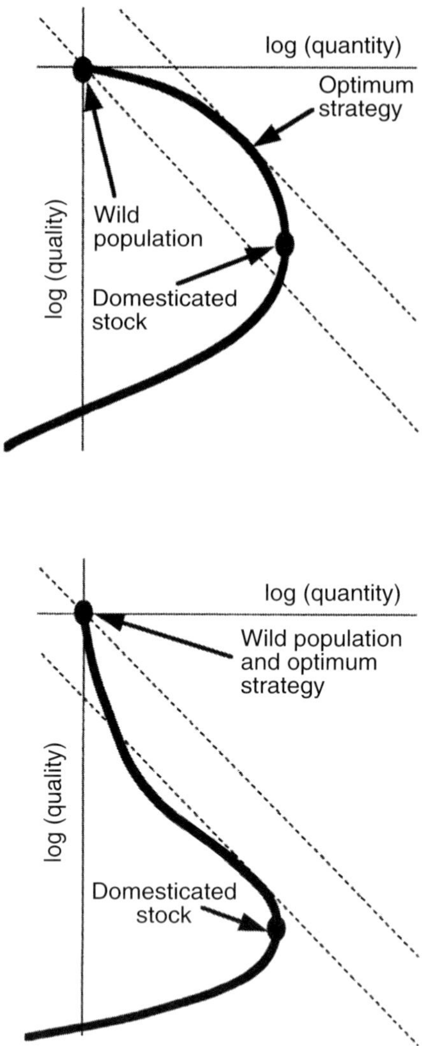

Fig. 6.1. The expected trade-off between the numbers produced in mass rearing (quantity) and field performance (quality). The solid curve defines the trade-off. The natural (wild) population is arbitrarily placed at the origin and the position of a population adapted to the captive-rearing facility over many generations (domesticated stock) is shown. The region of the trade-off curve below the point of domestication defines decreasing quality and quantity, due to the effect of long-term inbreeding depression. The dashed lines link combinations of quality and quantity that are equally effective (as defined by equation 1). The upper graph shows a trade-off curve with an optimum strategy of partial domestication; the lower graph shows a curve with no such optimum – the best strategy is to minimize domestication.

tions only about 1% of wild-caught females mated. Changes in mating behaviour apparently due to captive rearing have also been observed in houseflies (Fye and LaBrecque, 1966). In a laboratory simulation of SIT, sterile males from a 20-year-old laboratory population competed poorly with males from a wild-caught population, whereas sterile males from a newly established laboratory population were much more successful. In another example, Fletcher *et al.* (1968) showed significant differences between two captive populations of screw-worm fly. Males from both populations produced the male pheromone, but only females from one of the populations responded to the chemical. The authors suggested that differences in the captive rearing of the two populations may have selected for this difference.

Changes in mating behaviour are not only a problem for SIT. In classical biological control, the aim is to establish a self-perpetuating population of the natural enemy. If the mating system has been disrupted through domestication, the probability of establishment is inevitably reduced.

Life-history traits

Captive-rearing conditions almost inevitably select for faster development. This has been observed in medfly (Rössler, 1975; Wong and Nakahara, 1978; Vargas and Carey, 1989), oriental fruit fly (Foote and Carey, 1987), Caribbean fruit fly (Leppla *et al.*, 1976) and melon fly (Miyatake, 1993; Miyatake and Yamagishi, 1999). In the melon fly, these changes occurred during the first nine generations of captive rearing (Miyatake and Yamagishi, 1999).

Selection for faster development generally leads to a correlated decrease in adult size and lifetime female fecundity (Nunney, 1996); however, the expected correlations can break down when a population is introduced into a new environment (Service and Rose, 1985). Thus, although the adult size of melon fly decreased in response to selection for a shorter developmental period, lifetime fecundity did not, and captive melon-fly populations generally have a higher fecundity than wild-caught flies (Miyatake, 1998). Similarly, in both the Caribbean fruit fly (Leppla *et al.*, 1976) and the oriental fruit fly (Foote and Carey, 1987), the shorter development time of domesticated populations was associated with higher fecundity, relative to recently wild-caught flies.

Correlated responses can affect traits that we may not a priori expect to be influenced. Miyatake (1998) notes that selection for faster development in the melon fly results in individuals that have a shortened circadian period and that mate earlier in the day. These responses were not arbitrary; they were due to the pleiotropic effects of a single gene (Shimizu *et al.*, 1997). This result is an excellent illustration of how adaptation to the rearing facility (faster development) could have an unexpected negative effect on mating success in the field (due to flies attempting to mate at the wrong time of day).

General

We do not know which genetic loci are involved in the adaptation to a captive environment. The rapidity of adaptation is suggestive that relatively few loci are responsible for most of the change. For example, in tobacco budworm, it took only four generations for the oviposition pattern of a wild-caught population to converge on that of a laboratory culture (Raulston, 1975). More typically, significant adaptive change seems to occur over the first 6–10 generations (Raulston, 1975; Loukas *et al.*, 1985; Miyatake and Yamagishi, 1999).

In the screw-worm fly, Bush and Neck (1976) identified a candidate gene, the α-glycerophosphate dehydrogenase (α-GDH) locus. They found that one allele, rare in natural Texas populations, was very common in each of four large 'factory' populations. They argued that this was an adaptive change in response to the novel rearing conditions (constant high temperature, combined with selection for rapid development and reduced flight). Similarly, Loukas *et al.* (1985) found rapid changes at several allozyme loci when a population of the olive fly was reared in the laboratory. In only five generations, the commonest allele at the 6-phosphogluconate dehydrogenase (6-PGD) and alcohol dehydrogenase (ADH) loci declined from an initial frequency of 0.6–0.7 to close to 0.2.

Strong selection for adaptation to the captive-rearing environment can be expected to reduce the genetic variability of populations. Bush and Neck (1976) and Loukas *et al.* (1985) noted that the genetic (allozyme) variability of captive colonies decreased over time, and Miyatake and Yamagishi (1999) found that the heritability of larval development time in captive melon fly declined over time until it was not statistically different from zero.

Changes in Field Performance

There is no question that captive rearing results in a cascade of genetic changes as a population adapts to its new environment. But this leaves open the question of the extent to which these changes reduce field performance. As noted earlier, measuring field performance is almost always difficult and understandably many researchers have used simple laboratory tests to infer effectiveness in the field (see, for example, Cohen, 2000). However, data from the Japanese melon-fly SIT programme argue strongly against this approach.

The melon fly was successfully eradicated from the Japanese islands of Kume-zima in 1977 and Miyako-zima in 1987. This success came after several failed attempts in other parts of the world to use SIT to eradicate fruit flies. Ito (1988) reviewed the Japanese project and concluded that, contrary to prevailing practice, a relatively low sterile : wild-fly ratio is sufficient to achieve success provided that the harmful effects of domestication can be avoided. In particular, he emphasized that, although the negative effects of irradiation have always been a concern in SIT, mass rearing can cause a much greater reduction in the mating competitiveness of released males. This is a very important point. It suggests that the prevailing emphasis on quantity is misplaced. A failure to maintain quality can drive up the cost of biological control and can make successful control unlikely. Calkins and Ashley (1989) stressed this same point for medfly SIT. Using estimates from three medfly stocks, they calculated the dramatic increase in costs incurred when quality is compromised.

A decline in the field mating competitiveness of mass-reared melon flies became apparent after about 15 generations. No such decline was apparent under laboratory conditions (Fig. 6.2). At generation 18, mating

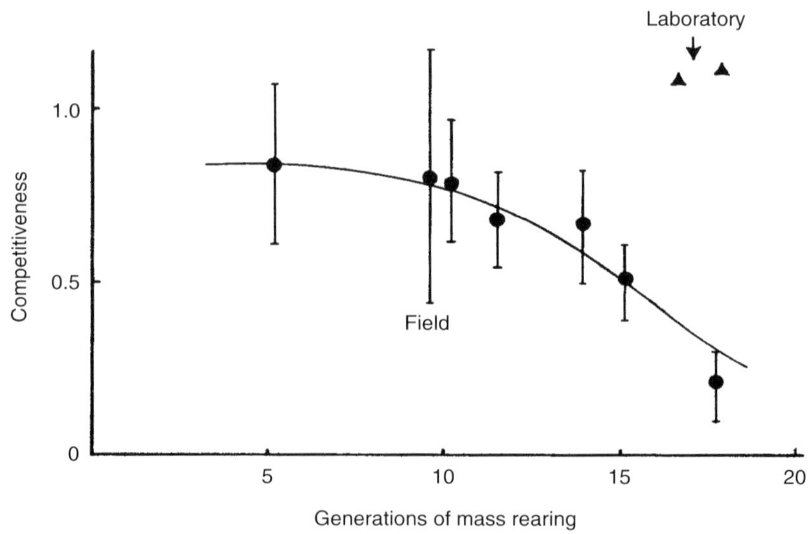

Fig. 6.2. The decline in the field mating competitiveness of sterile, captive-reared male melon flies, as a function of their time in mass production. Also shown are the results of laboratory trials at generations 16 and 17. The release programme on Kume-zima, Japan, started after generation 5. (Figure from Ito, 1988.)

competitiveness in the laboratory was very high; however, field performance had declined by about fourfold (relative to generation 5). The Japanese researchers believed that decreased flight ability and decreased mating success at lek sites (where males congregate to attract females) contributed to this decline. Field experiments demonstrated that dispersal distance of the captive-reared population was substantially less than that of wild-caught male melon flies, and this was confirmed in laboratory experiments on flight duration (see Ito, 1988). The differences between domesticated and wild stocks of melon fly became apparent after the first hour of flight, and Ito (1988) links this relatively subtle distinction in performance to the significant loss of quality. The 'low-quality' melon flies were generally able to fly for more than an hour. Contrast this level of flight ability with a criterion used for evaluating the quality of captive-reared medfly – the ability to successfully fly out of a 20 cm tall, 9 cm diameter cylinder (Boller *et al.*, 1981). A similar criterion is still being used (see Cayol and Zarai, 1999).

Since the quality of captive-reared populations declines over time, an important practical issue concerns when a population should be replaced. One useful approach is to compare the old strain with a newly established one, under field conditions. Ahrens *et al.* (1976) used recapture as a measure of quality in their comparison of two strains of sterilized screw-worm flies. The older strain had a relative recapture rate of 0.49 and had a lower mean dispersal distance. As a result, the older strain was phased out of production. The twofold difference in the performance of these two strains illustrates how the field environment can reveal gross inadequacies in a strain that performs well under captive conditions. A similar effect was observed with maize earworm (Young *et al.*, 1975). The field mating performance of sterile males from a laboratory colony was significantly improved by incorporating genetic material from a local wild population.

As noted earlier, Bush and Neck (1976) found evidence of directional selection favouring allele 2 at the α-GDH locus in captive populations of screw-worm flies. Whitten (1980) provided evidence that this genetic change adversely affected field performance. Specifically, in a field release, he compared prerelease and recaptured flies. The frequency of allele 2 decreased from 0.80 to 0.68 (the natural population had a frequency of 0.31). This shift in frequency suggested poor survival in the field of individuals carrying allele 2, even though they are favoured under captive rearing. Bush and Neck (1976) proposed that the constant temperature of the rearing facility was promoting the spread of allele 2 and Whitten (1980) provided support for this view by finding a significant association between the α-GDH heterozygosity of captured flies and the prevailing temperature.

Improving the Effectiveness of Mass Rearing

Boller (1972) had the insight to suggest that the first step in improving the effectiveness of mass rearing is a psychological one: we should stop thinking in terms of production efficiency (cost per individual), and instead think in terms of the cost to achieve a goal. This goal-directed approach would lead us to maximize effectiveness (the product of quantity and quality), as diagrammed in Fig. 6.1.

While recognizing that the optimum strategy is case-specific, there are some general guidelines that can be used to help maintain the quality of captive-reared populations. These guidelines can be considered under four general headings: colony founding; colony maintenance; colony replacement; and colony improvement.

Colony founding

It is important to avoid the detrimental effects of inbreeding during the first few generations of a new captive population. These detrimental effects include the random increase in the frequency of deleterious alleles and the random loss of potentially beneficial genetic variation. They can be avoided by ensuring that the effective (i.e. genetic) size of the population (N_e) is large. Unfortunately, 'large' is difficult to define. In the context of conservation of threatened species, Franklin (1980) suggested $N_e = 500$ as an acceptable minimum, whereas Lande (1995) pointed out that a target of $N_e = 5000$ is more appropriate. Furthermore, in any given generation, relationship between effective population size (N_e) and the number of adults (N) is a complex function; however, in the absence of exceptional circumstances, it will tend to be in the range $0.25 < N_e/N < 0.75$ (Nunney, 2000). From these figures it is clear that a reasonable goal is to found and maintain the population with $N > 1000$ (Pimentel, 1990). Practical constraints may preclude such a large founding population, but there is no question that several samples of $N > 100$ unrelated adults are necessary to reflect the genetic variation of the source population (Mackauer, 1976; Bartlett, 1994).

The size of a founding population is not the only parameter relevant to initiating a captive-rearing programme. It must also be decided which natural populations will be sampled. To maximize field adaptation, a source population should (if possible) be from a region that is climatically similar to the release sites (McDonald, 1976). But how many source populations should be used? Using more than one source population has a large potential advantage of increasing genetic variability. However, mixing individuals from different geographical locations can lead to the breakdown of geographically distinct coadapted gene complexes. Such breakdown results in a general loss of fitness (e.g. Burton *et al.*, 1999) and can cause an unpredictable change in some traits (see Carson and Templeton, 1984). Male mating traits have been shown to exhibit geographical genetic coadaptation (e.g. Aspi, 2000) and there could be a major problem if, in an SIT programme, there is a change in mating behaviour. We have no way of predicting when coadaptation is likely to be a problem. The best indicator of a potential problem is a large genetic distance between the populations. Since genetic distance is not necessarily well correlated with geographical distance (see Burton *et al.*, 1999), it is prudent to rear samples from different populations independently until genetic testing or other evaluation can be carried out.

Once the founding population has been introduced into the rearing facility, it is very important to avoid the 'crash' of the 'crash–recovery' cycle often seen in the initial stages of captive rearing (Leppla *et al.*, 1983). The crash occurs because the founding population is generally maladapted to the rearing environment. As a result, most genotypes fail to reproduce, but a few are successful. For example, Leppla *et al.* (1983) found that, during the establishment of a new medfly colony, fewer than half of the females were reproducing over the first ten generations.

The exclusive success of a few genotypes dramatically reduces N_e and creates a real danger of extremely rapid inbreeding. Temporarily dividing the founding sample into a large number of very small breeding units can minimize the problem. It may be necessary to expend significant effort to ensure the reproductive success of as many of these units as possible. The survival of many independent subpopulations ensures the reproductive success of a large number of genotypes and avoids large-scale genetic losses.

Colony maintenance

Bartlett (1994) lists some of the important variables that generally differ between the environment of natural and captive-reared populations. The most likely cause of declining quality is the inability of individuals adapted to the captive-rearing conditions to function under field conditions. Notably, in the natural environment, the physical parameters (e.g. temperature) are variable and the availability of resources (e.g. hosts and/or mates) is generally low. There are two qualitatively different ways of dealing with this problem.

First, stocks can be maintained as a large number of inbred (isofemale) lines (Roush and Hopper, 1995). This is the best solution for preventing adaptation to rearing conditions, since inbred lines cannot adapt, and is the optimal strategy when quality is rapidly lost upon domestication (see Fig. 6.1, lower graph). However, captive rearing based on inbred lines has many practical difficulties. Despite these difficulties, it may be particularly advantageous when animals are needed only at certain times a year or for animals that are particularly difficult to replace. This solution necessitates the maintenance of many inbred lines as the genetic reservoir. Roush and Hopper (1995) advocate 25–50 lines, although it is important to note that a substantially larger number ($>>$ 100) would be required both to have confidence that even moderately rare alleles ($P \approx 0.2$) would be retained and to provide a buffer against the loss of some difficult-to-maintain lines.

This method is obviously inappropriate for species that are difficult or impossible to maintain as inbred lines. Diploid animals generally carry a significant load of deleterious recessive alleles and exhibit marked inbreeding depression. This can make the captive rearing of inbred lines difficult. In haplodiploid animals, deleterious recessives are much less of a problem; however, in some groups of Hymenoptera problems are caused by the production of diploid males under inbreeding (see Luck *et al.*, 1999). On the other hand, the micro-Hymenoptera generally show negligible inbreeding depression, at least under laboratory conditions (e.g. Sorati *et al.*, 1996).

Inbred individuals are generally unsuitable for release because of their low fitness. Even when this is not the case, the number of distinct genotypes is limited to the number of lines. Thus, prior to release, the inbred lines should be crossed in some systematic way to create a population of F_1 hybrids. These F_1 hybrids (or better still their F_2 or F_3 offspring, creating recombinant genotypes) can then be released.

Maintaining inbred lines is often impractical, but other methodologies can be used to minimize detrimental adaptation to captive rearing. One possibility is to design the rearing facility to select for the maintenance of specific traits. Boller (1972) suggested adding 'luxury' stimuli, e.g. mating sites, and using 'suboptimal' conditions, e.g. temperature variation. Another approach is to reduce selection for early reproductive maturity by, for example, using older females to lay eggs (see Saul and McCombs, 1995). In addition, it is possible to intermittently select the population under more natural conditions. Mackauer (1976) suggested that readaptation to field conditions could be promoted by maintaining the colony for one or more generations in field cages. Another possibility, appropriate for natural enemies, is to recapture some of the released individuals and reintroduce them into the colony. This strategy is likely to be very beneficial, since recaptured individuals are a selected group of genotypes able to survive under field conditions.

Colony augmentation or replacement

A genetically variable captive population will inevitably adapt to its new environment, following a trajectory of the type shown in Fig. 6.1. The relatively stable 'domesticated stock' is the well-adapted end-point (although long-term inbreeding will eventually lead to a deterioration in the population). Unfortunately, this domesticated stock is extremely unlikely to be at the optimum that maximizes the effectiveness of a release programme. In general, the optimum will be either the non-adapted wild population (lower graph, Fig. 6.1) or, probably more usually, some intermediate between the wild and domesticated forms (upper graph, Fig. 6.1). As a result, there is a strong argument for either regularly augmenting the captive population with genotypes from the wild or replacing old populations with new ones after a defined number of generations.

This practice of augmentation and/or replacement is always necessary unless the captive population is maintained as a large number of inbred subpopulations. Roush and Hopper (1995) suggest that a mixed strategy may often be appropriate, with the inbred lines providing a backup for a large colony. Indeed, inbred lines can be used instead of wild-caught individuals to regularly augment a large colony, as outlined below.

Soemori and Nakamori (1981) proposed that melon-fly stocks should be replaced about every nine generations. Such frequent stock changes may be impractical in many cases; however, as noted earlier, this time period corresponds with the time it takes many captive populations to come close to their stable level of domestication. A satisfactory compromise may be possible with less frequent stock changes supplemented with augmentation from natural populations.

Catching and adding individuals from the wild may present little problem; however, this does not ensure that these new individuals contribute to the strain. Maladapted to the rearing environment, they may reproduce poorly or not at all. It may be necessary to hybridize the wild genotypes with the laboratory strain (Calkins, 1989), perhaps under semi-natural conditions, before the new genes can be successfully introduced. Haeger and O'Meara (1970) showed that wild-caught female *C. nigripalpus* (a mosquito) rarely bred in captivity; however, they could introduce wild genetic material into the captive population by crossing wild males with colony females.

Introductions should have a defined goal in terms of some measurable character. There should be a measurable change in the monitored character following a successful supplementation with wild-caught genotypes. For example, Young *et al.* (1975) demonstrated increased mating competitiveness in the field of a maize earworm population augmented by crossing to local wild-caught moths. In this example, the character (field mating success) was a direct measure of quality; however, the success of genetic introgression is more conveniently monitored in the rearing facility using a trait that shifts predictably in response to selection for domestication.

Saul and McCombs (1995) argued against the introduction of new genetic material into established colonies, using the generally correct, but misguided, argument that such introductions will reduce the fitness of individuals in a mass-rearing facility. In fact, this is the purpose of such introductions: the goal is to intentionally reduce fitness (i.e. quantity) in order to gain quality and shift the population closer to the point of maximum effectiveness (see Fig. 6.1).

Colony improvement

The evolutionary trajectory from a natural population to a domesticated one and then potentially to an inbred one (Fig. 6.1) can be modified, as noted earlier, by changing the captive rearing conditions. It can also be modified through selective breeding or genetic engineering. This is a potentially useful strategy whenever features of the release programme suggest potential improvements (Beckendorf and Hoy, 1985). For example, temperature extremes were implicated in the failure of the red-scale parasitoid *Aphytis lingnanensis* to become established in the inland areas of southern and central California. White *et al.* (1970) successfully selected *A. lingnanensis* for increased tolerance to temperature extremes. However, the effectiveness of this strategy was never tested, because in the meantime a congener, *Aphytis melinus*, became established in the area. Selection of a complex trait, such as temperature tolerance, with the goal of adapting a population to novel features of the release site is a strategy that has considerable merit. Furthermore, it is unlikely that complex traits will be amenable to genetic engineering in the foreseeable future.

Heilmann *et al.* (1994) list a number of genetically simple traits that they consider potential candidates for genetic engineering. This list includes such factors as sex ratio, diapause control and pesticide resistance. Both the elimination of diapause and pesticide resistance have been the subject of traditional selection experiments (e.g. Herzog and Phillips, 1974; Rosenheim and Hoy, 1988; Hoy *et al.*, 1989), but genetic engineering may improve efficiency and success.

The techniques of genetic engineering have been successfully applied to SIT eradication of medfly. An embryonic temperature-sensitive lethal allele is used to destroy female eggs (Franz and McInnis, 1995), so that only sterilized males are released. Cayol and Zarai (1999) showed that these flies were effective in the field. However, this effectiveness was probably far from its potential maximum. Even without irradiation and shipment, fewer than 50% of the pupae produced males that were able to fly!

The low vigour of the genetically engineered sexing strain of medfly highlights an important issue. Traditional captive-reared colonies are derived from natural populations, but it has proved difficult to maintain colonies that have acceptable field performance (quality). The problem of maintaining the quality of genetically engineered strains is much greater because the beneficial genetic changes are initially incorporated into inbred, laboratory-adapted stocks. Outcrossing the stocks into a higher-quality genetic background can be quite complicated. For example, in the medfly sexing strain, one autosome carries a Y translocation and its homologue carries two recessive mutations. The integrity of these arrangements must be maintained in any programme of outcrossing.

Significant effort should be made to ensure that genetically engineered strains are put in the highest-quality genetic backgrounds. Probably the best solution is to build up a genetic reservoir of independently derived isofemale lines from the original engineered strain. These isofemale lines can be maintained indefinitely and periodically combined to reinitiate the rearing colony.

Classical Biological Control

The methodologies discussed above for optimizing the trade-off between quality and quantity apply primarily when the continuous augmentative release of a natural enemy requires long-term captive rearing. However, this trade-off is also important when the goal is to establish a natural enemy for self-sustaining biological control. Domestication and loss of genetic variability during the initial rearing phase could contribute to the failure of an introduction. Potential causes of failure include inbreeding depression, lack of adaptation of released individuals to their new natural environment and lack of ability of a newly established population to adapt to changes in their environment.

Inbreeding is the random loss of genetic variation and results in a lack of individual genetic heterozygosity. It can also result in inbreeding depression. For example, the extinction risk facing natural populations of a northern European butterfly increases with their level of inbreeding (Saccheri *et al.*, 1998). However, biological control agents rarely exhibit significant inbreeding depression in the laboratory (Roush, 1990), but this generality must be treated with some caution since it is now recognized that inbreeding depression can increase dramatically when individuals are subjected to stressful conditions (Bijlsma *et al.*, 2000).

A lack of adaptation of released individuals to their new natural environment is likely to have a profound influence on the probability of successful colonization (McDonald, 1976; Tauber and Tauber, 1986; Roderick, 1992). For this reason, even when laboratory adaptation can be avoided, it is still important to plan carefully how the founding population is to be established (e.g. geographical origins; see González and Gilstrap, 1992). Ideally, the released population should both be preadapted and have the potential for further adaptation to the environment of the release site. As noted earlier, colony improvement for such traits as insecticide resistance can be important and may be amenable to genetic engineering, but the potential value of *in situ* evolution of more complex adaptations after release should not be underestimated. Releasing genetically depauperate stocks initiated from a handful of wild-caught ancestors does not guarantee failure, but it can be expected to minimize the chance of success.

The importance of postrelease adaptation is difficult to evaluate experimentally and such evaluation has rarely been attempted (Roderick, 1992). However, interest in the role of genetic variability in the adaptation and persistence of populations in the face of environmental fluctuation has been stimulated by the growth of conservation genetics since the 1980s (Soulé, 1987; Nunney, 2000). Many of the issues faced in classical biological control have parallels in conservation biology. Perhaps the most obvious is the necessity to re-establish populations of threatened species using individuals taken from other areas (see Hedrick, 2001).

Conclusions

The importance of maintaining quality has long been recognized as a problem in the rearing of natural enemies, but there has been little effort to consider the problem quantitatively. Here I argue that the paradox of captive rearing – the trade-off between rearing quantity and field quality – should be evaluated in terms of the parameter 'effectiveness'. Effectiveness is defined as the product of quantity and quality, and recognizing the importance of this parameter allows us to define how much of a decline in quality is acceptable (and necessary) to achieve the optimal strategy.

The shape of the trade-off curve between quantity and quality created by captive rearing defines the optimal strategy (Fig. 6.1). Unfortunately, we lack the data necessary to draw this curve for any specific case. Admittedly, getting these data is no easy task; however, it is very important to encourage measurement of these curves. The alternative is either to ignore the problem of adaptation to captive rearing or to attempt to manage such adaptation based on guesswork. The dollar costs of failing to maximize effectiveness can be extremely large.

Sometimes the best solution, when it can be employed, is to maintain natural enemies as isofemale lines and then hybridize these lines two or three generations prior to release. However, this approach does not justify maintaining only a handful of isofemale lines. In order to release a population that has levels of genetic variability comparable to a natural population, the minimum would be 100 or more lines; however, maintaining around 50 lines will perpetuate most of the common genetic alleles.

Most of the data documenting the genetic changes associated with captive rearing and documenting the effect of these changes on field performance are derived from tephritid fruit flies used in SIT. However, there is no reason to believe that large-scale captive rearing of predators or parasitoids is fundamentally different. Indeed, many species of natural enemy are difficult to rear, creating a high potential for laboratory adaptation and for a large trade-off with field performance. While none of these tephritid studies can be considered complete, the high-quality work carried out on the melon fly illustrates the kinds of data we should be gathering on commonly used biological control agents.

References

Ahrens, E.H., Hofmann, H.C., Goodenough, J.L. and Peterson, H.D. (1976) A field comparison of two strains of sterilized screwworm flies. *Journal of Medical Entomology* 12, 691–694.

Aspi, J. (2000) Inbreeding and outbreeding depression in male courtship song characters in *Drosophila montana*. *Heredity* 84, 273–282.

Bartlett, A.C. (1994) Maintaining genetic diversity and laboratory colonies of parasites and predators. In: Narang, S.K., Bartlett, A.C. and Faust, R.M. (eds) *Applications of Genetics to Arthropods of Biological Control Significance*. CRC Press, Boca Raton, Florida, pp. 133–145.

Beckendorf, S.K. and Hoy, M.A. (1985) Genetic improvement of arthropods natural enemies through selection, hybridization or genetic engineering techniques. In: Hoy, M.A. and Herzog, D.C. (eds) *Biological Control in Agricultural IPM Systems*. Academic Press, Orlando, Florida, pp. 167–187.

Bijlsma, R., Bundgaard, J. and Boerema, A.C. (2000) Does inbreeding affect the extinction risk of small populations? Predictions from *Drosophila*. *Journal of Evolutionary Biology* 13, 502–514.

Boller, E. (1972) Behavioral aspects of mass-rearing of insects. *Entomophaga* 17, 9–25.

Boller, E.F., Katsoyannos, B.I., Remund, U. and Chambers, D.L. (1981) Measuring, monitoring and improving the quality of mass-reared Mediterranean fruit flies, *Ceratatis capitata* Wied. 1. The RAPID quality control system for early warning. *Zeitschrift für angewandte Entomologie* 92, 67–83.

Burton, R.S., Rawson, P.D. and Edmands, S. (1999) Genetic architecture of physiological phenotypes: empirical evidence for coadapted gene complexes. *American Zoologist* 39, 451–462.

Bush, G.L. and Neck, R.W. (1976) Ecological genetics of the screwworm fly, *Cochliomyia hominivorax* (Diptera: Calliphoridae) and its bearing on the quality control of mass-reared insects. *Environmental Entomology* 5, 821–826.

Calkins, C.O. (1989) Quality control. In: Robinson, A.S. and Hooper, G. (eds) *Fruit Flies: Their Biology, Natural Enemies and Control*, Vol. 3B. Elsevier, Oxford, pp. 153–165.

Calkins, C.O. and Ashley, T.R. (1989) The impact of poor quality of mass-reared Mediterranean fruit flies on the sterile insects technique used for eradication. *Journal of Applied Entomology* 108, 401–408.

Carson, H L. and Templeton, A.R. (1984) Genetic revolutions in relation to speciation phenomena: the founding of new populations. *Annual Review of Ecology and Systematics* 15, 97–132.

Cayol, J.P. and Zarai, M. (1999) Field releases of two genetic sexing strains of the Mediterranean fruit fly (*Ceratitis capitata* Wied.) in two isolated oases of Tozeur governorate, Tunisia. *Journal of Applied Entomology* 123, 613–619.

Cohen, A.C. (2000) Feeding fitness and quality of domesticated and feral predators: effects of long-term rearing on artificial diet. *Biological Control* 17, 50–54.

Fletcher, L.W., Claborne, H.V., Turner, J.P. and Lopez, E. (1968) Difference in response of two strains of screw-worm flies to the male pheromone. *Journal of Economic Entomology* 61, 1386–1388.

Foote, D. and Carey, J.R. (1987) Comparative demography of a laboratory and a wild strain of the Oriental fruit fly, *Dacus dorsalis*. *Entomologia Experimentalis et Applicata* 44, 263–268.

Franklin, I.R. (1980) Evolutionary changes in small populations. In: Soulé, M.E. and Wilcox, B.A. (eds) *Conservation Biology: an Evolutionary-ecological Perspective*. Sinauer Associates, Sunderland, Massachusetts, pp. 135–149.

Franz, G. and McInnis, D.O. (1995) A promising new twist – a genetic sexing strain based on a temperature sensitive lethal (tsl) mutation. In: Morse, J.G., Metcalf, R.L., Carey, J.R. and Dowell, R.V. (eds) *The Medfly in California: Defining Critical Research*. Center for Exotic Pest Research, University of California, Riverside, California, pp. 187–198.

Fye, R.L. and LaBrecque, G.C. (1966) Sexual acceptability of laboratory strains of male houseflies in competition with wild strains. *Journal of Economic Entomology* 59, 538–540.

González, D. and Gilstrap, F.E. (1992) Foreign exploration: assessing and prioritizing natural enemies and consequences of reintroduction studies. In: Kauffman, W.C. and Necols, J.E. (eds) *Selection Criteria and Ecological Consequences of Importing Natural Enemies*. Entomological Society of America, Lanham, Maryland, pp. 53–70.

Haeger, J.S. and O'Meara, G.F. (1970) Rapid incorporation of wild genotypes of *Culex nigripalpus* (Diptera: Culicidae) into laboratory-adapted strains. *Annals of the Entomological Society of America* 63, 1390–1391.

Hedrick, P.W. (2001) Conservation genetics: where are we now? *Trends in Ecology and Evolution* 16, 629–636.

Heilmann, L.J., DeVault, J.D., Leopold, R.L. and Narang, S.K. (1994) Improvement of natural enemies for biological control: a genetic engineering approach. In: Narang, S.K., Bartlett, A.C. and Faust, R.M. (eds) *Applications of Genetics to Arthropods of Biological Control Significance*. CRC Press, Boca Raton, Florida, pp. 167–189.

Herzog, G.A. and Phillips, J.R. (1974) Selection for a nondiapause strain of the bollworm. *Environmental Entomology* 3, 525–527.

Hoy, M.A., Cave, F.E., Beede, R.H., Grant, J., Kreuger, W.H., Olson, W.H., Spollen, K.M., Barnett, W.W. and Hendricks, L.C. (1989) Release, dispersal, and recovery of a laboratory-selected strain of the walnut aphid parasite *Trioxys pallidus* (Hymenoptera: Aphidiidae) resistant to azinphosmethyl. *Journal of Economic Entomology* 83, 89–96.

Ito, Y. (1988) Role of basic ecological/ethological studies for measurement and control of the mating competitiveness of mass-reared melon fly males: a review. In: Vijaysegaran, S. and Ibrahim, A.G. (eds) *1st International Symposium on Fruit Flies in the Tropics*. Malaysian Plant Protection Society, Kuala Lumpur, pp. 251–266.

Lande, R. (1995) Mutation and conservation. *Conservation Biology* 9, 782–791.

Leppla, N.C., Huettel, M.D., Chambers, D.L. and Turner, W.K. (1976) Comparative life history and respiratory activity of 'wild' and colonized Caribbean fruit flies [Dipt.: Tephritidae]. *Entomophaga* 21, 353–357.

Leppla, N.C., Huettel, M.D., Chambers, D.L., Ashley, T.R., Miyashita, D.H., Wong, T.T.Y. and Harris, E.J. (1983) Strategies for colonization and maintenance of the Mediterranean fruit fly. *Entomologia Experimentalis et Applicata* 33, 89–96.

Loukas, M., Economopoulos, E., Zouros, E. and Vergini, Y. (1985) Genetic changes in artificially reared colonies of the olive fruit fly (Diptera: Tephritidae). *Annals of the Entomological Society of America* 78, 159–165.

Luck, R.F., Nunney, L. and Stouthamer, R. (1999) Factors affecting sex ratio and quality in the culturing of parasitic hymenoptera: a genetic and evolutionary perspective. In: Fisher, T.W., Bellows, T.S., Caltagirone, L.E., Dahlsten, D.L., Huffaker, C.B. and Gordh, G. (eds) *A Handbook of Biological Control*. Academic Press, New York, pp. 653–672.

McDonald, I.C. (1976) Ecological genetics and the sampling of insect populations for laboratory colonization. *Environmental Entomology* 5, 815–820.

Mackauer, M. (1976) Genetic problems in the production of biological control agents. *Annual Review of Entomology* 21, 369–385.

Miyatake, T. (1993) Difference in the larval and pupal periods between mass-reared and wild strain of the melon fly, *Bactrocera cucurbitae* (Coquillet) (Diptera: Tephritidae). *Applied Entomology and Zoology* 28, 577–581.

Miyatake, T. (1998) Genetic changes of life history and behavioral traits during mass-rearing in the melon fly, *Bactrocera cucurbitae* (Diptera: Tephritidae). *Researches on Population Ecology* 40, 301–310.

Miyatake, T. and Yamagishi, M. (1999) Rapid evolution of larval development time during mass-rearing in the melon fly, *Bactrocera cucurbitae*. *Researches on Population Ecology* 41, 291–297.

Nunney, L. (1996) The response to selection for fast larval development in *Drosophila melanogaster* and its effect on adult weight: an example of a fitness trade-off. *Evolution* 50, 1193–1204.

Nunney, L. (2000) The limits to knowledge in conservation genetics: the predictive value of effective population size. *Evolutionary Biology* 32, 179–194.

Nunney, L. (2002) The population genetics of mass-rearing. In: Leppla, N.C., Bloem, S. and Luck, R.F. (eds) *Quality Control for Mass-Reared Arthropods*. Kluwer, Dordrecht, The Netherlands.

Pimentel, D. (1990) Population dynamics and the importance of evolution in successful biological control. In: Pimentel, D. (ed.) *Handbook of Pest Management*, Vol. 2. CRC Press, Boca Raton, Florida, pp. 171–175.

Raulston, J.R. (1975) Tobacco budworm: observations on the laboratory adaptation of a wild strain. *Annals of the Entomological Society of America* 68, 139–142.

Roderick, G.K. (1992) Postcolonization evolution of natural enemies. In: Kauffman, W.C. and Necols, J.E. (eds) *Selection Criteria and Ecological Consequences of Importing Natural Enemies*. Entomological Society of America, Lanham, Maryland, pp. 71–86.

Rosenheim, J.A. and Hoy, M.A. (1988) Genetic improvement of a parasitoid biological control agent: artificial selection for insecticide resistance in *Aphytis melinus* (Hymenoptera: Aphelinidae). *Journal of Economic Entomology* 81, 1539–1550.

Rössler, Y. (1975) Reproductive differences between laboratory-reared and field-collected populations of the Mediterranean fruit fly, *Ceratitis capitata*. *Annals of the Entomological Society of America* 68, 987–991.

Roush, R.T. (1990) Genetic variation in natural enemies: critical issues for colonization in biological control. In: Mackauer, M., Ehler, L.E. and Hants, J. (eds) *Critical Issues in Biological Control*. Intercept, Andover, UK, pp. 263–288.

Roush, R.T. and Hopper, K.R. (1995) Use of single family lines to preserve genetic variation in laboratory colonies. *Annals of the Entomological Society of America* 88, 713–717.

Saccheri, I., Kuussaari, M., Kankare, M., Vikman, P., Fortelius, W. and Hanski, I. (1998) Inbreeding and extinction in a butterfly metapopulation. *Nature* 392, 491–494.

Saul, S.H. and McCombs, S.D. (1995) Genetics and ecology of colonization and mass rearing of Hawaiian fruit flies (Diptera: Tephritidae) for use in sterile insect control programs. *Proceedings of the Hawaiian Entomological Society* 32, 21–37.

Service, P.M. and Rose, M.R. (1985) Genetic covariation among life-history components: the effect of novel environments. *Evolution* 39, 943–944.

Shimizu, T., Miyatake, T., Watari, Y. and Arai, T. (1997) A gene pleiotropically controlling developmental and circadian periods in the melon fly, *Bactrocera cucurbitae* (Diptera: Tephritidae). *Heredity* 79, 600–605.

Soemori, H. and Nakamori, H. (1981) Production of successive generations of a new strain of the melon fly, *Dacus cucurbitae* Coquillett (Diptera: Tephritidae) and reproductive characteristics in mass rearing. *Japanese Journal of Applied Entomology and Zoology* 25, 229–235.

Sorati, M., Newman, M. and Hoffmann, A.A. (1996) Inbreeding and incompatibility in *Trichogramma* nr. *brassicae*: evidence and implications for quality control. *Entomologia Experimentalis et Applicata* 78, 283–290.

Soulé, M.E. (1987) Where do we go from here? In: Soulé, M.E. (ed.) *Viable Populations for Conservation*. Cambridge University Press, Cambridge, UK, pp. 175–183.

Tauber, C.A. and Tauber, M.J. (1986) Ecophysiological responses in life-history evolution: evidence for their importance in a geographically widespread insect species-complex. *Canadian Journal of Zoology* 64, 875–884.

Vargas, R.I. and Carey, J.R. (1989) Comparison of demographic parameters of wild and laboratory-adapted Mediterranean fly (Diptera: Tephritidae). *Annals of the Entomological Society of America* 82, 55–59.

Via, S., Gomulkiewicz, R., De Jong, G., Scheiner, S.M., Schlichting, C.D. and Van Tienderen, P.H. (1995) Adaptive phenotypic plasticity: consensus and controversy. *Trends in Ecology and Evolution* 10, 212–217.

White, E.B., DeBach, P. and Garber, M.J. (1970) Artificial selection for genetic adaptation to temperature extremes in *Aphytis lingnanensis* (Hymenoptera: Aphelinidae). *Hilgardia* 40, 161–192.

Whitten, C.J. (1980) Use of the isozyme technique to assess the quality of mass-reared sterilized screw-worm flies. *Annals of the Entomological Society of America* 73, 7–10.

Wong, T.T.Y. and Nakahara, L.M. (1978) Sexual development and mating response of laboratory-reared and native Mediterranean fruit flies. *Annals of the Entomological Society of America* 71, 592–596.

Young, J.R., Snow, J.W., Hamm, J.J., Perkins, W.D. and Haile, D.G. (1975) Increasing the competitiveness of laboratory-reared corn earworm by incorporation of indigenous moths from the area of sterile release. *Annals of the Entomological Society of America* 68, 40–42.

7 Adaptive Recovery after Fitness Reduction: the Role of Population Size

R.F. Hoekstra
Laboratory of Genetics, Wageningen University, Arboretumlaan 4, 6703 BD Wageningen, The Netherlands

Abstract

The transfer of natural enemies from the field to a mass-production facility may result in a sudden reduction in fitness. A sharp decrease in population size during the season in the mass-production facility can lead to further reduction in fitness and will, in addition, enhance the possibility of fixation of deleterious mutations in the population by genetic drift. Such reductions in fitness can be prevented by keeping very large populations in the mass-production unit and by regularly replacing laboratory populations with new, field-collected individuals.

Introduction: Deleterious Mutations

Organisms may experience sudden reductions of fitness. Perhaps the most common reason is a change in environmental conditions. Populations tend to consist of genotypes that are well adapted to the prevailing conditions. When these conditions change, previously well-adapted genotypes may no longer be advantageous or may even become deleterious. For example, a mutation conferring resistance to an antibiotic may have risen to high frequency in the presence of the antibiotic, but is likely to become disadvantageous in an environment without this antibiotic for reasons explained in the next paragraph. Another possible cause of a sudden reduction in fitness is the fixation of a deleterious mutation due to genetic drift in a (temporarily) very small population.

Genomes are subject to the inevitable occurrence of mutations. In the great majority of organisms having DNA genomes, mutations occur roughly at a rate of 10^{-5} per locus, in some viruses with RNA genomes, e.g. those causing influenza or AIDS, the mutation rate may be several orders of magnitude higher. Deleterious mutations are expected to disappear again from the population due to the action of natural selection. Occasionally, however, a deleterious mutation may reach a high frequency in the population as a consequence of genetic drift. The likelihood of such an event is very small in large populations, but may be relatively high in small populations – even when a normally large population experiences a 'bottleneck': a strong reduction in size for only one or two generations.

This chapter considers the micro-evolutionary adaptive recovery following a sudden reduction in fitness.

© CAB International 2003. *Quality Control and Production of Biological Control Agents: Theory and Testing Procedures* (ed. J.C. van Lenteren)

Fitness Reduction as Side Effect of Selection

Genes often simultaneously affect several functional aspects of an organism, a phenomenon called pleiotropy. This comes as no surprise to physiologists who are aware of the intricate network of metabolic machinery by which several functions may interact, or to medical geneticists who often observe that a disease phenotype associated with a single hereditary factor involves the malfunctioning of different organs. Pleiotropy is also frequently apparent in so-called major mutations: mutations with a big effect on one particular trait of the organism almost always display 'side-effects' on other traits.

Well-studied examples are mutations causing resistance to antibiotics, pesticides or other toxic compounds. For example, we recently isolated spontaneous *Aspergillus nidulans* mutants resistant to the fungicide fludioxonil. When tested on the standard medium without the fungicide, these resistant strains showed a 10 to 30% decrease in growth rate compared to the sensitive wild-type strain (Schoustra, unpublished results). Similarly, Björkman *et al.* (1998) isolated spontaneous *Salmonella typhimurium* mutants that were resistant to the antibiotics rifampicin and naladixic acid. In subsequent competition experiments against the wild-type strain after injection of the bacteria into mice that were not treated with the antibiotics, the mutants appeared to be at a general disadvantage. Therefore, the resistance mutations caused a reduction in virulence. Nevertheless, these mutations will tend to become common in an environment containing the toxin because of the protection they provide, despite their negative side-effects.

Do Deleterious Genes Disappear as a Result of Selection?

Of particular interest in the context of resistance mutations is the fate of resistant strains or genotypes after termination of the application of the relevant antibiotic or pesticide. On the basis of naïve reasoning one might expect that the now deleterious allele would quickly disappear, being selected against because of its negative side-effects. However, there is increasing evidence that this scenario is not very likely. A more probable course of events involves the selection of mutations that take away the negative side-effects of the resistance, thus enabling these strains to enhance their fitness while retaining their resistance. In a more general sense, the experiments discussed below suggest that evolution will rarely be reversible. Therefore the view that when a population adapted to one environment returns to a previous environment, evolution will (re)produce the original genotypic state, is unlikely to be correct.

A convincing illustration is provided by the study of Björkman *et al.* (1998) on antibiotic-resistant mutants of *Salmonella typhimurium*. They selected seven mutant strains, resistant to streptomycin, rifampicin, or nalidixic acid. From these, six had lost their virulence in mice, due to apparent side-effects of the resistance mutations. The mutant strains were then allowed to grow in mice in the absence of antibiotics and samples were regularly examined for restoration of virulence. After several growth cycles all mutant strains showed restored virulence. In one case, full restoration of virulence appeared to result from a true reversion, i.e. a precise back-mutation to the sensitive state. But all others had retained their resistance. In these cases, the restored virulence had resulted from so-called compensatory mutations – mutations at sites other than that of the resistance mutation and apparently taking away some of the side-effects of the resistance mutation that had caused the reduced virulence. The authors inferred from their results that reduction in the use of antibiotics might not result in the disappearance of the resistant bacteria already present. That this phenomenon is not restricted to prokaryotes is shown by recent work in our laboratory. We have observed compensatory evolution in fungicide-resistant *A. nidulans* strains, restoring growth rate to the original level of the parental sensitive strain while retaining the resistance (S.E. Schoustra, unpublished).

At least one study has demonstrated a similar phenomenon in insects. McKenzie *et al.* (1993) studied the establishment of resistance to diazinon in the sheep blowfly *Lucilla cuprina*. Initially, resistant genotypes were at a selective disadvantage relative to the susceptible genotypes in a diazinon-free environment. But, after the resistance became widespread, resistant genotypes were no longer at a selective disadvantage in the absence of the insecticide when they also contained a particular mutation located at a different chromosome.

The Role of Population Size

Population size plays an important role in the adaptive recovery process, in at least two ways. First, a sharp decrease (bottleneck) in population size will enhance the possibility of fixation of a deleterious mutation in the population by genetic drift. This would provide the starting-point of subsequent compensatory evolution. Secondly, the population size during the recovery process affects the probability of various types of fitness-restoring mutations. Basically, the available evidence suggests that instantaneous and complete recovery can only be achieved by very rare mutational events, such as the precise back-mutation that reverts the genotype to its original state. On the other hand, many different mutations are often possible that restore fitness to a lesser degree. In a very large population, the single unique back-mutation might perhaps be expected to occur within a reasonably short time period and its large selective advantage might cause its rapid spread. But, in smaller populations, this mutation is unlikely to occur and selection will promote the spread of the more common compensatory mutations of small effect.

Several experiments have shown this latter scenario to be realistic. Burch and Chao (1999) subjected the bacteriophage $\phi 6$ to intensified genetic drift and caused viral fitness to decline following the fixation of a deleterious mutation. They then propagated the mutated virus at a range of population sizes and allowed fitness to recover by natural selection. Typically, it was recovered in small steps. Step size during recovery was smaller with decreasing size of the recovery population. This result suggests that mutations improving fitness by a small amount are more common than those with bigger positive effects. Burch and Chao also demonstrated that the advantageous mutations of small effect were compensatory mutations whose advantage is conditional on the presence of the deleterious mutation that caused the fitness decline. Levin *et al.* (2000) performed similar experiments using streptomycin-resistant mutants of *Escherichia coli*. They showed that the fitness recovery is mediated primarily by intermediate-fitness compensatory mutations, rather than by high-fitness revertants, and that this result is dependent on the numerical bottlenecks associated with serial passage in their experiments.

Conclusions

The evidence discussed above suggests that recovery following a fitness reduction is often of a compensatory nature, in particular if populations are experiencing occasional bottlenecks in numbers. This means that fitness is restored not by removing the deleterious gene or genotype, but by the spread of mutations with a beneficial effect only in the presence of the gene that caused the fitness reduction. This gene or genotype will therefore remain (for longer) in the population. An example is provided by mutations causing resistance to antibiotics or pesticides that remain present after the application of the relevant toxins has been terminated. However, the principle may apply to other deleterious mutations as well, such as a virulence-reducing mutation in an agent used in biological control. Were such a mutation to be fixed in a population due to the passage through an extreme numerical bottleneck, fitness might be restored by compensating mutations that do not affect virulence directly and the reduction of virulence might thus become a trait that would be difficult to remove from the population.

What might the meaning of these findings be for mass production of biological control agents? If the aim is to rear natural enemies that are similar to the initial field-collected population, one should either prevent bottlenecks and always keep large populations or one should replace laboratory populations that have experienced a bottleneck with new, field-collected material.

References

Björkman, J., Hughes, D. and Andersson, D.I. (1998) Virulence of antibiotic-resistant *Salmonella typhimurium*. *Proceedings of the National Academy of Sciences, USA* 95, 3949–3953.

Burch, C.L. and Chao, L. (1999) Evolution by small steps and rugged landscapes in the RNA virus ϕ6. *Genetics* 151, 921–927.

Levin, B.R., Perrot, V. and Walker, N. (2000) Compensatory mutations, antibiotic resistance and the population genetics of adaptive evolution in bacteria. *Genetics* 154, 985–997.

McKenzie, J.A. (1993) Measuring fitness and intergenic interactions: the evolution of resistance to diazinon in *Lucilia cuprina*. *Genetica* 90, 227–237.

8 The Use of Unisexual Wasps in Biological Control

R. Stouthamer*
Laboratory of Entomology, Wageningen University, PO Box 8031, 6700 EH Wageningen, The Netherlands

Abstract

Unisexual reproduction has long been seen as a clear advantage for wasps to be applied in biological control projects. The discovery that the mode of reproduction in parasitoid wasps may be manipulated from sexual to unisexual and vice versa will allow biocontrol workers to test the advantage of either mode of reproduction for biological pest control. Here a review is presented of the cases of unisexual reproduction found in parasitoid wasps. Unisexual reproduction is not rare among parasitoids; at least 150 cases of unisexual reproduction have been reported. The literature is reviewed for cases where both unisexual and sexual forms are used in the same control project to determine if the theoretical advantage of unisexual reproduction indeed materializes. Few cases can be used to test the presumed advantage of unisexuals. Some evidence is found for two advantages of unisexual reproduction: unisexuals are cheaper to produce in mass rearing than sexuals, and in classical biocontrol projects they are more easily established.

Introduction

Biological control workers have long been fascinated by the phenomenon of unisexual reproduction. Timberlake and Clausen (1924) explained the possible advantage of a unisexual parasitoid over a form reproducing sexually by calculating the population increase of the sexual form compared with the unisexual form. The difference in population growth rate in such calculations can be astonishing (Fig. 8.1). In such calculations we assume that the unisexual wasps produce equal numbers of offspring to the sexual forms and therefore over time the unisexual population should outcompete the sexual form since all the offspring of the unisexual form consist of females only.

The concept of choosing a unisexual mode of reproduction for the wasps to be used in biological control is interesting; however, up until now, this can only be applied in those cases where two modes of reproduction are present in a species. In the not too distant future, it may be possible to render sexual forms unisexual by infection with parthenogenesis-inducing (PI) microorganisms. Natural infection with these bacteria is the

*Present address: Department of Entomology, University of California at Riverside, Riverside, CA 92521, USA.

© CAB International 2003. *Quality Control and Production of Biological Control Agents: Theory and Testing Procedures* (ed. J.C. van Lenteren)

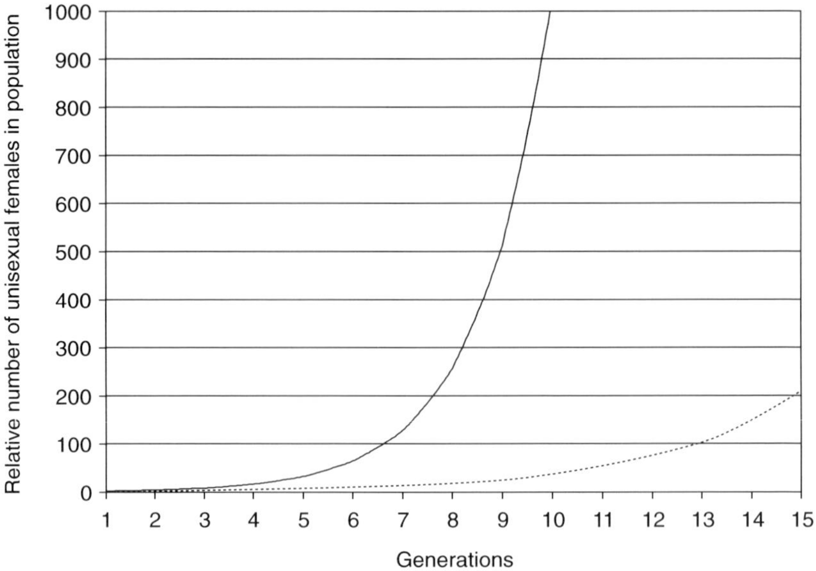

Fig. 8.1. Relative size of unisexual population (= number of unisexual females/number of sexual females in generation *n*) if at generation 1 equal numbers of sexual and unisexual wasps are released. The sexual wasps produce offspring sex ratio (% females) of either 50% ———— or of 70% ··············. No other differences are assumed between the sexual and unisexual form; populations in exponential growth phase.

cause of unisexual reproduction in many Hymenoptera (Stouthamer, 1997), and initial experiments have shown that in some cases inter- and intraspecific transfers of these bacteria are possible (Chapter 9; Grenier *et al.*, 1998; Huigens *et al.*, 2000).

Two papers published in the early 1990s discussed the use of sexual versus unisexual lines in biocontrol. The first paper, by Aeschlimann (1990), suggested initially releasing unisexual forms, because they may be easier to establish. Subsequently sexual forms could be released to introduce genetic variation in the population. The generality of that idea was questioned by Stouthamer (1993), who argued that the sequence in which these two forms should be released depends on: (i) the type of biological control the release is intended for; and (ii) the density of the hosts that are to be controlled.

In the following sections, I shall give an overview of the knowledge that we have gained about unisexual reproduction over the last 10 years and discuss work done specifically to test the merits of using either a sexual form or a unisexual form for biological control.

Causes of Unisexual Reproduction

Two classes of causes are known for unisexual reproduction in Hymenoptera: (i) microbial infection; and (ii) other genetic mechanisms that allow unfertilized eggs to develop into females.

Over the last 15 years many species have been discovered that are infected with PI *Wolbachia* (Stouthamer *et al.*, 1990b, 1993; Stouthamer, 1997). These bacteria allow infected females to produce daughters from both fertilized and unfertilized eggs. In many species where PI-*Wolbachia* infection is known, the infection has gone to fixation and all individuals in the 'fixed' population are infected females (Stouthamer, 1997). An example is the biocontrol icon *Encarsia formosa* (Zchori-Fein *et al.*, 1992; van Meer *et al.*, 1995). In a number of other species the infection with PI *Wolbachia* is restricted to a smaller part of the population and both infected and uninfected individuals co-occur and gene flow still takes place between these two subpopulations ('mixed populations') (Stouthamer and Kazmer, 1994). Only wasps in genus *Trichogramma* populations are known to consist of both forms. The best-studied species is *Trichogramma kaykai* from the Mojave Desert (Stouthamer and Kazmer, 1994; Pinto *et al.*, 1997; Huigens *et al.*, 2000; Stouthamer *et al.*, 2001). The distinction between these two classes – fixed versus mixed – is important in regard to the influence the *Wolbachia* may have on the life-history characters of the infected wasps.

In fixed populations we expect that there will be a selection for accommodation between the bacteria and their wasp hosts. The evolutionary interests of both host and *Wolbachia* are the same as long as the infection is only passed on from the mother to her offspring. The wasp–*Wolbachia* combination that produces the most daughters will be selected for. In the case of infected individuals occurring in a population with sexual individuals where gene flow between the sexual and unisexual individuals still occurs, the evolutionary interests of the *Wolbachia* and the nuclear genes of the wasp are not the same (Stouthamer, 1997; Stouthamer *et al.*, 2001). Under such circumstances the optimal sex ratio for the nuclear genes is a sex ratio involving at least some males, whereas for the PI *Wolbachia* the optimal sex ratio is 100% female. This conflict in evolutionary interest of these two genetic elements can lead to an arms race between these elements, which may have as a by-product a reduced offspring production of the infected females. Other differences also exist between infected females from fixed versus mixed populations. When females from fixed populations are fed antibiotics, the males they produce are often unable to successfully mate with the females of their own line (Zchori-Fein *et al.*, 1992; Stouthamer *et al.*, 1994; Pijls *et al.*, 1996; Arakaki *et al.*, 2000). The most likely cause of this is an accumulation of mutations in genes involved with sexual reproduction in these infected lines. Such mutations are assumed to accumulate because they are not selected against any longer. Therefore, in populations where the infection has been fixed for a prolonged period, sexual reproduction is no longer possible and sexual lines cannot be established from these forms.

Besides the influence of microbial infection on parthenogenesis, there are also many cases of unisexual reproduction where this mode of reproduction is not sensitive to antibiotic treatment and therefore we assume that there is a genetic cause of unisexual reproduction. Examples of such unisexual reproduction are *Venturia canescens* (Speicher *et al.*, 1965; Beukeboom and Pijnacker, 2000) and *Trichogramma cacoeciae* (Stouthamer *et al.*, 1990a).

Unisexuals in Biocontrol

The incidence of unisexual reproduction among wasps used in biological control appears to be very high. In a sample of wasps used for biological control the percentage of unisexuals was at least 15% (Luck *et al.*, 1999). The cause for this high proportion of species carrying unisexual forms may be: (i) the proportion is not extreme but simply reflects the proportion of unisexual forms in nature; or (ii) the proportion is high because the establishment of a species in quarantine, in mass rearing (Stouthamer and Luck, 1991) and in the field is much more successful for unisexual forms than for sexual forms (Hung *et al.*, 1988). If this latter reason is correct, then the high proportion of species that are unisexual in wasps used in biological control is simply a reflection of our inability to start cultures with only a few sexual individuals.

In Table 8.1, I give an overview of the cases of unisexual reproduction in parasitoid wasps. Most of these cases were discovered when the wasps were studied for biological control purposes. In some genera, such as *Aphytis*, *Encarsia* and *Trichogramma*, the number of unisexual species is particularly high. This is most probably due to the relative importance of these groups in biological control and the awareness of the workers in this field of unisexual reproduction (DeBach, 1969). In other cases, it appears that certain host species seem to have a disproportionately high number of unisexual wasps attacking them; an example of this is weevils. The

Table 8.1. Literature review of unisexual species or biotypes among parasitic Hymenoptera.

Taxa	References
Ichneumonoidea	
Braconidae	
Agathis stigmaterus	Hummelen (1974)
Apanteles cerialis	Wysoki and Izhar (1981)
Apanteles circumscriptus	Shaw and Askew (1976)
Apanteles pedias	Laing and Heraty (1981)
Apanteles thompsoni	Vance (1931)
Centistes excrusians	Loan (1963b)
Chelonus blackburni	Platner and Oatman (1972)
Meteorus japonicus	Clausen (1940); Li (1984); Fuester *et al.* (1993)
Microctonus brevicollis	Kunckel d'Herculais and Langlois (1891)
Microctonus hyperodae	Phillips and Baird (1996)
Microctonus sp.	Loan (1963a)
Microctonus vittatae	Nagasawa (1947)
Perilitus coccinellae	Balduf (1926); Wright (1978)
Peristenus conradi	Day *et al.* (1992)
Peristenus howardi	Day *et al.* (1999)
Pygostylus falcatus	Loan and Holdaway (1961); Loan (1963b); Milbrath and Weiss (1998)
Rogas unicolor	Dowden (1938)
Aphidiidae	
Aphidius colemani	Tardieux and Rabasse (1988)
Lysiphlebus ambiguus	Rosen (1967)
Lysiphlebus cardui	Nemec and Stary (1985)
Lysiphlebus confusus	Nemec and Stary (1985)
Lysiphlebus fabarum	Rosen (1967); Nemec and Stary (1985)
Lysiphlebus tritici	Kelly and Urbahns (1908)
Ichneumonidae	
Biolysia tristis	Puttler and Coles (1962)
Diadromus collaris	Kfir (1998)
Gelis tenellus	Muesenbeck and Dohanian (1927)
Mesochorus nigripes	Hung *et al.* (1988)
Polysphincta pallipes	Clausen (1940)
Sphecophaga burra	Schmieder (1939)
Sphecophaga vesparum	Reichert (1911)
Thersilochus parkeri	Kerrich (1961); Clancy (1969)
Trathala flavoorbitalis	Sandanayake and Edirisinghe (1992)
Venturia canescens	Speicher *et al.* (1965)
Chalcidoidea	
Pteromalidae	
Mesopolobus diffinis	Redfern (1976)
Muscidifurax uniraptor	Legner (1985)
Spalangia erythromera	Baker (1979)
Eupelmidae	
Anastatus pearsalli	Muesenbeck and Dohanian (1927)
Eupelmus vesiculari	Muesenbeck and Dohanian (1927); Phillips and Poos (1927)
Aphelinidae	
Aphelinus asychis	Hartley (1922); Force and Messenger (1964)
Aphelinus jucundus	Griswold (1929)

Continued

Table 8.1. *Continued.*

Taxa	References
Chalcidoidea *(Continued)*	
Aphelinidae *(Continued)*	
Aphytis aonidae	Rosen and DeBach (1979)
Aphytis chilensis	Rosen and DeBach (1979); Gottlieb *et al.* (1998)
Aphytis chrysomphali	Bartlett and Fisher (1950); Gottlieb *et al.* (1998)
Aphytis comperei	Rosen and DeBach (1979)
Aphytis diaspidis	Zchori-Fein *et al.* (1995); Gottlieb *et al.* (1998)
Aphytis hispanicus	Gerson (1968)
Aphytis holoxanthus	Rosen and DeBach (1979)
Aphytis lingnanensis	Zchori-Fein *et al.* (1995); Gottlieb *et al.* (1998)
Aphytis melinus	Rosen and DeBach (1979)
Aphytis mytilaspidis	Rosen and DeBach (1979)
Aphytis neuter	Rosen and DeBach (1979)
Aphytis opuntiae	Rosen and DeBach (1979)
Aphytis phoenicus	Rosen and DeBach (1979)
Aphytis proclia	Sumaroka (1967)
Aphytis simmondsiae	DeBach (1984)
Aphytis testaceus	Rosen and DeBach (1979)
Aphytis vandenboschi	Rosen and DeBach (1979); Titayavan and Davis (1988)
Aphytis yanonenesis	DeBach and Rosen (1982)
Azotus perspeciosus	Pedata and Viggiani (1991)
Azotus pulcherimus	Viggiani (1972)
Encarsia citrina	Flanders (1953a)
Encarsia formosa	Speyer (1926)
Encarsia hispida	Avilla *et al.* (1991)
Encarsia inquirenda	Gerson (1968)
Encarsia lounsburyi	Flanders (1953a)
Encarsia meritoria	Pedata and Hunter (1996)
Encarsia pergandiella	Hunter (1999)
Encarsia perniciosi	Flanders (1953b)
Eretmocerus mundus	de Barro *et al.* (2000)
Eretmocerus sp. Hawaii	Powell and Bellows (1992)
Eretmocerus sp. Hong Kong	McAuslane and Nguyen (1996)
Eretmocerus staufferi	Rose and Zolnerowich (1997)
Signiphoridae	
Signiphora borinquensis	Quezada *et al.* (1973)
Signiphora coquilletti	Woolley (1984)
Signiphora flavella	DeBach *et al.* (1958); Woolley (1984)
Signiphora merceti	DeBach *et al.* (1958)
Encyrtidae	
Achrysophagus modestus	Timberlake and Clausen (1924)
Adelencyrtus odonaspidis	Timberlake (1919)
Anagyrus subalbicornis	Timberlake and Clausen (1924)
Apoanagyrus diversicornis	Pijls *et al.* (1996)
Blepyrus mexicanus	Timberlake (1919)
Chrysopophagus flaccus	Timberlake (1919)
Clausenia purpurea	Rivnay (1942)
Compariella unifasciata	Clausen (1940)
Encyrtus fulginosus	Flanders (1943)
Encyrtus infelix	Embleton (1904)
Habrolepis dalmani	Clausen (1940)

Continued

Table 8.1. *Continued.*

Taxa	References
Chalcidoidea *(Continued)*	
Encyrtidae *(Continued)*	
Habrolepis rouxi	Flanders (1945, 1958)
Hambletonia pseudococcina	Carter (1937); Bartlett (1939)
Microterys speciosus	Ishii (1932)
Ooencyrtus fecundus	Laraichi (1978)
Ooencyrtus submetallicum	Wilson and Woolcock (1960a, b)
Pauridia peregrina	Timberlake (1919); Flanders (1959)
Plagiomerus diaspidis	Gordh and Lacey (1976)
Pseudoleptomastix squammulata	Timberlake and Clausen (1924)
Trechnites psyllae	Slobodchikoff and Daly (1971)
Tropidophryne melvillei	Doutt and Smith (1950)
Eulophidae	
Ceranisus americensis	Loomans and van Lenteren (1995)
Ceranisus menes	Clausen (1940); Loomans and van Lenteren (1995)
Ceranisus russelli	Russell (1911)
Ceranisus vinctus	Loomans and van Lenteren (1995)
Galeopsomyia fausta	Argov *et al.* (2000)
Nesolynx sp.	Bueno *et al.* (1987)
Pedobius nawaii	Muesenbeck and Dohanian (1927)
Tetrastichus asparagi	Russell and Johnston (1912)
Tetrastichus brevistigma	Berry (1938)
Tetrastichus cecidophagus	Wangberg (1977)
Tetrastichus nr. *venustus*	Teitelbaum and Black (1957)
Thripobius semiluteus	Hessein and McMurtry (1988)
Mymaridae	
Anagrus atomus	Perkins (1905b)
Anagrus delicates	Cronin and Strong (1996)
Anagrus ensifer	Walker (1979)
Anagrus flaveolus	Chandra (1980)
Anagrus frequens	Perkins (1905b)
Anagrus optabilis	Perkins (1905b)
Anagrus perforator	Perkins (1905b)
Anagrus sp. nov. 1	Claridge *et al.* (1987)
Anagrus takeyanus	Gordh and Dunbar (1977)
Anaphes diana	Aeschlimann (1986, 1990)
Polynema enchenopae	Kiss (1986)
Polynema euchariformis	Clausen (1940)
Trichogrammatidae	
Megaphragma deflectum	Takagi (1988); Loomans and van Lenteren (1995)
Megaphragma mymaripenne	Hessein and McMurtry (1988)
Trichogramma brevicappilum	Pinto (1998)
Trichogramma cacoeciae	Marchal (1936)
Trichogramma chilonis	Stouthamer *et al.* (1990a); Chen *et al.* (1992)
Trichogramma cordubensis	Cabello *et al.* (1985)
Trichogramma deion	Bowen and Stern (1966); Stouthamer *et al.* (1990a)
Trichogramma dianae	Pinto (1998)
Trichogramma embryophagum	Birova (1970)
Trichogramma evanescens	Marchal (1936); Voegele and Russo (1981)
Trichogramma flavum	Marchal (1936)

Continued

Table 8.1. *Continued.*

Taxa	References
Chalcidoidea *(Continued)*	
Trichogrammatidae *(Continued)*	
Trichogramma kaykai	Stouthamer and Kazmer (1994); Pinto *et al.* (1997)
Trichogramma oleae	Pointel *et al.* (1979)
Trichogramma pintoi	Wang and Zhang (1988)
Trichogramma platneri	Stouthamer *et al.* (1990a); Pinto (1998)
Trichogramma pretiosum	Orphanides and Gonzalez (1970); Rodriguez *et al.* (1996)
Trichogramma semblidis	Pintureau *et al.* (2000)
Trichogramma telengai	Sorakina (1987)
Leucospidae	
Leucospis gigas	Berland (1934)
Pelecinoidea	
Pelecinus polyturator	Brues (1928); Johnson and Musetti (1998)
Proctutropoidea	
Amitus bennetti	Viggiani and Evans (1992)
Amitus fuscipennis	Viggiani (1991); Manzano *et al.* (2000)
Platygaster virgo	Day (1971)
Telonomus dignus	van der Goot (1915)
Telenomus nakagawai	Hokyo and Kiritani (1966)
Telenomus nawai	Arakaki *et al.* (2000)
Cynipoidea	
Hexacola sp.	James (1928)
Hexacola sp. near *websteri*	Eskafi and Legner (1974)
Leptopilina austalis	Werren *et al.* (1995)
Leptopilina clavipes	Eijs and van Alphen (1999)
Phaenoglyphis ambrosiae	Andrews (1978)
Bethylidea	
Scleroderma immigrans	Bridwell (1929); Keeler (1929a, b)
Trigonalyidae	
Taeniogonalos venatoria	Weinstein and Austin (1996)
Dryinidae	
Gonatopus contortus	Perkins (1905a)
Gonatopus sepsoides	Waloff (1974)
Haplogonatopus hernandazae	Pilar Hernandez and Belloti (1984)
Haplogonatopus vitiensis	Clausen (1978)

interpretation of this list is difficult because it does not constitute an independent sample of all parasitoid species. One might expect that the frequency of unisexual reproduction would be high particularly for solitary species and species with extremely small individuals, because for them the encounter between the sexes might be the most difficult. Indeed several trichogrammatid and mymarid genera appear to be well represented. However, being small is not a prerequisite for unisexual reproduction, because some of the largest parasitic wasps species also have unisexual biotypes, e.g. *Pelecinus polyturator* (Johnson and Musetti, 1998).

Potential Advantages of Unisexuals

As summarized by Stouthamer (1993), the advantages of unisexual wasps in biocontrol are: (i) unisexual wasps have a potentially higher rate of increase than sexual wasps; (ii) unisexual wasps are cheaper to produce, all the wasps reared in mass rearing are females and only females are effective in biological control; (iii) unisexual forms should be easier to establish in classical biocontrol projects because they do not suffer from the Allee effect, i.e. a shortage of mating partners, which may limit the growth rate of sexual forms when wasp density is low; and (iv) for the same reason unisexual wasps may be able to reduce the host density to lower levels than the sexuals, since low wasp densities may cause a reduction in the ability of females to find mates and therefore to produce daughters for the next generation.

Do unisexual wasps indeed have a higher rate of increase than sexual forms?

This will depend entirely on the number of female offspring produced per unisexual female versus per sexual female. Little is known about the number of daughters produced by comparable sexual and unisexual females. In the case of *Wolbachia*-induced parthenogenesis, the relative offspring production of unisexual (infected) females differs from that of the sexual females. This has been studied extensively in *Trichogramma* species, where unisexual lines could be cured of their infection and rendered sexual (Stouthamer *et al.*, 1990a). When unisexual and sexual forms of the same line are compared, the offspring production of the sexual form is generally much higher when the unisexual form originated from a population where both sexuals and unisexuals co-occurred (mixed populations), while, if these comparisons were made using *Trichogramma* from populations where the infection has gone to fixation, no significant difference in offspring production could be found. In general, it appears that the influence of the infection on offspring production is much higher in those cases where the infected and uninfected wasps occur together (i.e. mixed populations) (van Meer, 1999). Similarly, there appears to be hardly any negative influence of the *Wolbachia* infection in species such as *E. formosa* and *Muscidifurax uniraptor* (Stouthamer *et al.*, 1994).

These comparisons have been made in the laboratory using conditions where the wasps were given a surplus of hosts. In the field, the situation may be entirely different. Even if the unisexual forms are capable of producing fewer offspring than the sexual forms in the laboratory, this may not be very important in the field. The number of hosts that a wasp encounters determines the number of offspring produced and, as long as this number is below the maximal egg production of the unisexual line, all hosts encountered by both forms will be parasitized (Stouthamer and Luck, 1993). Even when the hosts are more numerous than the maximum egg production (M_u) of the unisexual line, the number of daughters produced by a unisexual female will be higher until the number of hosts encountered reaches the threshold T. If we define the sex ratio produced by sexual females as S, expressed as the fraction of daughters in the offspring, T can be derived as follows:

$$M_u = T \times S,\ T = M_u / S$$

In the range of host densities of M_u to T, the sexual form will kill more hosts per female than the unisexual form and yet the growth rate of the unisexual population will be higher than that of the sexual population. These three zones of host-encounter rates ($< M_u$, M_u–T, $> T$) are useful values for making predictions about the relative usefulness of releasing unisexual versus sexual wasps for biocontrol (Fig. 8.2). As long as the host density is and remains such that the number encountered per female is larger than T, then it is more useful to release the sexual form. It will both have a faster rate of population growth and kill a higher number of hosts than the unisexual form. In the range of host densities between M_u and T, the sexual form will kill a higher fraction of the host population but the rate

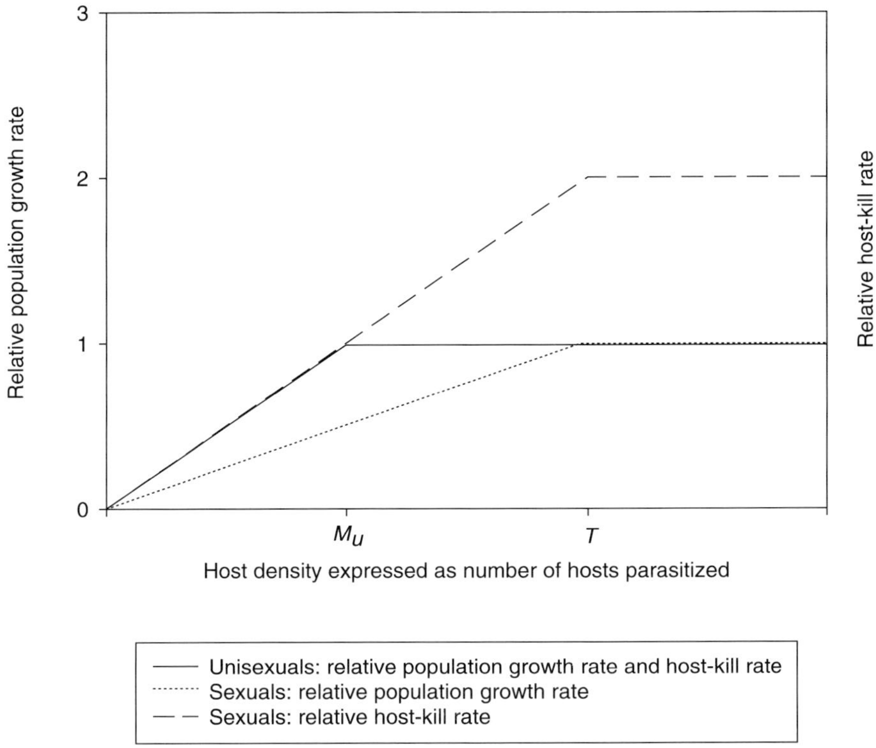

Fig. 8.2. Relative population growth rate expressed as number of daughters per mother and the relative host-kill rate expressed as the relative number of hosts killed per mother of a unisexual form and a sexual form. The unisexual form produces 100% daughters but can only parasitize M_u (maximum egg production of the unisexual female) hosts, while the sexual form can parasitize T hosts (maximum egg production of sexual host). Sex ratio of sexual form is assumed to be 50% females.

of increase of the sexual form will be lower than that of the unisexuals. Finally, below M_u, the unisexuals and sexuals will cause the same number of hosts to be killed per female but the population growth rate of the unisexuals will be higher ($1/S$ times as high per generation). If we assume that the wasp population is in an exponential growth phase, the criteria can be derived to determine the relative number of hosts killed over time in this tract; in the simplest case, the relative number of sexual females present of either form is given by $(SV/M_u)^n$. The number of hosts killed per female equals V/M_u; therefore the relative fraction of hosts killed by the sexual forms in generation n equals:

$$(SV/M_u)^n \times V/M_u = S^n V^{1+n}/M_u^{1+n}.$$

Are unisexuals cheaper to produce?

A major part of the cost of producing parasitoids for biological control is the cost of producing hosts. When unisexual wasps are used, all hosts result in female parasitoids and no hosts are wasted in the production of males. This should result in a reduction in production cost per female. While in species with a female-biased sex ratio the difference in production costs is not very large, in those species with sex ratios close to 50% the difference can be substantial. In addition, even in species that normally have a female-biased sex ratio, the sex ratio in mass rearing is often male-biased (Heimpel and Lundgren, 2000). Particularly for species used in inundative biological control, these production costs may form an important reason for choosing the unisexual form.

Are unisexual forms easier to establish in classical biocontrol projects?

Few papers directly address this issue. There are some indirect indications, summarized in the paper by Hung *et al.* (1988). They discuss a list of species where the mode of reproduction in the native area of the species is sexual, while in the area where they are released for biological control the mode of reproduction is unisexual. The most likely explanation is that in the native area the population consists of a mixture of both sexual and unisexual forms and in the release area the unisexuals are better able to colonize. Statistics on the mode of reproduction and the success of colonization are not available. However, Hopper and Roush (1993) show how important Allee effects in mate finding may be for the success of biological control. Some striking examples exist of the establishment of unisexual forms in biocontrol; for instance, Laing and Heraty (1981) report the successful establishment of *Apanteles pedias* after placing only two females in sleeve cages in the field.

Are unisexuals able to suppress the pest to a lower density than sexuals?

The same Allee effect as mentioned above would allow the unisexual to suppress the host density to a lower level than the sexual. Very low host densities and therefore low wasp densities would make mate finding difficult, but unisexuals would not suffer from such a drawback. This may not be very important because, if low host densities (and therefore low wasp densities) cause wasps to be unable to encounter each other, it may also be extremely difficult to find hosts. No evidence exists for this hypothesis.

Disadvantages of Unisexuals

Often the lack of sexual reproduction is seen as a dead end, because the wasps would be unable to adjust to changes in the environment and over time mutations in the genome of the wasps are assumed to accumulate (Muller's ratchet). This may indeed be a problem, but the time-scale in which these problems may manifest themselves is probably substantial. For instance, all of our knowledge of *E. formosa* indicates that parthenogenesis has been present in that species for a long time. Sexual reproduction is no longer possible between males and females of this species and no sexual populations are known. This species has been reared for biological control more or less continuously since the 1930s and shows no signs of losing its effectiveness.

A potential disadvantage of the unisexual reproduction of the parasitoid may be the evolution of resistance against the parasitoid by its sexual host species. In the case of *E. formosa*, we tried to find evidence for such resistance in its host, the greenhouse whitefly, *Trialeurodes vaporariorum*. Several factors made it unlikely that we would have to fear this development of resistance, the main reason being the method the insectaries use to grow the *E. formosa*. The whiteflies that are used to start the host population in the next generation are those that were not parasitized or that survived parasitization in the wasp rearing. The rearing method that the insectaries apply is the perfect experiment to determine if resistance is evolving. No clear evidence for the evolution of resistance to the whiteflies has been found.

Case-studies: Unisexuals in Classical Biological Control

Sexuals more successful than unisexuals

For several years, Neuffer (1962, 1964a, b) tried to establish the unisexual form of *Encarsia perniciosi* for the biological control of San José scale in Germany. This was without success. He subsequently established cultures of the sexual form of *E. perniciosi* (Neuffer, 1966, 1968, 1969, 1975, 1981, 1990), and was able both to establish the cultures and to control the San José scale. At the same time, the sexual form of *E. perniciosi* also seemed to have spread, at least in some areas of the USA, where the unisexual form prevailed earlier (Stouthamer and Luck, 1991).

The introduction of both the unisexual *Aphytis yanonensis* and the sexual *Coccobius fulvus* for the control of the arrowhead scale *Unaspis yanonensis* was studied in detail (Itioka *et al.*, 1997). In this introduction, it appeared that the control exerted by the sexual *C. flavius* was much more substantial than that of *A. yanonensis*. Only in the last few dates of the study, when the scale population density was substantially reduced did the importance of *A. yanonensis* increase. At the lower densities of the host and therefore of the wasps, the Allee effect may result in the lower efficiency of the sexual form. Apparently in this study this was not the case, either because the wasp density did not become that low or because the hypothesis is not correct. A study covering the next few years of this interaction showed that also in the years after the last reported date the importance of *A. yanonensis* did not increase substantially (T. Itioka, Nagoya, Japan, January 2001, personal communication).

The control exerted by the unisexual *Aphytis chrysomphali* on California red scale in the first half of the 20th century in California was considered variable and in some cases satisfactory. The parasitoids had spread throughout the red-scale-infested areas of California. In 1947 the sexual species *Aphytis lingnanensis* was introduced. This species replaced *A. chrysomphali* in a relatively short time in all areas except in a few small coastal areas (Clausen, 1978). For the control of the California red scale, several unisexual species have been released (*E. perniciosi, A. chrysomphali, Habrolepis rouxi*); although they do play a minor part in its control, in general the sexual species are more important.

Another case where the initially successful unisexual parasitoid was replaced by sexual species is in the biological control of the citrus blackfly in Mexico (Smith *et al.*, 1964; Flanders, 1969). In one of the most thorough biocontrol efforts ever, H.D. Smith collected large numbers of individuals of several species of whitefly parasitoids in India and Pakistan. These parasitoids were mass-reared in infested orchards throughout Mexico and subsequently distributed. Three sexual species were imported: *Encarsia opulenta, Encarsia clypealis* and *Amytis hesperidum* and a single unisexual species *Encarsia smithi*. Initially, *E. smithi* was very successful and a total of approximately 3,000,000 wasps had been dispersed throughout the country. In one grove where releases were carried out, the results were spectacular and in a short time the blackfly was controlled. However, in most other areas the sexual parasitoids did better and replaced *E. smithi* (Smith *et al.*, 1964; Flanders, 1969). This was not only because the other sexual *Encarsia* species produced males hyperparasitically, but also because the larvae of *A. hesperidum* proved competitively superior to those of *E. smithi* (Smith *et al.*, 1964). In subsequent infestations in Texas and Florida, only the sexual species were released. An accidental introduction into a rearing facility in Florida of *E. smithi* brought to light the fact that this species is most probably also a sexual species, whose males closely resemble those of *E. opulenta* (Nguyen and Sailer, 1987). Since these wasps originated in Mexico, either H.D. Smith wrongly classified the species or the species consisted of two different forms when released and the sexual form became the contaminant in the mass rearing.

For the control of the olive scale, a large number of different forms of *Aphytis maculicornis* were collected throughout the world. Several were introduced; the unisexual strain collected from Egypt was released in 1949 and it became established. Later on, a unisexual Spanish strain and sexual Persian and Indian forms were released. The sexual Persian strain soon established and was apparently the most effective (Clausen, 1978).

Inconclusive

Aeschlimann *et al.* (1989) tried to establish *Anaphes diana* in Australia from material collected around the Mediteranean consisting of both unisexual and sexual forms (Aeschlimann, 1990). The establishment in Australia failed (Aeschlimann *et al.*, 1989). In North America *A. diana* was released from the same source material; it is also assumed that it contained both unisexual and sexual forms. In total, approximately 12,500 individuals were released but only very few individuals were recovered: 3 years after the last release two males and one female were found, and 6 years after the last release another two males and a single female were found (Dysart, 1990).

Unisexual successful

Encarsia berlesei, a unisexual parasitoid of the white peach scale, proved extremely efficient in the control of this pest. The parasitoid was released in several European countries where the scale was a pest, and complete control was attained (Clausen, 1978).

Biological control of the citriculus mealy bug by the unisexual encyrtid *Clausenia purpurea* in Israel was very successful and no other parasitoids were released. Similarly, the unisexual parasitoid *Tetrastichus asparagi* is a successful imported parasitoid of the asparagus beetle in North America, where it became established (Clausen, 1978).

In a shipment of parasitoids for the lucerne weevil, a number of ichneumonids (*Biolysia tristis*) were present that did not end up in their intended locale (Utah) but were diverted to Washington, DC. There they were released and have since spread to many of the states on the east coast and in the Midwest, where they reduce the populations of their host, the clover-leaf weevil (*Hyperica punctata*). The unisexual sweet-clover-weevil parasitoid *Pygostolus falcatus* was established successfully in Canada; however, hardly any control was exerted by these parasitoids (Clausen, 1978).

The unisexual *Hexacola* sp. nr. *websteri* was imported together with several other sexual species for the control of eye gnats; although several parasitoids, including the *Hexacola* sp. nr. *websteri*, were established, none exerted substantial control of the pest (Clausen, 1978). The unisexual wasp *A. pedias* was established in Canada on the spotted tentiform leafminer from an initial collection of only two individuals and spread rapidly in a large area (Laing and Heraty, 1981). Several unisexual lines of the South American species *Microctonus hyperodae* were released for biological control of the weevil *Listronotus bonariensis*. Several populations were released, all originating from different parts of South America. Successful establishment took place and high parasitism rates have been reported (Phillips *et al.*, 1997; Goldson *et al.*, 1998).

Case-studies: Unisexuals in Seasonal Inoculative Biological Control

The best-known case of the use of unisexuals in seasonal inoculative biological control is the use of *E. formosa* for the biological control of the greenhouse whitefly (*T. vaporariorum*). This parasitoid is probably one of the most applied biological control agents in greenhouses and its use and biology have been extensively reviewed (see, for instance, van Lenteren *et al.*, 1997). No sexual forms of this species are known, so no comparative work has been done on the relative advantages of either form. Other unisexual species, such as *Eretmocerus staufferi* and *Amitus bennetti*, are also being considered for inoculative biological control of whiteflies in greenhouses (Drost *et al.*, 2000).

Case-studies: Unisexuals in Inundative Biological Control

The only controlled study that has been done thus far to test the potential advantage of unisexual forms over sexual forms of the same species is the study of Silva *et al.* (2000). In this study small greenhouses were used, in which unisexual and sexual forms of the same line were released, both of *Trichogramma deion* and of *Trichogramma cordubensis*. The sexual forms had been derived from the unisexual forms by antibiotic treatment. In the greenhouse, tomato plants were placed with egg cards attached to them. Either a mixture of unisexual and sexual wasps was released or the different forms were released in adjacent greenhouses. The location and the number of the parasitized egg patches was determined, as well as the mode of reproduction of the female that had found the patch. This experiment showed that females of both forms were equally capable of finding host patches, but that the sexual females parasitized more hosts per patch. Overall, the conclusion was that, under these circumstances, the use of unisexual wasps is more economic even when the hosts are found in patches. When the hosts are solitary, the use of the unisexual wasps becomes even more economically practical (Silva *et al.*, 2000).

Discussion

While in theory the use of unisexual wasps in biocontrol should result in advantages, in classical biological control this thesis has not been rigorously tested. In a number of cases, the unisexual form established when released in a new area, but whether the sexual form, had it been released in sufficient numbers, would have done worse remains untested (Hung *et al.*, 1988). The two studies where both sexual and unisexual forms of the same species were released do not give a clear result either. In the case of *E. perniciosi* the unisexual form was replaced by the sexual form (Neuffer, 1990), while in the case of *Anagyrus* the species either did not establish or established in such low frequency that no conclusion can be drawn (Aeschlimann *et al.*, 1989; Dysart, 1990). In the biological control effort against citrus red scale, the initial established unisexual species was competitively displaced by the sexual species, *A. lingnanensis* (Clausen, 1978). While for the control of the arrowhead scale the unisexual *A. yanonenis* coexists together with the sexual *Coccobius* (Itioka *et al.*, 1997), of these two species the sexual species appears to be the more effective parasitoid, both at high host densities and at low host densities. The predicted advantage of the unisexual form at low densities did not materialize. All in all, the present literature on classical biological control releases cannot be used to make statements about the superiority of unisexual forms over sexual forms. However, few studies are done that directly test this thesis and, in many cases, unisexual forms are effective classical biological control agents.

In seasonal inoculative biological control, a single extremely effective unisexual biological control agent stands out: *E. formosa*. In the genus *Encarsia* sexual species generally have a curious mode of reproduction, in which males develop as hyperparasitoids of female larvae, sometimes of conspecific female larvae (Walter, 1983). This mode of reproduction is called heteronomous hyperparasitism. The presence of only a heteronomous hyperparasitic species may hamper its own population growth rate and effectiveness, although in classical biological control such heteronomous species are often very effective. However, if the main biocontrol agent is a unisexual species the presence of a heteronomous species may influence the growth rate of the unisexual species, because sexual males are produced on unisexual female larvae (Vet and van Lenteren, 1981; Pedata and Hunter, 1996; Hunter and Kelly, 1998).

No large-scale use is made of unisexual lines in inundative biological control. Most inundative biological control programmes use species of the genus *Trichogramma*. In this genus large numbers of unisexual lines are known (Stouthamer, 1997), but no unisexual form is known in the species most commonly used in inundative programmes: *Trichogramma brassicae*. In other *Trichogramma* species used for biological control, unisexual forms are known and are sometimes applied in biological control programmes. For instance, *T. cacoeciae* is used in the USA and Europe for the control of codling moth (Dolphin *et al.*, 1972; Hassan and Rost, 1993), *Trichogramma* nr. *sibericum* is used for the control of cranberry pests (Li *et al.*, 1994), a unisexual form of *Trichogramma pintoi* is used in China (Wang and Zhang, 1988), and *Trichogramma chilonis* may be used in Taiwan (Chen *et al.*, 1992). However, among other commercially used species, unisexual forms exist but are not used on a large scale: these species include *Trichogramma platneri*, *Trichogramma pretiosum* and *Trichogramma evanescens* (Stouthamer, 1997). The most likely reason for the lack of application of these unisexuals is that they are only known in academic institutions and are unknown in the insectary industry.

In conclusion, we may state that the potential advantages of unisexual forms in classical biological control have not been tested rigorously. This will remain difficult to do because of the often limited time available for biological control projects. Once a good biocontrol agent has been found, funds generally dry up for doing additional work. In inundative releases, the potential for doing comparative work on the economy of applying unisexuals is better and this should be done. Finally, if at some point the technology of rendering sexual forms unisexual has progressed to the stage where it is easy to make sexual lines unisexual, it will also be possible to test the potential advantages of unisexuality in classical biological control.

References

Aeschlimann, J.P. (1986) Distribution and effectiveness of *Anaphes diana*, a parasitoid of *Sitona* spp. eggs in the Mediterranean region. *Entomophaga* 31, 163–172.

Aeschlimann, J.P. (1990) Simulatanious occurrence of thelytoky and bisexuality in hymenopteran species, and its implications for the biological control of pests. *Entomophaga* 35, 3–5.

Aeschlimann, J.P., Hopkins, D.C., Cullen, J.M. and Cavanaugh, J.A. (1989) Importation and release of *Anaphes diana* Girault (Hym., Mymaridae), a parasitoid of *Sitona discoideus* Gyllenhal (Col., Cuculionidae) eggs in Australia. *Journal of Applied Entomology* 107, 418–423.

Andrews, F.G. (1978) *Taxonomy and Host Specificity of Nearctic Alloxystinae with a Catalog of the World Species (Hym: Cynipidae).* Occasional Papers of the California Department of Food and Agriculture Laboratory Serving Entomology 25, Sacramento, California, 128 pp.

Arakaki, N., Noda, H. and Yamagishi, K. (2000) *Wolbachia*-induced parthenogenesis in the egg parasitoid *Telenomus nawai*. *Entomologia Experimentalis et Applicata* 96, 177–184.

Argov, Y., Gottlieb, Y., Amin, S.S. and Zchori-Fein, E. (2000) Possible symbiont-induced thelytoky in *Galeopsomyia fausta*, a parasitoid of the citrus leafminer *Phyllocnistis citrella*. *Phytoparasitica* 28, 212–218.

Avilla, J., Anadon, J., Sarasua, M.J. and Albajes, R. (1991) Egg allocation of the autoparasitoid *Encarsia tricolor* at different relative densities of the primary host and two secondary hosts. *Entomologia Experimentalis et Applicata* 59, 219–227.

Baker, R.H.A. (1979) Studies on the interactions between *Drosophila* parasites. PhD thesis, Oxford University.

Balduf, W.V. (1926) The bionomics of *Dinocampus coccinellae* Schrank. *Annals of the Entomological Society of America* 19, 465–498.

Bartlett, B.R. and Fisher, T.W. (1950) Laboratory propagation of *Aphytis chrysomphali* for release to control California red scale. *Journal of Economic Entomology* 43, 802–806.

Bartlett, K.A. (1939) Introduction and colonization of two parasites of the pineapple mealybug in Puerto Rico. *Journal of the Agricultural University of Puerto Rico* 23, 67–72.

Berland, L. (1934) Un cas probable de parthénogenèse géographique chez *Leucospis gigas*. *Bulletin Zoologique de France* 59, 172–175.

Berry, P.A. (1938) Tetrastichus brevistigma *Gahan, a Pupal Parasite of the Elm Leaf Beetle*. Circular 485, United States Department of Agriculture, Washington, DC, 11 pp.

Beukeboom, L.W. and Pijnacker, L.P. (2000) Automictic parthenogenesis in the parasitoid *Venturia canescens* revisited. *Genome* 43, 939–944.

Birova, H. (1970) A contribution to the knowledge of the reproduction of *Trichogramma embryophagum*. *Acta Entomologica Bohemoslovaca* 67, 70–82.

Bowen, W.R. and Stern, V.M. (1966) Effect of temperature on the production of males and sexual mosaics in uniparental race of *Trichogramma semifumatum*. *Annals of the Entomological Society of America* 59, 823–834.

Bridwell, J.C. (1929) Thelytoky and arrhenotoky in *Scleroderma immigrans*. *Psyche* 36, 119–120.

Brues, C.T. (1928) A note on the genus *Pelicinus*. *Psyche* 35, 205–209.

Bueno, V.H.P., Berti-Filho, E. and Matiolo, J.C. (1987) Aspects of the biology and behavior of *Nesolynx* sp. *Anais da ESA 'Luiz de Quiroz'* 44, 105–117.

Cabello, G.T., Vargas, P.P., Garcia, T.C. and Piqueras, P.V. (1985) Temperature as a factor influencing the form of reproduction of *Trichogramma cordubensis* Vargas & Cabello (Hym., Trichogrammatidae). *Zeitschrift für Angewandte Entomologie* 100, 434–441.

Carter, W. (1937) Importation and laboratory breeding of two chalcid parasites of *Pseudococcus brevipes*. *Journal of Economic Entomology* 30, 370–372.

Chandra, G. (1980) Dryinid parasitoids of rice leafhoppers and planthoppers in the Philippines. *Acta Oecologica* 1, 161–172.

Chen, B.H., Kfir, R. and Chen, C.N. (1992) The thelytokous *Trichogramma chilonis* in Taiwan. *Entomologia Experimentalis et Applicata* 65, 187–194.

Clancy, D.W. (1969) A new parasite of the vegetable weevil in southern California. *Journal of Economic Entomology* 62, 743–745.

Claridge, M.F., Claridge, L.C. and Morgan, J.C. (1987) *Anagrus* egg parasitoids of rice feeding planthoppers. In: *Proc. 6th Auchenorrhyncha Meeting, Turin, Italy, 7–11 September* 1987, pp. 617–621.

Clausen, C.P. (1940) *Entomophagous Insects*. McGraw-Hill, New York.

Clausen, C.P. (1978) *Introduced Parasites and Predators of Arthropod Pests and Weeds: a World Review*. Agricultural Handbook No. 480, ARS, USDA, Washington, DC.

Cronin, J.T. and Strong, D.R. (1996) Genetics of oviposition success of a thelytokous fairyfly parasitoid, *Anagrus delicatus*. *Heredity* 76, 43–54.

Day, M.C. (1971) A new species of *Platygaster* Latreille (Hym: Proctotrupoidae, Platygasteridae) reproducing by thelytokous parthenogenesis. *Entomological Gazette* 22, 37–42.

Day, W.H., Marsh, P.M., Fuester, R.W., Hoyer, H. and Dysart, R.J. (1992) Biology, initial effect, and description of a new species of *Peristenus*, a parasite of the alfalfa plant bug, recently established in the United States. *Annals of the Entomological Society of America* 85, 482–488.

Day, W.H., Baird, C.R. and Shaw, S.R. (1999) New, native species of *Peristenus* parasitizing *Lygus hesperus* in Idaho: biology, importance, and description. *Annals of the Entomological Society of America* 92, 370–375.

DeBach, P. (1969) Uniparental, sibling and semi-species in relation to taxonomy and biological control. *Israel Journal of Entomology* 4, 11–27.

DeBach, P. (1984) *Aphytis simmondsiae* n. sp., a parasite of jojoba scale, *Diaspis simmondsiae*. *Folia Entomologica Mexicana* 60, 103–112.

DeBach, P. and Rosen, D. (1982) *Aphytis yanonensis* n.sp. a parasite of *Unaspis yanonensis*. *Kontyu* 50, 626–634.

DeBach, P., Kennett, C.E. and Pence, R.J. (1958) Species of *Thysanus* as primary parasites. *Journal of Economic Entomology* 51, 114–115.

de Barro, P.J., Hart, P.J. and Morton, R. (2000) The biology of two *Eretmocerus* spp. (Haldeman) and three *Encarsia* spp. Forster and their potential as biological control agents of *Bemisia tabaci* biotype B in Australia. *Entomologia Experimentalis et Applicata* 94, 93–102.

Dolphin, R.E., Cleveland, M.L., Mouzin, T.E. and Morrison, R.K. (1972) Releases of *Trichogramma minutum* and *T. cacoeciae* in an apple orchard and the effects on populations of codling moths. *Environmental Entomology* 1, 481–484.

Doutt, R.L. and Smith, R.A. (1950) Males and intersexes in a normally thelytokous insect, *Tropidophryne melvillei* Compere. *Canadian Entomologist* 82, 165–170.

Dowden, P.B. (1938) *Rogas unicolor*, a braconid parasite of the Satin Moth. *Journal of Agricultural Research* 56, 523–535.

Drost, Y.C., Qui, Y.T., Postuma-Doodeman, C.J.A.M. and van Lenteren, J.C. (2000) Comparison of searching strategies of five parasitoid species of *Bemisia argentifolii*. *Journal of Applied Entomology* 124, 105–112.

Dysart, R.J. (1990) The introduction and recovery in the United States of *Anaphes diana*, an egg parasite of *Sitona* weevils. *Entomophaga* 35, 307–313.

Eijs, I.E.M. and van Alphen, J.J.M. (1999) Life history correlations: why are hymenopteran parasitoids an exception? *Ecology Letters* 2, 27–35.

Embleton, A.L. (1904) On the anatomy and development of *Comys infelix* Embleton, a hymenopterous parasite of *Lecanium haemisphaericum*. *Zoological Transactions of Linnean Society of London* 9, 231–254.

Eskafi, F.M. and Legner, E.F. (1974) Parthenogenetic reproduction in *Hexacola* sp. near *websteri*, a parasite of *Hippelates* eye gnats. *Annals of the Entomological Society of America* 67, 767–768.

Flanders, S.E. (1943) The role of mating in the reproduction of parasitic hymenoptera. *Journal of Economic Entomology* 36, 802–803.

Flanders, S.E. (1945) The bisexuality of uniparental hymenoptera, a function of the environment. *American Nauralist* 79, 122–141.

Flanders, S.E. (1953a) Hymenopterous parasites of three species of oriental scale insects. *Bollettino del Laboratorio di Entomologia Agraria 'Filippo Silvestri' Portici* 33, 10–28.

Flanders, S.E. (1953b) Aphelinid biologies with implications for taxonomy. *Annals of the Entomological Society of America*, 46, 84–94.
Flanders, S.E. (1958) Bisexuality in uniparental hymenoptera. In: *Internal report Department of Entomology*, No. 1702, University of California, Riverside, pp. 11–45.
Flanders, S.E. (1959) Differential host relations of the sexes in parasitic hymenoptera. *Entomologia Experimentalis et Applicata* 2, 125–142.
Flanders, S.E. (1969) Herbert D. Smith's observations on citrus blackfly parasites in India and Mexico and the correlated circumstances. *Canandian Entomologist* 101, 467–480.
Force, D.C. and Messenger, P.S. (1964) Duration of development, generation time and longevity of three hymenopterous parasites of *Therioaphis maculata*, reared at various constant temperatures. *Annals of the Entomological Society of America* 57, 405–413.
Fuester, R.W., Taylor, P.B., Peng, H. and Swan, K. (1993) Laboratory biology of a uniparental strain of *Meteorus pulchricornis* (Hymenoptera: Braconidae), an exotic larval parasite of the gypsy moth (Lepidoptera: Lymantriidae). *Annals of the Entomological Society of America* 86, 298–304.
Gerson, U. (1968) The comparative biologies of two hymenopterous parasites of the chaff scale, *Parlatoria pergandii*. *Entomophaga* 13, 163–173.
Goldson, S.L., Proffitt, J.R. and Baird, D.B. (1998) Establishment and phenology of the parasitoid *Microctonus hyperodae* (Hymenoptera: Braconidae) in New Zealand. *Environmental Entomology* 27, 1386–1392.
Gordh, G. and Dunbar, D.M. (1977) A new *Anagrus* important in the biological control of *Stephanitis takeyai* and a key to the North American species. *Florida Entomologist* 60, 85–95.
Gordh, G. and Lacey, L. (1976) Biological studies of *Plagiomerus diaspidis*, a primary internal parasite of diaspidid scale insects. *Proceedings of the Entomological Society of Washington* 78, 132–144.
Gottlieb, Y., Zchori-Fein, E., Faktor, O. and Rosen, D. (1998) Phylogenetic analysis of parthenogenesis-inducing *Wolbachia* in the genus *Aphytis* (Hymenoptera: Aphelinidae). *Insect Molecular Biology* 7, 393–396.
Grenier, S., Pintureau, B., Heddi, A., Lassabliere, F., Jager, C., Louis, C. and Khatchadourian, C. (1998) Successful horizontal transfer of *Wolbachia* symbionts between *Trichogramma* wasps. *Proceedings of the Royal Society of London Series B Biological Sciences* 265, 1441–1445.
Griswold, G.H. (1929) On the bionomics of a primary parasite and of two hyperparasites of the geranium aphid. *Annals of the Entomological Society of America* 22, 438–457.
Hartley, E.A. (1922) Some bionomics of *Aphelinus semiflavus* (Howard) chalcid parasite of aphids. *Ohio Journal of Science* 22, 209–236.
Hassan, S.A. and Rost, W.M. (1993) Mass rearing and utilization of *Trichogramma*: 13. Optimizing the control of the codling moth *Cydia pomonella* L. and the summer fruit tortrix moth *Adoxophyes orana* F.R. *Gesunde Pflanzen* 45, 296–300.
Heimpel, G.E. and Lundgren, J.G. (2000) Sex ratios of commercially reared biological control agents. *Biological Control* 19, 77–93.
Hessein, N.A. and McMurtry, J.A. (1988) Observations on *Megaphragma mymaripenne* Timberlake (Hymenoptera: Trichogrammatidae), an egg parasite of *Heliothrips haemorrhoidalis* Bouche (Thysanoptera: Thripidae). *Pan Pacific Entomologist* 64, 250–254.
Hokyo, N. and Kiritani, K. (1966) Oviposition behaviour of two egg parasites, *Asolcus mitsukurii* Ashmead and *Telenomus nakagawai* Watanabe. *Entomophaga* 11, 191–201.
Hopper, K.R. and Roush, R.T. (1993) Mate finding, dispersal, number released, and the success of biological control introductions. *Ecological Entomology* 18, 321–331.
Huigens, M.E., Luck, R.F., Klaassen, R.H.G., Maas, M.F.P.M., Timmermans, M.J.T.N. and Stouthamer, R. (2000) Infectious parthenogenesis. *Nature* 405, 178–179.
Hummelen, P.J. (1974) Relations between two rice borers in Surinam, *Rupela albinella* and *Diatraea saccharalis*, and their hymenopterous larval parasites. *Mededeleingen van de Landbouwhogeschool Wageningen* 74, 1–86.
Hung, A.C.F., Day, W.H. and Hedlund, R.C. (1988) Genetic variability in arrhenotokous and thelytokous forms of *Mesochorus nigripes*. *Entomophaga* 33, 7–15.
Hunter, M.S. (1999) The influence of parthenogenesis-inducing *Wolbachia* on the oviposition behaviour and sex-specific developmental requirements of autoparasitoid wasps. *Journal of Evolutionary Biology* 12, 735–741.
Hunter, M.S. and Kelly, S.E. (1998) Hyperparasitism by an exotic autoparasitoid: secondary host selection and the window of vulnerability of conspecific and native heterospecific hosts. *Entomologia Experimentalis et Applicata* 89, 249–259.

Ishii, T. (1932) The Encyrtidae of Japan. II Studies on morphology and biology. *Bulletin of the Imperial Central Agricultural Experiment Station* 3, 161–202.
Itioka, T., Inoue, T., Matsumoto, T. and Ishida, N. (1997) Biological control by two exotic parasitoids: eight-year population dynamics and life tables of the arrowhead scale. *Entomologia Experimentalis et Applicata* 85, 65–74.
James, H.C. (1928) On the life histories and the economic status of certain cynipid parasites of dipterous larvae, with descriptions of some new larval forms. *Annals of Applied Biology* 15, 287–316.
Johnson, N.F. and Musetti, L. (1998) Geographic variation of sex ratio in *Pelecinus polyturator* (Drury) (Hymenoptera: Pelecinidae). *Journal of Hymenoptera Research* 7, 48–56.
Keeler, C.E. (1929a) Thelytoky in *Scleroderma immigrans*. *Psyche* 36, 41–44.
Keeler, C.E. (1929b) Critical data upon thelytoky in *Scleroderma immigrans*. *Psyche* 36, 121–122.
Kelly, E.O. and Urbahns, T.D. (1908) Note read by the secretary. *Proceedings of the Entomological Society of Washington* 10, 163–164.
Kerrich, G.J. (1961) A study of the *Tersilochine* parasites of vegetable weevils of the genus *Listeroderes*. *EOS, Revista Espanola de Entomologia* 37, 497–503.
Kfir, R. (1998) Origin of the diamondback moth (Lepidoptera: Plutellidae). *Annals of the Entomological Society of America* 91, 164–167.
Kiss, A. (1986) Sympatric parthenogenetic and biparental races of the egg-parasite *Polynema enchenopae* (Hym.: Mymaridae). Unpublished manuscript, 23 pp.
Kunckel d'Herculais, J. and Langlois, C. (1891) *Perilitus brevicollis* Haliday, Hymenoptère braconide, Parasite de l'Altise de la Vigne. *Annales du Société Entomologique de France* 60, 457–466.
Laing, J.E. and Heraty, J.M. (1981) Establishment in Canada of the parasite *Apanteles pedias* on the Spotted Tentiform Leafminer, *Phyllonorycter blancardella* (Fabr.). *Environmental Entomology* 10, 933–935.
Laraichi, M. (1978) L'effet de hautes températures sur le taux sexuel de *Ooencyrtus fecundus* (Hym: Encyrtidae. *Entomologia Experimentalis et Applicata* 23, 237–242.
Legner, E.F. (1985) Natural and induced sex ratio changes in populations of thelytokous *Muscidifurax uniraptor*. *Annals of the Entomological Society of America* 78, 398–402.
Li, Q.T. (1984) Preliminary observations on thelytoky of *Meteorus cypris*. *Kunchong Zhishi* 21, 82–83.
Li, S.Y., Henderson, D.E. and Myers, J.H. (1994) Selection of suitable *Trichogramma* species for potential control of the blackheaded fireworm infesting cranberries. *Biological Control* 4, 244–248.
Loan, C.C. (1963a) Observations on the biology of *Centistes excrusians* Haliday. *Proceedings of the Entomological Society of Ontario* 94, 56–61.
Loan, C.C. (1963b) Parasitism of the dogwood flea-beetle *Altica corni*, in Ontario. *Journal of Economic Entomology* 56, 537–538.
Loan, C. and Holdaway, F.G. (1961) *Pygostolus falcatus*, a parasite of *Sitona* species. *Bulletin of Entomological Research* 52, 473–488.
Loomans, A.J.M. and van Lenteren, J.C. (1995) Biological control of thrips pests: a review on thrips parasitoids. In: Loomans, A.J.M., van Lenteren, J.C., Tommasini, M.G., Maini, S. and Riudavets, J. (eds) *Biological Control of Thrips Pests*. Wageningen University, Wageningen, pp. 92–201.
Luck, R.F., Nunney, L.P. and Stouthamer, R. (1999) Sex ratio and quality in the culturing of parasitic hymenoptera: a genetic and evolutionary perspective. In: Bellows, T.S., Jr and Fisher, T.W. (eds) *Handbook of Biological Control*. Academic Press, San Diego, California, pp. 653–672.
McAuslane, H.J. and Nguyen, R. (1996) Reproductive biology and behavior of a thelytokous species of *Eretmocerus* (Hymenoptera: Aphelinidae) parasitizing *Bemisia argentifolii* (Homoptera: Aleyrodidae). *Annals of the Entomological Society of America* 89, 686–693.
Manzano, M.R., van Lenteren, J.C., Cardona, C. and Drost, Y.C. (2000) Developmental time, sex ratio, and longevity of *Amitus fuscipennis* MacGown and Nebeker (Hymenoptera: Platygasteridae) on the greenhouse whitefly. *Biological Control* 18, 94–100.
Marchal, P. (1936) Recherches sur la biologie et développement des Hymenoptères parasites. Les Trichogrammes. *Annales des Epiphyties et de Phytogénétique* 2, 447–550.
Milbrath, L.R. and Weiss, M.J. (1998) Development, survival and phenology of the sweetclover weevil parasitoid, *Pygostolus falcatus* (Hymenoptera: Braconidae). *Great Lakes Entomologist* 31, 129–136.
Muesenbeck, C.F.W. and Dohanian, S.M. (1927) *A study of hyperparasitism, with particular reference to the parasites of* Apanteles melanoscelus. Bulletin 1487, United States Department of Agriculture, Muesenbeck, Washington, DC.
Nagasawa, S. (1947) Studies on life-history and bionomics of the fleabeetle parasite *Microctonus tittatae* Muesenbeck VI. On parthenogenesis, hyperparasitism and economic importance. *Seibutsu* 2, 48–52.

Nemec, V. and Stary, P. (1985) Population diversity in deuterotokous *Lysiphlebus* species, parasitiods of aphids (Hymenoptera, Aphidiidae). *Acta Entomologica Bohemoslovaca* 82, 170–174.
Neuffer, G. (1962) Zur Zucht und Verbreitung von *Prospaltella perniciosi* und anderen Parasiten der San-Jose-Schildlaus in Baden-Wurttemberg. *Nachrichtenblatt des Deutschen Pflanzenschutzdienstes* 14, 97–101.
Neuffer, G. (1964a) Bemerkungen zur Parasitenfauna von *Quadraspidiotus perniciosus* und zur Zucht bisexueller *Prospaltella perniciosi* im Insektarium. *Zeitschrift für Pflannzenkrankheiten und Pflanzenzucht* 71, 1–11.
Neuffer, G. (1964b) Zu den Aussetzenversuchen von *Prospaltella perniciosi* gegen die San Jose Schildlaus in Baden-Wurttemberg. *Entomophaga* 9, 131–136.
Neuffer, G. (1966) Zur Parasitenfauna von *Quadraspidiotus perniciosus* unter besonderer Beruchsichtigung der importierten *Prospaltella perniciosi*. *Entomophaga* 11, 383–393.
Neuffer, G. (1968) Die Wirksamkeit der Aphelinide *Prospaltella perniciosi* im sudwestdeutschen Befallsgebiet der San-Jose-Schildlaus *Quadraspidiotus perniciosus*. *Anzeiger für Schadlingskunde* 41, 97–101.
Neuffer, G. (1969) Biological control of the San Jose scale with *Prospaltella perniciosi* in south-western Germany. *European Plant Protection Organization Publications Series A* 48, 49–55.
Neuffer, G. (1975) Zur Situation in den Befallsgebieten der San-Jose-Schildlaus in Baden-Wurttemberg nach zwanzig Jahren Freilassungen von *Prospaltella perniciosi*. *Zeitschrift für Pflanzen Krankheiten* 82, 503–514.
Neuffer, G. (1981) Bemerkungen zur Befallsituation der San-Jose-Schildlaus, *Quadraspidiotus perniciosus* in Sudwestdeutschland 27 Jahre nach Beginn der biologischen Bekampfung mit *Prospaltella perniciosi*. *Miteilungenteilungen aus der BBA* 203, 145–146.
Neuffer, G. (1990) Zur Abundanz und Gradation der San-Jose-Schildlaus und deren Gegenspieler *Prospaltella perniciosi*. *Gesunde Pflanzen* 42, 89–96.
Nguyen, R. and Sailer, R.I. (1987) Facultative hyperparasitism and sex determination of *Encarsia smithi* (Silvestri) (Hymenoptera: Aphelinidae). *Annals of the Entomological Society of America* 80, 713–719.
Orphanides, G.M. and Gonzalez, D. (1970) Identity of a uniparental race of *Trichogramma pretiosum*. *Annals of the Entomological Society of America* 63, 1784–1786.
Pedata, P.A. and Hunter, M.S. (1996) Secondary host choice by the autoparasitoid *Encarsia pergandiella*. *Entomologia Experimentalis et Applicata* 81, 207–214.
Pedata, P.A. and Viggiani, G. (1991) Preliminary morpho-biological observations on *Azotus perspeciosus* (Girault) (Hymenoptera: Aphelinidae), hyperparasitoid of *Pseudaulacaspis pentagona* (Targioni, Tozzetti) (Homoptera: Diaspididae). *Redia* 74, 343–350.
Perkins, R.C.L. (1905a) Leaf-hoppers and their natural enemies (Dryinidae). *Hawaii Sugar Planters' Association Experimental Station Bulletin* 1, 1–69.
Perkins, R.C.L. (1905b) Leaf-hoppers and their natural enemies (Mymaridae, Platygasteridae). *Hawaii Sugar Planters' Association Experiment Station Bulletin* 1, 187–203.
Phillips, C.B. and Baird, D.B. (1996) A morphometric method to assist in defining the South American origins of *Microctonus hyperodae* Loan (Hymenoptera: Braconidae) established in New Zealand. *Biocontrol Science and Technology* 6, 189–205.
Phillips, C.B., Baird, D.B. and Goldson, S.L. (1997) South American origins of *Microctonus hyperodae* Loan (Hymenoptera: Braconidae) established in New Zealand as defined by morphometric analysis. *Biocontrol Science and Technology* 7, 247–258.
Phillips, W.J. and Poos, F.W. (1927) Two hymenopterous parasites of american jointworms. *Journal of Agricultural Research* 34, 473–488.
Pijls, J., van Steenbergen, H.J. and van Alphen, J.J.M. (1996) Asexuality cured: the relations and differences between sexual and asexual *Apoanagyrus diversicornis*. *Heredity* 76, 506–513.
Pilar Hernandez, M. and Belloti, A. (1984) Life cycle and behaviour of *Haplogonatopus hernandezae* Olmi (Hymenoptera: Dryinidae), a natural enemy of rice leafhoppers *Sogatodes orizicola* (Muir). *Revista Colombiana de Entomologia* 10, 3–4.
Pinto, J.D. (1998) Systematics of the North American species of *Trichogramma*. *Memoirs of the Entomological Society of Washington* 22, 1–287.
Pinto, J.D., Stouthamer, R. and Platner, G.R. (1997) A new cryptic species of *Trichogramma* (Hymenoptera: Trichogrammatidae) from the Mojave Desert of California as determined by morphological, reproductive and molecular data. *Proceedings of the Entomological Society of Washington* 99, 238–247.
Pintureau, B., Chaudier, S., Lassabliere, F., Charles, H. and Grenier, S. (2000) Addition of wsp sequences to the *Wolbachia* phylogenetic tree and stability of the classification. *Journal of Molecular Evolution* 51, 374–377.

Platner, G.R. and Oatman, E.R. (1972) Techniques for culturing and mass producting parasites of the potato tuberworm. *Journal of Economic Entomology* 65, 1336–1338.

Pointel, J.G., Jardak, T., Pintureau, B. and Voegele, J. (1979) The demonstration of a new species of *Trichogramma* (Hym. Trichogrammatidae). Intersexuality phenomenon, enzymatic study. *Annales de la Société Entomologique de France* 15, 635–642.

Powell, D.A. and Bellows, T.S., Jr (1992) Development and reproduction of two populations of *Eretmocerus* species (Hymenoptera: Aphelinidae) on *Bemisia tabaci* (Homoptera: Aleyrodidae). *Environmental Entomology* 21, 651–658.

Puttler, B. and Coles, L.W. (1962) Biology of *Biolysia tristis* and its role as a parasite of the clover leaf weevil. *Journal of Economic Entomology* 55, 831–833.

Quezada, J.R., DeBach, P. and Rosen, D. (1973) Biological and taxonomic studies of *Signophora borinquensis*, new species (Hym: Signiphoridae), a primary parasite of diaspine scales. *Hilgardia* 41, 543–604.

Redfern, M. (1976) The identity of *Mesoplobus* species (Hym: Pteromalidae) parasitizing *Taxomyia taxi* (Inchbold) (Dipt., Cecidomyiidae) on jew *Taxus baccata* L. *Entomologist's Monthly Magazine* 111, 201–204.

Reichert (1911) *Sphecophaga vesparum* Rtzb., eine Schlupfwespe im Wespenneste. *Entomologisches Jahrbuch* 20, 180–182.

Rivnay, E. (1942) *Clausenia purpurea* Ishii, a parasite of *Pseudococcus comstocki* Kuw. introduced into Palestine. *Bulletin de Société Fouad ler Entomologie* 26, 1–19.

Rodriguez, J., Pintureau, B. and Galan, M. (1996) On the identity of some Cuban *Trichogramma* Westwood (Hym.: Trichogrammatidae). *Boletin de Sanidad Vegetal, Plagas* 22, 585–599.

Rose, M. and Zolnerowich, G. (1997) *Eretmocerus* Haldeman (Hymenoptera: Aphelinidae) in the United States, with descriptions of new species attacking *Bemisia* (*tabaci* complex) (Homoptera: Aleyrodidae). *Proceedings of the Entomological Society of Washington* 99, 1–27.

Rosen, D. (1967) The hymenopterous parasites and hyperparasites of aphids on citrus in Israel. *Annals of the Entomological Society of America* 60, 394–399.

Rosen, D. and DeBach, P. (1979) *Species of* Aphytis *of the world*. Series Entomologica Vol. 17, Dr W. Junk BV, The Hague.

Russell, H.M. (1911) A true internal parasite of Thysanoptera. *Proceedings of the Entomological Society of Washington* 13, 235–238.

Russell, H.M. and Johnston, F.A. (1912) The life history of *Tetrastichus asparagi* Crawf. *Journal of Economic Entomology* 5, 429–433.

Sandanayake, W.R.M. and Edirisinghe, J.P. (1992) Aspects of reproductive biology of *Trathala flavoorbitalis* (CAM.): a parasitoid of *Leucinodes orbonalis* (Guen.). *Entomon* 17, 159–168.

Schmieder, R.G. (1939) The dimorphism of coccoons of *Sphecophaga burra*. *Entomological News* 50, 91–97.

Shaw, M.R. and Askew, R.R. (1976) Ichneumonoidae parasitic upon leaf-mining insects of the orders Lepidoptera, Hymenoptera and Coleoptera. *Ecological Entomology* 1, 127–133.

Silva, I., Van, M.M.M.M., Roskam, M.M., Hoogenboom, A., Gort, G. and Stouthamer, R. (2000) Biological control potential of *Wolbachia*-infected versus uninfected wasps: laboratory and greenhouse evaluation of *Trichogramma cordubensis* and *T. deion* strains. *Biocontrol Science and Technology* 10, 223–238.

Slobodchikoff, C.N. and Daly, H.V. (1971) Systematic and evolutionary implications of parthenogenesis in the Hymenoptera. *American Zoologist* 11, 273–282.

Smith, H.D., Maltby, H.L. and Jimenez, J.J. (1964) *Biological Control of the Citrus Blackfly in Mexico*. Technical Bulletin 1311, United States Department of Agriculture, Washington, DC, 30 pp.

Sorakina, A.P. (1987) Biological and morphological substantiation of the specific distinctness of *Trichogramma telengai* sp. n. *Entomological Review* 66 (4), 20–34.

Speicher, B.R., Speicher, K.G. and Roberts, F.L. (1965) Genetic segregation in the unisexual wasp *Devorgilla*. *Genetics* 52, 1035–1041.

Speyer, E.R. (1926) An important parasite of the greenhouse white-fly (*Trialeurodes vaporariorum*). *Bulletin of Entomological Research* 17, 301–308.

Stouthamer, R. (1993) The use of sexual versus asexual wasps in biological control. *Entomophaga* 38, 3–6.

Stouthamer, R. (1997) *Wolbachia*-induced parthenogenesis. In: O'Neill, S.L., Hoffmann, A.A. and Werren, J.H. (eds) *Influential Passengers: Inherited Microorganisms and Arthropod Reproduction*. Oxford University Press, New York, pp. 102–124.

Stouthamer, R. and Kazmer, D.J. (1994) Cytogenetics of microbe-associated parthenogenesis and its consequences for gene flow in *Trichogramma* wasps. *Heredity* 73, 317–327.

Stouthamer, R. and Luck, R.F. (1991) Transition from bisexual to unisexual cultures in *Encarsia perniciosi* (Hymenoptera: Aphelinidae): new data and a reinterpretation. *Annals of the Entomological Society of America* 84, 150–157.

Stouthamer, R. and Luck, R.F. (1993) Influence of microbe-associated parthenogenesis on the fecundity of *Trichogramma deion* and *T. pretiosum*. *Entomologia Experimentalis et Applicata* 67, 183–192.

Stouthamer, R., Luck, R.F. and Hamilton, W.D. (1990a) Antibiotics cause parthenogenetic *Trichogramma* to revert to sex. *Proceedings of the National Academy of Sciences of the USA* 87, 2424–2427.

Stouthamer, R., Pinto, J.D., Platner, G.R. and Luck, R.F. (1990b) Taxonomic status of thelytokous forms of *Trichogramma* (Hymenoptera: Trichogrammatidae). *Annals of the Entomological Society of America* 83, 475–481.

Stouthamer, R., Breeuwer, J.A.J., Luck, R.F. and Werren, J.H. (1993) Molecular identification of microorganisms associated with parthenogenesis. *Nature* 361, 66–68.

Stouthamer, R., Luko, S. and Mak, F. (1994) Influence of parthenogenesis *Wolbachia* on host fitness. *Norwegian Journal of Agricultural Sciences Supplement* 16, 117–122.

Stouthamer, R., Van Tilborg, M., de Jong, J.H., Nunney, L. and Luck, R.F. (2001) Selfish element maintains sex in natural populations of a parasitoid wasp. *Proceedings of the Royal Society London B* 268, 617–622.

Sumaroka, A.F. (1967) Factors affecting the sex ratio of *Aphytis proclia*, external parasite on the San Jose scale. *Entomological Review* 46, 179–185.

Takagi, M. (1988) Natural enemies of thrips. In: Umeya, I., Kudo, I. and Miyazaki, M. (eds) *Pest Thrips in Japan*. Zenkoku Noson Kyoiku Kyokai Publishers, Tokyo, Japan.

Tardieux, I. and Rabasse, J.M. (1988) Induction of a thelytokous reproduction in the *Aphidius colemani* (Hym., Aphidiidae) complex. *Journal of Applied Entomology* 106, 58–61.

Teitelbaum, S.S. and Black, L.M. (1957) The effect of phytophagous species of *Tetrastichus*, new to the United States, on sweet clover infected with wound-tumor virus. *Phytopathology* 44, 548–550.

Timberlake, P.H. (1919) Observations on the sources of Hawaiian Encyrtidae. *Hawaiian Entomological Society Proceedings* 4, 183–196.

Timberlake, P.H. and Clausen, C.P. (1924) The parasites of *Pseudococcus maritimus* (Ehrhorn) in California. *University of California Publications Technical Bulletin Entomology* 3, 223–292.

Titayavan, M. and Davis, D.W. (1988) Studies of a uniparental form of *Aphytis vandenboschi*, a parasite of the San Jose scale in northern Utah. *Great Basin Naturalist* 48, 388–393.

Vance, A.M. (1931) Apanteles thompsoni *Lyle, a Braconid Parasite of the European Corn Borer*. Technical Bulletin 233, United States Department of Agriculture, Washington, DC, 28 pp.

van der Goot, P. (1915) Over boorderparasieten en boorderbestrijding. *Archief voor de suikerindustrie in Nederlandsch-Indie* 1915, 407–460.

van Lenteren, J.C., Drost, Y.C., van Roermond, H.J.W. and Postuma-Doodeman, C.J.A.M. (1997) Aphelinid parasitoids as sustainable biological control agents in greenhouses. *Journal of Applied Entomology* 121, 473–485.

van Meer, M.M.M. (1999) *Phylogeny and Host–Symbiont Interactions of Thelytoky Inducing* Wolbachia *in Hymenoptera*. Department of Plant Sciences, Laboratory of Entomology, Wageningen University, Wageningen, 118 pp.

van Meer, M.M.M., van Kan, F., Breeuwer, J.A.J. and Stouthamer, R. (1995) Identification of symbionts associated with parthenogenesis in *Encarsia formosa* (Hymenoptera: Aphelinidae) and *Diplolepis rosae* (Hymenoptera: Cynipidae). *Proceedings of the Section Experimental and Applied Entomology of the Netherlands Entomological Society* 6, 81–86.

Vet, L.E.M. and van Lenteren, J.C. (1981) The parasite–host relationship between *Encarsia formosa* Gah. (Hymenoptera: Aphelinidae) and *Trialeurodes vaporariorum* (Westw.) (Homoptera: Aleyrodidae). X. A comparison of three *Encarsia* spp. and one *Eretmocerus* sp. to estimate their potentialities in controlling whitefly on tomatoes in greenhouses with a low temperature regime. *Zeitschrift für Angewandte Entomologie* 91, 327–348.

Viggiani, G. (1972) Richerche sugli Hymenoptera Chalcidoidea. XL. Observazioni morfo-bioloche sull'*Azotus pulcherrimus* Merc. *Bollettino del Laboratorio di Entomologia Agraria 'Filippo Silvestri' Portici* 30, 300–311.

Viggiani, G. (1991) Redescription of *Amitus fuscipennis* Macg. and Neb. (Hymenoptera: Platygastridae), exotic parasitoid of *Trialeurodes vaporariorum* (Westw.), with preliminary notes on its introduction into Italy. *Redia* 74, 177–184.

Viggiani, G. and Evans, G.A. (1992) Descriptions of three new species of *Amitus* Haldeman (Hymenoptera, Platygasteridae), parasitoids of known whiteflies from the New World. *Bollettino del Laboratorio di Entomologia Agraria 'Filippo Silvestri'* 49, 189–194.

Voegele, J. and Russo, J. (1981) *Trichogramma* spp. Vc. Discovery in Alsace of two new species of *Trichogramma, Trichogramma schuberti* and *T. rhenana* (Hym. Trichogrammatidae) in egg-masses of *Ostrinia nubilalis* Hubn., (Lepid. Pyralidae). *Annales de la Société Entomologique de France* 17, 535–541.

Walker, I. (1979) Some British species of *Anagrus*. *Zoological Journal of the Linnean Society* 67, 181–202.

Waloff, N. (1974) Biology and behavior of some species of Dryinidae. *Journal of Entomology* 49, 97–109.

Walter, G.H. (1983) 'Divergent male ontogenies' in Aphelinidae: a simplified classification and a suggested evolutionary sequence. *Biological Journal of the Linnean Society* 19, 63–82.

Wang, F. and Zhang, S. (1988) Studies on *Trichogramma pintoi*, its deuterotokous reproduction, artificial propagation and field releases. *Chinese Journal of Biological Control* 4, 149–151.

Wangberg, J.K. (1977) A new *Tetrastichus* parasitising tephritid gall-formers on *Chrysothamnus* in Idaho (Hymenoptera: Eulophidae). *Pan Pacific Entomologist* 53, 237–240.

Weinstein, P. and Austin, A.D. (1996) Thelytoky in *Taeniogonalos venatoria* Riek (Hymenoptera: Trigonalyidae), with notes on its distribution and first description of males. *Australian Journal of Entomology* 35, 81–84.

Werren, J.H., Zhang, W. and Guo, L.R. (1995) Evolution and phylogeny of *Wolbachia*: reproductive parasites of arthropods. *Proceedings of the Royal Society of London Series B Biological Sciences* 261, 55–63.

Wilson, F. and Woolcock, L.T. (1960a) Environmental determination of sex in a parthenogenetic parasite. *Nature* 186, 99–100.

Wilson, F. and Woolcock, L.T. (1960b) Temperature determination of sex in a parthenogenetic parasite, *Ooencyrtus submetallicus*. *Australian Journal of Zoology* 8, 153–169.

Woolley, J.B. (1984) Higher classification of the signiphoridae (Hymenoptera: Chalcidoidea) and revision of *Signiphora* Ashmead of the New World. PhD thesis, Department of Biological Control, University of California, Riverside.

Wright, E.J. (1978) Observations on the copulatory behaviour of *Perilitus coccinellae* (Hymenoptera: Braconidae). *Proceedings of the Entomological Society of Ontario* 109, 22.

Wysoki, M. and Izhar, Y. (1981) Biological data on *Apanteles cerialis* Nixon (Hymenoptera: Braconidae), a parasite of *Boarmia (Ascotis) selenaria* Schiff. (Lepidoptera; Geometridae). *Phytoparasitica* 9, 19–25.

Zchori-Fein, E., Roush, R.T. and Hunter, M.S. (1992) Male production induced by antibiotic treatment in *Encarsia formosa* (Hymenoptera: Aphelinidae), an asexual species. *Experientia* 48, 102–105.

Zchori-Fein, E., Faktor, O., Zeidan, M., Gottlieb, Y., Czosnek, H. and Rosen, D. (1995) Parthenogenesis-inducing microorganisms in *Aphytis* (Hymenoptera: Aphelinidae). *Insect Molecular Biology* 4, 173–178.

9 Comparison of Artificially vs. Naturally Reared Natural Enemies and Their Potential for Use in Biological Control

S. Grenier[1] and P. De Clercq[2]
[1]UMR INRA/INSA de Lyon, Biologie Fonctionnelle, Insectes et Interactions, Institut National des Sciences Appliquées, Bât. Pasteur, 20 av. A. Einstein, 69621 Villeurbanne Cedex, France; [2]Laboratory of Agrozoology, Department of Crop Protection, Faculty of Agricultural and Applied Biological Sciences, Ghent University, Coupure Links 653, B-9000 Ghent, Belgium

Abstract

Quality assessment of entomophagous arthropods used in augmentative biological control is one of the main concerns after their mass production. The quality-testing procedures for natural enemies reared on artificial diets largely remain to be defined. As a first approach, comparisons of some relevant parameters between *in vivo*- and *in vitro*-reared entomophages are presented in this chapter. Results from experiments with different kinds of diets with or without insect components, developed for parasitoids and predators, are examined. Morphological traits as well as development and reproduction parameters used for comparisons between *in vivo*- and *in vitro*-grown arthropods are discussed. Morphological characters include body size and weight and the occurrence of abnormalities. Immature development is assessed by measuring duration and survival rates of the different stages. Sex ratio and symbiont association, fecundity–fertility and longevity are compared as reproduction parameters. It is important to consider biochemical parameters, such as protein, lipid and carbohydrate content, for quality control. These parameters may also indicate the deficiency or excess in a particular nutritional component. Behavioural and genetic parameters are considered as well. The establishment of relationships between certain parameters, e.g. between body size and fecundity or longevity, may help in simplifying quality control procedures. The ultimate quality criterion for an artificially reared natural enemy is its capacity to reduce pest populations, which can be evaluated by measuring the predation efficiency or the parasitization rate under laboratory or field conditions.

Introduction

The main concerns for artificial mass culture of entomophagous insects, after production of effective natural enemies at acceptable cost, are the maintenance of insect quality and reliable methods for measuring quality. Relevant and robust quality control programmes are necessary for mass-produced natural enemies, particularly if they are reared on factitious or artificial diets (Sorati *et al.*, 1996). Cost-effective production of

© CAB International 2003. *Quality Control and Production of Biological Control Agents: Theory and Testing Procedures* (ed. J.C. van Lenteren)

high-quality natural enemies is a prerequisite for increasing the use of entomophagous insects for pest control. For biological control strategies, the ultimate quality criterion for a mass-produced natural enemy is excellent field performance against the target pest insect (Thompson and Hagen, 1999). However, very few field evaluations have been done for artificially reared beneficials. Egg parasitoids, especially *Trichogramma*, are used worldwide for biological control on many crops (van Lenteren, 2000). In China, *Trichogramma* spp. and *Anastatus* spp. produced at an industrial scale on artificial eggs have been released on thousands of hectares of different crops, with parasitism rates of usually more than 80% (Dai *et al.*, 1991; Liu *et al.*, 1995). In the USA, biocontrol companies have started to integrate artificial diets in their production process, and some predators are at least partially being reared on various artificial diets (P.D. Greany, Gainesville, USA, 2001, personal communication), but there are no reports on their field performance. Field assessments of artificially reared natural enemies are only useful, however, when basic parameters during the production process in the laboratory are strictly controlled and changes in the rearing procedure are carefully monitored.

Quality control of mass-produced entomophagous insects can be based on procedures developed earlier for non-entomophagous insects (Chapter 2; Leppla and Ashley, 1989; Leppla and Fischer, 1989; Williams and Leppla, 1992). For example, quality control during storage, packaging and shipment of all kinds of mass-reared insects could be similar. There are already quality control processes recommended for *in vivo* production that could be used as a guideline for *in vitro* production (Bigler, 1994). Comparative analyses of the reproductive attributes of commercially *in vivo*-produced *Trichogramma* species were described by Kuhlmann and Mills (1999). Both these latter authors and Liu and Smith (2000) chose fecundity, longevity, sex ratio and adult size as quality parameters. Parasitization rate is another key parameter tested for quality evaluation (Losey and Calvin, 1995).

For quality control of insects no absolute criteria exist, but different criteria could be defined in relation to the objectives for which they are produced. The degree of difference between insects produced *in vivo* and *in vitro* that we can accept will depend on the goals of the production (Moore *et al.*, 1985), in this case the pest-control efficiency to be obtained in the field. For *in vitro* rearing of entomophages we are only at the very first steps of developing quality control, except for a few species of parasitoids and some predators (Table 9.1). The main question will be: what kinds of parameters are reliable enough to be used in the assessment of quality? In this chapter we shall not develop a quality control scheme. Instead, we shall conduct *in vivo–in vitro* comparisons, which form the starting-point for establishment of quality control parameters for artificially reared entomophages.

Different Kinds of Artificial Diets for Parasitoids and Predators

It would be convenient if the different kinds of diets used in mass rearing could be typified by simple terms. Long ago some terms were used based on the presence or absence of complex components, but they were not very clearly defined: holidic media (chemical structure of all ingredients known), meridic media (holidic base to which at least one substance or preparation of unknown structure or uncertain purity is added) or oligidic media (crude organic materials). However, we are of the opinion that these distinctions are not very relevant, because only a complete description of the composition of a diet would be able to characterize it. Nevertheless, for practical considerations, a critical characteristic is the presence or the absence of insect components. Thus, considering that synthetic diets were supposed to replace the insect host or prey, it is worth distinguishing two main kinds of media: those including and those excluding insect components. Addition of insect materials implies the necessity to culture not only the host but often also the host's food plant, rendering entomophage

Table 9.1. Proposed parameters for quality control of parasitoids and predators produced under artificial conditions.

Morphological parameters
Size or weight of last larval instar/pupa/adult
Percentage abnormalities (deformation of wings/abdomen)
Development and reproduction parameters
Duration of egg/larval/pupal stage
Survival rate of egg/larval/pupal/adult stage
Sex ratio – presence/absence of symbionts influencing sex ratio
Fecundity/fertility
Duration of preoviposition/oviposition/postoviposition period
Longevity
Biochemical parameters
Protein, lipid, carbohydrate content
Hormone titre
Behavioural parameters
Predation or parasitization efficiency
Locomotion/flight activity
Host/prey localization capability
Genetic parameters
Genetic variability
Homozygosity rate

production more expensive. But we have to emphasize that in some parts of the world, especially in China and some other Asian countries and in Latin America, insect components, such as haemolymph, could be by-products of the silk industry and thus cheap and easy to obtain.

Diets with insect additives

Insect additives can be used in different ways. Sometimes almost the whole host contents are used as scarcely diluted extracts. The main elements used are whole-body tissue extracts or haemolymph from lepidopterous pupae in artificial diets for parasitoids. This is the case for larval parasitoids, such as the chalcidid *Brachymeria intermedia* (Dindo *et al.*, 1997), the ichneumonid *Diapetimorpha introita* (Ferkovich *et al.*, 1999, 2000) or the tachinid *Exorista larvarum* (Dindo *et al.*, 1999), and for oophagous parasitoids, such as *Trichogramma* spp. (for a review, see Grenier, 1994). Usually silkworm species (*Antheraea pernyi, Philosamia cynthia*) and easily reared insects like *Galleria mellonella* are used for these extracts.

Bee extracts or even whole pulverized bees or bee brood have been added in diets for coccinellid predators (Smirnoff, 1958; Niijima *et al.*, 1977, 1986).

Some diets for *Trichogramma* contain egg juice from a natural host (Consoli and Parra, 1996). For the egg parasitoid *Edovum puttleri* a homogenate of host eggs (Colorado potato beetle) was used (Hu *et al.*, 1998).

In hymenopterous parasitoids, teratocytes play various roles (Dahlman, 1990), mainly in the exploitation of the host by the parasitoid larva, through secretion of digestive enzymes attacking host tissues or proteins as food for the parasitoid larva (Falabella *et al.*, 2000). *In vitro*, cell products or cell cultures were also used in lieu of haemolymph or host factors (Grenier *et al.*, 1994).

Diets devoid of insect components

Very few diets are chemically defined. The first defined diet concerning a true parasitoid species was that for *Itoplectis conquisitor* (Yazgan, 1972). Diets whose entire chemical composition is known, even if the structure of some components is not fully defined (nucleic acids, proteins), can be considered as chemically defined. A small number of diets that fit such a definition were tested successfully for rearing entomophagous insects (Grenier *et al.*, 1994). In such diets, many complex or 'crude' components can be added as host substitutes. Irrespective of the species reared, whether parasitoids or predators, the most commonly used components are hen's egg yolk, chicken embryo extract, calf fetal serum, bovine serum, cow's milk, yeast extract or hydrolysate, crude proteins or protein hydrolysates, meat or liver extracts and seed oils. For recent reviews of such diets, see Thompson (1999) and Thompson and Hagen (1999).

Success in Development of Some Species in Artificial Conditions

The main successes in artificial mass rearing have been obtained with hymenopterous egg and pupal parasitoids, tachinid larval parasitoids and some polyphagous predators. Extensive general reviews of artificial diets for entomophagous arthropods have been published by Grenier *et al.* (1994), Thompson (1999) and Thompson and Hagen (1999).

Koinobiontic endoparasitic Hymenoptera (parasitoids that do not immediately kill their hosts and where the parasitoid larvae develop in the still living host) are the most difficult species to rear *in vitro* because the parasitoid has a close relationship with its living host, which probably supplies the parasitoid with some specific growth factors necessary for normal development of the parasitoid larva (Greany *et al.*, 1989). Moreover, endoparasitoids, for which the diet is not only their food but also their environment for larval development, have special requirements compared with ectoparasitoids or predators. Thus, special attention has to be paid to factors such as osmotic pressure and pH (Grenier *et al.*, 1994).

Comparison of Artificially and Naturally Reared Natural Enemies

Many parameters used as quality criteria are linked, such as adult body weight and longevity, fecundity, flight activity and searching ability (Kazmer and Luck, 1995). Quality control procedures could be simplified and could thus be made less costly if we were able to use one parameter that is easily measured (e.g. size) to predict the value of another trait that is more complex or time-consuming to determine (e.g. fecundity or field performance). In parasitoids, body size may be related to fecundity, longevity, rate of search and flight ability (Kazmer and Luck, 1995). Bigler (1994) pointed out that the female body size of a parasitoid could be used as an index of fitness or a quality parameter, as in *Trichogramma*. But female size is not a reliable parameter for predicting field performance when the parasitoids are reared on factitious or artificial hosts. In *Trichogramma*, large-sized wasps developed from *in vitro* rearing showed characteristic abnormalities called 'big belly'. Despite their large body size, such adults usually have a low viability. The size of a normally shaped *Trichogramma* adult produced *in vitro* is also larger than that of a wasp that developed in the natural host (Nordlund *et al.*, 1997). This is often found in oophagous parasitoids and is the result of a low number of parasitoid eggs developing in the large amount of food that is available to them (Grenier *et al.*, 1995).

In general, the size of *Trichogramma* and other oophagous parasitoids varies according to the number of adults developing in the same host, which consume all the available host material. Remains of the host prevent proper pupation of parasitoids, and parasitoid larvae that are excessively large cannot pupate. In a natural situation with a great many *Trichogramma* larvae in one host, adult parasitoid size will be reduced accordingly. Under artificial rearing conditions, however, the quantity of food in the artificial host egg is usually very large compared with a natural host egg, and the number of parasitoid eggs laid is often too low for the development of normal-sized *Trichogramma* (Grenier *et al.*, 2001).

All parameters related to reproduction are important, and sometimes reproduction capacity can be estimated by a simple measurement, such as the body size of the parasitoid, as in *Encarsia formosa* (van Lenteren, 1999). In predators as well, body size is often believed to be a good predictive index of fecundity, but the relationship between the two parameters is not always clear. For instance, females of the predatory pentatomid *Podisus maculiventris* reared on an artificial diet were significantly smaller than those fed larvae of *Tenebrio molitor*, but their fecundities were similar (De Clercq *et al.*, 1998a). Rojas *et al.* (2000) obtained females of *Perillus bioculatus* on artificial diet with similar size to that of those offered *Leptinotarsa decemlineata* larvae, but their fecundity was only 10% of that of prey-fed controls. Establishing a relationship between the size and predation capacity of a laboratory-produced predator has been shown to be even more problematic, even when it is produced on live prey (e.g. De Clercq *et al.*, 1998b). Cohen (2000a) reported that *Geocoris punctipes* reared for over 6 years on artificial diet were significantly smaller than feral specimens but had similar predation capacities. Chocorosqui and De Clercq (1999) found that, despite their smaller size, artificially reared nymphs of *P. maculiventris* even showed significantly greater predation rates than prey-fed controls.

Several morphological traits and developmental and reproductive parameters that have been used to assess the quality of artificially reared parasitoids and predators will be reviewed below. For clarity we shall treat the most widely used parameters separately, but, where applicable, links between different traits will be discussed.

Morphological parameters

Size or weight of different developmental stages

Size or weight of eggs deposited by *in vitro*-reared natural enemies has only rarely been monitored. Fresh weights of eggs laid by females of the heteropteran predators *G. punctipes* and *P. maculiventris* cultured on bovine meat diets were similar to those of eggs from females maintained on insect prey (Cohen, 1985; De Clercq and Degheele, 1992).

Only occasionally have workers monitored fresh weights of larvae during their development on artificial diets. Nevertheless, such studies may reveal where nutritional deficiencies are most crucial. Larval weights of predators fed on artificial diets have often been found to be inferior to those of their prey-fed peers (Hagen and Tassan, 1965; Hussein and Hagen, 1991; Chocorosqui and De Clercq, 1999; Wittmeyer and Coudron, 2001).

Puparial weight was surprisingly higher in the tachinid fly *E. larvarum* reared *in vitro* on artificial diets than *in vivo* in *G. mellonella* (Dindo *et al.*, 1999). Availability of food *ad libitum* in artificial rearing might explain this result. Vanderzant (1969) also reported greater pupal weights for artificially reared lacewings, *Chrysoperla carnea*, compared with controls fed *Sitotroga cerealella* eggs. Hassan and Hagen (1978) succeeded in rearing larvae of *C. carnea* on an artificial diet and obtained pupae with similar weights to those in controls fed moth eggs. Likewise, no difference was observed between pupal weight of *Chrysoperla rufilabris* fed a meat-based artificial diet and those fed *Ephestia kuehniella* eggs (Cohen and Smith, 1998). Pupal weights of the colydiid beetle *Dastarcus helophoroides* reared on artificial diet were not different from those of predators reared on live prey (Ogura *et al.*, 1999), but pupal weights of *in vitro*-reared *Trogossita japonica* were only half of those of beetles reared *in vivo* (Ogura and Hosada, 1995).

In both parasitoids and predators, adult body weight and size have often been used to evaluate the effectiveness of a diet (see above). Adult weights of *D. introita* developed *in vitro* were lower than those of individuals fed host pupae (*Spodoptera frugiperda*), and they were not significantly improved by adding some conditioned tissue-culture media (Ferkovich *et al.*, 1999), but they were improved when the diet was supplemented with certain host lipid extracts (Ferkovich *et al.*, 2000). Usually larger hosts produce larger wasps showing a greater egg load and a higher longevity, as described in the encyrtid *Metaphycus* sp. (Bernal *et al.*, 1999). As in many insects, puparial weight in tachinid flies is closely correlated with adult weight (Grenier and Bonnot, 1983). As a consequence, differences in pupal weights between *in vitro*- and *in vivo*-reared *E. larvarum* are reflected in the adult stage (see also below).

In many cases, adults of predatory insects from various orders reared on artificial food are smaller than those obtained on insect prey (e.g. Racioppi *et al.*, 1981; Hussein and Hagen, 1991; De Clercq and Degheele, 1992; Hattingh and Samways, 1993; Ogura and Hosada, 1995; Zanuncio *et al.*, 1996; De Clercq *et al.*, 1998a; Cohen, 2000a). However, Rojas *et al.* (2000) and Arijs and De Clercq (2001) obtained female adults of the predatory bugs *P. bioculatus* and *Orius laevigatus*, respectively, with equal weights to those of predators maintained on insect prey.

Abnormalities

Abnormalities observed in artificial rearing are usually deformations of wings (expanded or not) and of the abdomen. The percentage of deformed females, i.e. mainly unexpanded wings, is a criterion for measuring abnormalities in *Trichogramma* (Cerutti and Bigler, 1995). Bloated larvae with unexpanded wings are produced *in vitro* due to too few eggs per artificial host egg (Grenier *et al.*, 1995). Wing and abdominal deformations occurred in *Trichogramma* reared on artificial diet, but no differences in the morphology of their genital apparatus were detected (Consoli and Parra, 1996). The same wing and body anomalies were recorded after *Trichogramma* was produced for ten generations *in vitro* (Nordlund *et al.*, 1997). Suboptimal diets have also been reported to cause deformation of wings and abnormal pigmentation in some predatory coccinellids (Atallah and Newsom, 1966; Racioppi *et al.*, 1981).

Development and reproduction parameters

Duration of different developmental stages

Many workers have assessed developmental rates of immature stages to evaluate the nutritional value of a diet. Usually the immature development of parasitoids (from egg or first-instar larva to adult emergence) is longer under artificial conditions than in natural hosts. This was observed in the tachinid fly *E. larvarum* (Dindo *et al.*, 1999) and in the ichneumonid *D. introita* (Greany and Carpenter, 1998; Ferkovich *et al.*, 1999). The development from egg to adult on artificial diets was delayed for *Trichogramma pretiosum* compared with natural hosts, but no differences were observed for *Trichogramma galloi* (Consoli and Parra, 1996). The development of the egg stage was not very often tested, but for *Trichogramma australicum* it took longer in artificial diet vs. natural-host eggs (Dahlan and Gordh, 1998). In *T. australicum* also the larval stages develop more slowly in artificial diet vs. natural-host eggs (Dahlan and Gordh, 1998). Total development time is not different in *Bracon mellitor* and *Catolaccus grandis* reared *in vivo* vs. *in vitro* (Guerra *et al.*, 1993; Guerra and Martinez, 1994). Usually, when the whole preimaginal development is longer, only the larval development is delayed, the embryonic and pupal development being less dependent on the diet composition.

Likewise, developmental times in predators are usually longer on artificial diets. Some workers have even reported total developmental periods two to three times longer than those observed in controls on natural food (Hagen and Tassan, 1965; Vanderzant, 1969; Racioppi *et al.*, 1981; Hussein and Hagen, 1991). Whereas in a number of studies differences in developmental times were consistent across all larval instars, differences in some species became more important in later instars (Racioppi *et al.*, 1981; Zanuncio *et al.*, 1996; Wittmeyer and Coudron, 2001). Because Wittmeyer and Coudron (2001) found differences in developmental durations between *in vitro*- and *in vivo*-reared *P. maculiventris* to be most marked in the fifth (final) nymphal stage, they suggested selecting this stage to evaluate the suitability of a diet. Similar developmental periods for artificially and naturally reared predators have occasionally been reported, e.g. by Grenier *et al.* (1989) for *Macrolophus caliginosus*, by De Clercq *et al.* (1998a) for individually reared *P. maculiventris* and by Arijs and De Clercq (2001) for *O. laevigatus*.

Survival rates of the different developmental stages

The survival rates of both *T. pretiosum* and *T. galloi* reared on artificial diets are lower than those of individuals reared on natural hosts (Consoli and Parra, 1996). The larval density could be critical in influencing general quality parameters, such as size, or survival in the case of cannibalism. Oophagous parasitoids cannot regulate the quantity of food ingested, so the number of parasitoids developing per unit of food is a key parameter for development success (see above; Grenier *et al.*, 1995; Dahlan and Gordh, 1997).

Although rearing on artificial food has variable effects on the fecundity of insect predators, hatchability of eggs produced by artificially reared individuals is generally similar to that of conspecifics fed insect prey (Hassan and Hagen, 1978; De Clercq and Degheele, 1992; Adams, 2000b; Arijs and De Clercq, 2001). Only Rojas *et al.* (2000) and Wittmeyer *et al.* (2001) have reported lower viability of eggs laid by females of *P. bioculatus* and *P. maculiventris*, respectively, on artificial diet. When assessing egg viability of predators, however, one should take cannibalistic activities of adults into account. Both Cohen (1985) and De Clercq and Degheele (1992) have noted egg cannibalism by diet-fed adults of *G. punctipes* and *Podisus* spp., respectively. The degree of egg cannibalism will evidently depend on adult rearing density, but in *Podisus* even females reared in isolation have been seen to feed on their own eggs.

Nymphal or larval survival was compared among diets in a number of predatory insects and results often depended upon predators being reared individually or in groups. In *Podisus* spp. survival of nymphs fed either artificial diet or live food was equally good when the predators were reared individually (> 90%), but when they were reared in groups cannibalistic activities reduced nymphal survival to 55–75% (De Clercq and Degheele, 1992; De Clercq *et al.*, 1998a). Hassan and Hagen (1978) and Cohen and Smith (1998) reported survival rates of *Chrysoperla* spp. that were similar on artificial diet and on flour-moth eggs. In both of these studies, however, larvae were reared individually. Despite individual rearing of *P. bioculatus* nymphs in multiwell tissue-culture plates, nymphal survival on artificial food was inferior to that on frozen Colorado potato-beetle eggs (Rojas *et al.*, 2000). Group rearing of *O. laevigatus* on a number of meat-based diets yielded similar to slightly lower survival rates compared with controls fed *E. kuehniella* eggs (Arijs and De Clercq, 2001).

Percentages of cocoons of *D. introita* developed *in vitro* were not different from those obtained on pupae of *S. frugiperda* (Greany and Carpenter, 1998), and could not be increased by adding some conditioned tissue-culture media (Ferkovich *et al.*, 1999).

The emergence rate of adults is a quality control criterion used for *in vivo* comparison in *Trichogramma* (Cerutti and Bigler, 1995). The emergence rate of both *Trichogramma dendrolimi* and *Trichogramma chilonis* reared *in vitro* is 90% of that reared *in vivo* (Feng *et al.*, 1999). The percentage of adult emergence of the tachinid fly *E. larvarum* obtained either *in vivo* (in *G. mellonella*) or *in vitro* is quite similar (Dindo *et al.*, 1999).

Emergence rates of *T. dendrolimi* are lower in artificial diets without insect components than in diets with haemolymph, but results are variable according to the strain tested (Grenier *et al.*, 1995). Emergence rates of *D. introita* on an artificial diet developed by Greany and Carpenter (1998) were similar to those on host pupae (Carpenter and Greany, 1998). Supplementing the diet with conditioned tissue-culture media or certain host lipid extracts did not improve emergence (Ferkovich *et al.*, 1999, 2000).

Sex ratio and association with symbionts

The sex ratio is a quality criterion used in *Trichogramma* (Cerutti and Bigler, 1995). *Wolbachia* is a well-known Rickettsiaceae that strongly interferes with 'normal' sex-ratio determination in many arthropod species (Werren, 1997; Stouthamer *et al.*, 1999). For entomophages the main effects are cytoplasmic incompatibility and thelytokous parthenogenesis in Hymenoptera and male killing in Coccinellidae. Contrary to some statements (Consoli and Parra,

1999), the use of antibiotics in artificial diets does not always prevent the production of entomophagous insects harbouring symbionts belonging to alpha Proteobacteria (Rickettsiaceae). The antibiotics used in diets for preventing bacterial contamination, mainly by gamma Proteobacteria, are not effective in removing endosymbiotic *Wolbachia* from infected strains (Stouthamer, 1990). In fact, penicillin and streptomycin are used for bacterial control and tetracycline and rifampicin for *Wolbachia* elimination. Pure female lines with a *Wolbachia* infection could be an advantage for biological control, because only females kill pest insects. But *Wolbachia* infections may modify the fecundities (see below and Chapter 8). The presence or absence of a symbiont could be a key quality criterion for some insects. Symbionts transferred from one species to another could be used to improve the quality of entomophages released for biological control (Grenier *et al.*, 1998), but for a full discussion of this topic see also Chapter 8.

In general, sex ratios of adult predators obtained on artificial diets are not different from those observed in prey-fed controls.

Fecundity and fertility

Fecundity is influenced by female size (see above) and also by the quality and quantity of the adult food (Bernal *et al.*, 1999). The fecundity of *Trichogramma* females produced on *E. kuehniella* eggs is proposed as a quality control parameter (Dutton *et al.*, 1996). *In vitro*-reared females of *T. australicum* produced significantly more progeny and parasitized more host eggs than the females reared on natural or factitious hosts (Nurindah *et al.*, 1997). *Wolbachia* symbionts inducing thelytokous reproduction could also modify the fecundity in *T. pretiosum* even when grown in artificial diet (Grenier *et al.*, 2002). The fecundity of the tachinid fly *E. larvarum* is slightly, but not significantly, higher in individuals reared in artificial diets than *in vivo* (Dindo *et al.*, 1999). *D. introita* females laid significantly fewer eggs when reared on artificial diet than when reared on live hosts (Greany and Carpenter, 1998).

Although several diets have been developed for predatory insects that have yielded only slightly inferior to similar developmental success, most often the fecundity of adults maintained on such diets is clearly inferior to that on insect prey. Fecundity appears to be less of an issue when the adult stage does not feed at all, only does so to a limited extent or has different feeding habits from those of the larval stage (e.g. honeydew feeders). For instance, artificial larval diets for *Chrysoperla* spp. have been reported to yield adults with similar reproductive capacities to those obtained on flour-moth eggs (Hassan and Hagen, 1978; Cohen and Smith, 1998). However, adult lacewings can be maintained on a much simpler diet than that for larval feeding, usually consisting of water, sugar and yeast. Only rarely have artificial diets yielded similar total fecundities when the adult stage also has to be maintained on artificial food.

Arijs and De Clercq (2001) found that females of the anthocorid *O. laevigatus* fed different bovine meat and liver diets had a similar fecundity to those fed *E. kuehniella* eggs (120–200 eggs per female). Similar results were reported for the mirid *Dicyphus tamaninii* (Iriarte and Castañe, 2001). De Clercq *et al.* (1998a) reported that the fecundity of *P. maculiventris* on a meat diet (*c.* 550 eggs per female) was similar to that on *T. molitor* larvae (*c.* 460 eggs per female), but lower than that on *G. mellonella* larvae (*c.* 1000 eggs per female). The fecundity of other arthropod predators reared on artificial diets is reportedly half (Kennett and Hamai, 1980; Abou-Awad *et al.*, 1992; De Clercq and Degheele, 1992) to less than one-tenth (Kennett and Hamai, 1980; Chain-Ing *et al.*, 1993; Hattingh and Samways, 1993; Rojas *et al.*, 2000) of that of predators fed natural or factitious food. Very few workers have calculated demographic statistics to evaluate an insect diet. Wittmeyer and Coudron (2001) estimated life–fertility-table parameters for *P. maculiventris* reared on artificial diet or on *Trichoplusia ni* larvae. Net reproductive rate (R_O) and intrinsic rate of increase (r_m) were significantly lower, even when the predator was only partially reared on the artificial food.

Determining growth factors responsible for the full expression of the reproductive capacities of arthropod predators remains a major challenge. Results from recent studies on the pentatomid predator *P. maculiventris* suggest that adult nutrition is more critical for fecundity than the diet of the immature stages. In so called 'rescue' experiments, *P. maculiventris* fed artificial diet as nymphs but transferred to live prey upon adult emergence showed normal fecundity. Conversely, nymphs fed prey and given the diet as adults exhibited reduced fecundity (P.D. Greany, unpublished data). However, Wittmeyer and Coudron (2001) and Wittmeyer *et al.* (2001) reported for the same species that switching to live food in the adult stage still yielded lower fecundity compared with predators reared on insect prey in both nymphal and adult stages. Adams (2000b) and Wittmeyer *et al.* (2001) showed that predatory pentatomids fed on artificial diet do not have impaired ovipositional abilities but lay fewer eggs because they produce fewer mature follicles than prey-fed controls.

Finally, measurement of fecundity is a tedious and time-consuming activity if one has to collect eggs over a long span of time. Adams (2000a) and Wittmeyer *et al.* (2001), and Shapiro *et al.* (2000) proposed methods to reduce assay time by looking at first-cycle oocyte development or at vitellin levels, respectively. Further, instead of testing fecundity during the entire female life, it could be evaluated during a shorter period of time, but this may not always be very reliable, as observed *in vivo* with different *Trichogramma* spp. (Silva, 1999; Grenier *et al.*, 2002).

Longevity

The lifespan of mated females is a quality criterion used in *Trichogramma* for *in vivo* rearing in different conditions (Cerutti and Bigler, 1995; Dutton *et al.*, 1996). Longevity is reduced in *T. pretiosum* and *T. galloi* reared *in vitro* compared with parasitoids reared in natural hosts (Consoli and Parra, 1996). On the other hand, longevity is similar for the tachinid fly *E. larvarum* obtained either *in vivo* (on *G. mellonella*) or *in vitro* on artificial diets (Dindo *et al.*, 1999). No difference in longevity of diet-reared and host-reared *D. introita* was noted by Greany and Carpenter (1998).

The longevity of female *P. maculiventris* was not affected by diet (De Clercq and Degheele, 1992; De Clercq *et al.*, 1998a). Whereas Adams (2000b) did not observe a difference in longevity between diet-fed and prey-fed females of *P. bioculatus*, a difference was reported by Rojas *et al.* (2000).

Finally, it should be pointed out that the relationships between longevity (or fecundity) and the kind of food supplied (natural vs. artificial) in parasitoids is completely different from that in most predators, because the adults of the former group are non-carnivorous, even if they may occasionally show some host-feeding. It is also necessary to separate pro- and synovigenic parasitoid species for their more or less close dependence on their larval food.

Biochemical parameters

The composition of key components in living organisms, such as proteins, lipids, carbohydrates, enzymes and hormones, might be a good criterion for estimating the quality of artificially reared natural enemies.

Proteins

Proteins are key constituents of living organisms. Total amino acid content is a good indication of the viability of *in vitro*-reared insects, but free amino acid level is a better criterion for detecting unbalanced food. Total amino acid content of *T. dendrolimi* grown in artificial diet is strongly improved and reaches nearly the level of that of parasitoids grown *in vivo* when proteins were added in the diet. A better composition in amino acids is correlated with improvements of biological parameters. Free amino acid levels for *T. dendrolimi* grown *in vitro*, especially on an artificial diet with haemolymph, revealed high excess in some amino acids, notably serine (Grenier *et al.*, 1995). The total amino acid content of *M. caliginosus* reared on artificial diets showed less than 20% variation for each amino acid compared with control bugs reared on natural prey. The *in vitro*-grown bugs developed normally into fecund adults (Grenier *et al.*, 1989).

Salivary protein profiles of *P. maculiventris* reared on artificial diet were not different from those reared on caterpillar prey (Habibi *et al.*, 2001).

Some proteins are directly related to reproduction capacity, such as vitellogenin and vitellin. In *P. maculiventris* and *Orius insidiosus* vitellin content is determined in haemolymph samples or whole-body homogenates to quickly estimate the influence of diet on fecundity (Shapiro *et al.*, 2000; J.P. Shapiro and S.M. Ferkovich, unpublished data). Vitellus content is a good parameter for estimating fitness and discovering problems of quality in mass-reared insects also in *Lygus hesperus* and *C. rufilabris* (Cohen, 2000b).

Lipids and carbohydrates

Most of the energy sources are constituted by lipids (mainly triglycerides) and carbohydrates (glycogen, for example). Triglycerides are reserve compounds that can be synthesized from carbohydrates. The concentrations in total fatty acids in the tachinid parasitoid *Phryxe caudata* were quite similar when reared on different host species compared with artificial diet, but the detailed patterns are very different according to the food. Tachinid larvae are able to modify the host pattern towards their own pattern, but the parasitoid pattern is not necessarily reached in all situations (Delobel and Pageaux, 1981). *Exeristes roborator* reared *in vitro* contain half the total lipids of those reared on the host insect, mainly due to a lack of triglyceride synthesis and deposition in parasitoids reared on artificial diets (Thompson and Johnson, 1978). The total amount of fatty acids as neutral lipids as well as phospholipids could be a good estimation of the quality of a parasitoid.

Hormones

Ecdysteroids added to an artificial diet, mainly as 20-hydroxyecdysone, promote the growth and development of different parasitoid larvae (Grenier, 1987; Fanti, 1990; Nakahara *et al.*, 2000). In *D. introita*, the ecdysteroid titre in the haemolymph of host-reared parasitoids was higher than in diet-reared parasitoids. The insufficient hormonal level in diet-reared parasitoids could in part be responsible for mortality and lower percentages of pupation and emergence *in vitro* (Gelman *et al.*, 2000). On the other hand, the presence of juvenile hormones exerts no beneficial effect and may even have a negative influence (including the inhibition of moulting) on the development of different larval parasitoids (Beckage, 1985). Juvenile-hormone mimetics, such as fenoxycarb, may have strong negative effects, such as a delay in or inhibition of moulting, morphological/physiological abnormalities and even death, on many entomophagous insects (Grenier and Grenier, 1993).

Behaviour

The effectiveness of entomophages produced in or on artificial diets may be reduced when they are directly released in the field, but artificially reared natural enemies may be readapted to their natural host or prey through a process of learning (see also Chapters 3 and 4). For instance, exposure of the newly emerging adults of *C. grandis* to its natural host will 'train' female wasps and stimulate their egg production (Rojas *et al.*, 1999). Moreover, methods of artificial rearing allowing the incorporation of semiochemicals also open possibilities for preimaginal conditioning of parasitoids and predators (Greany *et al.*, 1984; Grenier, 1994; De Clercq *et al.*, 1998a).

Predation efficiency

Predation rates and food preferences have been measured for a number of predators reared for successive generations on artificial diets. After 60 generations and over 6 years on a bovine meat diet, there were no or only slight differences in predatory capabilities, as measured by handling times and consumption rates, between domesticated *G. punctipes* and feral predators (Cohen, 2000a; Hagler, 2000). In paired-choice experiments, *in vitro*-reared *G. punctipes* showed similar prey preferences to those of their wild counterparts

(Hagler and Cohen, 1991). When *P. maculiventris* and *Podisus sagitta* (*Podisus nigrispinus*) were returned to a diet of live prey after more than 15 generations on artificial food, the predators did not show any difference in their predation capacity on *Spodoptera exigua* larvae and had similar developmental and reproductive rates to those of predators maintained on live prey (De Clercq and Degheele, 1993). After three to four generations on a meat-based artificial diet, *P. nigrispinus* nymphs and adults exhibited prey-capture abilities comparable to those reared on housefly larvae, in both laboratory and greenhouse assays (Saavedra *et al.*, 1997). Some workers have even reported greater predation rates for diet-reared predators than for their prey-fed peers. Chocorosqui and De Clercq (1999) found that, despite their smaller size, nymphs of *P. maculiventris* reared on artificial diet had killed significantly more *S. exigua* larvae on sweet-pepper plants after 24 h than had control insects fed live prey. After 96 h, however, this difference had disappeared. Greater aggressiveness of diet-reared predators has also been noted for nymphs of *P. nigrispinus* by Saavedra *et al.* (1997) and for adults of *D. tamaninii* by Castañe *et al.* (2002). It is worth noting that the above predation studies were mostly done using simple laboratory arenas, in some cases not even involving host plants. Given the complexity of the field situation, predation rates measured under unrealistic laboratory conditions should not be extrapolated to the field. In this respect, more research is needed to evaluate the influence of tritrophic factors on the searching abilities of predators (and parasitoids) reared on unnatural foods.

Parasitization rate (see also above)

The parasitization rates of *T. dendrolimi* and *T. chilonis* reared *in vitro* are similar, 65 and 68%, respectively (Feng *et al.*, 1999), but these rates are lower in artificially than in naturally reared *T. pretiosum* and *T. galloi* (Consoli and Parra, 1996), as well as in *Anastatus japonicus* (Han *et al.*, 1988). After ten generations cultured *in vitro*, *Trichogramma minutum* females parasitize more host eggs than *in vivo*-reared insects (Nordlund *et al.*, 1997). Under field conditions, mortality induced by *in vitro*- and *in vivo*-reared *C. grandis* are similar (Morales-Ramos *et al.*, 1998). Field efficiency is usually evaluated by the number of hosts parasitized, but it is also necessary to include host stinging and host feeding, which could be an important mortality factor in specific host–parasitoid complexes, such as in *Trichogramma* spp. (Shipp *et al.*, 1998). Nevertheless, parasitization rates obtained in the field are also closely related to the conditions in which the releases were conducted (Bigler, 1994).

Host localization, walking and flying

Flight activity and walking behaviour (walking speed) were tested to compare different strains of *Trichogramma* developed *in vivo* (Cerutti and Bigler, 1995; Dutton and Bigler, 1995; Dutton *et al.*, 1996). Dispersal activity and host localization are a key parameter for *Trichogramma* quality, which could be tested in a special chamber or maze (Honda *et al.*, 1999). These parameters still need to be tested for *in vitro*-produced insects.

Genetic parameters

Long-term rearing of *Lixophaga diatraeae* in the laboratory on the alternative host *G. mellonella* induced modifications of some biological characteristics, such as puparial size and developmental durations (Pintureau and Grenier, 1992) and of the capability to develop on artificial diet (Grenier and Pintureau, 1991). Long-term rearing under laboratory conditions, especially when using artificial diets, could lead to genetic bottleneck effects and impose severe selection pressures on the population of an entomophagous arthropod. The concomitant loss of genetic variability is often believed to lead to a loss of quality in the natural enemy. In sensitive species, strong genetic (and non-genetic) effects are expected to arise during the very first generations grown in artificial conditions. The size and heterogeneity of the founder population usually declines during the colony initiation, as described in fruit flies by Leppla (1989) (see also Chapters 6 and 7). In particular, elevated selection pressure associated with the use of unnatural food implies a greater risk that unadapted genotypes (i.e. those that will be inefficient in controlling a pest in the field) will eventually dominate the laboratory population. Unfortunately, little effort has been made to quantify genetic variability in entomophages bred on artificial media, e.g. by allozymic analyses or by direct analysis of DNA polymorphisms via PCR-based techniques.

Further, inbreeding can often increase homozygosity and may thus increase the risk that recessive deleterious alleles are expressed. In species where there is no inbreeding depression, crosses among different inbred strains of the same species could be a useful tool for minimizing the effects of laboratory adaptation, as in *Trichogramma* nr. *brassicae*. The persistence of genetic variation for several traits within this species indicates heritable variation, which could be manipulated, when necessary, to optimize the efficiency of the strain used in biological control. Thus, inbreeding might provide a way to increase the useful life of a strain in mass-release programmes (Sorati *et al.*, 1996). Given the lack of inbreeding depression even after many generations of inbreeding, the same was also believed to be true for the predator *P. maculiventris* produced *in vivo* (De Clercq *et al.*, 1998b). Conversely, outbreeding (or heterosis), as in the crossing of *Cotesia flavipes* strains from different origins, may improve the performance of a parasitoid, including female sex ratio, rate of development and size of females (Gu and Dorn, 2000).

Conclusions

Tests for quality comparisons between natural enemies that were reared artificially or on their natural host were mainly conducted on the first generations after *in vitro* culture, but on rare occasions effects of continuous culture for several generations have been tested (e.g. Hassan and Hagen, 1978; Gao *et al.*, 1982; De Clercq and Degheele, 1992, 1993; Nordlund *et al.*, 1997; Cohen, 2000a). We suggest that it may not be advisable to maintain entomophagous insects on synthetic diets for many generations, because they may suffer from non-intentional selection, inducing a reduction in genetic variability and finally a deterioration in performance. On the other hand, the frequent introduction of new strains to initiate *in vitro* mass production could generate inconveniences, such as the necessity for a few generations of laboratory adaptation, the risk of misidentification of the introduced strain or species and the danger of introducing pathogens or hyperparasitoids (see Chapters 1 and 10).

The ultimate test for quality of entomophagous insects is the assessment of their field efficiency measured as the rate of parasitism or predation. However, besides being expensive and time-consuming, the complexity of a field setting may obscure the actual causes for the failure or success of natural-enemy releases. Therefore, the first assessment of the quality of an *in vitro*- or *in vivo*-produced beneficial will preferably be done in a laboratory setting.

Currently, quality control of *in vitro*-reared entomophagous insects has been done for the major part only by comparing selected characteristics between *in vitro*- and *in vivo*-grown insects in the laboratory. Obviously, such comparisons should be done in a fair way, with artificial diets being compared with the best natural-rearing protocols. Further, it is important to try to define which parameters should be considered as key criteria to be tested in a first quality assessment of entomophages. Fecundity and the rate of parasitization in parasitoids and the predation capacity in predators are probably the most relevant criteria for estimating the ultimate quality of a natural enemy (Table 9.1).

At the laboratory level, however, such biological parameters could be associated with biochemical parameters, as we demonstrated above. We believe that it is worthwhile assessing these biochemical parameters because, contrary to biological traits, they can be used to suggest modifications of the *in vitro*-rearing system, eventually leading to an improvement of the insects produced. Excess or deficiencies of some elements could be balanced by deletion or supplementation of nutritional components in the diet, based on a better understanding of the nutritional physiology of an insect. One could say that the insect protein content as a structural element mainly reflects the identity of the species, and the carbohydrate/lipid content, as an energy reserve, gives an indication of its life potential or fitness.

Acknowledgements

We thank Norman Leppla, Patrick Greany, Steve Ferkovich, Jeff Shapiro and Joop van Lenteren for their helpful suggestions.

References

Abou-Awad, B.A., Reda, A.S. and Elsawi, S.A. (1992) Effects of artificial and natural diets on the development and reproduction of two phtyoseiid mites *Amblyseius gossipi* and *Amblyseius swirskii* (Acari: Phytoseiidae). *Insect Science and its Application* 13, 441–445.

Adams, T.S. (2000a) Effect of diet and mating status on ovarian development in a predacious stink bug *Perillus bioculatus* (Hemiptera: Pentatomidae). *Annals of the Entomological Society of America* 93, 529–535.

Adams, T.S. (2000b) Effect of diet and mating on oviposition in the twospotted stink bug *Perillus bioculatus* (F.) (Heteroptera : Pentatomidae). *Annals of the Entomological Society of America* 93, 1288–1293.

Arijs, Y. and De Clercq, P. (2001) Factitious prey and artificial diets for rearing *Orius laevigatus*. *53rd International Symposium on Crop Protection*, Ghent, Belgium May 8, Abstract book p. 37.

Atallah, Y.H. and Newsom, L.D. (1966) Ecological and nutritional studies on *Coleomegilla maculata* De Geer (Coleoptera: Coccinellidae). I. The development of an artificial diet and a laboratory rearing technique. *Journal of Economic Entomology* 59, 1173–1179.

Beckage, N.E. (1985) Endocrine interactions between endoparasitic insects and their hosts. *Annual Review of Entomology* 30, 371–413.

Bernal, J.S., Luck, R.F. and Morse, J.G. (1999) Host influences on sex ratio, longevity, and egg load of two *Metaphycus* species parasitic on soft scales: implications for insectary rearing. *Entomologia Experimentalis et Applicata* 92, 191–204.

Bigler, F. (1994) Quality control in *Trichogramma* production. In: Wajnberg, E. and Hassan, S.A. (eds) *Biological Control with Egg Parasitoids*. CAB International, Wallingford, UK, pp. 93–111.

Carpenter, J.E. and Greany, P.D. (1998) Comparative development and performance of artificially reared versus host-reared *Diapetimorpha introita* (Cresson) (Hymenoptera: Ichneumonidae) wasps. *Biological Control* 11, 203–208.

Castañe, C., Iriarte, J. and Lucas, E. (2002) Comparison of prey consumption by *Dicyphus tamaninii* reared conventionally, and on a meat-based diet. *BioControl* 47, 657–666.

Cerutti, F. and Bigler, F. (1995) Quality assessment of *Trichogramma brassicae* in the laboratory. *Entomologia Experimentalis et Applicata* 75, 19–26.

Chain-Ing, T.S., Chang, H.Y., Hsu, P.H. and Hwang, Y.F. (1993) Responses of *Amblyseius ovalis* (Evans) (Acarina: Phytoseiidae) to natural food resources and two artificial diets. *Experimental and Applied Acarology* 17, 503–519.

Chocorosqui, V.R. and De Clercq, P. (1999) Developmental and predatory performance of *Podisus maculiventris* (Say) (Heteroptera: Pentatomidae) reared on a meat-based diet. *Mededelingen Faculteit Landbouwkundige en Toepgepaste Biologische Wetenschappen Universiteit Gent* 64, 229–234.

Cohen, A.C. (1985) Simple method for rearing the insect predator *Geocoris punctipes* (Heteroptera: Lygaeidae) on a meat diet. *Journal of Economic Entomology* 78, 1173–1175.

Cohen, A.C. (2000a) Feeding fitness and quality of domesticated and feral predators: effects of long-term rearing on artificial diet. *Biological Control* 17, 50–54.

Cohen, A.C. (2000b) Using biochemical approaches for quality control of insects reared on artificial diets. In: Gazzoni, D.L. (ed.) *Abstracts of the XXI International Congress of Entomology*, Vol. I. EMBRAPA, Londrina, Brazil, p. 381.

Cohen, A.C. and Smith, L.K. (1998) A new concept in artificial diets for *Chrysoperla rufilabris*: the efficacy of solid diets. *Biological Control* 13, 49–54.

Consoli, F.L. and Parra, J.R.P. (1996) Biology of *Trichogramma galloi* and *T. pretiosum* (Hymenoptera: Trichogrammatidae) reared *in vitro* and *in vivo*. *Annals of the Entomological Society of America* 89, 828–834.

Consoli, F.L. and Parra, J.R.P. (1999) *In vitro* rearing of parasitoids: constraints and perspectives. *Trends in Entomology* 2, 19–32.

Dahlan, A.N. and Gordh, G. (1997) Development of *Trichogramma australicum* (Hym.: Trichogrammatidae) at low and high population density in artificial diet. *Entomophaga* 42, 525–536.
Dahlan, A.N. and Gordh, G. (1998) Development of *Trichogramma australicum* Girault (Hymenoptera: Trichogrammatidae) in eggs of *Helicoverpa armigera* (Hübner) (Lepidoptera: Noctuidae) and in artificial diet. *Australian Journal of Entomology* 37, 254–264.
Dahlman, D.L. (1990) Evaluation of teratocyte functions: an overview. *Archives of Insect Biochemistry and Physiology* 13, 159–166.
Dai, K.J., Ma, Z.J., Zhang, L.W., Cao, A.H., Zhan, Q.X., Xu, K.J., Pan, D.S. and Zhang, J.L. (1991) Research on technology of industrial production of the artificial host egg of Trichogramma. In: Wajnberg, E. and Vinson, S.B. (eds) *Trichogramma and Other Egg Parasites*. Les Colloques de l'INRA, 56, Paris, pp. 137–139.
De Clercq, P. and Degheele, D. (1992) A meat-based diet for rearing the predatory stinkbugs *Podisus maculiventris* and *Podisus sagitta* (Het.: Pentatomidae). *Entomophaga* 37, 149–157.
De Clercq, P. and Degheele, D. (1993) Quality assessment of the predatory bugs *Podisus maculiventris* (Say) and *Podisus sagitta* (Fab.) (Heteroptera: Pentatomidae) after prolonged rearing on a meat-based artificial diet. *Biocontrol Science and Technology* 3, 133–139.
De Clercq, P., Merlevede, F. and Tirry, L. (1998a) Unnatural prey and artificial diets for rearing *Podisus maculiventris* (Heteroptera: Pentatomidae). *Biological Control* 12, 137–142.
De Clercq, P., Vandewalle, M. and Tirry, L. (1998b) Impact of inbreeding on performance of the predator *Podisus maculiventris*. *Biocontrol* 43, 299–310.
Delobel, B. and Pageaux, J.F. (1981) Influence de l'alimentation sur la composition en acides gras totaux de Diptères Tachinaires. *Entomologia Experimentalis et Applicata* 29, 281–288.
Dindo, M.L., Sama, C., Fanti, P. and Farneti, R. (1997) *In vitro* rearing of the pupal parasitoid *Brachymeria intermedia* (Hym.: Chalcididae) on artificial diets with and without host components. *Entomophaga* 42, 445–453.
Dindo, M.L., Farneti, R., Scapolatempo, M. and Gardenghi, G. (1999) *In vitro* rearing of the parasitoid *Exorista larvarum* (L.) (Diptera: Tachinidae) on meat homogenate-based diets. *Biological Control* 16, 258–266.
Dutton, A. and Bigler, F. (1995) Flight activity assessment of the egg parasitoid *Trichogramma brassicae* (Hym.: Trichogrammatidae) in laboratory and field conditions. *Entomophaga* 40, 223–233.
Dutton, A., Cerutti, F. and Bigler, F. (1996) Quality and environmental factors affecting *Trichogramma brassicae* efficiency under field conditions. *Entomologia Experimentalis et Applicata* 81, 71–79.
Falabella, P., Tremblay, E. and Pennacchio, F. (2000) Host regulation by the aphid parasitoid *Aphidius ervi*: the role of teratocytes. *Entomologia Experimentalis et Applicata* 97, 1–9.
Fanti, P. (1990) Fattori ormonali inducenti la prima muta larvale del parassitoids *Pseudogonia rufifrons* Wied. (Diptera: Tachinidae) in substrati di crescita *in vivo* e *in vitro*. *Bolletino d'ell Istituto di Entomologia della Universita degli studi di Bologna* 45, 47–59.
Feng, J.G., Tao, X., Zhang, A.S., Yu, Y. and Zhang, C.W. (1999) Study on using *Trichogramma* spp. on artificial host egg to control corn pests. *Chinese Journal of Biological Control* 15, 97–99.
Ferkovich, S.M., Moralesramos, J.A., Rojas, M.G., Oberlander, H., Carpenter, J.E. and Greany, P. (1999) Rearing of ectoparasitoid *Diapetimorpha introita* on an artificial diet: supplementation with insect cell line-derived factors. *BioControl* 44, 29–45.
Ferkovich, S.M., Shapiro, J. and Carpenter, J.E. (2000) Growth of a pupal ectoparasitoid, *Diapetimorpha introita*, on an artificial diet: stimulation of growth rate by a lipid extract from host pupae. *BioControl* 45, 401–413.
Gao, Y.G., Dai, K.J. and Shong, L.S. (1982) *Trichogramma* sp. and their utilization in People's Republic of China. In: INRA (ed.) *Les Trichogrammes*. Les Colloques de l'INRA 9, Paris, p. 181.
Gelman, D.B., Carpenter, J.E. and Greany, P.D. (2000) Ecdysteroid levels/profiles of the parasitoid wasp, *Diapetimorpha introita*, reared on its host, *Spodoptera frugiperda* and on an artificial diet. *Journal of Insect Physiology* 46, 457–465.
Greany, P.D. and Carpenter, J.E. (1998) Culture medium for parasitic and predacious insects. US Patent no. 5,799,607, USPTO, USA.
Greany, P.D., Vinson, S.B. and Lewis, W.J. (1984) Insect parasitoids: finding new opportunities for biological control. *BioScience* 34, 690–696.
Greany, P.D., Ferkovich, S.M., and Clark, W.R. (1989) Progress towards development of an artificial diet and an *in vitro* rearing system for *Microplitis croceipes*. *Southwestern Entomologist* 12, 89–94.
Grenier, S. (1987) Developmental relationships between the tachinid parasitoid *Pseudoperichaeta nigrolineata* and two host species – hormonal implications. In: Bouletreau, M. and Bonnot, G. (eds) *Parasitoid Insects*. INRA 48, Paris, pp. 87–89.

Grenier, S. (1994) Rearing of *Trichogramma* and other egg parasitoids on artificial diets. In: Wajnberg, E. and Hassan, S.A. (eds) *Biological Control with Egg Parasitoids*. CAB International, Wallingford, UK, pp. 73–92.

Grenier, S. and Bonnot, G. (1983) Evolution pondérale du parasitoïde *Lixophaga diatraeae* (Dipt. Tachinidae) de la sortie de l'hôte à l'émergence. Estimations des poids par mensurations de la pupe. *Entomophaga* 28, 259–270.

Grenier, S. and Grenier, A.M. (1993) Fenoxycarb, a fairly new insect growth regulator: a review of its effects on insects. *Annals of Applied Biology* 122, 369–403.

Grenier, S. and Pintureau, B. (1991) Essai d'élevage en milieux artificiels de *Lixophaga diatraeae* (Dipt. Tachinidae), parasitoïde des Lépidoptères foreurs de la canne à sucre. Importance de la souche. In: Pavis, C. and Kermarrec, A. (eds) *Rencontres Caraïbes en Lutte Biologique*. Les Colloques de l'INRA 58, Paris, pp. 451–458.

Grenier, S., Guillaud, J., Delobel, B. and Bonnot, G. (1989) Nutrition et élevage du prédateur *Macrolophus caliginosus* (Heteroptera, Miridae) sur milieux artificiels. *Entomophaga* 34, 77–86.

Grenier, S., Greany, P.D. and Cohen, A.C. (1994) Potential for mass release of insect parasitoids and predators through development of artificial culture techniques. In: Rosen, D., Bennett, F.D. and Capinera, J.L. (eds) *Pest Management in the Subtropics: Biological Control – A Florida Perspective*. Intercept 10, Andover, pp. 181–205.

Grenier, S., Yang, H., Guillaud, J. and Chapelle, L. (1995) Comparative development and biochemical analyses of *Trichogramma* (Hymenoptera: Trichogrammatidae) grown in artificial media with haemolymph or devoid of insect components. *Comparative Biochemistry and Physiology* 111B, 83–90.

Grenier, S., Pintureau, B., Heddi, A., Lassablière, F., Jager, C., Louis, C. and Khatchadourian, C. (1998) Successful horizontal transfer of *Wolbachia* symbionts between *Trichogramma* wasps. *Proceedings of the Royal Society of London Series B* 265, 1441–1445.

Grenier, S., Grillé, G., Basso, C. and Pintureau, B. (2001) Effects of the host species and the number of parasitoids per host on the size of some *Trichogramma* species (Hymenoptera: Trichogrammatidae). *Biocontrol Science and Technology* 11, 23–28.

Grenier, S., Gomes, S.M., Pintureau, B., Lassablière, F. and Bolland, P. (2002) Use of tetracycline in larval diet to study the effect of *Wolbachia* on host fecundity and clarify taxonomic status of *Trichogramma* species in cured bisexual lines. *Journal of Invertebrate Pathology* 80, 13–21.

Gu, U. and Dorn, S. (2000) Consequences of inbreeding and hybridization in the gregarious parasitoid *Cotesia glomerata*. In: Gazzoni, D.L. (ed.) *Abstracts of the XXI International Congress of Entomology*, Vol. I. EMBRAPA, Londrina, Brazil, p. 389.

Guerra, A.A. and Martinez, S. (1994) An *in vitro* rearing system for the propagation of the ectoparasitoid *Catolaccus grandis*. *Entomologia Experimentalis et Applicata* 72, 11–16.

Guerra, A.A., Robacker, K.M. and Martinez, S. (1993) *In vitro* rearing of *Bracon mellitor* and *Catolaccus grandis* with artificial diets devoid of insect components. *Entomologia Experimentalis et Applicata* 68, 303–307.

Habibi, J., Backus, E.A., Coudron, T.A. and Brandt, S.L. (2001) Effect of different host substrates on hemipteran salivary protein profiles. *Entomologia Experimentalis et Applicata* 98, 369–375.

Hagen, K.S. and Tassan, R.L. (1965) A method of providing artificial diets to *Chrysopa* larvae. *Journal of Economic Entomology* 58, 999–1000.

Hagler, J.R. (2000) Biological control of insects. In: Rechcigl, J.E. and Rechcigl, N.A. (eds) *Insect Pest Management*. Agriculture and Environment Series 3, Lewis Publishers, Boca Raton, Florida, pp. 207–241.

Hagler, J.R. and Cohen, A.C. (1991) Prey selection by *in vitro*- and field-reared *Geocoris punctipes*. *Entomologia Experimentalis et Applicata* 59, 201–205.

Han, S.C., Chen, Q.X., Xu, X., Zhang, M.L., Zhu, D.F. and Liu, W.H. (1988) *In vitro* rearing *Anastatus japonicus* Ashmead (Hym.: Eupelmidae) for controlling litchi stink bug, *Tessaratoma papillosa* Drury (Hem.: Pentatomidae). *Natural Enemies of Insects* 10, 170–173.

Hassan, S.A. and Hagen, K.S. (1978) A new artificial diet for rearing *Chrysopa carnea* larvae (Neuroptera, Chrysopidae). *Zeitschrift fürAngewandte Entomologie* 86, 315–320.

Hattingh, V. and Samways, M.J. (1993) Evaluation of artificial diets and two species of natural prey as laboratory food for *Chilocorus* spp. *Entomologia Experimentalis et Applicata* 69, 13–20.

Honda, J.Y., Silva, I.M.M.S., Vereijssen, J. and Stouthamer, R. (1999) Laboratory bioassay and greenhouse evaluation of *Trichogramma cordubensis* strains from Portugal. *BioControl* 44, 1–11.

Hu, J.S., Gelman, D.B., Bell, R.A. and Loeb, M.J. (1998) *In vitro* rearing of *Edovum puttleri*, an egg parasitoid of the Colorado potato beetle – development from egg through the pupal stage. *BioControl* 43, 1–16.

Hussein, M.Y. and Hagen, K.S. (1991) Rearing of *Hippodamia convergens* on artificial diet of chicken liver, yeast and sucrose. *Entomologia Experimentalis et Applicata* 59, 197–199.

Iriarte, J. and Castañe, C. (2001) Artificial rearing of *Dicyphus tamaninii* (Heteroptera: Miridae) on a meat based diet. *Biological Control* 22, 98–102.

Kazmer, D.J. and Luck, R.F. (1995) Field tests of the size-fitness hypothesis in the egg parasitoid *Trichogramma pretiosum*. *Ecology* 76, 412–425.

Kennett, C.E. and Hamai, J. (1980) Oviposition and development in predacious mites fed with artificial and natural diets (Acari: Phytoseiidae). *Entomologia Experimentalis et Applicata* 28, 116–122.

Kuhlmann, U. and Mills, N.J. (1999) Comparative analysis of the reproductive attributes of three commercially-produced *Trichogramma* species (Hymenoptera: Trichogrammatidae). *Biocontrol Science and Technology* 9, 335–346.

Leppla, N.C. (1989) Laboratory colonization of fruit fies. In: Robinson, A.S. and Hooper, G. (eds) *World Crop Pests, Fruit Flies, Their Biology, Natural Enemies and Control*, 3B. Elsevier Science Publishers, Amsterdam, pp. 91–102.

Leppla, N.C. and Ashley, T.R. (1989) Quality control in insect mass production: a review and model. *Bulletin of the Entomological Society of America* 35, 33–44.

Leppla, N.C. and Fischer, W.R. (1989) Total quality control in insect mass production for insect pest management. *Journal of Applied Entomology* 108, 452–461.

Liu, F.H. and Smith, S.M. (2000) Measurement and selection of parasitoid quality for mass-reared *Trichogramma minutum* Riley used in inundative release. *Biocontrol Science and Technology* 10, 3–13.

Liu, Z.C., Liu, J.F., Wang, C.X., Yang, W.H. and Li, D.S. (1995) Mechanized production of artificial host egg for mass-rearing of parasitic wasps. In: Wajnberg, E. (ed.) *Trichogramma and Other Egg Parasitoids*. INRA 73, Paris, pp. 163–164.

Losey, J.E. and Calvin, D.D. (1995) Quality assessment of four commercially available species of *Trichogramma* (Hymenoptera: Trichogrammatidae). *Journal of Economic Entomology* 88, 1243–1250.

Moore, R.F., Odell, T.M. and Calkins, C.O. (1985) Quality assessment in laboratory-reared insects. In: Singh, P. and Moore, R.F. (eds) *Handbook of Insect Rearing*, Vol. 1. Elsevier, Amsterdam, pp. 107–135.

Morales-Ramos, J.A., Rojas, M.G., Coleman, R.J. and King, E.G. (1998) Potential use of *in vitro*-reared *Catolaccus grandis* (Hymenoptera: Pteromalidae) for biological control of the boll weevil (Coleoptera: Curculionidae). *Journal of Economic Entomology* 91, 101–109.

Nakahara, Y., Hiraoka, T. and Iwabuchi, K. (2000) Growth-promoting effects of ecdysteroids and juvenile hormone on *in vitro* development of the larval endoparasitoid, *Venturia canescens* (Hymenoptera : Ichneumonidae). *Journal of Insect Physiology* 46, 467–476.

Niijima, K., Nishimura, R. and Matsuka, M. (1977) Nutritional studies of an aphidophagous coccinellid, *Harmonia axyridis*. 3. Rearing of larvae using a chemically defined diet and fractions of drone honeybee powder. *Bulletin of the Faculty of Agriculture of Tamagawa University* 17, 45–51.

Niijima, K., Matsuka, M. and Okada, I. (1986) Artificial diets for an aphidophagous coccinellid, *Harmonia axyridis*, and its nutrition. In: Hodek, I. (ed.) *Ecology of Aphidophaga*. Academia, Prague, pp. 37–50.

Nordlund, D.A., Wu, Z.X. and Greenberg, S.M. (1997) *In vitro* rearing of *Trichogramma minutum* Riley (Hymenoptera: Trichogrammatidae) for ten generations, with quality assessment comparisons of *in vitro* and *in vivo* reared adults. *Biological Control* 9, 201–207.

Nurindah, Gordh, G. and Cribb, B.W. (1997) Oviposition behaviour and reproductive performance of *Trichogramma ustralicum* Girault (Hymenoptera: Trichogrammatidae) reared in artificial diet. *Australian Journal of Entomology* 36, 87–93.

Ogura, N. and Hosada, R. (1995) Rearing of a coleopterous predator, *Trogossita japonica* (Col.: Trogossitidae), on artificial diets. *Entomophaga* 40, 371–378.

Ogura, N., Tabata, K. and Wang, W. (1999) Rearing of the colydiid beetle predator, *Dastarcus helophoroides*, on artificial diet. *Biocontrol* 44, 291–299.

Pintureau, B. and Grenier, S. (1992) Variability of morphological and biological characteristics in two strains of *Lixophaga diatraeae* (Diptera: Tachinidae). *Biological Control* 2, 176–180.

Racioppi, J.V., Burton, R.L. and Eikenbary, R. (1981) The effects of various oligidic synthetic diets on the growth of *Hippodamia convergens*. *Entomologia Experimentalis et Applicata* 30, 68–72.

Rojas, M.G., Morales-Ramos, J.A. and King, E.G. (1999) Synthetic diet for rearing the hymenopterous ectoparasitoid, *Catolaccus grandis*. US patent no. 5,899,2168, USPTO, USA.

Rojas, M.G., Morales-Ramos, J.A. and King, E.G. (2000) Two meridic diets for *Perillus bioculatus* (Heteroptera: Pentatomidae), a predator of *Leptinotarsa decemlineata* (Coleoptera : Chrysomelidae). *Biological Control* 17, 92–99.

Saavedra, J.L.D., Zanuncio, J.C., Zanuncio, T.V. and Guedes, R.N.C. (1997) Prey capture ability of *Podisus nigrispinus* (Dallas) (Het., Pentatomidae) reared for successive generations on a meridic diet. *Journal of Applied Entomology* 121, 327–330.

Shapiro, J.P., Wasserman, H.A., Greany, P.D. and Nation, J.L. (2000) Vitellin and vitellogenin in the soldier bug, *Podisus maculiventris*: identification with monoclonal antibodies and reproductive response to diet. *Archives of Insect Biochemistry and Physiology* 44, 130–135.

Shipp, J.L., Wang, K. and Ferguson, G. (1998) Evaluation of commercially produced *Trichogramma* spp. (Hymenoptera: Trichogrammatidae) for control of tomato pinworm, *Keiferia lycopersicella* (Lepidoptera: Gelechiidae), on greenhouse tomatoes. *Canadian Entomologist* 130, 721–731.

Silva, I.M.M.S. (1999) *Identification and Evaluation of* Trichogramma *Parasitoids for Biological Pest Control.* Wageningen University, Wageningen, 151 pp.

Smirnoff, W.A. (1958) An artificial diet for rearing coccinellid beetles. *Canadian Entomologist* 90, 563–565.

Sorati, M., Newman, M. and Hoffmann, A.A. (1996) Inbreeding and incompatibility in *Trichogramma* nr. *brassicae*: evidence and implications for quality control. *Entomologia Experimentalis et Applicata* 78, 283–290.

Stouthamer, R. (1990) Effectiveness of several antibiotics in reverting thelytoky to arrhenotoky in *Trichogramma* spp. In: Wajnberg, E. and Vinson, S.B. (eds) Trichogramma *and Other Egg Parasitoids.* INRA 56, Paris, pp. 119–122.

Stouthamer, R., Breeuwer, J.A.J. and Hurst, G.D.D. (1999) *Wolbachia pipientis*: microbial manipulator of arthropod reproduction. *Annual Review of Microbiology* 53, 71–102.

Thompson, S.N. (1999) Nutrition and culture of entomophagous insects. *Annual Review of Entomology* 44, 561–592.

Thompson, S.N. and Hagen, K.S. (1999) Nutrition of entomophagous insects and other arthropods. In: Bellows, T.S. and Fisher, T.W. (eds) *Handbook of Biological Control: Principles and Applications.* Academic Press, San Diego, California, pp. 59–652.

Thompson, S.N. and Johnson, J.A. (1978) Further studies on lipid metabolism in the insect parasite, *Exeristes roborator* (Fabricius). *Journal of Parasitology* 64, 731–740.

Vanderzant, E.S. (1969) An artificial diet for larvae and adults of *Chrysopa carnea*, an insect predator of crop pests. *Journal of Economic Entomology* 62, 25–257.

van Lenteren, J.C. (1999) Fundamental knowledge about insect reproduction: essential to develop sustainable pest management. *Invertebrate Reproduction and Development* 36, 1–15.

van Lenteren, J.C. (2000) Measures of success in biological control of arthropods by augmentation of natural enemies. In: Gurr, G.M. and Wratten, S.D. (eds) *Measures of Success in Biological Control.* Kluwer Academic Publishers, Dordrecht, pp. 77–103.

Werren, J.H. (1997) Biology of *Wolbachia. Annual Review of Entomology* 42, 587–609.

Williams, D.W. and Leppla, N.C. (1992) The future of augmentation of beneficial arthropods. In: Kauffman, W.C. and Nechols, J.E. (eds) *Selection Criteria and Ecological Consequences of Importing Natural Enemies.* Entomological Society of America, Thomas Say Publications, Lanham, Maryland, pp. 87–102.

Wittmeyer, J.L. and Coudron, T.A. (2001) Life table parameters, reproductive rate, intrinsic rate of increase and realized cost of rearing *Podisus maculiventris* (Say) (Heteroptera: Pentatomidae) on an artificial diet. *Journal of Economic Entomology* 94, 1344–1352.

Wittmeyer, J.L., Coudron, T.A. and Adams, T.S. (2001) Ovarian development, fertility and fecundity in *Podisus maculiventris* Say (Heteroptera: Pentatomidae): an analysis of the impact of nymphal, adult, male and female nutritional source on reproduction. *Invertebrate Reproduction and Development* 39, 9–20.

Yazgan, S. (1972) A chemically defined synthetic diet and larval nutritional requirements of the endoparasitoid *Itoplectis conquisitor* (Hymenoptera). *Journal of Insect Physiology* 18, 2123–2141.

Zanuncio, J.C., Saavedra, J.L.D., Oliveira, H.N., Degheele, D. and De Clercq, P. (1996) Development of the predatory stinkbug *Brontocoris tabidus* (Signoret) (Heteroptera: Pentatomidae) on different proportions of an artificial diet and pupae of *Tenebrio molitor* L. (Coleoptera: Tenebrionidae). *Biocontrol Science and Technology* 6, 619–625.

10 Pathogens of Mass-produced Natural Enemies and Pollinators

S. Bjørnson[1] and C. Schütte[2]

[1]*Department of Biology, Saint Mary's University, 923 Robie Street, Halifax, Nova Scotia, Canada B3H 3C3;* [2]*Laboratory of Entomology, Wageningen Agricultural University, PO Box 8031, 6700 EH Wageningen, The Netherlands*

Abstract

Pathogens are found in both field-collected and mass-reared natural enemies. This raises concern regarding their quality and efficacy in biological pest-control programmes. Some pathogens affect the performance of natural enemies and crop pollinators by lowering their efficacy, whereas others alter their reproduction. It is therefore important to screen individuals for pathogens on a routine and continual basis. Special attention should be given to field-collected arthropods and those exchanged among rearing facilities. In order to obtain high efficacy in pest control, the release of pathogen-free natural enemies in biological control programmes is of the utmost importance. This chapter first provides an overview of the types of pathogens to which beneficial arthropods are susceptible and describes the symptoms that may be associated with infection. Next, diseases of natural enemies and insect pollinators are presented according to host genus.

Introduction

About 125 natural enemies are reared for inundative and seasonal inoculative forms of biological control (Chapters 1 and 11; van Lenteren *et al.*, 1997). Several species are used throughout the world. The success of a biological control programme is dependent on many factors, including the quality of the beneficial arthropods that are used. In several cases, reports of pathogens in mass-reared natural enemies have raised questions regarding their quality and efficacy.

Pathogens have been reported in natural enemies collected from the field (Lipa *et al.*, 1975; Geden *et al.*, 1995), from those currently mass-produced for biological pest control (Beerling and van der Geest, 1991; Becnel and Geden, 1994; Bjørnson and Keddie, 2000) and from those being evaluated for their potential as biological control agents (Kluge and Caldwell, 1992). The origin of many of these pathogens has not been demonstrated and it is therefore difficult to determine if pathogens originate from field-collected natural enemies or arise in mass-rearing systems as a result of intense and continuous rearing under laboratory conditions. Temporary starvation and localized overcrowding are often unavoidable in mass-rearings and these conditions place stresses on individuals, presumably making

© CAB International 2003. *Quality Control and Production of Biological Control Agents: Theory and Testing Procedures* (ed. J.C. van Lenteren)

them more susceptible to disease (Goodwin, 1984; Maddox, 1987). The following summary of pathogens in natural enemies and pollinators will include those reported from both field-collected and mass-reared individuals. An overview of pathogen presence in mite and insect genera that include species used in biological control or pollination is presented in Table 10.1.

Causes of Diseases in Natural Enemies and Pollinators

Insect diseases may be broadly categorized as either infectious or non-infectious, based on the respective presence or absence of a transmissible, living organism. Diseases classified as non-infectious may be caused by mechanical injury, adverse physical environmental factors, chemical toxins or poisons, injuries made by predators and parasitoids, genetic abnormalities or nutritional deficiencies (Thomas, 1974; Gaugler, 1987). In contrast, infectious diseases among insects are caused by a variety of pathogens, including viruses, bacteria, fungi, protozoa and nematodes (Thomas, 1974; Tanada and Kaya, 1993).

Diseases caused by viruses

Viruses have been isolated from more than 1000 species of insects, representing at least 12 insect orders, and have also been reported from mites, ticks and marine crustaceans (Martignoni and Iwai, 1981; Tanada and Kaya, 1993).Viruses are obligate, intracellular pathogens that consist of one or more nucleic acid template molecules, normally encased in a protective coat or coats of protein or

Table 10.1. Pathogen presence in mite and insect genera that include species used for biological control or pollination.

	Viruses	Bacteria	Protozoa	Fungi	Nematodes	Unidentified disease
Steinernema	X		X	X		
Heterorhabditis	X		X	X		
Metaseiulus		X				X
Neoseiulus	X	X	X	X		X
Phytoseiulus	X	X	X			X
Aphidoletes				X		
Chrysoperla	X		X	X		
Adalia		X	X	X		
Coccinella			X	X	X	
Coleomegilla		X		X		
Harmonia		X	X	X		
Hippodamia		X	X	X		
Aphidius				X		
Aphytis		X				
Cotesia	X		X	X		
Dacnusa				X		
Encarsia		X	X			
Eretmocerus		X				X
Lysiphlebus		X				
Muscidifurax		X	X			
Nasonia		X				
Opius		X				
Pediobius			X			
Trichogramma		X	X			
Bombus	X	X	X	X	X	

X, pathogen from respective group present.

lipoprotein (capsid) (Tanada and Kaya, 1993). The simplest viruses can be seen only with the aid of the electron microscope; however, some viruses are occluded in proteinaceous bodies that can be detected by light microscopy (Thomas, 1974).

More than 20 groups of viruses consist of known insect pathogens (Martignoni and Iwai, 1981). The Baculoviridae are the most common and widely studied group of insect viruses and are found exclusively in arthropods. This family consists of two major subgroups: the nuclear polyhedrosis viruses and the granulosis viruses. The former group has been reported from more than 400 insect species, whereas the latter have been found exclusively in lepidopteran hosts. The Polydnaviridae are associated with parasitic wasps and are categorized into two major groups based on viral structure and wasp-family association. Some polydnaviruses are pathogenic, whereas others form mutualistic relationships with their hosts (Tanada and Kaya, 1993).

In general, infection occurs after viruses are ingested but transmission may occur transovarially or through spiracles or wounds (Thomas, 1974; Tanada and Kaya, 1993). Viruses are able to organize their own replication within suitable host cells. This involves the adsorption, uptake and uncoating of virus particles, followed by the replication of the viral genome and production of viral progeny (Tanada and Kaya, 1993). Diagnostic features that are considered to be general characteristics of infection include: dead larvae found hanging or lying on leaf or plant surfaces, a very fragile cuticle, which may rupture easily when touched to release the liquefied body contents, and white masses of fat body, which are visible through the cuticle. The host body has no filamentous structures on the cuticle, which may acquire a bluish or bluish-purple iridescence (Thomas, 1974). Infection may induce prominent behavioural changes in the host, including changes in the level of activity (wandering behaviour) and/or changes of microhabitat preference (elevation-seeking behaviour or 'tree-top' diseases, movement to exposed locations and diurnal behaviour of nocturnal insects) (Horton and Moore, 1993).

Viruses are reported infrequently in natural enemies and are documented for *Bombus* (MacFarlane *et al.*, 1995), *Cotesia* (Hamm *et al.*, 1985) and *Phytoseiulus* (Šut'áková and Rüttgen, 1978).

Diseases caused by bacteria

Bacteria are the most common microorganisms associated with insects (Tanada and Kaya, 1993); however, relatively few bacteria are capable of infecting or killing their insect hosts. Most bacteria are non-pathogenic but may become pathogenic if conditions are favourable (Thomas, 1974; Tanada and Kaya, 1993). Such facultative pathogens are mainly found in the genera *Enterobacter*, *Serratia*, *Pseudomonas* and *Proteus* (Tanada and Kaya, 1993). Transmission of bacterial pathogens occurs mostly through the mouth and digestive tract. Less often, they are transmitted through the egg or through wounds in the integument (Thomas, 1974; Andreadis, 1987; Tanada and Kaya, 1993). Upon ingestion by a susceptible host, pathogenic bacteria multiply and produce enzymes or toxins that degrade the cells of the gut. Bacterial pathogens may cause septicaemia, whereby they invade the haemocoel, multiply, and produce toxins that kill the host. Some bacteria may cause toxaemia, a condition resulting from the production of bacterial toxins, while the bacteria themselves are often confined to the gut lumen (Tanada and Kaya, 1993).

In general, insects that are killed by bacteria rapidly become dark in colour and are often very soft. Symptoms are usually more pronounced in larval stages. While the integument remains intact, internal tissues and organs degenerate and the body becomes viscid. The infected host may stop feeding and excrete diarrhoea-like faeces. Advanced infection is often accompanied by a putrid odour. The cadaver often shrivels and becomes dry and hard (Thomas, 1974; Tanada and Kaya, 1993).

Reports of bacterial diseases among natural enemies used for biological control are scarce. However, the facultative pathogen *Serratia marcescens* is common among laboratory-reared insects and can be responsible for parasitoid death if it is transmitted by female parasitoids during oviposition (Brooks, 1993). The effect of *S. marcescens* and a second facultative pathogen, *Pseudomonas fluorescens*, has been tested on *Metaseiulus occidentalis* (Nesbitt) (Lighthart *et al.*, 1988) and *Hippodamia convergens* Guérin-Méneville (James and Lighthart, 1992).

Rickettsiae are obligate intracellular symbionts. Many are non-pathogenic and occur as commensals or mutualists (Tanada and Kaya, 1993). Entomopathogenic rickettsiae (family Rickettsiaceae, tribe Wolbachiae) belong to two recognized genera. Members of the genus *Rickettsiella* are common pathogens, whereas those of the genus *Wolbachia* are seldom pathogenic in the true sense but have evolved various means of manipulating their hosts in order to enhance their own transmission (see Stouthamer *et al.*, 1999). Rickettsiae have complex developmental cycles and a diversity of forms. These range from typical rod-, coccoid and kidney-shaped structures to spherical or giant forms.

Rickettsiella spp. have been isolated from many different insect orders. They generally produce chronic infections in their hosts. Following the ingestion of the small, dense, bacterial rods by a susceptible host, the rickettsiae pass through the gut into the haemocoel and infect the cells of other tissues and organs, especially the fat body and gonads. *Rickettsiella* spp. generate a variety of structures, including bipyramidal crystalline bodies. Most tissues become infected, with the exception of the gut epithelium. Infected cells usually lyse and release their contents into the haemolymph. After the infected host dies, infective stages are released into the soil, where rod-shaped forms can survive for 1 or more years. *Rickettsiella* spp. that produce sublethal infections are usually transovarially transmitted (Tanada and Kaya, 1993). Infection may induce prominent behavioural changes in the host, including elevation-seeking behaviour and changes in temperature preference (Horton and Moore, 1993).

Wolbachia spp. are common cytoplasmic symbionts of insects, crustaceans, mites and filarial nematodes (see Stouthamer *et al.*, 1999). *Wolbachia* spp. are rarely pathogenic but may manipulate the host biology by inducing parthenogenesis (whereby infected females exclusively produce daughters), feminization (whereby infected genetic males reproduce as females), male-killing (whereby infected male embryos die while female embryos develop into infected females), cytoplasmic incompatibility (unidirectional in its simplest form: whereby the crossing of an uninfected female and infected male result in embryo mortality) or by enhancing host fecundity (Stouthamer *et al.*, 1999).

Wolbachia spp. may be present in various tissues but are predominantly present in gonadal tissue (Stouthamer *et al.*, 1999). The symbionts are transmitted vertically through the egg. Therefore, infected mothers give rise to infected offspring. Phylogenetic studies of *Wolbachia* indicate that horizontal transmission must have taken place rather frequently. An intraspecific horizontal transfer of *Wolbachia* has recently been reported in *Trichogramma kaykai* (Huigens *et al.*, 2000). *Wolbachia*-infected arthropods may be cured by treatment with several antibiotics and by rearing in elevated temperatures (Stouthamer *et al.*, 1999). As culturing *Wolbachia* outside their hosts has been successful in only one case, molecular techniques, such as the polymerase chain reaction (PCR), are used in detecting *Wolbachia* infections (Stouthamer *et al.*, 1993). However, a recent study demonstrated that *Wolbachia* infections can be simply established, stably maintained and stored *in vitro* using standard tissue-culture techniques (Dobson *et al.*, 2002). These findings will facilitate the development of a *Wolbachia* stock centre and permit novel approaches for the study of *Wolbachia*.

Wolbachia spp. are known to infect a wide array of natural enemies and are common among parasitoids, including *Aphytis, Encarsia, Lysiphlebus, Muscidifurax, Nasonia* and *Trichogramma*. *Wolbachia* have also been reported to infect predaceous natural enemies, including *Adalia, Phytoseiulus, Neoseiulus* and *Metaseiulus* (Stouthamer *et al.*, 1999). Although *Wolbachia* spp. may affect host reproduction, they may have little or no measurable effect on host fitness (Zchori-

Fein *et. al.*, 2000). Therefore, *Wolbachia* spp. are not always considered to be detrimental to natural enemies that are used for biological control (Chapter 8).

Diseases caused by protozoa

Mass-reared beneficial arthropods are known hosts of several types of entomopathogenic or symbiotic protozoa, including eugregarines, neogregarines, microsporidia and trypanosomes.

Eugregarines have been observed in invertebrates for many years (Brooks, 1974) and a large number have been described from arthropods. However, few studies regarding eugregarine pathogenicity and host–parasite relationships have been presented (Brooks, 1974; Tanada and Kaya, 1993).

The life cycle of eugregarines can vary greatly and is often complex (Tanada and Kaya, 1993). Transmission of gregarines is achieved *per os* (orally) when a susceptible host ingests gregarine spores. Shortly afterward, sporozoites are released in the digestive tract where they become partially embedded in, and invade, the epithelial cells of the intestinal wall. During the initial phase of development, the parasite (trophozoite) develops intracellularly and may increase greatly in size. These eventually emerge from the host cells, destroying them in the process.

Trophozoites obtain nourishment in two ways: through the host epithelial cells to which they remain attached by a specialized organelle (epimerite), or by the absorption of fluids from the gut lumen after they detach from these cells (Brooks, 1974; Tanada and Kaya, 1993).

Most species of eugregarines are relatively harmless parasites or commensals that live in the intestinal tract or haemocoel of their hosts. The absence of asexual reproduction (merogony) in their life cycle is considered to be a major factor for their apparent lack of pathogenicity (Brooks, 1974; Tanada and Kaya, 1993). Eugregarines have been reported in *Bombus* (MacFarlane *et al.*, 1995), *Coccinella* and *Hippodamia* (Lipa, 1967; Lipa *et al.*, 1975). Conversely, neogregarines are known to produce lethal infections in insects. These pathogens have been reported in *Bombus* (Liu *et al.*, 1975; Lipa and Triggiani, 1996).

Microsporidia are spore-forming protozoa that infect a wide range of hosts from all major animal phyla. Fish and arthropods are their most common hosts (Tanada and Kaya, 1993). Microsporidia are obligate, intracellular parasites that lack mitochondria. These pathogens are dependent on host cells for energy in the form of adenosine triphosphate (Canning and Hollister, 1990), which they use for their own development. Microsporidia can cause either acute infections, resulting in the death of the host, or chronic infections, which produce sublethal and debilitating disease (Tanada and Kaya, 1993).

The microsporidian life cycle is complex and consists of two phases. The vegetative phase is responsible for the proliferation of the pathogen, whereas the sporulation phase results in the production of transmissible spores. These are somewhat resistant to unfavourable environmental conditions and provide the pathogen with a means of dispersal (Maddox, 1973).

Microsporidia may invade the host tissues when spores are ingested, when the pathogen is transmitted from parent to progeny or occasionally through wounds in the integument (Tanada and Kaya, 1993). Microsporidian spores are structurally unique and contain a characteristic tube-like polar filament through which an infective stage (sporoplasm) is injected into an adjacent host cell. This begins the infective cycle of the pathogen.

Over 50 cases of microsporidioses have been reported in beneficial arthropods (Bjørnson, 1998), including *Bombus*, *Coccinella*, *Cotesia*, *Encarsia*, *Muscidifurax* and *Phytoseiulus*. Microsporidia often cause a significant reduction in host fecundity and longevity, increased mortality, developmental problems (delayed development, deformations, moulting problems) and reduced adult emergence of infected parasitoids (Geden *et al.*, 1995; Bjørnson and Keddie, 1999). Therefore, microsporidia may adversely affect the establishment and efficacy of mass-reared natural enemies (Kluge and Caldwell, 1992). Infected hosts may show no physical symptoms; however, infected tissues appear white in some hosts. Infected hosts may have soft bodies or bodies that dry to a scale that yields a milky mass in water (Thomas, 1974). Microsporidia-infected insects may also exhibit behavioural changes, including changes in temperature preference (Horton and Moore, 1993).

Diseases caused by fungi

Classification of fungi is largely based on the morphology of reproductive structures and associated processes. Many entomopathogenic fungi have wide host ranges (e.g. *Beauveria bassiana, Metarhizium anisopliae*) but different isolates may demonstrate host specificity. Although fungi may infect all stages of the host, they are infrequent pathogens of pupae and rarely infect insect eggs (Tanada and Kaya, 1993).

Most pathogenic fungi penetrate their insect hosts through the cuticle or spiracles; however, some penetrate through the gut. Once in the host, pathogenic fungi proliferate and compete for soluble nutrients in the haemocoel, where they release mycotoxins. They then invade and digest tissues and cause premature death of the host (Thomas, 1974; Tanada and Kaya, 1993).

Adhesion and germination of fungal spores on the host cuticle are highly dependent on relative humidity and temperature, but light conditions and nutritional requirements are also important factors (Tanada and Kaya, 1993). In soft-bodied insects, fungal penetration can occur anywhere on the body. On larger, more heavily sclerotized insects, penetration often occurs at the thinner arthrodial membranes that are characteristic of joints. In many insects, early instars may be more susceptible to fungal penetration because the cuticle becomes thicker in successive instars. Insects that have recently moulted may also be more susceptible to fungal pathogens (Bell, 1974; Thomas, 1974; Tanada and Kaya, 1993).

In some cases, behavioural changes occur prior to death. Symptoms may include loss of coordination and body tremors, reproductive behaviour of castrated hosts and changes in microhabitat preference. The latter include elevation-seeking behaviour (fungal 'summit disease'), movement to exposed locations, changes in oviposition or foraging sites and change in temperature preference (Horton and Moore, 1993).

The most common and easily recognized signs of fungal infection are the presence of filamentous hyphae and characteristic reproductive structures (fruiting structures, spores), usually produced on the external surface of the dead host (Thomas, 1974). The body of the infected host may discolour. It may become mummified, hard or cheese-like and will not disintegrate in water (Thomas, 1974). Fungi are important pathogens of *Bombus* (MacFarlane *et al.*, 1995) and several coccinellid beetles (Lipa *et al.*, 1975).

Diseases caused by nematodes

Entomopathogenic nematodes parasitize their hosts by directly penetrating through the cuticle into the haemocoel or by entering through natural openings, such as the spiracles, mouth or anus (Tanada and Kaya, 1993). Nematode-infected insects generally show few external symptoms. The expression of external abnormalities in the host is dependent on nematode number, host age and the time of infection. Such anomalies may include colour changes, distorted abdomens, deformed wings or the formation of intersexes (Tanada and Kaya, 1993). Internal morphological and physiological changes as a result of infection may result in host sterility or death (Tanada and Kaya, 1993). Behavioural changes that may occur after nematode infection include changes in activity level and reproductive behaviour of castrated hosts, whereby parasitized females release nematodes instead of eggs (Horton and Moore, 1993). Nematodes are uncommon pathogens of natural enemies; however, they are known to infect *Coccinella* (Rhamhalinghan, 1992) and to sterilize *Bombus* queens (MacFarlane *et al.*, 1995).

Diseases of Beneficial Arthropods Presented Following Host Systematics

Pathogens that infect field or laboratory populations of beneficial arthropods are presented in the following sections. Unidentified diseases and descriptions of (potential) pathogens whose effects are not yet clear have also been included. However, the side-effects of entomopathogens used as microbial pesticides on beneficial arthropods are not discussed (for a review, see Lipa and Smits, 1999). Often, host specificity for a pathogen or group of pathogens has not been studied in great detail and the range of susceptible hosts is unknown. Therefore, a pathogen of a particular natural enemy may have a broader host range and be virulent towards members of the same and related genera. As a result, pathogens of natural enemies are presented according to host genus.

Entomopathogenic nematodes (Nematoda: Heterorhabditidae, Steinernematidae)

Heterorhabditis, Steinernema*: general parasites of arthropods*

Several pathogens reduce the effectiveness of entomopathogenic nematodes. These pathogens attack either the nematode or its bacterial symbionts (for a review, see Kaya *et al.*, 1998).

VIRUSES. Lysogenic phages have been isolated from *Photorhabdus luminescens* and *Xenorhabdus* sp., bacterial symbionts of entomopathogenic nematodes (cited by Kaya *et al.*, 1998). These phages lyse the symbiotic bacteria and thereby reduce the food supply for developing nematodes. Phages may become problematic in nematode mass-production systems if they become established. Future research is needed to find a means of detecting and eliminating them (Kaya *et al.*, 1998).

PROTOZOA. *Pleistophora schulbergi* and *Nosema mesnili* infect *Steinernema carpocapsae* (Weiser) (cited by Kaya *et al.*, 1998); however, the effect of these microsporidia on the efficacy of *S. carpocapsae* is unknown. An undescribed microsporidium isolated from *Steinernema glaseri* has detrimental effects on its host that range from slight damage to mortality, depending on the severity of infection. Furthermore, infective *S. glaseri* juveniles are smaller and do not live as long as uninfected ones (cited by Kaya *et al.*, 1998).

Neoplectana glaseri (Steiner) infected with an unidentified microsporidium produce very few infective juveniles. The microsporidium infects all stages of *N. glaseri*, including the eggs and infective juveniles. Microsporidian developmental stages and spores are observed throughout the intestine and reproductive system. The intestinal lumen of infective juveniles is frequently packed with spores. The effects of the microsporidium on *N. glaseri* are dependent upon the severity of infection and range from partial or complete castration in both sexes to reduced survival or mortality. The inadvertent introduction of microsporidia to mass cultures of *N. glaseri* can cause problems in mass rearings (Poinar, 1988).

FUNGI. Nematophagous fungi are common in soil and can reduce steinernemid and heterorhabdid populations. Predatory fungi trap their prey using specialized hyphae that penetrate into the body of the nematode, whereas endoparasitic fungi produce spores that infect their hosts.

There are several species of trapping fungi that successfully trap steinernemids and heterorhabdids, including *Monacrosporium eudermatum*, *Monacrosporium cionopagum*, *Monacrosporium elipsosporum*, *Geniculifera paucispora*, *Nematoctonus concurrens*, *Arthrobotrys robusta*, *Arthrobotrys oligospora* and *Dactylaria* sp. (cited by Kaya *et al.*, 1998). In addition to their presence in the soil, trapping fungi may also be present on the insect integument, where they serve to protect against nematode infection. Trapping fungi may survive as saprophytes, whereas endoparasitic fungi are obligate parasites. Spores of endoparasitic fungi attach to the nematode cuticle, germinate and penetrate into the body cavity. An example of an entomopathogenic fungus known to infect nematodes under natural conditions is *Hirsutella rhossiliensis* (cited by Kaya *et al.*, 1998).

Entomopathogenic nematodes show differential susceptibilities to nematophagous fungi; therefore, knowledge of the types of fungi that are present in the soil may be of assistance when selecting the appropriate entomopathogenic nematode for controlling a particular pest (Kaya and Koppenhöfer, 1996).

Predatory mites (Acari: Phytoseiidae)

Mass-reared phytoseiid mites may harbour *Wolbachia* and microsporidia (for recent reviews, see Poinar and Poinar, 1998; van der Geest *et al.*, 2000). Moreover, several cases of unidentified diseases have been reported.

Metaseiulus*: predators of mites*

BACTERIA. *Metaseiulus occidentalis* is host to unidentified microorganisms and *Wolbachia* endosymbionts. In some cases, *M. occidentalis* exhibits cream to pink rectal plugs that extrude from the anus and occasionally cause mites to become firmly attached to their substrate (Hess and Hoy, 1982). Rectal plug formation in *M. occidentalis* is common in older females and is associated with motor disfunction, reduced oviposition and eventual death. Affected immatures and males rarely exhibit rectal plugs. In some cases, the immatures and females become pale, thin and translucent.

Hess and Hoy (1982) described two unidentified microorganisms in *M. occidentalis*. One of these was found in the midgut, Malpighian tubules and epidermis but not in the ovarian or nervous tissues. This microorganism was present in all mites in varying numbers and was not detrimental to *M. occidentalis*. The second microorganism is rickettsia-like and was present in the majority (two-thirds) of asymptomatic and symptomatic mites. These rickettsia-like organisms were observed in the haemocoel and Malpighian tubules. Their presence within ovarian tissue and on egg surfaces suggests that they are transovarially transmitted. These microorganisms are thought to cause pathology in some circumstances, particularly when *M. occidentalis* is reared under crowded laboratory conditions (Hess and Hoy, 1982).

By using molecular methods (PCR with *Wolbachia*-specific primers), *Wolbachia* endosymbionts were detected in eight of nine laboratory populations of *M. occidentalis* and in four laboratory populations of *Tetranychus urticae* Koch that served as food for *M. occidentalis* (Johanowicz and Hoy, 1996). Therefore, it is probable that the rickettsia-like microorganisms described by Hess and Hoy (1982) are *Wolbachia* (for a discussion, see van der Geest *et al.*, 2000).

In *M. occidentalis*, *Wolbachia* spp. cause non-reciprocal reproductive incompatibilities between infected males and uninfected females. Uninfected females crossed with infected males produce few eggs and no female progeny. Many of the eggs that are produced are shrivelled (Johanowicz and Hoy, 1998b). The mechanisms by which *Wolbachia* spp. cause reproductive incompatibilities in *M. occidentalis* are unknown. However, *Wolbachia* infection seems to be associated with fitness costs, as the number of female progeny is lower in infected control crosses than in uninfected control crosses. These fitness costs may have prevented the rapid spread of *Wolbachia* in three laboratory populations of *M. occidentalis* (Johanowicz and Hoy, 1998a). *Wolbachia* spp. are eliminated from *M. occidentalis* when they are reared at an elevated temperature (33°C) (Johanowicz and Hoy, 1998a,b).

Lighthart *et al.* (1988) tested the effect of several stress factors on the susceptibility of *M. occidentalis* to the weak bacterial pathogen *Serratia marcescens*. A high preinoculation temperature pulse in relatively uncrowded conditions was most effective in enhancing susceptibility, higher mortality being the only disease symptom. Remarkably, starvation did not have such an effect.

Neoseiulus *(formerly* Amblyseius*): predators of mites and thrips*

VIRUSES. Unidentified, non-occluded virus-like particles were observed in the yolk of developing eggs inside *Neoseiulus cucumeris* (Oudemans) females. The effects of these virus-like particles on predator efficacy are not known (Bjørnson *et al.*, 1997).

BACTERIA. *Wolbachia* spp. have been detected in a population of *Neoseiulus barkeri* (Hughes) collected in The Netherlands and a population of *Neoseiulus bibens* Blommers from Madagascar (Breeuwer and Jacobs, 1996). Effects of *Wolbachia* spp. on these mites have not yet been investigated, but it is likely that *Wolbachia* are associated with non-reciprocal reproductive incompatibilities (for a discussion, see Breeuwer and Jacobs, 1996).

PROTOZOA. In 1991, microsporidia were reported in *Amblyseius cucumeris* and *Amyblyseius barkeri*, predatory mites that are used for controlling western flower thrips (*Frankliniella occidentalis* (Pergande)) and onion thrips (*Thrips tabaci* (Lindeman)), respectively (Beerling and van der Geest, 1991). This was the first report of microsporidia in mass-reared predatory mites. Microsporidia reduce the fecundity of *A. cucumeris* and *A. barkeri* and lower their predation capacity (Ramakers *et al.*, 1989; Beerling and van der Geest, 1991). During advanced stages of the disease, infected mites are swollen, white and lethargic and spores are so numerous that they often conceal the internal tissues (Beerling and van der Geest, 1991). Microsporidia may be transmitted vertically (from parent to progeny) or horizontally (from one infected individual to another) (Beerling and van der Geest, 1991). However, the mechanisms of pathogen transmission have not been determined for this system. Three types of microsporidian spores have been found in *A. cucumeris* and *A. barkeri* (Beerling *et al.*, 1993) but it is unclear if these represent one species of microsporidia with three different spore types or three distinct species. Beerling *et al.* (1993) developed a monoclonal-antibody ELISA as a bioassay for the detection of microsporidia in mass-reared *A. cucumeris* and *A. barkeri*. Further work is needed to determine the sensitivity of this test as a suitable screening method for microsporidia in mites.

FUNGI. A pathogenic fungus has been reported from phytoseiid mites collected in Brazil (Furtado *et al.*, 1996; Keller, 1997). Field-collected *Euseius citrifolius* (formerly *Amblyseius citrifolius*) (Denmark and Muma) were heavily infected by the fungus *Neozygites* sp. (Furtado *et al.*, 1996) and showed a high rate of mortality. Some cadavers carried near-white hyphae that produced pear-shaped conidia. However, *Amblyseius idaeus* Denmark and Muma and *Amblyseius limonicus* Garman and McGregor were not infected by *Neozygites* sp. isolated from the cassava green mite in laboratory tests (De Moraes and Delalibera, 1992). *E. citrifolius* collected in Brazil on two subsequent occasions contained viable resting spores and hyphal bodies of two distinct fungal species identified as *Neozygites acaricida* and *Neozygites* cf. *acaridis* (Keller, 1997).

UNIDENTIFIED DISEASE. In some cases, *Amblyseius hibisci* (Chant) exhibit dark-red occlusions within the alimentary tract near the distal opisthosoma. These are thought to be associated with the incomplete digestion of their prey, the citrus red mite (*Panonychus citri* (McGregor)). Symptomatic mites appear dorsal–ventrally flattened and are lethargic; females do not produce eggs and complete mortality of immatures occurs at 32 and 35°C. Symptoms are not observed in mites fed a diet of pollen from the ice plant, *Malephora crocea* (Jacq.) (Tanigoshi *et al.*, 1981).

Phytoseiulus: predators of mites

VIRUSES. Gravid *Phytoseiulus persimilis* Athias-Henriot females may carry unidentified, non-occluded virus-like particles in the yolk of developing eggs. The effects of these virus-like particles on predator efficacy are not known (Bjørnson *et al.*, 1997). Mites infected with *Rickettsiella phytoseiuli* contain virus-like particles that are abundant and visible in the dorsal part of the body, immediately below the cuticle (Šut'áková and Rüttgen, 1978).

BACTERIA. *R. phytoseiuli* reported in *P. persimilis* from the Ukraine (Šut'áková and Rüttgen, 1978) are observed throughout the tissues, with the exception of the gut diverticula, salivary glands and Malpighian tubules. *R. phytoseiuli* has no known effect on the development and longevity of *P. persimilis.*

Breeuwer and Jacobs (1996) detected *Wolbachia* in a commercial population of *P. persimilis* from The Netherlands. The effects of *Wolbachia* on the efficacy of *P. persimilis* have not been established.

In a microscopic study of the digestive tract of *P. persimilis*, bacteria-like entities detected in the gut lumen were thought to have entered the digestive tract during feeding (Arutunyan, 1985). However, these bacteria bear a marked similarity to the birefringent dumb-bell-shaped crystals that are frequently observed in the caecae and rectum of some phytoseiid mites (see Unidentified disease, below; Schütte *et al.*, 1995; Bjørnson *et al.*, 1997).

PROTOZOA. Three distinct species of microsporidia have been reported from *P. persimilis* from three commercial sources (North America, Europe and Israel; Bjørnson and Keddie, 2000). Microsporidia are not restricted to specific tissues, and spores are found in muscle fibres, the super- and suboesophageal ganglia, ovaries, eggs, cells underlying the cuticle and cells lining the caecal lumen and Malpighian tubules. Early development of all three microsporidia occurs in cells of the lyrate organ. The lyrate organ occupies a significant portion of the body and is thought to be involved in oogenesis or embryogenesis. Each microsporidium occupies a specific site within these cells. Infection of the lyrate organ may be necessary for the efficient vertical transmission of microsporidia in *P. persimilis* (vertical transmission is 100%). Males do not contribute to infection of the progeny.

Microsporidium phytoseiuli does not infect two-spotted spider mites (*T. urticae*). Therefore, prey mites do not contribute to pathogen transmission among *P. persimilis* mites. Horizontal transmission of microsporidia does not readily occur when uninfected adult predators are placed in proximity to infected *P. persimilis* females or solutions of microsporidian spores. Horizontal transmission is low (about 15%) when uninfected immatures develop in proximity to infected adult and immature mites (Bjørnson and Keddie, 2001). Although microsporidia are horizontally and vertically transmitted, little is known regarding the mechanisms of transmission.

Microsporidia-infected *P. persimilis* do not exhibit any obvious external symptoms; therefore, routine monitoring is necessary to detect microsporidia when disease prevalence is low (Bjørnson and Keddie, 1999). Microsporidia-infected *P. persimilis* produce fewer eggs than uninfected predators and the longevity and prey consumption of infected females are reduced (Bjørnson and Keddie, 1999). Decreases in fecundity, longevity and prey consumption of microsporidian-infected mites may have a profound effect on the performance of *P. persimilis* when released on crops for pest control.

To further complicate matters, microsporidia may alter the sex ratio of *P. persimilis*. Adult females infected with *M. phytoseiuli* produce fewer female progeny than those produced by uninfected females (Bjørnson and Keddie, 1999). This may affect the intrinsic rate of increase of the population and have an adverse effect on the establishment of mite colonies.

UNIDENTIFIED DISEASE. Birefringent dumb-bell-shaped crystals have been observed in *P. persimilis* from several sources (Schütte *et al.*, 1995; Bjørnson *et al.*, 1997, 2000). Excessive crystal formation is associated with white discoloration of the opisthosoma. Symptoms appear as two white stripes down the lateral sides of the body in the region of the Malpighian tubules or as a U-shaped discoloration in the distal opisthosoma. A rectal plug may be observed when symptoms are more pronounced. Rectal plugs often disrupt normal excretion and may cause the affected individual to become stuck to the leaf surface (Bjørnson *et al.*, 1997). The frequent occurrence of a prominent white dot in the opisthosoma of *P. persimilis* is correlated to reduced fecundity in mites examined following shipment (Bjørnson *et al.*, 2000). Crystals are observed in immature and adult *P. persimilis* (Bjørnson *et al.*, 1997); therefore, non-excessive crystal formation is probably a normal physiological process (Schütte *et al.*, 1995; Bjørnson *et al.*, 1997). An examination of *P. persimilis* from 14 commercial and research sources showed that there is no correlation between the occurrence of crystals and the presence of microsporidia, rickettsia or virus-like particles in *P. persimilis* (Bjørnson *et al.*, 1997).

A permanent change in the foraging behaviour has been observed in a laboratory and a commercial population of *P. persimilis* (Dicke *et al.*, 2000). This change consisted of a reduced level of attraction to odours emanating from prey-infested lima-bean plants. As this odour response plays an important role in host location in the field, it is expected that predators with a lower odour response are less effective for controlling spider-mite populations. Several hypotheses that may explain this behavioural change have been tested. Strong evidence has been found to support the hypothesis that the behavioural change is a symptom of an infectious disease (Schütte *et al.*, 1998; Dicke *et al.*, 2000). The behavioural change is a contagious phenomenon. Contact with dead conspecifics and their products originating from a population with a low odour-response level induces a reduced odour response and higher mortality in predators originating from a normal population (Schütte *et al.*, 1998). Maintenance of a population with a high odour response in the laboratory is possible only when strict hygienic rearing protocols are followed (Dicke *et al.*, 2000). A prominent symptom in adult female *P. persimilis* from a population with a low odour-response level is the presence of birefringent excretory crystals outside the excretory and digestive tract, predominantly in the legs. The presence of crystals in the legs is associated with reduced fecundity, a lower degree of attraction to prey-infested plants, and predator death (Schütte *et al.*, 1995).

Predatory insects (various unrelated taxonomic groups)

Aphidoletes *(Cecidomyiidae): predators of aphids*

FUNGI. The entomopathogenic fungus *Entomophthora apiculata* was recorded from *Aphidoletes aphidimyza* (Rondani) in a greenhouse in the southern part of Finland (Kariluoto, 1981). The midges were used to control *Myzus persicae* (Sulzer). The same fungus was isolated from *A. aphidimyza* released in experimental cages outside the glasshouse. Primary conidia measure 27.3 × 32.4 μm.

During a field study in southern Germany and Switzerland, three entomopathogenic fungi were isolated from dead individuals of *Aphidoletes thomsonii*. *B. bassiana* and *Isaria farinosa* were isolated from larvae, whereas *Entomophthora* sp. was isolated from larvae and adults (Smirnoff and Eichhorn, 1970).

Chrysoperla *(formerly* Chrysopa*) (Neuroptera: Chrysopidae): predators of aphids*

VIRUSES. *Chrysoperla* (L.) is a known host of two viruses: a cytoplasmic polyhedrosis and a nuclear polyhedrosis virus (Martignoni and Iwai, 1981). The effects of these viruses on the efficacy of *Chrysoperla* is not known.

PROTOZOA. The microsporidium *Pleistophora californica* has detrimental effects on the lacewing, *Chrysoperla californica* Coquillett. The lifespan and egg production of infected individuals are greatly reduced. However, the microsporidium has no effect on the lacewing's host, the potato tuber moth, *Gnorimoschema operculella* (Zeller) (Finney, 1950).

FUNGI. Several entomopathogenic fungi affect *Chrysoperla carnea* (Stevens). *C. carnea* is susceptible to *B. bassiana* when placed under temperature, starvation or nutritional stress or a combination of these stresses. The effect that *B. bassiana* has on *C. carnea* varies according to the age and gender of the host (Donegan and Lighthart, 1989).

When nitrogen is removed from the diet, the resulting nutritional stress causes a disruption of larval development and a significant increase in adult and larval mortality. Under these circumstances, the effects of *B. bassiana* on *C. carnea* are intensified. Starvation has a greater effect on the larval stages and it is thought that differences in susceptibility are due to larval feeding behaviour. Early-instar larvae require a continuous food supply, whereas adults feed little during the first few days after pupal emergence. Temperature stress affects *C. carnea* mortality to *B. bassiana* only in combination with nutrition or starvation stress (Donegan and Lighthart, 1989).

Fungi may also infect *C. carnea* larvae when they are not subjected to stress. *Verticillium lecanii*, *B. bassiana* and *Paecilomyces fumoso-roseus* cause larval mortality, and toxic effects on surviving individuals may be traced until the fifth generation after inoculation. Effects of these fungi on *C. carnea* include decreased adult emergence, development of teratogenic forms, hampered cocoon formation and decreased fecundity and predatory rates (Pavlyushin, 1996).

Lady bird beetles *(Coleoptera: Coccinellidae): predators of aphids*

Several groups of invertebrate pathogens' symbionts have been recorded from the Coccinellidae, including gregarines, microsporidia, fungi, nematodes and symbionts' bacteria (Drea and Gordon, 1990; Majerus and Hurst, 1997). A single pathogen may be found within several coccinellid species, bringing the host specificity of these pathogens into question. Additional information is needed to determine if native pathogens from field-collected beetles are encountered in mass-rearing systems.

During the past decade, male-killing symbionts (maternally inherited bacteria that kill male hosts during embryogenesis) have been intensively studied in aphidophagous coccinellids (for a review, see Majerus and Hurst, 1997). Within the Coccinellidae, there is more than one agent of male lethality. Male-killing is caused by a diverse array of inherited bacteria from three different groups (Mollicutes, Proteobacteria, Flavobacteria-Bacterioides). Maternally inherited male-killing symbionts are associated with coccinellids of six different genera, including *Adalia bipunctata* (L.), *Adonia variegata* (Goeze), *Coleomegilla maculata* (De Geer), *Harmonia axyridis* Pallas, *Hippodamia convergens* and *Menochilus sexmaculatus* Fabricius (cited by Majerus and Hurst, 1997).

The habits of ladybirds (production of eggs in clutches, sibling cannibalism, high level of neonate larval mortality due to starvation and feeding on ephemeral prey) are thought to have made them particularly prone to male-killing symbionts. The presence of male-killing bacteria results in resource allocation among siblings. By consuming their dead male siblings, newly hatched female larvae increase their chances of survival following dispersal. Furthermore, low egg hatch in infected clutches reduces the risk of cannibalism of female larvae by their siblings (Majerus and Hurst, 1997).

Adalia*: predators of aphids*

BACTERIA. Some strains of the two-spot ladybird beetle, *A. bipunctata*, and a strain of the ten-spot ladybird beetle, *Adalia decempunctata* L., are known to produce broods with a strong bias towards females (Hurst *et al.*, 1992; von der Schulenburg *et al.*, 2001) or produce no male offspring at all (cited by Matsuka *et al.*, 1975; von der Schulenburg *et al.*, 2001). Egg hatches of *A. bipunctata* broods with distorted sex ratios were about half that of broods with non-biased sex ratios due to the death of male offspring early in their development (Hurst *et al.*, 1992). In both *Adalia* species, the male-killing trait is heritable only through the female line and is efficiently transmitted from the adult female to her female progeny. Following treatment with tetracycline, adult females revert to producing clutches with high hatch rates and a normal 1:1 sex ratio (Hurst *et al.*, 1992; von der Schulenburg *et al.*, 2001). These findings indicate that the male-killing element is probably a vertically transmitted bacterium.

At least four different bacteria cause sex-ratio distortion in *A. bipunctata*: a *Rickettsia* (Werren *et al.*, 1994), a *Spiroplasma* (Hurst *et al.*, 1999b) and two strains of *Wolbachia* (Hurst *et al.*, 1999a). In one case, all four bacteria were present in a single sample of *A. bipunctata* (Majerus *et al.*, 2000). So far, only one male-killing *Rickettsia* has been reported in *A. decempunctata* (von der Schulenburg *et al.*, 2001). Phylogenetic analysis of *Rickettsia* DNA sequences from different populations of both *Adalia* species revealed a single origin of male-killing in the genus *Rickettsia* and indicated possible horizontal transfer of *Rickettsia* between the two hosts (von der Schulenburg *et al.*, 2001).

In *A. bipunctata*, little is known regarding the mechanism by which male offspring are killed early in embryogenesis. Infection has no effect on larval emergence or development rate; however, infected females produce fewer progeny and do not live as long as uninfected females (Hurst *et al.*, 1994).

PROTOZOA. The gregarine *Gregarina ruszkowski* and the microsporidium *Nosema coccinellae* were detected in *A. bipunctata* collected in Poland. The microsporidium was found to infect the gut of only one of 168 beetles (Lipa *et al.*, 1975). For a description of this gregarine and microsporidium, see the section on *Coccinella* below.

FUNGI. During a field study in Poland, *B. bassiana* was detected in hibernating adult beetles from nine coccinellid species, including *A. bipunctata* (Lipa *et al.*, 1975). Several specimens collected in spring were totally covered with mycelium, suggesting that *B. bassiana* plays an important role in the mortality of overwintering beetles (Lipa *et al.*, 1975).

Two other parasitic fungi have been recorded from *Adalia*. These include a species of *Laboulbenia* in *Adalia* spp. in France (cited by Drea and Gordon, 1990) and *Hesperomyces virescens* from hibernating *A. bipunctata* in England (Weir and Beakes, 1996). The effect of these fungi on beetle mortality during hibernation has not yet been studied.

Coccinella: *predators of aphids*

PROTOZOA. Several species of gregarines have been reported from *Coccinella* collected in Asia and Eastern Europe, suggesting that gregarines are common in these hosts. However, little is known regarding the effect, if any, that gregarines have on the predation efficiency of infected coccinellids. *Gregarina coccinellae* (synonym: *Gregarina barbarara*) was originally described from *Coccinella septempunctata* L. and *Myrrha octodecimguttata* (L.) collected in Poland and the Soviet Union. It was later found infecting *Exochomus quadripustulatus* L. and *Harmonia quadripunctata* (Pontoppidan) (Lipa *et al.*, 1975). Immature stages develop within the midgut epithelial cells and mature sporonts are found in the gut lumen, where as many as 100 individual gregarines are observed. This gregarine species is reported to exert some harmful, but unspecified, effects on the life processes of the host. *Gregarina hyashii* is morphologically similar to *G. coccinellae*. This gregarine was reported in *Coccinella transversalis* (Fabricius). Early development occurs within the midgut epithelial cells and an increase in basal lamina suggests that there is a change in the secretory habits of the infected cells (Sengupta and Haldar, 1996). The gregarine *G. ruszkowski* Lipa (synonym: *Gregarina katherina*) was reported in *C. septempunctata* and *C. quinquepunctata* L. from Poland but has also been found in *A. bipunctata* and *E. quadripustulatus* (Lipa *et al.*, 1975). A fourth species, *Gregarina dasguptai*, is found in *C. septempunctata* collected from tea plants in West Bengal, India (Mandal *et al.*, 1986). A fifth species, *Anisolobus indicus*, is morphologically similar to gregarines of the genus *Gregarina* and is reported in *C. septempunctata* collected in India (Haldar *et al.*, 1988). Further work is needed to determine the effects of gregarines on *Coccinella* and other related beetle genera.

The microsporidium *N. cocinnellae* was originally described from the coccinellids *C. septempunctata*, *M. octodecimguttata* and *Hippodamia tredecimpunctata* L. (Lipa, 1968). This microsporidium was first detected in coccinellids from Poland in 1963 and was subsequently detected in specimens from the Soviet Union. These two localities are more than 2000 km apart, giving Lipa (1968) reason to speculate that the distribution of *N. coccinellae* could cover the whole of Europe. *N. coccinellae* was later found to infect *A. bipunctata*, *C. quinquepunctata* and *E. quadripustulatus*. This microsporidium parasitizes the gut epithelium, Malpighian tubules, muscles, nerves and gonads but is not found in the fat body (Lipa, 1968; Lipa *et al.*, 1975). In the case of *C. septempunctata*, only part of the gut and susceptible organs are infected (infection is focal). The most intensive infection is observed in the Malpighian tubules.

Infected portions of the organs are hypertrophied, due to the large number of spores that occupy the cells (Lipa, 1968). In *M. octodecimguttata*, infection is observed in the gut epithelium, which is frequently destroyed by the pathogen (Lipa, 1968). Spores of *N. coccinellae* are ellipsoidal and measure 4.4–6.7 μm long by 2.3–3.4 μm wide in fresh smears (Lipa, 1968). *N. coccinellae* spores are about 1 μm larger than those of *N. hippodamiae*, a microsporidium that infects the fat body of *H. convergens* Guérin (Lipa, 1968). Depending on the host species, up to 25% of the beetle population may be infected with *N. coccinellae*. There are no data, however, that describe the effects of this microsporidium on the efficacy of its hosts. The life cycle of the microsporidium *Nosema tracheophila* has been investigated in adult *C. septempunctata* of the USA (Cali and Briggs, 1967). Spores of this species are ovoid and measure 4.0–5.3 μm long by 2.2–3.1 μm wide. Infected hosts do not show external signs of the disease. Spores are present in the haemocytes, tracheal epithelium and connective tissue. Even when heavily infected, spores are not present in midgut cells or the fat body of *C. septempunctata*, as is the case for *N. hippodamiae*.

FUNGI. The fungal pathogen *B. bassiana* is an important mortility factor in several overwintering coccinellid species in Poland, including *A. bipunctata*, *C. quinquepunctata*, *C. septempunctata* and *H. quadripunctata* (Lipa *et al.*, 1975; Hemptinne, 1988).

NEMATODES. Female *C. septempunctata* from India may be parasitized by the solitary mermithid nematode, *Coccinellimermis*. This nematode may constitute 27% of the total body weight (Rhamhalinghan, 1987). After infection, juvenile worms curl up and occupy the host haemocoel, where they may inflict injuries on the heart (Rhamhalinghan, 1992). Obvious symptoms of nematode infection include a lower feeding rate, resulting in malnutrition, a lower level of activity before parasite emergence and paralysation after parasite emergence. Secondary effects include a lower haemocoel volume and heart-beat frequency (Rhamhalinghan, 1992).

Coleomegilla*: predators of aphids*

BACTERIA. A male-killing trait in *C. maculata* is heritable only through the female line and is tetracycline-sensitive (Hurst *et al.*, 1996). The trait, however, is frequently absent early in the reproductive cycle of the host. Some females first produce both male and female offspring but only female offspring are produced later on. Early clutches from such females show good hatchability and females emerging from these clutches do not show the male-killing trait. The progressive nature of the male-killing trait in *C. maculata* may be caused by a host–parasite conflict over transmission (for a discussion, see Hurst *et al.*, 1996). The male-killing bacterium of *C. maculata* differs from the male-killer present in *A. bipunctata*. The causal agent is most closely related to *Blattabacterium* (Flavobacteria–Bacterioides group), a host-beneficial symbiont found in cockroaches and some termites (Hurst *et al.*, 1997). This first record of a male-killing agent in the Flavobacteria–Bacterioides group indicates that this trait has evolved independently in different symbionts and in many host species.

FUNGI. Infection experiments have shown that the entomopathogenic fungus *B. bassiana* may be pathogenic to *C. maculata*. In one study, beetles were susceptible to *B. bassiana* when they were directly contaminated by sprayed fungal conidia, whereas no infection occurred when prey insects were consumed from leaves that had been sprayed (Lord *et al.*, 1988).

Harmonia*: predators of aphids*

BACTERIA. Several populations of *H. axyridis* have been shown to occasionally produce highly female-biased sex ratios (Matsuka *et al.*, 1975; Gotoh, 1982; Majerus *et al.*, 1998). This trait may be maintained within a population over several generations and is maternally inherited (Matsuka *et al.*, 1975). Moreover, several types of abnormal sex ratio (SR) are present: the 'complete SR' (hatching rate half as high as for the normal egg batches; only female adults develop), the 'incomplete SR' (hatching rate somewhat higher; few male adults develop) and the 'progressive SR' (normal hatching rate; male and female offspring develop from egg batches laid during the initial oviposition period, whereas egg batches deposited later on yield only females with a low hatching rate). In two cases, 'cured' females developed spontaneously from female-biased lines. In another study, three populations of *H. axyridis* showed a high variance in the presence of the male-killing trait (Majerus *et al.*, 1998). This trait was absent in a Russian population, whereas approximately 2% and 49% of the females carried the trait in a Japanese and a Mongolian population, respectively. The trait has high vertical-transmission efficiency and is antibiotic-sensitive. Antibiotic treatments (tetracycline and chloramphenicol) fail to produce an increase in egg hatch rates in *H. axyridis* exhibiting abnormal sex ratios. However, antibiotic treatment results in a significant increase in the proportion of eggs showing embryonic development. Treatment of *H. axyridis* with abnormal sex ratios does not usually effect a complete cure (Majerus *et al.*, 1998).

As is the case for *A. bipunctata*, several distinct bacteria may cause heterogeneous male-killing in *H. axyridis*. However, only one causal agent has been detected. Molecular analysis of a Russian line bearing a maternally inherited male-killing trait revealed the presence of a bacterium belonging to the genus *Spiroplasma* (Zakharov *et al.*, 1999). The *Spiroplasma* from *H. axyridis* is not identical to the one detected in *A. bipunctata*. A more detailed study of geographically distant Japanese populations revealed the presence of an identical *Spiroplasma* in two populations of *H. axyridis* (Majerus *et al.*, 1999). A spiroplasm-specific PCR verified the presence of *Spiroplasma* in all-female and predominantly female *H. axyridis* populations, whereas it was absent in individuals from normal populations. Bacteria are present in lymph and haemocytes only in populations showing the male-killing trait. Phylogenetic analysis revealed that the bacterium is most closely related to *Spiroplasma ixodetis*, a *Spiroplasma* from ticks, and the male-killing *Spiroplasma* from *A. bipunctata* (Majerus *et al.*, 1999).

PROTOZOA. The gregarine *G. coccinellae* was recorded in *H. quadripunctata* during a field study on hibernating coccinellids in Poland (Lipa *et al.*, 1975). The immature stages of this gregarine develop in the epithelial cells of the midgut, whereas the mature sporonts live in the gut lumen (see section on *Coccinella*, above).

FUNGI. The entomopathogenic fungus *B. bassiana* was detected on hibernating *H. quadripunctata* (Lipa *et al.*, 1975) and is thought to be an important mortality factor in hibernating coccinellids (Lipa *et al.*, 1975).

Hippodamia: *predators of aphids*

BACTERIA. Abnormal sex ratios have been reported in *Hippodamia quinquesignata* (Kirby) collected in Utah. The trait is maternally inherited (Shull, 1948). It is not known if the sex-ratio distortion is due to the presence of bacteria, as is the case for *A. bipunctata*, *C. maculata* and *H. axyridis*.

James and Lighthart (1992) tested the susceptibility of *H. convergens* to the weak bacterial pathogen *Pseudomonas fluorescens*. This facultative pathogen may invade the haemocoel of the host and cause septicaemia, eventually leading to the death of the host. Fourth-instar larvae are more resistant to *P. fluorescens* than earlier-instar larvae. A poor diet offered at 25°C decreases the susceptibility of first-instar larvae to *P. fluorescens*, whereas starvation at a high temperature (30°C) increases susceptibility to the bacterium (James and Lighthart, 1992).

Two species of ice-nucleating active (INA) bacteria, *Enterobacter agglomerans* and *Enterobacter taylorae*, were isolated from the gut of field-collected *H. convergens* (Lee *et al.*, 1991). INA bacteria are unique in their capacity to catalyse ice nucleation at temperatures as high as 1–2° below 0°C. Ingestion of INA bacteria cause freezing of *H. convergens* at temperatures as high as −1.5°C, whereas unfed beetles and those fed sterile water or the non-INA bacterium *Escherichia coli* freeze at −16°C to −17°C. The effects of INA bacteria on hibernating *H. convergens* under field conditions are not yet known.

PROTOZOA. Beetles of the genus *Hippodamia* are hosts to at least two described species of microsporidia. *Nosema hippodamiae* was the first microsporidian pathogen described from *H. convergens* collected in California (Lipa and Steinhaus, 1959). This microsporidium infects the midgut epithelium and pupal fat body. Infected pupae show no conspicuous external symptoms and the adults and larvae are not infected. *N. hippodamiae* occurs in localized epizootics. There is no evidence that the disease is severe in local populations. Spores are ovoid and measure 3.3–5.4 μm long × 2.2–2.7 μm in width. Specimens infected with *N. hippodamiae* were also infected with an unidentified eugregarine (Lipa and Steinhaus, 1959).

The microsporidium *N. coccinellae* was found to infect *H. tredecimpunctata* from Poland. Fresh spores are slightly larger than those of *N. hippodamiae* and measure 4.4–6.7 μm long × 2.3–3.4 μm wide. This microsporidium causes localized infections of the midgut epithelium, Malpighian tubules, gonads, muscles and nerves (Lipa *et al.*, 1975).

An undescribed microsporidium found in *H. convergens* from California has spores that are similar in size to those of *N. hippodamiae*. The pathogen causes lesions in the midgut epithelium and Malpighian tubules. The fat body of infected beetles lacks vacuoles and cell boundaries and has cell nuclei that are hypertrophied. The disease begins in the fat body and progresses to other tissues. The gut is the last tissue to become infected (Sluss, 1968). Little is known regarding the effects of microsporidia on the life-history characteristics of *Hippodamia*.

H. tredecimpunctata, like *C. septempunctata*, *M. octodecimguttata*, *E. quadripustulatus* and *H. quadripunctata*, is host to the gregarine parasite *G. coccinellae* (Lipa, 1967). For more information on this disease, see *Coccinella*.

FUNGI. Three entomopathogenic fungi (*B. bassiana*, *M. anisopliae* and *P. fumoso-roseus*) with a broad host range show some degree of virulence towards *H. convergens* (James and Lighthart, 1994). First-instar beetle larvae exposed to different concentrations of fungal preparations in the laboratory exhibit high mortality. However, the effects of these fungi on *Hippodamia* under field conditions are not known.

Parasitoids (Hymenoptera)

Endoparasitoid larvae develop within their host where they may confront host pathogens. As a result, parasitoids may be affected by diseases from the host directly or indirectly (Brooks, 1993). Most host–parasitoid–pathogen interactions are detrimental to the parasitoid. Parasitoids may be adversely affected due to the premature death of the host, the presence of pathogen-produced toxins in the host or the alteration of host nutrition or physiology as a result of infection. Furthermore, the parasitoid may itself be susceptible to the host pathogen (Brooks, 1993). We shall present only reports of the latter case. For more information on the other cases, see Brooks (1993).

Parasitoids may also be infected independently of their host. *Wolbachia* infections represent such a case. They are present in several parasitoids used for biological pest control and induce parthenogenesis (*Aphytis*, *Encarsia*, *Lysiphlebus*, *Muscidifurax* and *Trichogramma*), cytoplasmic incompatibility (*Nasonia*) or elevated fecundity (*Trichogramma*). These symbionts may cause little or no measurable effect on parasitoid fitness (Zchori-Fein *et al.*, 2000); therefore, *Wolbachia* spp. are not always detrimental to natural enemies (Chapter 8).

Aphidius (Aphidiidae): parasitoids of aphids

FUNGI. Potato aphids (*Macrosiphum euphorbiae* (Thomas)) parasitized by *Aphidius nigripes* Ashmead are susceptible to infection by *V. lecanii*, an opportunistic fungus that is commercially produced for aphid and whitefly control in commercial cropping systems. The application of *Verticillium lecanii* for aphid control can have an adverse affect on the survival of *A. nigripes*. When developing inside an infected host, successful parasitoid development is dependent on the timing of host infection. Parasitoid survival increases if host exposure to the fungus is delayed following parasitism (Askary and Brodeur, 1999). When aphids are heavily infected by *V. lecanii* under laboratory conditions, dense aggregations of hyphae invade the host tissues and the fungus penetrates the *A. nigripes* larvae. Under these circumstances, infection of parasitoid tissues is localized. In some cases, abundant hyphae are found in the gut of *A. nigripes* larvae, suggesting that parasitoid larvae consume fungal spores and hyphae when feeding on host haemolymph and other tissues. However, there is no evidence of invasion of the parasitoid tissues by the fungus (Askary and Brodeur, 1999). Further work is necessary to determine the suitability of infected hosts for parasitoid development and to quantify the effects of *V. lecanii* on *A. nigripes* and other biological control agents under greenhouse conditions.

Aphidius spp. that develop within cereal aphids are infected by an entomophthoraceous fungus, probably *Erynia neoaphidis* (cited by Brooks, 1993). However, it is not known if the fungus invades the parasitoid before or after death. Moreover, only two of 26 examined parasitoid larvae were colonized by the fungus (cited by Brooks, 1993).

Aphytis *(Aphelinidae): parasitoids of scale insects*

BACTERIA. About one-quarter of all *Aphytis* species exhibit thelytokous parthenogenesis (cited by Zchori-Fein *et al.*, 1995), whereby unfertilized females produce predominantly female offspring. In these species, males are produced regularly but usually at a low rate. *Wolbachia* spp. are associated with parthenogenesis induction in several thelytokous *Aphytis* species that exhibit reversible parthenogenesis (*A. chilensis* Howard, *A. chrysomphali* (Mercet), *A. diaspidis* Howard, *A. lingnanensis* (Comp) and *A. yanonensis* DeBach) (Zchori-Fein *et al.*, 1995, 1998; Gottlieb *et al.*, 1998). *Wolbachia* spp. are found in the ovaries and eggs of thelytokous lines, suggesting that these microorganisms are transmitted transovarially (Zchori-Fein *et al.*, 1998). Treatment with rifampicin has been successful in eliminating *Wolbachia* from these lines (Zchori-Fein *et al.*, 1995, 1998). A phylogenetic analysis of *Wolbachia* spp. from four *Aphytis* species belonging to distant phylogenetic groups place these *Wolbachia* spp. in group 'A', together with the parthenogenesis-inducing *Wolbachia* found in *Muscidifurax raptor* Girault and Saunders. The sequence analysis of two different *Wolbachia* genes (16S ribosomal DNA (rDNA) and *ftsZ*) results in the same phylogenetic tree (Zchori-Fein *et al.*, 1995, 1998). Virtually identical *Wolbachia* spp. are found in *Aphytis* species belonging to distant phylogenetic groups. This evidence suggests that the horizontal transmission of *Wolbachia* occurred recently in *Aphytis*.

Cotesia *(formerly Apanteles) (Braconidae): parasitoids of Lepidoptera*

VIRUSES. The survival or development of *Cotesia* may be affected when their lepidopteran hosts are infected with viruses. *Cotesia marginiventris* (Cresson) failed to develop within non-occluded virus-infected *Spodoptera frugiperda* (J.E. Smith) that are infected before or during parasitization. The virus does not kill the host before *C. marginiventris* is able to successfully complete its development. Rather, it is thought that either the parasitoid eggs or the larvae are killed when the host is inoculated (Hamm *et al.*, 1985). *Cotesia melanoscelus* (Ratzeburg) that develop in *Lymantria dispar* (L.) larvae infected with a nuclear polyhedrosis virus carry viral polyhedra in their haemocytes and fat body. However, parasitoid larvae develop successfully in gypsy-moth larvae infected with nuclear polyhedrosis virus (cited by Brooks, 1993).

PROTOZOA. Parasitoids of the genus *Cotesia* are host to several species of microsporidia. For example, *Cotesia glomerata* is susceptible to *Nosema aporivora*, a pathogen of its host *Aporia crataegi* L., *C. marginiventris* is host to *Vairimorpha* sp., a pathogen of its host *Heliothis zea* (Boddie), and *Cotesia rubecula* (Marshall) is host to *N. mesnili*, a pathogen of its host *Pieris brassicae* (cited by Brooks, 1993).

Progeny of *C. glomerata* (L.) are host to *Nosema polyvora*, a microsporidium that infects *P. brassicae* (L.). Female parasitoids that are heavily infected with *N. polyvora* produce few progeny. The ovaries of infected females contain eggs with many microsporidian spores. These eggs are incapable of developing within *P. brassicae*; however, they are able to infect the host with microsporidia (Issi and Maslennikova, 1966). The microsporidium *N. mesnili* also infects *C. glomerata* and its host, *P. brassicae*. Infected parasitoid larvae may die prematurely if the host is heavily infected (Tanada, 1955; Hostounský, 1970).

The microsporidium *Nosema bordati* is highly pathogenic to *Cotesia flavipes* Cameron and its host, the cereal stem borer *Chilo partellus* (Swinhoe). High parasitoid mortality is observed in all stages of infected *C. flavipes*, and dead larvae and pupae contain high numbers of spores. *N. bordati* is transmitted transovarially (Kfir and Walters, 1996). It multiplies in all stages of *C. flavipes* and may be horizontally transmitted by infected parasitoids (Bordat *et al.*, 1994).

Encarsia *(Aphelinidae): parasitoids of mainly whiteflies*

BACTERIA. All known populations of *Encarsia formosa* Gahan are parthenogenetic (Zchori-Fein *et al.*, 1992). However, male production can be induced by exposing infected females to high temperatures (31°C) for two or more generations (Kajita, 1993) or by treating them with certain antibiotics (Zchori-Fein *et al.*, 1994). Resulting male progeny produce sperm and may mate but insemination of females does not occur (Zchori-Fein *et al.*, 1992). It is possible that *E. formosa* has been parthenogenetic for such a long time that behavioural, physical and/or mechanical prezygotic barriers exist in either males or females that prevent fertilization (Zchori-Fein *et al.*, 1992). Thelytoky (parthenogenetic reproduction in which only females are produced) in *E. formosa* is mediated by *Wolbachia* symbionts (van Meer *et al.*, 1995). Sequence analysis of 16S rDNA reveals that the symbiont of *E. formosa* is closely related to *Wolbachia pipiens*, which is associated with cytoplasmic incompatibility in *Culex pipiens* Pallens (van Meer *et al.*, 1995). The effect of *Wolbachia* on parasitoid fitness is unclear. In one study, female parasitoids treated with antibiotics produced more offspring than untreated females (Zchori-Fein *et al.*, 1994), whereas the opposite results were found in another study (Stouthamer *et al.*, 1994). However, *Wolbachia* spp. have clear effects on oviposition behaviour and sex-specific developmental requirements (Hunter, 1999). Most sexual *Encarsia* are obligate autoparasites: females lay fertilized female-producing eggs in hosts, while unfertilized male-producing eggs are laid in immature parasitoids. Cured *E. formosa* females show a very different behaviour. They deposit most of their unfertilized eggs in their hosts and males emerge exclusively from them (Hunter, 1999). Antibiotic treatment induced male production in two thelytokous populations of *Encarsia meritoria* Gahan (synonym *Encarsia hispida* DeSantis). After treatment with tetracycline, females of an Italian population produced 81% male progeny (Giorgini, 2001), whereas females of a Spanish population produced almost entirely male offspring (Hunter, 1999).

PROTOZOA. An undescribed microsporidium (*Nosema* sp.) found in the ovaries of *Encarsia* nr. *pergandiella* from Brazil is associated with a noted decline in fecundity in production systems (Sheetz *et al.*, 1997). The effects of this microsporidium on the efficacy of *Encarsia* nr. *pergandiella* are yet to be determined. The authors report that adult *Encarsia* nr. *pergandiella* were effectively cured of microsporidia when they were given a single treatment of rifampicin administered in solution. However, it is not clear if the infection status of the treated and untreated individuals were verified before the rifampicin was administered. Nevertheless, this is the first report of microsporidia in *Encarsia*, suggesting that mass-produced *Encarsia* may be susceptible to microsporidia.

Eretmocerus *(Aphelinidae): parasitoids of whiteflies*

BACTERIA. The thelytokous parasitoid *Eretmocerus staufferi* Rose and Zolnerowich is host to parthenogenesis-inducing *Wolbachia* (van Meer *et al.*, 1999). Sequencing of the *wsp* gene revealed that these bacteria are closely related to *Wolbachia* from *Leptopilina australis* (Belizin) but not to those in *E. formosa* (van Meer *et al.*, 1999). *Wolbachia* spp. were also detected in an Australian parthenogenic form of *Eretmocerus mundus* (De Barro and Hart, 2001). Parasitoids treated with the antibiotic rifampicin produced males that were partially functional, as they produced sperm, copulated with females and inseminated them. However, no sexual line could be established. For each of the three generations following antibiotic treatment, infected females produced more progeny than cured females, due to a greater egg mortality in the cured line (De Barro and Hart, 2001). Most descendants of cured females were free of detectable *Wolbachia* infection.

UNIDENTIFIED DISEASE. Varying numbers of laboratory-reared *E. mundus* Mercet females were found to be sterile (Gerling and Fried, 1997). At a lower rate, sterility also occurs in field populations. Sterility in *E. mundus* is significantly correlated with crowding of female parasitoids during oviposition and is variable among laboratory populations (Gerling and Fried, 1997). Sterile females show pronounced behavioural changes and appear more restless. Sterility is thought to be induced by an unknown pathogen (Gerling and Fried, 1997).

Lysiphlebus *(Aphidiidae): parasitoids of aphids*

BACTERIA. *Wolbachia*-induced thelytoky has been reported in *Lysiphlebus cardui* (Marshall), *Lysiphlebus confusus* (Tremblay and Eady) and *Lysiphlebus fabarum* Marshall (cited by Starý, 1999). Thelytokous populations of these three species consist almost exclusively of females, although males are occasionally produced. Arrhenotokous and thelytokous populations coexist in the same localized areas (Starý, 1999). However, the distribution of thelytokous populations is restricted to the West Palaearctic species of *Lysiphlebus*. *Wolbachia* spp. have not been detected in *Lysiphlebus* populations of North America, despite an extensive study of this group (cited by Starý, 1999). Thelytokous virgin females of the three *Lysiphlebus* species display two peculiar behavioural features. First, they refuse the mating attempts of biparental males and, second, they form aggregations whereby they display aberrant oviposition behaviour by attempting to oviposit in each other. The thelytokous trait is maternally transmitted (Starý, 1999).

Muscidifurax *(Pteromalidae): parasitoids of flies*

BACTERIA. *Wolbachia* endosymbionts present in a parthenogenetic strain of *Muscidifurax uniraptor* Kogan and Legner were not found in the closely related sexual species *M. raptor* (Stouthamer *et al.*, 1993). Sequence analysis of the 16S rDNA revealed that *Wolbachia* spp. of *M. uniraptor* are most closely related to *Wolbachia* spp. that induce cytoplasmic incompatibility in different *Nasonia* species (Stouthamer *et al.*, 1993). *Wolbachia* densities are reduced when adult females are treated with the antibiotic rifampicin, resulting in an increase in the proportion of male offspring. Both *Wolbachia* density and the number of male offspring are dependent on antibiotic dose (Zchori-Fein *et al.*, 2000). Results indicate that *Wolbachia* spp. have no effect on the fecundity, longevity or reproductive rate of *M. uniraptor* (Zchori-Fein *et al.*, 2000). Two additional studies have failed to detect any negative effects of *Wolbachia* on host fitness (Stouthamer *et al.*, 1994; Horjus and Stouthamer, 1995). These findings are in accordance with the hypothesis that a negative effect of *Wolbachia* on host fitness is not expected in species where the symbiont infection has gone to fixation, as in *M. uniraptor* (Stouthamer *et al.*, 1994). Gottlieb and Zchori-Fein (2001) found three reproductive barriers between antibiotic-induced males and conspecific females of *M. uniraptor*: males do not produce mature sperm, females are reluctant to mate and a major muscle is absent from the spermatheca. These findings support the hypothesis that thelytokous reproduction is irreversible in *M. uniraptor*.

PROTOZOA. Microsporidia are common pathogens of pteromalid parasitoids associated with livestock and poultry houses, including *Muscidifurax* spp., *Nasonia vitripennis* (Walker), *Pachycrepoideus vindemiae* (Rondani), *Spalangia cameroni* Perkins, and *Spalangia endius* Walker (Dry *et al.*, 1999). Host specificity of microsporidia that infect pteromalid parasitoids is not known. Microsporidia are found in wild parasitoids collected on dairy farms; however, disease prevalence is often greater on farms where commercially reared parasitoids have been previously released (Becnel and Geden, 1994; Dry *et al.*, 1999). The microsporidium *Nosema muscidifuracis* was described from *M. raptor* collected from dairy farms in central New York. Developmental stages of the pathogen are detected in the midgut epithelium, Malpighian tubules, ovaries and fat body of infected larvae and adults. Two morphologically distinct spores are found in larvae, pupae and adults. One of these is thought to be responsible for autoinfection (reinfection of the same host) and another for pathogen transmission from one host to the next. A third spore type, found in eggs of *M. raptor*, is thought to either initiate infection at eclosion or to transmit the pathogen to a new host when infected eggs are cannibalized (Becnel and Geden, 1994).

Horizontal transmission of *N. muscidifuracis* occurs when uninfected parasitoid immatures cannibalize infected immatures in superparasitized pupal hosts or when uninfected adults feed on infected parasitoid immatures within host puparia. *N. muscidifuracis* infects the oocytes of female parasitoids, resulting in vertical transmission that is 100% efficient (Geden *et al.*, 1995). Parasitoid crowding, which leads to superparasitism, can greatly amplify the prevalence of this microsporidium in mass rearings (Geden *et al.*, 1995).

Infected *M. raptor* take longer to complete development, live half as long and produce one-tenth as many progeny as uninfected parasitoids (Zchori-Fein *et al.*, 1992; Geden *et al.*, 1995). Therefore, microsporidia may cause a dramatic reduction in the efficacy of *M. raptor* as a biological control agent.

Treatment of infected adult parasitoids with fumagillin has no effect on the disease; however, disease prevalence is reduced when fly puparia that contain parasitoid eggs are immersed in a 47°C water-bath for 30–60 min. The most practical way to eliminate *N. muscidifuracis* from *M. raptor* colonies with a 100% disease rate is to use heat to reduce disease prevalence and then isolate uninfected females and use them to start uninfected colonies (Geden *et al.*, 1995).

The presence of microsporidia in *M. raptor* raises questions regarding the quality of mass-produced parasitoids used for biological control. The presence of disease may also explain some of the inconsistencies reported for *M. raptor* life-history traits (Zchori-Fein *et al.*, 1992).

Nasonia *(Pteromalidae): parasitoids of flies*

BACTERIA. *Nasonia vitripennis* is known to have a female-biased sex ratio, resulting from a low (20%) hatch rate of male eggs (cited by Balas *et al.*, 1996). This son-killer trait is caused by the bacterium *Arsenophonus nasoniae* (Gherna *et al.*, 1991). The bacterial cells are non-motile, non-spore-forming, long rods, which are occasionally filamentous in young cultures (Gherna *et al.*, 1991). *A. nasoniae* infects several tissues, including the brain, fat body, muscles, eyes, haemocytes and reproductive tract (Huger *et al.*, 1985). Infections originate in the larval midgut, suggesting that infection takes place *per os* (Huger *et al.*, 1985). Bacteria are apparently transferred to fly pupae during parasitoid oviposition. *A. nasoniae* is maternally transmitted; horizontal transmission occurs when progeny from an uninfected female develop within a host parasitized by an infected female parasitoid (Werren *et al.*, 1986). *A. nasoniae* has also been reported in field-collected *Nasonia longicornis* Darling, a sibling species of *N. vitripennis* (Balas *et al.*, 1996). Over 2 years, the son-killer trait was expressed in 5.4 and 7.8% of natural *Nasonia* populations, respectively (Balas *et al.*, 1996).

Wolbachia-induced cytoplasmic incompatibility is a well-documented phenomenon in *Nasonia* species. *Wolbachia* spp. are found in *Nasonia giraulti* Darling, *N. longicornis* and *N. vitripennis* (Breeuwer *et al.*, 1992; Werren *et al.*, 1995). Cytoplasmic incompatibility is typically unidirectional: crosses between infected males and uninfected females yield no offspring or only sterile males, whereas the cross of infected females with uninfected males yields viable offspring (cited in Breeuwer and Werren, 1990). In *Nasonia*, *Wolbachia* spp. cause paternal chromosome loss in incompatible crosses, whereby the paternal chromosomes fail to condense properly, become fragmented and are lost in fertilized eggs. This results in the production of all-male lines (Breeuwer and Werren, 1990). *Nasonia* may be cured of *Wolbachia* endosymbionts with antibiotics or by prolonging larval diapause (Breeuwer and Werren, 1990; Perrot-Minnot *et al.*, 1996).

Wolbachia spp. are associated with the reproductive isolation of *N. vitripennis* and *N. giraulti* (Breeuwer and Werren, 1990). These two species mate readily but only male progeny are produced. However, elimination of *Wolbachia* by antibiotic treatment allows the two species to produce viable hybrids (Breeuwer and Werren, 1990).

Two independent studies on the phylogeny of *Wolbachia* using the 16S rDNA gene and the *ftsZ* gene showed that *Nasonia* spp. may be infected by two different species of *Wolbachia* (Breeuwer *et al.*, 1992; Werren *et al.*, 1995). Males from double-infected strains show typical unidirectional cytoplasmic incompatibility with single-infected females. Males of single-infected strains, however, show bidirectional incompatibility, meaning that all crosses are incompatible (Perrot-Minnot *et al.*, 1996).

Reports on the effect of *Wolbachia* on host fitness are controversial. Stolk and Stouthamer (1996) found a positive fitness effect in one strain of *N. vitripennis* and no effect in another strain. Here fitness was measured as lifetime fecundity. Preliminary tests done by Bordenstein and Werren (2000) also suggest a positive effect of *Wolbachia*, but no consistent positive effect was found once host genetic background was controlled for.

UNIDENTIFIED DISEASE. A field-collected strain of *N. vitripennis* produced few or no male offspring. This factor, termed *msr*, causes 97% female families and is maternally inherited (Skinner, 1982). Experimental results suggest that the observed sex ratios are not caused by the mortality of male offspring, as is the case with *Arsenophonus nasoniae*.

Opius *(formerly* Biosteres*) (Braconidae): parasitoids of fruit flies*

BACTERIA. Severe mortality of premature adult parasitoids may occur in laboratory cultures of *Opius longicaudatus* (Ashmead). Parasitoid mortality may exceed 80% (Greany *et al.*, 1977). The most common bacteria isolated from these parasitoids are *Serratia marcescens*, *Proteus mirabilis*, *Pseudomonas aeruginosa* and *Enterobacter cloacae*. Although these bacteria are also present in the host (the Caribbean fruit fly, *Anastrepha suspensa* (Loew)), the incidence of bacteria is highest in newly deceased parasitoids (Greany *et al.*, 1977). Parasitoid mortality is probably caused by stress-induced septicaemia of several species of opportunistic bacterial pathogens. Parasitoid mortality may be controlled by rearing *B. longicaudatus* under optimal conditions without the use of antibiotics (Greany *et al.*, 1977).

Pediobius *(Eulophidae): parasitoids of Mexican bean beetles*

PROTOZOA. *Pediobius foveolatus* (Crawford) is highly susceptible to *Nosema epilachnae* and *Nosema varivestis*, two microsporidia that also infect its host, the Mexican bean beetle (*Epilachna varivestis* Mulsant). Both *N. epilachnae* and *N. varivestis* are detrimental to *P. foveolatus* (Own and Brooks, 1986; Chapman and Hooker, 1992).

In beetles that are heavily infected with *N. epilachnae*, most progeny (96%) become infected and many of them are unable to complete development. Parasitoid mortality occurs primarily in the pupal stage. Of the parasitoids that are able to emerge from their infected hosts, many appear normal. However, about one-third exhibit obvious malformations, including malformed wings, greatly distended abdomens or both. Microsporidian spores are observed in all adult tissues. *N. epilachnae* does not affect the developmental period of emerging adults but adult longevity is significantly reduced. Adult parasitoids are susceptible to spores when administered *per os* and *N. epilachnae* is transmitted mechanically during oviposition (Own and Brooks, 1986).

P. foveolatus is also highly susceptible to *N. varivestis*, particularly when it parasitizes late-instar beetle larvae that are infected as first instars. Under these circumstances, all parasitoid progeny are infected and about half are unable to emerge. Spores are present in midgut epithelial cells, the ventral nerve cord and adipose tissue. *N. varivestis* has no effect on parasitoid development. However, infected females produce fewer eggs and a higher proportion of male progeny than uninfected females. *N. varivestis* is vertically transmitted and, although it is less virulent than *N. epilachnae*, disease prevalence is close to 100% for both species (Own and Brooks, 1986). Most infected adults appear normal; only a small proportion (12%) has malformed wings and greatly distended abdomens (Own and Brooks, 1986). Because most infected parasitoids appear normal, microsporidia may remain undetected in *P. foveolatus* colonies, ultimately having an adverse effect on their efficacy as biological control agents (Chapman and Hooker, 1992).

Trichogramma *(Trichogrammatidae): parasitoids of Lepidoptera*

BACTERIA. *Wolbachia*-induced parthenogenesis is a well-studied phenomenon in the genus *Trichogramma*. Molecular evidence for the presence of *Wolbachia* has been found in at least nine *Trichogramma* species: *T. brevicapillum* Pinto and Platner, *T. chilonis* Ishii, *T. cordubensis* Vargas and Cabello, *T. deion* Pinto and Oatman, *T. kaykai* Pinto and Stouthamer, *T. nubilale* Ertle and Davis, *T. oleae*, *T. pretiosum* Riley and *T. sibericum* Sokorina (Chapter 8; Stouthamer *et al.*, 1993; cited in Stouthamer, 1997, and van Meer, 1999). Four completely parthenogenetic species (*T. deion*, *T. chilonis*, *T. platneri* Nagarkatti, and *T. pretiosum*) have permanently reverted to producing male and female progeny when treated with antibiotics (tetracycline hydrochloride, sulphamethoxazole or rifampicin) or when reared at temperatures above 30°C (Stouthamer *et al.*, 1990). These revertible lines carry microorganisms within their eggs, whereas the eggs of non-revertible lines or arrhenotokous (bisexual) lines do not (Stouthamer and Werren, 1993).

In most known cases of *Wolbachia*-induced thelytoky, only infected thelytokous lines are known (for example, *E. formosa and M. uniraptor*). However, in *Trichogramma*, most species are infected at a low level (Stouthamer, 1997) and uninfected arrhenotokous and infected thelytokous forms co-occur and interbreed in mixed populations in the field. Under such conditions, horizontal transmission of *Wolbachia* may occur between individuals that share the same host. For example, infected *T. kaykai* larvae transmit *Wolbachia* to conspecific uninfected larvae, after which the transferred *Wolbachia* may be vertically transmitted to their offspring (Huigens *et al.*, 2000). However, *T. kaykai* populations in the field have infection levels of 6–26% (cited by Stouthamer *et al.*, 2001). In this case, the infection is kept at a low level by a parasitic B chromosome that causes females to produce only male offspring. The B chromosome destroys the paternal chromosomes with the exception of itself, thereby converting the diploid fertilized egg into a male haploid egg (Stouthamer *et al.*, 2001).

Thelytokous and arrhenotokous conspecifics co-occur in many field populations of *Trichogramma*; therefore, one can choose either form for use in biological control programmes. In the case of *Trichogramma*, the question arises whether one form is more effective for biological control than the other (for a detailed discussion, see Chapter 8; Aeschlimann, 1990; Stouthamer, 1993). For *T. cordubensis* and *T. deion*, *Wolbachia*-infected (thelytokous) lines were compared with uninfected (arrhenotokous) lines (Silva *et al.*, 2000). Arrhenotokous lines have a higher fecundity rate and greater dispersal than thelytokous lines under laboratory conditions. However, dispersal is not affected or reduced when arrhenotokous lines are released in the greenhouse. Therefore, thelytokous lines have a greater potential for biological control, despite their low fecundity (Silva *et al.*, 2000).

Wolbachia symbionts of the A subdivision are present in several strains of the arrhenotokous species of *Trichogramma bourarachae* Pintureau and Babault (Vavre *et al.*, 1999). Strains infected with this type of *Wolbachia* have higher fecundity than uninfected strains. Moreover, antibiotic treatments can reduce fecundity in the high-fecundity line (30 eggs per female per 5 days in the treated line versus 60 eggs in the untreated line), suggesting that *Wolbachia* spp. are associated with an increase in parasitoid fecundity (Vavre *et al.*, 1999).

PROTOZOA. The microsporidium *Nosema pyrausta* causes chronic disease in the European cornborer, *Ostrinia nubilalis* (Hübner). *Trichogramma evanescens* Westwood is also susceptible to the microsporidium and *N. pyrausta* may persist in field populations where *T. evanescens* is released for cornborer control. Stained preparations of infected *T. evanescens* show heavy infection of several tissues, including the alimentary tract, fat body, Malpighian tubules and muscle and nervous tissue. In some cases, the abdomen is filled with microsporidian spores. Microsporidia-infected *T. evanescens* females produce about half the progeny that are produced by uninfected females. Transovarial transmission of *N. pyrausta* does not occur in *T. evanescens*. However, the incidence of infection in developing parasitoid larvae is dependent on the number of *O. nubilalis* eggs that are infected with the microsporidium (Huger, 1984).

T. nubilale that develop in *N. pyrausta*-infected *O. nubilalis* eggs are directly affected by the microsporidium. Microsporidian spores are confined to the gut lumen of *T. nubilale* larvae but are found in the gut epithelium, and in muscle and nervous tissues of adult and pupal stages. The microsporidium impairs larval–pupal development and reduces adult emergence and fecundity (Sajap and Lewis, 1988). *T. nubilale* does not discriminate between healthy hosts and those infected with *N. pyrausta*. Fewer and smaller adult parasitoids emerge from infected host eggs than from uninfected eggs (Saleh *et al.*, 1995).

A microsporidium of the genus *Vairimorpha* infects the cabbage moth *Plutella xylostella* (L.). This microsporidium is transovarially transmitted between generations of the host but has little impact on its fitness. However, *Vairimorpha* sp. has a detrimental impact on *T. chilonis*, a parasitoid of *P. xylostella*. The microsporidium develops in various tissues, including the flight muscles and nervous tissues. Microsporidian-infected parasitoids show abnormal development and their longevity and fecundity are reduced. Successful biological control strategies are dependent on rearing these parasitoids on disease-free host eggs and making several releases in areas where endemic hosts are known to be infected (Schuld *et al.*, 1999).

Pollinators (Hymenoptera)

Bombus *(Hymenoptera, Apidae): pollinators*

A large number of pathogens have been described from bumblebees (for recent reviews, see MacFarlane *et al.*, 1995; Schmid-Hempel, 1998). However, the effects and epidemiology of many of these pathogens are unknown.

VIRUSES. Two types of viruses are recorded from bumblebees. An acute bee paralysis virus is pathogenic to both the honeybee (*Apis mellifera* L.) and several *Bombus* species. The effects of an entomopoxvirus isolated from *Bombus fervidus* Fabricius, *Bombus impatiens* Cresson and *Bombus pennsylvanicus* DeGeer from the USA are not known (cited by MacFarlane *et al.*, 1995).

BACTERIA. *Spiroplasma melliferum* causes May disease or pollen intoxication in honeybees. Although this spiroplasma is isolated from the haemolymph of *B. impatiens* and *B. pennsylvanicus* from North America, the effects of *S. melliferum* on bumblebees have not been determined. *Aerobacter cloacae* causes B-melanosis, a disease affecting the ovaries of honeybee queens. Although *A. cloacae* has been isolated from bumblebee queens, the effects of this bacterium on bumblebees are not clear (Schmid-Hempel, 1998). Gram-positive, spore-forming bacteria associated with diseased *Bombus melanopygus* Nylander larvae from the USA cause infected larvae to harden and be susceptible to fungal infection (cited by MacFarlane *et al.*, 1995).

FUNGI. Several fungi have been isolated from bumblebees. Known and suspected pathogens include *Paecilomyces farinosus*, *B. bassiana*, *V. lecanii*, *M. anisopliae*, *Aspergillus candidus*, *Hirsutella* spp., *Acrostalagmus* spp., *Candida* spp., *Doratomyces putredinis* and *Chrysosporium pannorum* (cited by MacFarlane *et al.*, 1995, and Schmid-Hempel, 1998). Fungi pathogenic to bumblebees can be distinguished by the colour of their respective mycelia; *P. farinosus* is pale yellow to orange; *B. bassiana* mycelium is white; *V. lecanii* is snow-white and *C. pannorum* is dirty yellow-white to brown. *D. putredinis* is an apparent saprophyte that produces dirty white mycelium that has a strong, characteristic ammonia-like odour. Fungal infections can cause overwintering losses of queens. Furthermore, honey contaminated with yeast (*Candida* sp.) may be unsuitable as stored nectar for bumblebees (cited by MacFarlane *et al.*, 1995).

PROTOZOA. Bumblebees are hosts to several protozoans, including gregarines, microsporidia and trypanosomes (MacFarlane *et al.*, 1995; Schmid-Hempel, 1998). The neogregarine *Apicystis* (formerly *Mattesia*) *bombi* infects several species of bumblebees from North America and Europe (Liu *et al.*, 1974; Lipa and Triggiani, 1996). Neogregarine infection in *Bombus* is common, but disease prevalence is low. Oocysts are ingested and emerging sporozoites penetrate the midgut wall and infect the fat body. Heavily infected fat body is strikingly white but reduced in size (Lipa and Triggiani, 1996). Normal fat body is yellow to brown, depending on the age of the queen (MacFarlane *et al.*, 1995). Microscopic examination for the presence of oocysts is necessary to diagnose infection. Oocysts are elongate with rounded ends and measure 16–21 μm long by 5 μm wide in fresh preparations. Oocysts in fixed and stained preparations measure 11–14 μm long by 3.5–5 μm wide (Lipa and Triggiani, 1996). *A. bombi* causes premature death of bumblebee queens in spring.

The microsporidium *Nosema bombi* infects bumblebees from Europe, North America and New Zealand. This pathogen affects adult bees, developing primarily in the Malpighian tubules. Infection extends secondarily to the midgut, tracheal matrix, connective tissue and eventually the fat body and occurs in larvae and pupae (MacFarlane *et al.*, 1995; McIvor and Malone, 1995). Spores measure 4.2–5.9 μm by 2.1–3.5 μm in unstained preparations (McIvor and Malone, 1995). *N. bombi* isolated from *Bombus terrestris* L. is less infective in *Bombus lapidarius* (L.) and *Bombus hypnorum* (L.) but more virulent, causing greater mortality and early death in infected hosts of these two species (Schmid-Hempel and Loosli, 1998). Natural infections of *N. bombi* have no detectable effects on colony performance but they are correlated with an increased production of sexuals, particularly males (Imhoof and Schmid-Hempel, 1999). *N. bombi*-infected bumblebees have distended abdomens, are unable to mate successfully and excrete diarrhoea-like faeces. *N. bombi* can be eliminated from bumblebee colonies by treating them with fumagillin (MacFarlane *et al.*, 1995).

Although the microsporidium *Nosema apis* is a pathogen of the honeybee *Apis mellifera*, some reports suggest that *N. apis* is pathogenic to several species of bumblebees, where it is confined to the midgut of adult males and workers (as cited in MacFarlane *et al.*, 1995). *N. apis* reduces the longevity of worker honeybees and their ability to forage. The effects of *N. apis* on bumblebees are expected to be similar to those manifested in honeybees (MacFarlane *et al.*, 1995). It is assumed that there is little interaction between bumblebees and honeybees in nature, limiting the opportunity for the dissemination of microsporidian pathogens between them. In rearing facilities, however, the potential for pathogen dissemination may be increased if honeybee pollen is used as food for rearing bumblebee colonies or if honeybees are used to stimulate bumblebee queens to start producing brood (MacFarlane *et al.*, 1995).

Two trypanosomes, *Crithidia bombi* and *Leptomonas* sp., are found in bumblebee intestines (MacFarlane *et al.*, 1995). *C. bombi* is a prevalent parasite of bumblebees in Italy and Switzerland (Shykoff and Schmid-Hempel, 1991a; MacFarlane *et al.*, 1995). The pathogen passes through the alimentary tract and is occasionally excreted (Gorbunov,

1996). Pathogenic effects of *C. bombi* appear to be slight, resulting in reduced ovarian growth in infected individuals and slower colony growth at the start of the season (Shykoff and Schmid-Hempel, 1991b). Moreover, field-collected bumblebee workers infected with *C. bombi* are less likely to forage for pollen than uninfected workers (Shykoff and Schmid-Hempel, 1991a). Bumblebee mortality may increase if the host is subjected to periods of starvation (Imhoof and Schmid-Hempel, 1998; Brown *et al.*, 2000).

NEMATODES. The parasitic nematode, *Sphaerularia bombi* is found in several *Bombus* species from Europe, North America and New Zealand (cited by MacFarlane *et al.*, 1995; McCorquodale *et al.*, 1998). Mated female *S. bombi* (1–3 mm long) enter bumblebee queens during the autumn and undergo developmental changes during queen hibernation. Each female nematode is capable of producing thousands of eggs, which develop into larvae in the host haemocoel (McCorquodale *et al.*, 1998). Parasitized queens fly close above the ground and occasionally dig small holes into the soil, where they expel juvenile nematodes (cited by Horton and Moore, 1993). Infection sterilizes the queen by arresting development of the corpora allata. Infection does not affect queen longevity; however, *S. bombi*-infected queens usually fail to start colonies and do not reproduce. Infected queens forage for longer than uninfected ones, the latter return to their colonies sooner and forage activity is taken over by workers (MacFarlane *et al.*, 1995).

Management of pathogens in *Bombus* should involve routine screening of individuals for pathogens. Preventative measures and sanitation are important means of preventing the spread of pathogens. Soil for overwintering queens should be free of fungi and dead workers suspected of being killed prematurely by pathogens should be discarded (MacFarlane *et al.*, 1995). Although many of the pathogens that infect *Bombus* have been described from field-collected specimens, high concentrations of bumblebees within rearing facilities may predispose them to disease.

Interestingly, bumblebees are not susceptible to some of the more common honeybee diseases, including viral sac brood (*Moratovirus*), bacterial foul brood (*Bacillus* and *Streptococcus*) and some fungal pathogens (*Ascosphaera*). It is thought that the evolution and behaviour of bumblebees have helped them to develop resistance to some of the pathogenic microorganisms that infect honeybees. However, adult bumblebees and honeybees share some viral and protozoan pathogens (MacFarlane *et al.*, 1995).

Summary

Reports of pathogens from both field-collected and mass-reared natural enemies raise concerns regarding the quality and efficacy of natural enemies used for biological pest control (Kluge and Caldwell, 1992; Zchori-Fein *et al.*, 1992). Some pathogens affect the performance of natural enemies by lowering their efficacy (Geden *et al.*, 1995; Bjørnson and Keddie, 1999), whereas others alter arthropod reproduction (Majerus and Hurst, 1997; Stouthamer *et al.*, 1999; Zchori-Fein *et al.*, 2000). Therefore, the release of pathogen-free natural enemies in biological control programmes is of the utmost importance.

Sanitation and screening procedures are important for controlling pathogens in arthropod mass rearings (see Lacey, 1997). Although the process of eliminating a particular pathogen from the rearing system may seem time-consuming and expensive, preventive sanitation and exclusion methods will help minimize the time and expense of such programmes (Dunn and Andres, 1980). It is important to be familiar with the types of pathogens to which each natural enemy is susceptible and any symptoms that may be associated with infection. Some pathogens produce chronic and debilitating effects and are able to remain undetected in arthropod populations for extended periods (Dunn and Andres, 1980; Chapman and Hooker, 1992; Bjørnson and Keddie, 1999). Therefore, individuals should be screened for pathogens on a routine and continual basis. Special attention should be given to field-collected arthropods and those exchanged among rearing facilities.

The addition of wild stock from native populations of natural enemies may result in the inadvertent introduction of endemic pathogens. In some cases, native populations of natural enemies are known to harbour pathogens (Geden *et al.*, 1995) but little is known regarding the prevalence and effects of endemic pathogens in most cases. It may be argued that endemic pathogens are of little concern in arthropod mass rearings. However, pathogens in natural field populations may cause problems in mass rearings if diseased wild stock is inadvertently added to the rearing system.

Infected arthropods that are used as food or hosts in mass rearings may be unsuitable for the completion of predator or parasitoid development (Laigo and Tamashiro, 1967; Beerling *et al.*, 1993; Brooks, 1993). Host arthropods should be examined routinely to ensure that they are free of pathogens. Furthermore, to ensure the success of a biological control programme, releases should be made more frequently in areas where endemic hosts are known to be infected (Sajap and Lewis, 1988; Schuld *et al.*, 1999).

The release of infected natural enemies into a previously disease-free environment may result in the introduction and establishment of the disease in field populations where it did not exist previously (MacFarlane *et al.*, 1995). Therefore, natural enemies should be screened to ensure that they are free of pathogens prior to their release (Geden *et al.*, 1995; Bjørnson and Keddie, 1999). Once released, there is no guarantee that pathogen-free natural enemies will remain free of pathogens. It is possible that they may encounter endemic pathogens native to the site in which they are released and become infected with them. However, pathogen-free natural enemies are more likely to survive and become established following release than are infected natural enemies (Dunn and Andres, 1980).

In experimental trials involving life-history studies, it is important to ensure that test arthropods are free of pathogens if the interpretation of data is to be meaningful. In some cases, the presence of pathogens in experimental animals may explain some of the discrepancies in the literature in respect of life-history traits (Zchori-Fein *et al.*, 1992). Furthermore, every effort should be made to test and release pathogen-free arthropods in preliminary trials when natural enemies are being evaluated for biological control potential. Mortality and reduced performance of diseased individuals may result in the misinterpretation of test results (Dunn and Andres, 1980).

References

Aeschlimann, J.P. (1990) Simultaneous occurrence of thelytoky and bisexuality in hymenopteran species, and its implications for the biological control of pests. *Entomophaga* 35, 3–5.

Andreadis, T.G. (1987) Transmission. In: Fuxa, J.R. and Tanada, Y. (eds) *Epizootiology of Insect Diseases.* John Wiley & Sons, New York, pp. 159–176.

Arutunyan, E.S. (1985) Structural peculiarities of the digestive tract of phytoseiid mites. *Biologicheskii Zhurnal Armenii* 35, 394–400 (in Russian).

Askary, H. and Brodeur, J. (1999) Susceptibility of larval stages of the aphid parasitoid *Aphidius nigripes* to the entomopathogenic fungus *Verticillium lecanii. Journal of Invertebrate Patholology* 73, 129–132.

Balas, M.T., Lee, M.H. and Werren, J.H. (1996) Distribution and fitness effects of the son-killer bacterium in *Nasonia. Evolutionary Ecology* 10, 593–607.

Becnel, J.J. and Geden, C.J. (1994) Description of a new species of microsporidia from *Muscidifurax raptor* (Hymenoptera: Pteromalidae), a pupal parasitoid of muscoid flies. *Journal of Eukaryotic Microbiology* 41, 236–243.

Beerling, E.A. and van der Geest, L.P. (1991) Microsporidiosis in mass-rearings of the predatory mites *Amblyseius cucumeris* and *A. barkeri* (Acarina: Phytoseiidae). *Proceedings of the Section Experimental and Applied Entomology of the Netherlands Entomological Society (NEV), Amsterdam* 2, 157–162.

Beerling, E.A., van der Voort, R.J. and Kwakman, P. (1993) Microsporidiosis in mass rearings of predatory mites: development of a detection method. *Proceedings of the Section Experimental and Applied Entomology of the Netherlands Entomological Society (NEV), Amsterdam* 4, 199–204.

Bell, J.V. (1974) Mycoses. In: Cantwell, G.E. (ed.) *Insect Diseases*. Marcel Dekker, New York, pp. 185–236.

Bjørnson, S.E. (1998) Morphology and pathology of the predatory mite, *Phytoseiulus persimilis* Athias-Henriot (Acari: Phytoseiidae). PhD thesis, University of Alberta, Edmonton, Alberta, Canada.

Bjørnson, S. and Keddie, B.A. (1999) Effects of *Microsporidium phytoseiuli* (Microsporidia) on the performance of the predatory mite, *Phytoseiulus persimilis* (Acari: Phytoseiidae). *Biological Control* 15, 153–161.

Bjørnson, S. and Keddie, B.A. (2000) The development and pathology of two undescribed species of Microsporidia (Microspora) found infecting the predatory mite, *Phytoseiulus persimilis* Athias-Henriot. *Journal of Invertebrate Pathology* 76, 293–300.

Bjørnson, S. and Keddie, B.A. (2001) Disease prevalence and transmission of *Microsporidium phytoseiuli* infecting the predatory mite, *Phytoseiulus persimilis* Athias-Henriot (Acari: Phytoseiidae). *Journal of Invertebrate Pathology* 77, 114–119.

Bjørnson, S., Steiner, M.Y. and Keddie, B.A. (1997) Birefringent crystals and abdominal discoloration in the predatory mite *Phytoseiulus persimilis* (Acari: Phytoseiidae). *Journal of Invertebrate Pathology* 69, 85–91.

Bjørnson, S., Raworth, D.A. and Bédard, C. (2000) Abdominal discoloration and the predatory mite *Phytoseiulus persimilis* Athias-Henriot: prevalence of symptoms and their correlation with short-term performance. *Biological Control* 19, 17–27.

Bordat, D., Goudegnon, A.E. and Bouix, G. (1994) Relationships between *Apanteles flavipes* (Hym.: Braconidae) and *Nosema bordati* (Microspora, Nosematidae) parasites of *Chilo partellus* (Lep.: Pyralidae). *Entomophaga* 39, 21–32.

Bordenstein, S.R. and Werren, J.H. (2000) Do *Wolbachia* influence fecundity in *Nasonia vitripennis*? *Heredity* 84, 54–62.

Breeuwer, J.A. and Jacobs, G. (1996) *Wolbachia*: intracellular manipulators of mite reproduction. *Experimental and Applied Acarology* 20, 421–434.

Breeuwer, J.A. and Werren, J.H. (1990) Microorganisms associated with chromosome destruction and reproductive isolation between two insect species. *Nature* 346, 558–560.

Breeuwer, J.A., Stouthamer, R., Barns, S.M., Pelletier, D.A., Weisburg, W.G. and Werren, J.H. (1992) Phylogeny of cytoplasmic incompatibility microorganisms in the parasitoid wasp genus *Nasonia* (Hymenoptera: Pteromalidae) based on 16S ribosomal DNA sequences. *Insect Molecular Biology* 1, 25–36.

Brooks, W.M. (1974) Protozoan infections. In: Cantwell, G.E. (ed.) *Insect Diseases*. Marcel Dekker, New York, pp. 254–266.

Brooks, W.M. (1993) Host–parasitoid–pathogen interactions. In: Beckage, N.E., Thompson, S.N. and Federici, B.A. (eds) *Parasites and Pathogens of Insects*, Vol. 2, *Pathogens*. Academic Press, New York, pp. 231–272.

Brown, M.J.F., Loosli, R. and Schmid-Hempel, P. (2000) Condition-dependent expression of virulence in a trypanosome infecting bumblebees. *Oikos* 91, 421–427.

Cali, A. and Briggs, J.D. (1967) The biology and life history of *Nosema tracheophila* sp. n. (Protozoa: Cnidospora: Microsporidea) found in *Coccinella septempunctata* Linnaeus (Coleoptera: Coccinellidae). *Journal of Invertebrate Pathology* 9, 515–522.

Canning E.U. and Hollister, W.S. (1990) *Enterocytozoon bieneusi* (Microspora): prevalence and pathogenicity in AIDS patients. *Transactions of the Royal Society of Tropical Medicine and Hygiene* 84, 181–186.

Chapman, G.B. and Hooker, M.E. (1992) A light and electron microscopic investigation of the occurrence of *Nosema* sp. (Microsporida: Nosematidae) in the abdomen of the parasitic wasp *Pediobius foveolatus* (Hymenoptera: Eulophidae). *Transactions of the American Microscopical Society* 111, 314–326.

De Barro, P.J. and Hart, P.J. (2001) Antibiotic curing of parthenogenesis in *Eretmocerus mundus* (Australian parthenogenic form). *Entomologia Experimentalis et Applicata* 99, 225–230.

DeMoraes, G.J. and Delalibera, I., Jr (1992) Specificity of a strain of *Neozygites* sp. (Zygomycetes: Entomophthorales) to *Mononychellus tanajoa* (Acari: Tetranychidae). *Experimental and Applied Acarology* 14, 98–94.

Dicke, M., Schütte, C. and Dijkman, H. (2000) Change in behavioural response to herbivore-induced plant volatiles in a predatory mite population. *Journal of Chemical Ecology* 26, 1497–1514.

Dobson, S.L., Marsland, E.J., Veneti, Z., Bourtzis, K. and O'Neill, S.L. (2002) Characterization of *Wolbachia* host cell range via the *in vitro* establishment of infections. *Applied Environmental Microbiology* 68, 656–660.

Donegan, K. and Lighthart, B. (1989) Effect of several stress factors on the susceptibility of the predatory insect, *Chrysoperla carnea* (Neuroptera: Chrysopidae), to the fungal pathogen *Beauvaria bassiana*. *Journal of Invertebrate Pathology* 54, 79–84.

Drea, J.J. and Gordon, R.D. (1990) Coccinellidae. In: Rosen, D. (ed.) *Armored Scale Insects. Their Biology, Natural Enemies and Control*, Vol. B. Elsevier Science Publishers, Amsterdam, pp. 19–40.

Dry, F.E., Steinkraus, D.C. and McNew, R.W. (1999) Microsporidia (Protozoa: Microspora) in pteromalids (Hymenoptera: Pteromalidae) in broiler-breeder houses of Northwest Arkansas. *Biological Control* 16, 164–169.

Dunn, P.H. and Andres, L.A. (1980) Entomopathogens associated with insects used for biological control of weeds. In: *Proceedings of the Fifth International Symposium for the Biological Control of Weeds, Brisbane*, pp. 241–246.

Finney, G.L. (1950) Mass-culturing *Chrysopa californica* to obtain eggs for field distribution. *Journal of Economic Entomology* 43, 97–100.

Furtado, I.P., De Moraes, G.J. and Keller, S. (1996) Infection of *Euseius citrifolius* (Acari: Phytoseiidae) by an entomophthoralean fungus in Brazil. *Ecossistema* 21, 85–86.

Gaugler, R.R. (1987) Noninfectious diseases. In: Fuxa, J.R. and Tanada, Y. (eds) *Epizootiology of Insect Diseases*. John Wiley & Sons, New York, pp. 245–255.

Geden, C.J., Long, S.J., Rutz, D.A. and Becnel, J.J. (1995) *Nosema* disease of the parasitoid *Muscidifurax raptor* (Hymenoptera: Pteromalidae): prevalence, patterns of transmission, management, and impact. *Biological Control* 5, 607–614.

Gerling, D. and Fried, R. (1997) Density related sterility in *Eretmocerus mundus*. *Entomologia Experimentalis et Applicata* 84, 33–39.

Gherna, R.L., Werren, J.H., Weisberg, W., Cote, R., Woese, C.R., Mandelco, L. and Brenner, D.J. (1991) *Arsenophonus nasoniae* gen. nov., sp. nov., the causative agent of the son-killer trait in the parasitic wasp *Nasonia vitripennis*. *International Journal of Systematic Bacteriology* 41, 563–565.

Giorgini, M. (2001) Induction of males in thelytokous populations of *Encarsia meritoria* and *Encarsia protransvena*: a systematic tool. *BioControl* 46, 427–438.

Goodwin, R.H. (1984) Recognition and diagnosis of diseases in insectaries and the effects of disease agents on insect biology. In: King, E.G. and Leppla, N.C. (eds) *Advances and Challenges in Insect Rearing*. US Department of Agriculture, Agricultural Research Service, Southern Region, New Orleans, Louisiana, pp. 96–129.

Gorbunov, P.S. (1996) Peculiarities of life cycle in flagellate *Crithidia bombi* (Protozoa, Trypanosomatidae). *Zoological Zhurnal* 75, 803–810.

Gotoh, T. (1982) Experimental transfer of abnormal 'sex-ratio' in the lady-bird beetle, *Harmonia axyridis* (Pallas) (Coleoptera: Coccinellidae). *Applied Entomology and Zoology* 17, 319–324.

Gottlieb, Y. and Zchori-Fein, E. (2001) Irreversible thelytokous reproduction in *Muscidifurax uniraptor*. *Entomologia Experimentalis et Applicata* 100, 271–278.

Gottlieb,Y., Zchori-Fein, E., Faktor, O. and Rosen, D. (1998) Phylogenetic analysis of parthenogenesis-inducing *Wolbachia* in the genus *Aphytis* (Hymenoptera: Aphelinidae). *Insect Molecular Biology* 7, 393–396.

Greany, P.D., Allen, G.E., Webb, J.C., Sharp, J.L. and Chambers, D.L. (1977) Stress-induced septicemia as an impediment to laboratory rearing of the fruit fly parasitoid *Biosteres* (= *Opius*) *longicaudatus* (Hymenoptera: Braconidae) and the Caribbean fruit fly *Anastrepha suspensa* (Diptera: Tephritidae). *Journal of Invertebrate Pathology* 29, 153–161.

Haldar, D.P., Ray, S.K. and Bose, R. (1988) *Anisolobus indicus* sp. n. of septate gregarines (Apicomplexa: Sporozoea) from the beetle, *Coccinella septempunctata* (Coleoptera: Coccinellidae). *Acta Protozoologica* 27, 165–170.

Hamm, J.J., Nordlung, D.A. and Marti, O.G. (1985) Effects of a nonoccluded virus of *Spodoptera frugiperda* (Lepidoptera: Noctuidae) on the development of a parasitoid, *Cotesia marginiventris* (Hymenoptera: Braconidae). *Environmental Entomology* 14, 258–261.

Hemptinne, J.L. (1988) Ecological requirements for hibernating *Propylea quatuordecimpunctata* (L.) and *Coccinella septempunctata* (Col.: Coccinellidae). *Entomophaga* 33, 505–515.

Hess, R.T. and Hoy, M.A. (1982) Microorganisms associated with the spider mite predator *Metaseiulus* (= *Typhlodromus*) *occidentalis*: electron microscope observations. *Journal of Invertebrate Pathology* 40, 98–106.

Horjus, M. and Stouthamer, R. (1995) Does infection with thelytoky-causing *Wolbachia* in the pre-adult and adult life stages influence the adult fecundity of *Trichogramma deion* and *Muscidifurax uniraptor*? *Proceedings of the Section Experimental and Applied Entomology of the Netherlands Entomological Society (NEV), Amsterdam* 6, 35–40.

Horton, D.R. and Moore, J. (1993) Behavioural effects of parasites and pathogens in insect hosts. In: Beckage, N.E., Thompson, S.N. and Federici, B.A. (eds) *Parasites and Pathogens of Insects*, Vol. 1, *Parasites*. Academic Press, New York, pp. 107–124.

Hostounský, Z. (1970) *Nosema mesnili* (Paill.) a microsporidian of the cabbageworm, *Pieris brassicae* (L.), in the parasites *Apanteles glomeratus* (l.), *Hyposoter ebeninus* (grav.) and *Pimpla instigator* (F.). *Acta Entomologica Bohemoslovaca* 67, 1–5.

Huger, A.M. (1984) Susceptibility of the egg parasitoid *Trichogramma evanescens* to the microsporidium *Nosema pyrausta* and its impact on fecundity. *Journal of Invertebrate Pathology* 44, 228–229.

Huger, A.M., Skinner, S.W. and Werren, J.H. (1985) Bacterial infections associated with the son-killer trait in the parasitoid wasp *Nasonia (= Mormoniella) vitripennis* (Hymenoptera: Pteromalidae). *Journal of Invertebrate Pathology* 46, 272–280.

Huigens, M.E., Luck, R.F., Klaassen, R.H., Maas, M.F., Timmermans, M.J. and Stouthamer, R. (2000) Infectious parthenogenesis. *Nature* 405, 178–179.

Hunter, M.S. (1999) The influence of parthenogenesis-inducing *Wolbachia* on the oviposition behaviour and sex-specific developmental requirements of autoparasitoid wasps. *Journal of Evolutionary Biology* 12, 735–741.

Hurst, G.D., Majerus, M.E. and Walker, L.E. (1992) Cytoplasmic male killing elements in *Adalia bipunctata* (Linnaeus) (Coleoptera: Coccinellidae). *Heredity* 69, 84–91.

Hurst, G.D., Purvis, E.L., Sloggett, J.J. and Majerus, M.E. (1994) The effect of infection with male-killing *Rickettsia* on the demography of female *Adalia bipunctata* L. (two spot ladybird). *Heredity* 73, 309–316.

Hurst, G.D., Hammarton, T.C., Obrycki, J.J., Majerus, T.M., Walker, L.E., Bertrand, D. and Majerus, M.E. (1996) Male-killing bacterium in a fifth ladybird beetle, *Coleomegilla maculata* (Coleoptera: Coccinellidae). *Heredity* 77, 177–185.

Hurst, G.D., Hammarton, T.C., Bandi, C., Majerus, T.M., Bertrand, D. and Majerus, M.E. (1997) The diversity of inherited parasites of insects: the male-killing agent of the ladybird beetle *Coleomegilla maculata* is a member of the Flavobacteria. *Genetical Research (Cambridge)* 70, 1–6.

Hurst, G.D., Jiggins, F.M., von der Schulenburg, J.H., Bertrand, D., West, S.A., Goriacheva, I.I., Zakharov, I.A., Werren, J.H., Stouthamer, R. and Majerus M.E. (1999a) Male-killing *Wolbachia* in two species of insect. *Proceedings of the Royal Society of London, Series B: Biological Sciences* 266, 735–740.

Hurst, G.D., von der Schulenburg, J.H., Majerus, T.M., Bertrand, D., Zakharov, I.A., Baungaard, J., Völkl, W., Stouthamer, R. and Majerus, M.E. (1999b) Invasion of one insect species, *Adalia bipunctata*, by two different male-killing bacteria. *Insect Molecular Biology* 8, 133–139.

Imhoof, B. and Schmid-Hempel, P. (1998) Patterns of local adaptation of a protozoan parasite to its bumblebee host. *Oikos* 82, 59–65.

Imhoof, B. and Schmid-Hempel, P. (1999) Colony success of the bumble bee, *Bombus terrestris*, in relation to infections by two protozoan parasites, *Crithidia bombi* and *Nosema bombi*. *Insectes Sociaux* 46, 233–238.

Issi, I.V. and Maslennikova, V.A. (1966) The role of the parasite *Apanteles glomeratus* L. (Hymenoptera: Braconidae) in transmission of *Nosema polyvora* Blunk (Protozoa, Microsporidia). *Entomological Review (English Translation of Entomologicheskoye Obozreniye)* 45, 275–277.

James, R.R. and Lighthart, B. (1992) The effect of temperature, diet and larval instar on the susceptibility of an aphid predator, *Hippodamia convergens* (Coleoptera: Coccinellidae), to the weak bacterial pathogen *Pseudomonas fluorescens*. *Journal of Invertebrate Pathology* 60, 215–218.

James, R.R. and Lighthart, B. (1994) Susceptibility of the convergent lady beetle (Coleoptera: Coccinellidae) to four entomogenous fungi. *Environmental Entomology* 23, 190–192.

Johanowicz, D.L. and Hoy, M.A. (1996) *Wolbachia* in a predator–prey system: 16S ribosomal DNA analysis of two phytoseiids (Acari: Phytoseiidae) and their prey (Acari: Tetranychidae). *Annals of the Entomological Society of America* 89, 435–441.

Johanowicz, D.L. and Hoy, M.A. (1998a) Experimental induction and termination of non-reciprocal reproductive incompatibilities in a parahaploid mite. *Entomologia Experimentalis et Applicata* 87, 51–58.

Johanowicz, D.L. and Hoy, M.A. (1998b) The manipulation of arthropod reproduction by *Wolbachia* endosymbionts. *Florida Entomologist* 81, 310–317.

Kajita, H. (1993) Induction of males in the thelytokous wasp *Encarsia formosa* Gahan (Hymenoptera: Aphelinidae). *Applied Entomology and Zoology* 28, 115–117.

Kariluotu, K.T. (1981) New records of the genus *Entomophthora* Fres. infesting insects in Finland. *Annales Entomologici Fennici* 48, 49–50.

Kaya, H.K. and Koppenhöfer, A.M. (1996) Effect of microbial and other antagonistic organism and competition on entomopathogenic nematodes. *Biocontrol Science and Technology* 6, 357–371.

Kaya, H.K., Koppenhöfer, A.M. and Johnson, M. (1998) Natural enemies of entomopathogenic nematodes. *Japanese Journal of Nematology* 28, 13–21.

Keller, S. (1997) The genus *Neozygites* (Zygomycetes, Entomophtorales) with special reference to species found in tropical regions. *Sydowia* 49, 118–146.

Kfir, R. and Walters, H.S. (1996) Transovarial transmission of a microsporidian pathogen in hymenopterous parasitoids of the stem borer *Chilo partellus* (Swinhoe) (Lepidoptera: Pyralidae). *African Entomology* 4, 287–290.

Kluge, R.L. and Caldwell, P.M. (1992) Microsporidian diseases and biological weed control agents: to release or not to release? *Biocontrol News and Information* 13, 43N–47N.

Lacey, L.A. (ed.) (1997) *Manual of Techniques in Insect Pathology*. Academic Press, New York, 409 pp.

Laigo, F.M. and Tamashiro, M. (1967) Interactions between a microsporidian pathogen of the lawn-armyworm and the hymenopterous parasite *Apanteles marginiventris*. *Journal of Invertebrate Pathology* 9, 546–554.

Lee, R.E., Jr, Strong-Gunderson, J.M., Lee, M.R., Grove, K.S. and Riga, T.J. (1991) Isolation of ice nucleating active bacteria from insects. *Journal of Experimental Zoology* 257, 124–127.

Lighthart, B., Sewall, D. and Thomas, D.R. (1988) Effect of several stress factors on the susceptibility of the predatory mite, *Metaseiulus occidentalis* (Acari: Phytoseiidae), to the weak bacterial pathogen *Serratia marcescens*. *Journal of Invertebrate Pathology* 52, 33–42.

Lipa, J.J. (1967) Studies on gregarines (Gregarinomorpha) of arthropods in Poland. *Acta Protozoologica* 5, 97–179.

Lipa, J.J. (1968) *Nosema coccinellae* sp. n., a new microsporidian parasite of *Coccinella septempunctata, Hippodamia tredecimpunctata* and *Myrrha octodecimguttata*. *Acta Protozoologica* 5, 369–374.

Lipa, J.J. and Smits, H. (1999) Microbial control of pests in greenhouses. In: Albajes, R., Gullino, M.L., van Lenteren, J.C. and Elad, Y. (eds) *Integrated Pest and Disease Management in Greenhouse Crops*. Kluwer Academic Publishers, Dordecht, The Netherlands, pp. 295–309.

Lipa, J.J. and Steinhaus, E.A. (1959) *Nosema hippodamiae* n. sp., a microsporidian parasite of *Hippodamia convergens* Guérin (Coleoptera, Coccinellidae). *Journal of Insect Pathology* 1, 304–308.

Lipa, J.J. and Triggiani, O. (1996) *Apicystis* gen nov and *Apicystis bombi* (Liu, Macfarlane & Pengelly) comb nov (Protozoa: Neogregarinida), a cosmopolitan parasite of *Bombus* and *Apis* (Hymenoptera: Apidae). *Apidologie* 27, 29–34.

Lipa, J.J., Pruszyński, S. and Bartkowski, J. (1975) The parasites and survival of the lady bird beetles (Coccinellidae) during winter. *Acta Parasitologica Polonica* 23, 453–461.

Liu, H.J., MacFarlane, R.P. and Pengelly, D.H. (1974) *Mattesia bombi* n. sp. (Neogregarinida: Ophrocystidae), a parasite of *Bombus* (Hymenoptera: Apidae). *Journal of Invertebrate Pathology* 23, 225–231.

Lord, J.C., Wraight, S.P., Daoust, R.A. and Roberts, D.W. (1988) Pathogenicity of *Beauveria bassiana* and *Zoophthora radicans* to the coccinellid predators *Coleomegilla maculata* and *Eriopis connexa*. *Journal of Invertebrate Pathology* 52, 471–473.

MacFarlane, R.P., Lipa, J.J. and Liu, H.J. (1995) Bumble bee pathogens and internal enemies. *Bee World* 76, 130–148.

McCorquodale, D.B., Beresford, R.G., Francis, J.M., Thomson, C.E. and Bartlett, C.M. (1998) Prevalence of *Sphaerularia bombi* (Nematoda: Tylenchida: Sphaerulariidae) in bumble bee queens (Hymenoptera: Apidae) on Cape Breton Island, Nova Scotia, Canada. *Canadian Entomologist* 130, 877–882.

McIvor, C.A. and Malone, L.A. (1995) *Nosema bombi*, a microsporidian pathogen of the bumble bee *Bombus terrestris* (L.). *New Zealand Journal of Zoology* 22, 25–31.

Maddox, J.V. (1973) The persistence of the Microsporida in the environment. *Entomological Society of America Miscellaneous Publication* 9, 99–104.

Maddox, J.V. (1987) Protozoan diseases. In: Fuxa J.R. and Tanada, Y. (eds) *Epizootiology of Insect Diseases*. John Wiley & Sons, New York, pp. 417–452.

Majerus, M.E. and Hurst, G.D. (1997) Ladybirds as a model system for the study of male-killing symbionts. *Entomophaga* 42, 13–20.

Majerus, M.E., Hinrich, J., Schulenburg, G.V. and Zakharov, I.A. (2000) Multiple causes of male-killing in a single sample of the two-spot ladybird, *Adalia bipunctata* (Coleoptera: Coccinellidae) from Moscow. *Heredity* 84, 605–609.

Majerus, T.M., Majerus, M.E., Knowles, B., Wheeler, J., Bertrand, D., Kuznetzov, V.N., Ueno, H. and Hurst, G.D. (1998) Extreme variation in the prevalence of inherited male-killing microorganisms between three populations of *Harmonia axyridis* (Coleoptera: Coccinellidae). *Heredity* 81, 683–691.

Majerus, T.M., Schulenburg, G.V., Majerus, M.E. and Hurst, G.D. (1999) Molecular identification of a male-killing agent in the ladybird *Harmonia axyridis* (Pallas) (Coleoptera: Coccinellidae). *Insect Molecular Biology* 8, 551–555.

Mandal, D., Rai, M., Pradhan, B., Gurung, D., Sharma, S., Rai, P. and Mandal, T. (1986) On a new septate gregarine from *Coccinella septempunctata* L. of a tea garden, Darjeeling. *Archiv für Protistenkunde* 131, 299–302.

Martignoni, M.E. and Iwai, P.J. (1981) A catalogue of viral diseases of insects, mites and ticks. In: Burges, H.D. (ed.) *Microbial Control of Pests and Plant Diseases 1970–1980.* Academic Press, New York, pp. 897–911.

Matsuka, M., Hashi, H. and Okada, I. (1975) Abnormal sex-ratio found in the lady beetle, *Harmonia axyridis* Pallas (Coleoptera: Coccinellidae). *Applied Entomology and Zoology* 10, 84–89.

Own, O.S. and Brooks, W.M. (1986) Interactions of the parasite *Pediobius foveolatus* (Hymenoptera: Eulophidae) with two *Nosema* spp. (Microsporida: Nosematidae) of the Mexican bean beetle (Coleoptera: Coccinellidae). *Environmental Entomology* 15, 32–39.

Pavlyushin, V.A. (1996) Effect of entomopathogenic fungi on entomophagous arthropods. *Bulletin OILB/SROP* 19, 247–249.

Perrot-Minnot, M.-J., Guo, L.R. and Werren, J.H. (1996) Single and double infections with *Wolbachia* in the parasitic wasp *Nasonia vitripennis*: effects on compatibility. *Genetics* 143, 961–972.

Poinar, G.O., Jr (1988) A microsporidian parasite of *Neoplectana glaseri* (Steinernematidae: Rhabditida). *Revue de Nématologie* 11, 359–361.

Poinar G., Jr and Poinar, R. (1998) Parasites and pathogens of mites. *Annual Review of Entomology* 43, 449–469.

Ramakers, P.M., Dissevelt, M. and Peeters, K. (1989) Large scale introductions of phytoseiid predators to control thrips on cucumber. *Mededelingen van der Faculteit Landbouwwetenschappen Rijksuniversiteit Gent* 54, 923–929.

Rhamhalinghan, M. (1987) Host–parasite relationship in *Coccinella septempunctata* L. (Coleoptera: Coccinellidae) and *Coccinellimermis* Rubtzov (Nematoda: Mermithidae). *Journal of Advanced Zoology* 8, 94–99.

Rhamhalinghan, M. (1992) Pathophysiological changes in haemolymph volume and pulse rate in mermithid parasitized lady beetles *Coccinella septempunctata* L. (Coleoptera: Coccinellidae). *Indian Journal of Entomology* 54, 454–460.

Sajap, A.S. and Lewis, L.C. (1988) Effects of the microsporidium *Nosema pyrausta* (Microsporida: Nosematidae) on the egg parasitoid, *Trichogramma nubilale* (Hymenoptera: Trichogrammatidae). *Journal of Invertebrate Pathology* 52, 294–300.

Saleh, M.E., Lewis, L.C. and Obrycki, J.J. (1995) Selection of *Nosema pyrausta* (Microsporidia: Nosematidae)-infected *Ostrinia nubilalis* (Lepidoptera: Pyralidae) eggs for parasitization by *Trichogramma nubilale* (Hymenoptera: Trichogrammatidae). *Crop Protection* 14, 327–330.

Schmid-Hempel, P. (1998) The parasites and their biology. In: Schmid-Hempel, P. (ed.) *Parasites in Social Insects.* Princeton University Press, Princeton, New Jersey, pp. 24–76.

Schmid-Hempel, P. and Loosli, R. (1998) A contribution to the knowledge of *Nosema* infections in bumble bees, *Bombus* spp. *Apidologie* 29, 525–535.

Schuld, M., Madel, G. and Schmuck, R. (1999) Impact of *Vairimorpha* sp. (Microsporidia: Burnellidae) on *Trichogramma chilonis* (Hymenoptera, Trichogrammatidae), a hymenopteran parasitoid of the cabbage moth, *Plutella xylostella* (Lepidoptera, Yponomeutidae). *Journal of Invertebrate Pathology* 74, 120–126.

Schütte, C., Hulshof, J., Dijkman, H. and Dicke, M. (1995) Change in foraging behaviour of the predatory mite *Phytoseiulus persimilis*: some characteristics of a mite population that does not respond to herbivore-induced synomones. *Proceedings of the Section Experimental and Applied Entomology of the Netherlands Entomological Society (NEV), Amsterdam* 6, 133–139.

Schütte, C., van Baarlen, P., Dijkman, H. and Dicke, M. (1998) Change in foraging behaviour of the predatory mite *Phytoseiulus persimilis* after exposure to dead conspecifics and their products. *Entomologia Experimentalis et Applicata* 88, 295–300.

Sengupta, T. and Haldar, D.P. (1996) Three new species of septate gregarines (Apicomplexa: Sporozoea) of the genus *Gregarina* Dufour, 1828 from insects. *Acta Protozoologica* 35, 77–86.

Sheetz, R., Goolsby, J. and Poprawski, T. (1997) Antibiotic treatment of a *Nosema* sp. (Protozoa: Microsporida) infecting the ovaries of a parasitic *Encarsia* wasp (Hymenoptera: Aphelinidae). *Subtropical Plant Science* 49, 50–52.

Shull, A.F. (1948) An all-female strain of lady beetles with reversions to normal sex ratios. *American Naturalist* 82, 241–251.

Shykoff, J.A. and Schmid-Hempel, P. (1991a) Incidence and effects of four parasites in natural populations of bumble bees in Switzerland. *Apidologie* 22, 117–126.

Shykoff, J.A. and Schmid-Hempel, P. (1991b) Parasites delay worker reproduction in bumblebees: consequences for eusociality. *Behavioral Ecology* 2, 242–248.

Silva, I.M., van Meer, M.M., Roskam, M.M., Hoogenboom, A., Gort, G. and Stouthamer, R. (2000) Biological control potential of *Wolbachia*-infected versus uninfected wasps: laboratory and greenhouse evaluation of *Trichogramma cordubensis* and *T. deion* strains. *Biocontrol Science and Technology* 10, 223–238.

Skinner, S.W. (1982) Maternally inherited sex ratio in the parasitoid wasp *Nasonia vitripennis*. *Science* 215, 1133–1134.

Sluss, R. (1968) Behavioural and anatomical responses of the convergent lady beetle to parasitism by *Perilitus coccinellae* (Shrank) (Hymenoptera: Braconidae). *Journal of Invertebrate Pathology* 10, 9–27.

Smirnoff, W.A. and Eichhorn, O. (1970) Diseases affecting predators of *Adelges* spp. on fir trees in Germany, Switzerland and Turkey. *Journal of Invertebrate Pathology* 15, 6–9.

Starý, P. (1999) Biology and distribution of microbe-associated thelytokous populations of aphid parasitoids (Hym., Braconidae, Aphidiinae). *Journal of Applied Entomology* 123, 231–235.

Stolk, C. and Stouthamer, R. (1996) Influence of a cytoplasmic incompatibility-inducing *Wolbachia* on the fitness of the parasitoid wasp *Nasonia vitripennis*. *Proceedings of the Section Experimental and Applied Entomology of the Netherlands Entomological Society (NEV), Amsterdam* 7, 33–37.

Stouthamer, R. (1993) The use of sexual versus asexual wasps in biological control. *Entomophaga* 38, 3–6.

Stouthamer, R. (1997) *Wolbachia*-induced parthenogenesis. In: O'Neill, S.L., Hoffmann, A.A. and Werren, J.H. (eds) *Influential Passengers: Inherited Microorganisms and Arthropod Reproduction*. Oxford University Press, New York, pp. 102–124.

Stouthamer, R. and Werren, J.H. (1993) Microbes associated with parthenogenesis in wasps of the genus *Trichogramma*. *Journal of Invertebrate Pathology* 61, 6–9.

Stouthamer, R., Luck, R.F. and Hamilton, W.D. (1990) Antibiotics cause parthenogenetic *Trichogramma* (Hymenoptera/Trichogrammatidae) to revert to sex. *Proceedings of the National Academy of Sciences, USA* 87, 2424–2427.

Stouthamer, R., Breeuwer, J.A., Luck, R.F. and Werren, J.H. (1993) Molecular identification of microorganisms associated with parthenogenesis. *Nature* 361, 66–68.

Stouthamer, R., Lükö, S. and Mak, F. (1994) Influence of parthenogenesis *Wolbachia* on host fitness. *Norwegian Journal of Agricultural Sciences Supplement* 16, 117–122.

Stouthamer, R., Breeuwer, J.A. and Hurst, G.D. (1999) *Wolbachia pipientis*: microbial manipulator of arthropod reproduction. *Annual Review of Microbiology* 53, 71–102.

Stouthamer, R., van Tilborg, M., de Jong, J.H., Nunney, L. and Luck, R.F. (2001) Selfish element maintains sex in natural populations of a parasitoid wasp. *Proceedings of the Royal Society of London, Series B: Biological Sciences* 268, 617–622.

Šut'áková G. and Rüttgen, F. (1978) *Rickettsiella phytoseiuli* and virus-like particles in *Phytoseiulus persimilis* (Gamasoidea: Phytoseiidae) mites. *Acta Virologica* 22, 333–336.

Tanada, Y. (1955) Field observations on a microsporidian parasite of *Pieris rapae* (L.) and *Apanteles glomeratus* (L.). *Proceedings of the Hawaiian Entomological Society* 15, 609–616.

Tanada, Y. and Kaya, H.K. (1993) *Insect Pathology*. Academic Press, San Diego, California, 666 pp.

Tanigoshi, L.K., Fargerlund, J. and Nishio-Wong, J.Y. (1981) Significance of temperature and food resources to the developmental biology of *Amblyseius hibisci* (Chant) (Acarina, Phytoseiidae). *Zeitschrift für Angewandte Entomologie* 92, 409–419.

Thomas, G.M. (1974) Diagnostic techniques. In: Cantwell, G.E. (ed.) *Insect Diseases*. Marcel Dekker, New York, pp. 1–48.

van der Geest, L.P., Elliot, S.L., Breeuwer, J.A. and Beerling E.A. (2000) Diseases of mites. *Experimental and Applied Acarology* 24, 497–560.

van Lenteren, J.C., Roskam, M.M. and Timmer, R. (1997) Commercial mass production and pricing of organisms for biological control of pests in Europe. *Biological Control* 10, 143–149.

van Meer, M.M. (1999) Phylogeny and host–symbiont interactions of thelytoky inducing *Wolbachia* in Hymenoptera. PhD thesis, Wageningen University, The Netherlands.
van Meer, M.M., van Kan, F.J., Breeuwer, J.A. and Stouthamer, R. (1995) Identification of symbionts associated with parthenogenesis in *Encarsia formosa* (Hymenoptera: Aphelinidae) and *Diplolepis rosae* (Hymenoptera: Cynipidae). *Proceedings of the Section Experimental and Applied Entomology of the Netherlands Entomological Society (NEV), Amsterdam* 6, 81–86.
van Meer, M.M., Witteveldt, J. and Stouthamer, R. (1999) Phylogeny of the arthropod endosymbiont *Wolbachia* based on the *wsp* gene. *Insect Molecular Biology* 8, 399–408.
Vavre, F., Girin, C. and Boulétreau, M. (1999) Phylogenetic status of a fecundity-enhancing *Wolbachia* that does not induce thelytoky in *Trichogramma*. *Insect Molecular Biology* 8, 67–72.
von der Schulenburg, J.H., Habig, M., Sloggett, J.J., Webberley, K.M., Bertrand, D., Hurst, G.D. and Majerus, M.E. (2001) Incidence of male-killing *Rickettsia* spp. (α-Proteobacteria) in the ten-spot ladybird beetle *Adalia decempunctata* L. (Coleoptera: Coccinellidae). *Applied and Environmental Microbiology* 67, 270–277.
Weir, A. and Beakes, G.W. (1996) Correlative light- and scanning electron microscope studies on the developmental morphology of *Hesperomyces virescens*. *Mycologia* 88, 677–693.
Werren, J.H., Skinner, S.W. and Huger, A.M. (1986) Male-killing bacteria in a parasitic wasp. *Science* 231, 990–992.
Werren, J.H., Hurst, G.D., Zhang, W., Breeuwer, J.A., Stouthamer, R. and Majerus, M.E. (1994) Rickettsial relative associated with male killing in the ladybird beetle (*Adalia bipunctata*). *Journal of Bacteriology* 176, 388–394.
Werren, J.H., Zhang, W. and Guo, L.R. (1995) Evolution and phylogeny of *Wolbachia*: reproductive parasites of arthropods. *Proceedings of the Royal Society of London, Series B: Biological Sciences* 261, 55–71.
Zakharov, I.A., Zinkevich, N.S., Shaikevich, E.V., Vysotskaya, L.V., Dorzhu, C.M. and Majerus, M.E. (1999) Sex ratio and male killing in Siberian populations of *Harmonia axyridis* Pall. *Russian Journal of Genetics* 35, 653–657.
Zchori-Fein, E., Geden, C.J. and Rutz, D.A. (1992) Microsporidioses of *Muscidifurax raptor* (Hymenoptera: Pteromalidae) and other pteromalid parasitoids of muscoid flies. *Journal of Invertebrate Pathology* 60, 292–298.
Zchori-Fein, E., Roush, R.T. and Hunter, M.S. (1994) Male production induced by antibiotic treatment in *Encarsia formosa* (Hymenoptera: Aphelinidae), an asexual species. *Experientia* 48, 102–105.
Zchori-Fein, E., Faktor, O., Zeidan, M., Gottlieb, Y., Czosnek, H. and Rosen, D. (1995) Parthenogenesis-inducing microorganisms in *Aphytis* (Hymenoptera: Aphelinidae). *Insect Molecular Biology* 4, 173–178.
Zchori-Fein, E., Roush, R.T. and Rosen, D. (1998) Distribution of parthenogenesis-inducing symbionts in ovaries and eggs of *Aphytis* (Hymenoptera: Aphelinidae). *Current Microbiology* 36, 1–8.
Zchori-Fein, E., Gottlieb, Y. and Coll, M. (2000) *Wolbachia* density and host fitness components in *Muscidifurax uniraptor* (Hymenoptera: Pteromalidae). *Journal of Invertebrate Pathology* 75, 267–272.

11 Commercial Availability of Biological Control Agents

J.C. van Lenteren
Laboratory of Entomology, Wageningen University, PO Box 8031, 6700 EH Wageningen, The Netherlands

Abstract

The commercial use of biological control has seen a very fast development during the past 30 years. Currently, about 85 companies worldwide produce more than 125 species of natural enemies. The largest variety of commercially produced species of natural enemies is available in Europe, mainly as a result of a much larger greenhouse industry in Europe, although many species are also available in North America. Emerging markets are Latin America, Asia and (South) Africa. The most commonly sold natural enemies are discussed in this chapter. The recommended release rates, the unit of sale and the target pest(s) are specified. In addition, a list of the commercially available biocontrol agents is provided, together with the target pests and the year of first use.

Introduction

Although the biological control of pests has been applied since around 1870, the large-scale commercial use of natural enemies of pests spans a period of less than 40 years. In some areas of agriculture, such as apple orchards, maize, cotton, sugarcane, soybean, vineyards and greenhouses, it has been a very successful, environmentally and economically sound, alternative for chemical pest control (van Lenteren *et al.*, 1992; van Lenteren, 2000). Inundative and seasonal inoculative releases of natural enemies are commercially applied primarily in annual field crops and greenhouse cultures and have increased considerably over the last 25 years (van Lenteren, 2000). Success of biological control in these crops is primarily dependent on the quality of the natural enemies, which are produced by commercial mass-rearing companies.

Today, more than 125 natural-enemy species are on the market for biological pest control (Table 11.1). Worldwide, there are about 85 commercial producers of natural enemies for augmentative forms of biological control with a turnover of about US$50 million in 2000 and an annual growth of 15–20% (Bolckmans, 1999; K. Bolckmans, Berkel and Rodenrijs, The Netherlands, personal communication). In addition there are hundreds of state- or farmer-funded production units that may sell natural enemies (Chapter 1; van Lenteren, 2000).

© CAB International 2003. *Quality Control and Production of Biological Control Agents: Theory and Testing Procedures* (ed. J.C. van Lenteren)

Table 11.1. Commercially available natural enemies (parasitic insects, predatory insects, predatory mites and entomopathogenic nematodes, fungi, bacteria and viruses) of insects, mites and other invertebrate pests in Europe (situation in the year 2000).

Natural enemy (endemic/exotic)	Pest (endemic/exotic)	In use since
**Adalia bipunctata* (en)	*Toxoptera aurantii* (en)	1998
**Adoxophyes orana* granulosis virus (en)	*Adoxophyes orana* (en)	1995
**Aleochara bilineata* (en)	*Delia* root flies (en)	1995
Amblyseius barkeri (en)	*Thrips tabaci* (en)	1981
	Frankliniella occidentalis (ex)	1986
Amblyseius (*Neioseiulus*) *degenerans* (ex)	Thrips (en, ex)	1993
Amblyseius fallacis (ex)	Mites (ex)	1997
**Amblyseius largoensis* (ex)	Mites (ex)	1995
**Amblyseius lymonicus* (ex)	Thrips (en, ex)	1997
**Ampulex compressa* (ex)	Blattidae (en, ex)	1990
**Anagrus atomus* (en)	Cicadellidae (en, ex)	1990
**Anagyrus fusciventris* (ex)	Pseudococcidae (en,ex)	1995
**Anagyrus pseudococci* (en)	Pseudococcidae (en,ex)	1995
**Anthocoris nemorum* (en)	Thrips (en, ex)	1992
Aphelinus abdominalis (en)	*Macrosiphum euphorbiae* (en)	1992
	Aulacorthum solani (en)	1992
**Aphelinus mali* (ex)	*Eriosoma lanigerum* (ex)	1980
Aphidius colemani (ex)	*Aphis gossypii, Myzus persicae* (ex, en)	1992
Aphidius ervi (en)	*Macrosiphum euphorbiae* (en)	1996
	Aulacorthum solani (en)	1996
Aphidius matricariae (en)	*Myzus persicae* (en)	1990
**Aphidius urticae* (en)	*Aulacorthum solani* (en)	1990
Aphidoletes aphidimyza (en)	Aphids (en, ex)	1989
**Aphytis holoxanthus* (ex)	Diaspididae (ex)	1996
**Aphytis melinus* (ex)	Diaspididae (en, ex)	1985
**Aprostocetus hagenowii* (ex)	Blattidae (en, ex)	1990
Bacillus thuringiensis (en, ex)	Lepidoptera (en, ex)	1972
Beauveria brongniartii (en)	*Melolontha* (en)	1985
**Bracon hebetor* (ex)	Lepidoptera (en)	1980
**Cales noacki* (ex)	*Aleurothrixus floccosus* (ex)	1970
**Chilocorus baileyi* (ex)	Diaspididae (en, ex)	1992
**Chilocorus circumdatus* (ex)	Diaspididae (en, ex)	1992
**Chilocorus nigritus* (ex)	Diaspididae, Asterolecaniidae (en, ex)	1985
**Chrysoperla carnea* (en, ex)	Aphids (en, ex) and others	1987
**Chrysoperla rufilabris* (ex)	Aphids (en, ex) and others	1987
**Clitostethus arcuatus* (en)	Aleyrodidae (ex)	1997
**Coccinella septempunctata* (en)	Aphids (en)	1980
**Coccophagus lycimnia* (ex)	Coccidae (en, ex)	1988
**Coccophagus rusti* (ex)	Coccidae (en, ex)	1988
**Coccophagus scutellaris* (en)	Coccidae (en, ex)	1986
**Coenosia attenuata* (en)	Diptera (en), Sciaridae (en)	1996
	Agromyzidae (en, ex), Aleurodidae (ex)	1996
**Comperiella bifasciata* (ex)	Diaspididae (ex)	1985
**Cryptolaemus montrouzieri* (ex)	Pseudococcidae, Coccidae (en, ex), *Planococcus citri* (ex)	1992
**Cydia pomonella* granulosis virus (en)	*Cydia pomonella* (en)	1995
Dacnusa sibirica (en)	*Liriomyza bryoniae* (en)	1981
	Liriomyza trifolii (ex)	1981
	Liriomyza huidobrensis (ex)	1990
Delphastus pusillus (ex)	*Trialeurodes vaporariorum* (ex)	1993
	Bemisia tabaci/argentifolii (ex)	1993

Continued

Table 11.1. *Continued.*

Natural enemy (endemic/exotic)	Pest (endemic/exotic)	In use since
Dicyphus tamaninii (en)	Whiteflies (ex), thrips (en, ex)	1996
Diglyphus isaea (en)	*Liriomyza bryoniae* (en)	1984
	Liriomyza trifolii (ex)	1984
	Liriomyza huidobrensis (ex)	1990
**Diomus* sp. (ex)	*Phenacoccus manihoti* (ex)	1990
**Encarsia citrina* (ex)	Diaspididae (en, ex)	1984
Encarsia formosa (ex)[a]	*Trialeurodes vaporariorum* (ex)	1970 (1926)
	Bemisia tabaci/argentifolii (ex)	1988
Encarsia tricolor (en)	*Trialeurodes vaporariorum* (ex)	1985
**Encyrtus infelix* (ex)	Coccidae (en, ex)	1990
**Encyrtus lecaniorum* (en)	Coccidae (en, ex)	1985
**Episyrphus balteatus* (en)	Aphids (en, ex)	1990
Eretmocerus californicus (ex)	*Bemisia tabaci/argentifolii* (ex)	1995
Eretmocerus mundus (en)	*Bemisia tabaci/argentifolii* (ex)	1995
**Franklinothrips vespiformis* (ex)	Thrips (ex)	1990
**Gyranusoidea* spp. (ex)	Pseudococcidae (en, ex)	1990
**Harmonia axyridis* (ex)	Aphids (en)	1995
Heterorhabditis bacteriophora	*Otiorrhynchus sulcatus* and other spp. (en)	1984
Heterorhabditis megidis and other spp. (en, ex)	*Otiorrhynchus sulcatus* and other spp. (en)	1984
**Hippodamia convergens* (ex)	Aphids (en, ex)	1993
**Hungariella peregrina* (ex)	Pseudococcidae (en, ex)	1990
**Hypoaspis aculeifer* (en)	Sciaridae, *Rhizoglyphus echinopus* (en)	1996
	Rhizoglyphus rolini (en), thrips (en, ex)	1996
**Hypoaspis miles* (en)	Sciaridae, *Rhizoglyphus echinopus* (en)	1994
**Kampimodromus aberrans* (en)	Mites (*Panonychus ulmi*) (en)	1960
**Leptomastidea abnormis* (en)	Pseudococcidae (en, ex)	1984
**Leptomastix dactylopii* (ex)	*Planococcus citri* (en, ex)	1984
**Leptomastix epona* (en)	Pseudococcidae (en, ex)	1992
**Lysiphlebus fabarum* (en)	*Aphis gossypii* (ex)	1990
**Lysiphlebus testaceipes* (ex)	*Aphis gossypii* (ex)	1990
Macrolophus caliginosus (en)	Whiteflies (ex)	1994
**Macrolophus pygmaeus* (*nubilis*) (en)	Whiteflies (ex)	1994
**Metaphycus bartletti* (ex)	Coccidae (en, ex)	1997
**Metaphycus helvolus* (ex)	Coccidae (en, ex)	1984
**Metaseiulus occidentalis* (ex)	Mites (en)	1993
**Microterys flavus* (ex)	Coccidae (en, ex)	1987
**Microterys nietneri* (en)	Coccidae (en, ex)	1987
**Muscidifurax zaraptor* (ex)	Stable flies (en)	1982
**Nasonia vitripennis* (en)	Stable flies (en)	1982
**Neoseiulus barkeri* (en)	Mites (en), thrips (en, ex)	1990
Neoseiulus (*Amblyseius*) *californicus* (ex)	Mites (en, ex)	1995
Neoseiulus (*Amblyseius*) *cucumeris* (en, ex)	*Thrips tabaci* (en)	1985
	Frankliniella occidentalis (ex)	1986
	Mites (en, ex)	1990
Neoseiulus (*Amblyseius*) *cucumeris* (ex, non-diapause strain)	Thrips (en, ex)	1993
**Nephus reunioni* (ex)	Pseudococcidae (en,ex)	1990
**Ooencyrtus kuwanae* (ex)	Moth (*Lymantria dispar*) (en)	1980
**Ooencyrtus pityocampae* (ex)	*Thaumetopoea pityocampa* (ex)	1997
**Ophyra aenescens* (ex)	Stable flies (en 2 spp.)	1995
Opius pallipes (en)	*Liriomyza bryoniae* (en)	1980

Continued

Table 11.1. *Continued.*

Natural enemy (endemic/exotic)	Pest (endemic/exotic)	In use since
Orius spp. (en, ex)	*F. occidentalis*/*T. tabaci* (ex, en)	
**Orius albidipennis* (en)		1991
Orius insidiosus (ex)		1991
Orius laevigatus (en)		1995
**Orius majusculus* (en)		1991
**Orius minutus* (en)		1991
**Orius tristicolor* (ex)		1995
**Paecilomyces fumosoroseus* (en)	Whiteflies (ex)	1997
**Phasmarhabditis hermaphrodita* (en)	Snails (en)	1994
**Phytoseiulus longipes* (ex)	*Tetranychus urticae* (en)	1990
Phytoseiulus persimilis (ex)	*Tetranychus urticae* (en)	1968
**Picromerus bidens* (en)	Lepidoptera (en)	1990
**Podisus maculiventris* (ex ?)	Lepidoptera (en)	1996
	Leptinotarsa decemlineata (ex)	1996
**Praon volucre* (en)	Aphids (en)	1990
**Pseudaphycus angelicus* (ex)	Pseudococcidae (en, ex)	1990
**Pseudaphycus flavidulus* (en)	Pseudococcidae (en, ex)	1990
**Pseudaphycus maculipennis* (en)	*Pseudococcus* spp. (en)	1980
**Rhyzobius chrysomeloides* (ex)	*Matsococcus feytaudi* (ex)	1997
**Rhyzobius* (*Lindorus*) *lophanthae* (ex)	Diaspididae (en, ex), *Pseudalacapsis pentagona*	1980
**Rodolia cardinalis* (ex)	*Icerya purchasi* (ex)	1990
**Rumina decollata* (en)	Snails (en)	1990
**Scolothrips sexmaculatus* (en)	Mites, thrips (en, ex)	1990
**Scutellista caerulea* (*cyanea*) (ex)	Coccidae (en, ex)	1990
**Scymnus rubromaculatus* (en)	Aphids (en)	1990
**Spodoptera* NPVirus (en)	*Spodoptera exigua* (ex)	1994
**Steinernema carpocapsae* (en)	*Otiorrhynchus sulcatus* and other spp. (en)	1984
Steinernema feltiae (en)	Sciaridae and other spp. (en)	1984
**Stethorus punctillum* (en)	Mites (en)	1995
**Stratiolaelaps miles* (en)	Sciaridae, *Rhizoglyphus echinopus* (en)	1994
**Sympherobius* sp. (en)	Pseudococcidae (en, ex)	1990
**Therodiplosis* (=*Feltiella*) *persicae* (en)	Mites in open fields (en)	1990
**Thripobius semiluteus* (ex)	Thrips (ex)	1995
**Trichogramma brassicae* (en)	Lepidoptera, several spp. (en)	1980
**Trichogramma cacoeciae* (en)	Lepidoptera, orchards, several spp. (en)	1980
**Trichogramma dendrolimi* (en)	Lepidoptera, orchards, several spp. (en)	1985
Trichogramma evanescens (en)	*Ostrinia nubilalis* in maize (en)	1975
	Lepidoptera in greenhouses (en, ex)	1992
**Typhlodromus pyri* (en)	Mites in apple, pear, grapes (en)	1985
**Verticillium lecanii* (en)	Whitefly/aphids (ex, en)	1990

* Small market products.
en, endemic: occurs in European Union countries; ex, exotic: originates from outside European Union countries, but may have been in Europe for 50 years or more; NPVirus, nucleopolyhedrovirus.
[a] *Encarsia* was already used for biological control of whitefly between 1926 and 1945. Its use was terminated when synthetic pesticides became popular.

The commercial availability of natural enemies is changing continuously, although several of the larger producers have been on the market for a period of 30 years now, which guarantees the permanent presence of the most important agents. Updated versions of commercially available biological control organisms, companies and suppliers are published on a regular basis in the *IPM Practitioner* (Anon., 2000) and on the web (e.g. www.koppert.nl, www.biobest.be etc.). Fewer than 30 beneficial species make up 90% of the total sales (Table 11.2; van Lenteren, 1997; Bolckmans, 1999). Extensive reviews of the availability of commercially produced biological control agents were not compiled until the mid-1990s, although some data are given in van Lenteren and Woets (1988). Cranshaw *et al.* (1996) correctly state that such information is essential for making calculations on the cost-effectiveness of using such biological control organisms. Cranshaw *et al.* (1996) reviewed the 1994 pricing and marketing by suppliers of organisms for biological control of arthropods in the USA. The same was done for Europe (van Lenteren *et*

Table 11.2. Most commonly used biological control organisms in Europe and North America, and life stage in which the organism is shipped from producer.

Biological control organism	Used in	Species life stage shipped
Amblyseius degenerans	Europe	Mixed life stages
Aphelinus abdominalis	Europe	Adult
Aphidius colemani	Europe	Pupa
Aphidius ervi	Europe	Pupa
Aphidoletes aphidimyza	Europe + America	Pupa
Aphytis melinus	America + Europe	Adult
Chrysoperla carnea and *Chrysoperla rufilibris*	America + Europe	Eggs, larvae, adults or mixed
Cryptoleamus montrouzieri	America + Europe	Adult
Dacnusa sibirica	Europe	Adult
Delphastus pusillus	Europe	Adult
Diglyphus isea	Europe	Adult
Encarsia formosa	Europe + America	Pupa
Eretmocerus californicus	America + Europe	Pupa
Eretmocerus mundus	Europe	Pupa
Galendromus occidentalis	America	Mixed life stages
Harmonia axyridis	Europe	Adult
Heterohabditis megides	Europe	Juveniles
Hippodamia convergens	America + Europe	Adult
Hypoaspis aculeifer	Europe	Mixed life stages
Hypoaspis miles	Europe	Mixed life stages
Leptomastix abnormis	Europe	Adult
Leptomastix dactylopii	Europe	Adult
Leptomastix epona	Europe	Adult
Macrolophus caliginosus	Europe	Adult
Mesoseiulus longipes	America	Mixed life stages
Metaphycus helvolus	America + Europe	Adult
Neoseiulus (Amblyseius) californicus	Europe + America	Mixed life stages
Neoseiulus (Amblyseius) cucumeris	Europe + America	Mixed life stages
Orius insidiosus	America	Adult
Orius laevigatus	Europe	Adult
Orius majusculus	Europe	Adult
Phytoseiulus persimilis	Europe + America	Mixed life stages
Steinernema carpocapsae	America	Juveniles
Steinernema feltiae	Europe + America	Juveniles
Trichogramma brassicae	Europe	Parasitized host egg
Trichogramma evanescens	America + Europe	Parasitized host egg
Trichogramma spp.	America + Europe	Parasitized host egg
Fly parasitoids	America + Europe	Pupa

al., 1997). The most commonly sold species in Europe and North America are discussed in this chapter (Table 11.2). These species are divided into four groups: parasitoids, predatory insects, predatory mites and entomopathogenic nematodes. Most natural enemies are used for biological control in greenhouses, with the exception of *Harmonia* sp. and *Trichogramma* spp., which are also used in the open field.

Parasitoids

Aphelinus abdominalis Dalman

This parasitic wasp is used to control the potato aphid *Macrosiphum euphorbiae* (Thomas) and the greenhouse potato aphid *Aulacorthum solani* Kaltenbach. It is sold in units of 100 or 250 adults or pupae. Since *A. abdominalis* does not easily spread over the crop, it should be introduced where aphid foci occur (Koppert, 1994). The recommended rates for release range from two to four wasps m^{-2} in the focal point of the infestation. Usually, A. *abdominalis* is used in combination with *Aphidoletes aphidimyza* Rondani.

Aphidius colemani Viereck

This species is a parasitoid of the cotton aphid *Aphis gossypii* Glover and the green peach aphid *Myzus persicae* Sulzer. The recommended rate for release ranges from 0.15 m^{-2} for preventive introductions to 1.5 m^{-2} in heavily infected areas.

Aphidius ervi Halliday

This parasitoid is used to control the potato aphid *M. euphorbiae*. The recommended release rate ranges from 0.15 m^{-2} to 1 m^{-2}.

Aphytis melinus DeBach

This species is sold for control of armoured scales (e.g. California red scale, yellow scale, and oleander scale). The suggested release rates are in the order of five to 100 adults per infested plant.

Dacnusa sibirica Telenga

This species is used to control leafminers such as the tomato leafminer *Liriomyza bryoniae* (Kaltenbach), the American serpentine leafminer *Liriomyza trifolii* (Burgess) and the pea leafminer *Liriomyza huidobrensis* (Blanchard). *D. sibirica* is supplied in units of 250 adults. This parasitoid is advised for use mainly during the winter in northern Europe.

Diglyphus isea (Walker)

This parasitoid is also used for controlling leafminers and it is sometimes sold with *D. sibirica* in mixed culture. It is sold in units of 250 adults. The recommended introduction rate (mostly mixed with *D. sibirica*) ranges from 0.25 m^{-2} for preventative introductions to 2 m^{-2} in heavily infested areas.

Encarsia formosa Gahan

This is a parasitoid of the greenhouse whitefly *Trialeurodes vaporariorum* (Westwood) and the whiteflies *Bemisia tabaci* (Gennadius) and *Bemisia argentifolii* Bellows and Perring, and is one of the most remarkable examples of the potential of a biological control agent (van Lenteren *et al.*, 1996). The small parasitoid was accidentally imported into Europe – as was greenhouse whitefly – and was discovered as a parasitoid of whiteflies by Speyer (1927). Shortly after its discovery, it became very popular in the 1930s, both in Europe and elsewhere. After the second world war, its use declined because of the availability of broad-spectrum insecticides. Nowadays, *E. formosa* is one of the most used biological control agents in protected crops, and also the most important natural enemy when expressed in monetary value for producers (van Lenteren, 1995; Bolckmans, 1999). *E. formosa* is produced in enormous numbers. The annual production by large companies is about half a billion parasitoids.

E. formosa is usually sold in units of 1000 as parasitized whitefly pupae, either on the leaf of its whitefly host plant, as pupae glued to different substrates or as pupae that are removed from a leaf. The recommended rate of release ranges from 1 m^{-2}, every 1–2 weeks, during a certain period for preventive use, to 9 m^{-2}, five times a week, for a severe infestation.

Eretmocerus californicus Howard

The recommended rate of release of this parasitoid of *Bemisia* spp. varies from 1.5 m^{-2} every 1–2 weeks for preventive use to 9 m^{-2} three times a week in a severe infestation of whiteflies.

Eretmocerus mundus Mercet

This parasitoid is also used for control of *Bemisia* spp., particularly in the Mediterranean area of Europe. Advice for releases is the same as for *E. californicus*.

Fly parasitoids

Fly parasitoids include various species of parasitic wasps that develop on pupae of manure-breeding flies, such as *Muscidifurax* spp., *Spalangia* spp. and *Nasonia* spp. They are used to reduce nuisance-fly problems by owners of horses and to control flies in feedlots and other animal rearing facilities, as well as around composting areas. Release rates and moments vary a lot; one commonly suggested release rate was 500 parasitoids per large animal at biweekly or monthly intervals, another release rate was related to the area to be protected, such as 1000 parasitoids 100 ft^{-2}. The most common unit package marketed included 8000–10,000 fly parasitoids (all information from Cranshaw *et al.*, 1996).

Leptomastix abnormis Girault

This is a parasitic wasp of mealybugs; see under *Leptomastix dactylopii*.

Leptomastix dactylopii (Howard)

This species parasitizes the citrus mealybug *Planococcus citri* (Risso). It is recommended to introduce one wasp m^{-2} in a light infestation and two wasps m^{-2} in a severe infestation of *P. citri*. Wasps should be introduced every 2 weeks.

Leptomastix epona (Walker)

This species is also a parasitoid of *P. citri*; see under *L. dactylopii* above.

Metaphycus helvolus (Compere)

This is the most common parasitoid sold for control of soft scales (e.g. hemispherical scale, black scale, nigra scale and soft brown scale). Recommended use rates: two to three releases at 2–3-week intervals of five to ten parasitoids per infested plant (Cranshaw *et al.*, 1996).

Trichogramma spp.

These species parasitize different lepidopteran pests. *Trichogramma* spp. are egg parasitoids and are supplied as black pupae attached to cardboard cards, as loose pupae in containers, in capsules or in other formulations. A multitude of species are produced and sold. The recommended rate of release in greenhouses ranges from 5 m^{-2} for preventive use to 20 m^{-2} in a severe infestation. In greenhouses, pupae should be introduced weekly. In maize with one generation of *Ostrinia nubilalis* (Hubner) in northern Europe, one release of 100,000 pupae of different ages per hectare is advised; pupae will emerge during a 2-week period. In North America, the numbers released in field crops are similar to or higher than those advised in Europe (Cranshaw *et al.*, 1996).

Predatory Insects

Aphidoletes aphidimyza (Rondani)

This gall midge is used for controlling aphids. The adult gall midges are active at night. The female gall midge deposits small orange eggs near the aphids. The emerging larvae paralyse the aphids and suck them dry (Koppert, 1994). *A. aphidimyza* is sold as black pupae in a vermiculite carrier. The recommended rate of release varies from 2 m^{-2} to 10 m^{-2}, depending on the seriousness of an infestation.

Chrysoperla spp.

Two species of green lacewings, *Chrysoperla carnea* Stehphens and *Chrysoperla rufilibris* (Burmeister), are produced and used as predators of aphids. The recommended rate of release for *C. carnea*, the most common one, ranges from 10 m^{-2} in a light infestation to 20 m^{-2} in a severe infestation.

Cryptolaemus montrouzieri Mulsant

This is a predator of mealybugs. Both the adult beetle and its larvae can kill all growth stages of the mealybug (Koppert, 1994). The beetle is particularly useful for controlling large mealybug populations. The recommended rate of release in Europe varies from 2 m^{-2} when introduced twice with an interval of 2 weeks, to 10 m^{-2} when introduced once in mealybug hot spots. In North America, the release rates are highly variable; an example is a release rate of two to five beetles per infested plant (Cranshaw *et al.*, 1996).

Delphastus pusillus (LeConte)

This predatory beetle is used for controlling *T. vaporariorum* and *Bemisia* spp. It is usually sold in units of 100 adults. As it is a rather new natural enemy, release rates are still highly variable.

Ladybirds

Two species of ladybird are commercially used as biological control agents against aphids. Adult beetles of the species *Hippodamia convergens* Guerin-Meneville are collected in enormous quantities in the USA. This species is not produced in Europe. It is introduced once during the production season. The recommended release rate ranges from 25 m^{-2} to 50 m^{-2}. *Harmonia axyridis* (Pallas) is a ladybird originating from Asia and advised for use against various aphids.

Macrolophus caliginosus Wagner

M. caliginosus is a predatory bug used for controlling the tobacco/silverleaf whitefly *B. tabaci/argentifolii* and the greenhouse whitefly *T. vaporariorum*. This predatory bug pierces the prey with its sucking mouthparts and sucks out the body fluids. It preys on all whitefly stages, but prefers eggs and larvae. An adult bug may feed on 30–40 whitefly eggs a day (Koppert, 1994). *M. caliginosus* is sold as adults in a vermiculite carrier. It is recommended for release of 0.5 m^{-2} to 5 predators m^{-2} twice, at an interval of 2 weeks.

Orius spp.

These predatory species are used to control thrips, such as the onion thrips, *Thrips tabaci* Lindeman, and the western flower thrips, *Frankliniella occidentalis* (Pergande). *Orius* spp. attack larvae and adult thrips. An adult predatory bug can eat five to 20 thrips a day (Koppert, 1994). The recommended rate of release ranges from 1 m^{-2} to 10 m^{-2}, depending on the level of pest infestation. Currently, there are about five *Orius* species on the commercial market. The most popular species in Europe is *Orius laevigatus* Fieber; in North America, *Orius insidiosus* (Say) is sold most often.

Predatory Mites

Amblyseius (Neoseiulus) cucumeris (Oudemans)

A. cucumeris is marketed as a predator of western flower thrips (*F. occidentalis*). This species of thrips is currently the most problematic pest in various greenhouse crops, including vegetables and ornamentals, and its control is difficult to achieve with either chemical or biological methods. *A. cucumeris* can be used also as a predator of the onion thrips, *T. tabaci*. The efficacy of *A. cucumeris* and *Amblyseius barkeri* Hughes in winter depends highly on their diapause characteristics. From research of Rodriguez *et al.* (1994) it is known that, at a night temperature of 12°C and below, the diapause incidence of *A. cucumeris* is 100%, whereas *A. barkeri* was not affected by a low night temperature. The mites are usually sold in units of 50,000.

Amblyseius (*Neoseiulus*) *barkeri* Hughes

This phytoseiid mite, like *A. cucumeris*, is marketed as a predator of the western flower thrips. It can also be used as a predator of the two-spotted spider mite, *Tetranychus urticae* Koch.

Amblyseius degenerans Berlese

This species is also marketed as a predator of thrips (*F. frankliniella* and *T. tabaci*) and is sold in units of 200 to 2000. The predatory mite also survives on pollen and can therefore be introduced preventively in pollen-bearing crops (Koppert, 1994).

Amblyseius (Neoseiulus) californicus (McGregor)

This predacious mite is employed as a predator of the two-spotted spider mite, *T. urticae*.

Amblyseius spp.

All the above mentioned *Amblyseius* species are released in very high numbers, often up to 50 or even more per plant. These predatory mites are distributed in cups or shaker bottles in a carrier such as bran. Some species are also sold in sachets, which can be affixed to plants and which also contain stored product mites as an additional food to sustain establishment. Such sachets provide predatory mites over a period of several weeks.

Galendromus occidentalis (Nesbitt)

This predatory mite, earlier named *Metaseiulus occidentalis* (Nesbitt), is sold for the control of various spider-mite species in outdoor crops, such as fruit trees and strawberries (Cranshaw *et al.*, 1996).

Hypoaspis miles (Berlese)

This mesostigmatic mite can be used as a predator of sciarid fly larvae and the bulb mite (*Rhizoglyphus echinopus* Fumouzze et Robin). Wright and Chambers (1994) found that all larval instars of sciarids (*Bradysia paupera* Tuom.) are attacked, although the numbers consumed are dependent on the size of the larvae. *H. miles* is sold mostly in units of 25,000.

Hypoaspis aculeifer (Canestrini)

This mesostigmatic mite is also produced as a predator of the bulb mite and sciarid fly larvae.

Mesoseiulus longipes Evans

This is a predator of spider mites that is usually sold for environments of lower humidity than is tolerated by the predatory mite *Phytoseiulus persimilis* (below). It is sold for interior/greenhouse use (Cranshaw *et al.*, 1996).

Phytoseiulus persimilis Athias–Henriot

P. persimilis is the most widely available species of all marketed natural enemies in our survey, and is released in even larger populations than *E. formosa*. It is used as a predator of the two-spotted spider mite, *T. urticae* Koch, but the carmine spider mite *Tetranychus cinnabarinus* (Boisduval) can also be controlled with this predator. This predatory mite is often mixed with vermiculite and put in shaker bottles. They need to be sprinkled on the leaves. Extra predatory mites should be placed in spider-mite hot spots (Koppert, 1994). For optimal development of the predator population, the relative humidity should not be under 75% and the temperature needs to be regularly above 20°C. *P. persimilis* is supplied in units of 1000 or 2000 mites, although a few producers offer a smaller quantity of 100 or 250 mites.

Entomopathogenic Nematodes

Members of the families of Steinernematidae and Heterorhabditidae are important entomopathogenic nematodes of scarabaeid larvae or lepidopteran caterpillars in soil. Desirable assets, such as ease of mass production, efficacy and safety for non-target organisms, have evoked commercial interest in these nematodes (Parwinder and Georgis, 1994). The nematodes kill their host by releasing symbiotic bacteria (*Xenorhabdus* spp.), which are carried in their alimentary tract. These bacteria multiply rapidly inside the insect's haemolymph and kill their host within 48 h. The symbiotic bacteria convert host tissue into products that can easily be taken up by the nematodes. Inside the beetle cadaver, the nematodes reproduce. As soon as the nematodes reach the infectious third-juvenile stage, they leave the old host and start searching for new weevil larvae. Usually billions of nematodes should be released per hectare.

Steinernema carpocapsae Weiser

This is the most widely available entomopathogenic nematode and is easy to rear, store and handle. It is sold to control a wide range of insects.

Steinernema feltiae (Filipjev)

This species is an entomopathogen of Sciaridae, but other target pests are also under investigation. This entomopathogenic nematode infects larval/nymphal, prepupal and pupal stages of insects that spend (part) of their life cycle in the soil (Tomalak, 1994), so they may be used for controlling thrips and leafminers.

Heterorhabditis megides Poinar

This nematode is used for control of the vine weevil *Otiorhynchus sulcatus* (F.). One producer supplies the nematodes, together with inert carrying material, in a 250 ml package, containing 50 million nematodes. This should be sufficient for 100 m^2 (for a blanket treatment) or for 10 m^3 soil (Koppert, 1994).

Entomopathogenic Fungi, Bacteria and Viruses

Currently, a number of entomopathogenic fungi, bacteria and viruses are on the market and can be used for control of pests. Because of their special character, way of production and application compared with the 'macrobiological' control agents, such as predators and parasitoids, they are not treated in detail here. To date, there are five nucleopolyhedroviruses (NPViruses), five *Bacillus* species/strains and ten species of fungi available for insect control in greenhouses (Lipa and Smits, 1999). For a review of these microbial natural enemies, refer to Lipa and Smits (1999).

Discussion and Conclusions

One way to express the importance of the different categories of natural enemies is by specifying the surface area on which they are applied. Although this is not an easy task (see Chapter 1 and van Lenteren, 2000), the roughly estimated areas under augmentative forms of biological control are given in Fig. 11.1.

The relative importance of the different natural enemies can be also expressed by their monetary value. Reliable data are available for biological control agents used in greenhouses (Bolckmans, 1999), but are lacking for field applications, although it is esti-

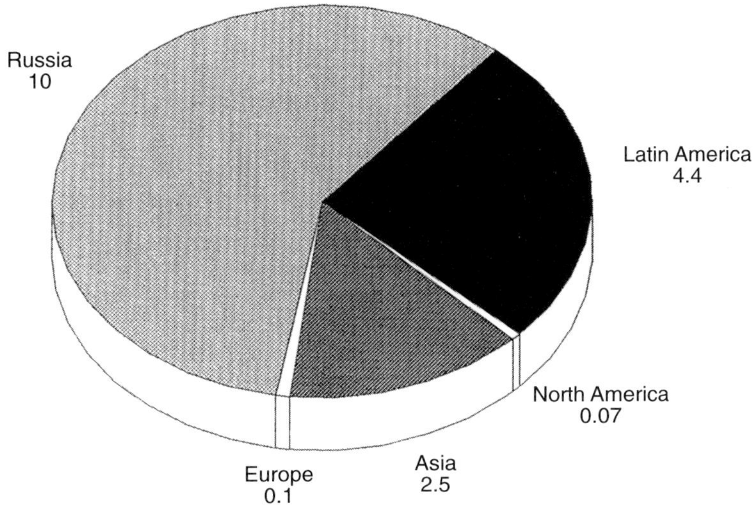

Fig. 11.1. Estimated areas (in million hectares) under augmentative forms of biological control (after van Lenteren, 2000).

mated that – expressed in monetary value – 80% of the commercial natural enemies are used in greenhouses. The vast amount of natural enemies used on about 16 million ha of field crops mainly consist of non-commercial products that are reared in state-funded laboratories. For these biological control agents cost estimates are often lacking. The most applied natural enemies in greenhouses are *E. formosa*, accounting for 25% of the total market, *P. persimilis*, accounting for 12%, and *A. cucumeris*, also accounting for 12%. Another good indicator of the significance of groups of natural enemies is the investment in money for control of the various groups of pests (Fig. 11.2). Four groups of pests – whiteflies, thrips, spider mites and aphids – account for 84% of the costs of biological pest control.

Large differences in prices for biological control agents exist among the commercial companies (for details, see van Lenteren *et al.*, 1997). A general observation is that there are many more species of natural enemies commercially available in Europe than in the USA, as a result of the much larger greenhouse industry in Europe. In comparison with the USA, it can also be concluded that commercial biological control suppliers in Europe are of larger size than their partners in the USA. When comparing the data of Cranshaw *et al.* (1996) with European data, a number of specific conclusions can be drawn. Comparisons between other continents are difficult to make because of lack of data.

Parasitoids

It appears that parasitoids of Diptera are hardly used in Europe, whereas they are widely available in the USA. The main reason for this is the large differences in poultry production between the two continents. A second difference is that most natural enemies in the USA are sold for application in open-field crops, while in Europe the major part of commercial biological control takes place in greenhouses. Sales of *Trichogramma* spp. are more common in the USA than in Europe, where these parasitoids are used on fewer than 10,000 ha for control of European cornborer in maize. Also, parasitoids of scale species seem to be marketed on a much larger scale in the USA than in Europe. In the USA, scale parasitoids are applied, for example, in citrus orchards, whereas, in Europe, they are applied in greenhouses and interior plantscapes. Natural enemies are used for controlling pests in orchards in Europe, primarily for inoculative releases with a long-term effect instead of inundative releases. In

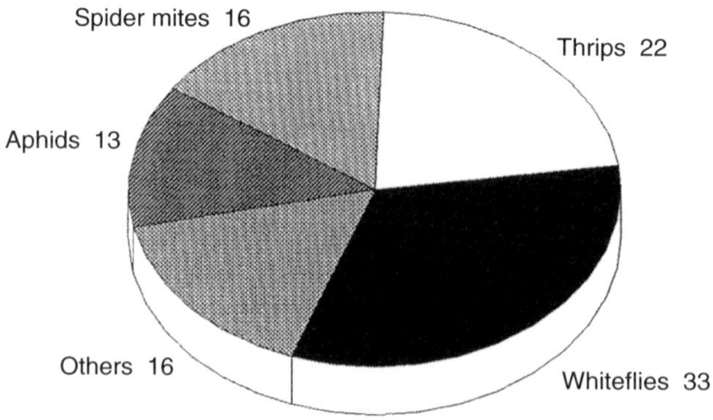

Fig. 11.2. Investments in money (expressed as percentages) in natural enemies used for control of different greenhouse pests (after Bolckmans, 1999).

Europe, many more aphid and leafminer parasitoids are marketed, both in total number as well as in the diversity of species. Also, the number of *E. formosa* sold in Europe is much higher than in the USA. The reason for this difference is that these natural enemies are applied in greenhouses, and the number of large greenhouses for commercial production of vegetables and ornamentals is much larger in Europe.

Predatory insects

It seems that, in the USA, *Chrysoperla* spp. are used more often than in Europe. Their high price and the observation that they are mainly effective at a high pest density make them not so attractive for the European greenhouse market, where only a very low pest density can be tolerated. Also *Cryptolaemus* sp. is more popular in the USA than in Europe, but the use of *Orius* spp. is more prevalent in Europe. Most of the predatory species are available both in the USA and Europe, so there is less difference in the predator-species composition than was found for parasitoids.

Predatory mites and entomophagous nematodes

The spectrum of available predatory mites and entomophagous nematode species is similar in the USA and Europe, with the exception of *Hypoaspis* species, which are used mainly in Europe. The quantity of predatory mites sold in Europe is much larger because predatory mites are used primarily in greenhouses.

A final conclusion that we can draw from the current data on mass production is that in Europe most natural enemies are used by professionals, whereas in North America biological control agents are also often used in home gardens and can be bought at garden centres or via garden catalogues (Cranshaw *et al.*, 1996).

References

Anon. (2000) 2001 directory of least-toxic pest control products. *IPM Practitioner* 22, 1–38.

Bolckmans, K.J.F. (1999) Commercial aspects of biological pest control. In: Albajes, R., Gullino, M.L., van Lenteren, J.C. and Elad, Y. (eds) *Integrated Pest and Disease Management in Greenhouse Crops*. Kluwer Academic Publishers, Dordrecht, pp. 310–338.

Cranshaw, W., Sclar, D.C. and Cooper, D. (1996) A review of 1994 pricing and marketing by suppliers of organisms for biological control of arthropods in the United States. *Biological Control* 6, 291–296.

Koppert (1994) *Koppert Products with Directions for Use*. Koppert, Berkel en Rodenrijs, 28 pp.
Lipa, J.J. and Smiths, P.H. (1999) Microbial control of pests in greenhouses. In: Albajes, R., Gullino, M.L., van Lenteren, J.C. and Elad, Y. (eds) *Integrated Pest and Disease Management in Greenhouse Crops*. Kluwer Academic Publishers, Dordrecht, pp. 295–309.
Parwinder, G. and Georgis, R. (1994) Fundamental research on entomopathogenic nematodes: an industrial perspective. In: *Proceedings XXVIIth Annual Meeting of the Society for Invertebrate Pathology*, Vol. 1, pp. 126–130.
Rodriguez, R.J.M., Ferragut, F., Carnero, A. and Pena, M.A. (1994) Diapause in the predacious mites *Amblyseius cucumeris* (Oud.) and *Amblyseius barkeri* (Hug.): consequences of use in integrated control programmes. *Journal of Applied Entomology* 118, 44–50.
Speyer, E.R. (1927) An important parasite of the greenhouse whitefly (*Trialeurodes vaporariorum* Westwood). *Bulletin of Entomological Research* 17, 301–308.
Tomalak, M. (1994) Genetic improvement of *Steinernema feltiae* for integrated control of the western flower thrips, *Frankliniella occidentalis*. *Bulletin IOBC/WPRS* 17, 17–20.
van Lenteren, J.C. (1995) Integrated pest management in protected crops. In: Dent, D. (ed.) *Integrated Pest Management*. Chapman & Hall, London, pp. 311–343.
van Lenteren, J.C. (1997) Benefits and risks of introducing exotic macro-biological control agents into Europe. *Bulletin OEPP/EPPO* 27, 15–27.
van Lenteren, J.C. (2000) Measures of success in biological control of arthropods by augmentation of natural enemies. In: Gurr, G. and Wratten, S. (eds) *Measures of Success in Biological Control*. Kluwer Academic Publishers, Dordrecht, pp. 77–103.
van Lenteren, J.C. and Woets, J. (1988) Biological and integrated pest control in greenhouses. *Annual Review of Entomololgy* 33, 239–269.
van Lenteren, J.C., Benuzzi, M., Nicoli, G. and Maini, S. (1992) Biological control in protected crops in Europe. In: van Lenteren, J.C., Minks, A.K. and de Ponti, O.M.B. (eds) *Biological Control and Integrated Crop Protection: Towards Environmentally Safer Agriculture*. Pudoc, Wageningen, pp. 77–84.
van Lenteren, J.C., van Roermund, H.J.W. and Suetterlin, S. (1996) Biological control of greenhouse whitefly (*Trialeurodes vaporariorum*): how does it work? *Biological Control* 6, 1–10.
van Lenteren, J.C., Roskam, M.M. and Timmer, R. (1997) Commercial mass production and pricing of organisms for biological control of pests in Europe. *Biological Control* 10, 143–149.
Wright, E.M. and Chambers, R.J. (1994) The biology of the predatory mite *Hypoaspis miles* (Acari: Laelapidae), a potential biological control agent of *Bradysia paupera* (Dipt.: Sciaridae). *Entomophaga* 39, 225–235.

12 Mass Production, Storage, Shipment and Release of Natural Enemies

J.C. van Lenteren[1] and M.G. Tommasini[2]
[1]*Laboratory of Entomology, Wageningen University, PO Box 8031, 6700 EH Wageningen, The Netherlands;* [2]*CRPV (Centro Ricerche Produzioni Vegetali), Via Vicinale Monticino 1969, 47020 – Diegaro di Cesena (FC), Italy*

Abstract

Mass production of natural enemies started in the 1940s and quickly developed thereafter. These developments were mainly triggered by trying to economize rearing and making biological control more competitive when compared with other pest-control methods. Storage of natural enemies is only possible for very short periods, with the exception of species for which it is known how to start and terminate diapause. Initially, the collection, shipment and release of biological control agents were rather amateurish, but enormous progress has also been made in this area. Many natural-enemy species can now be produced at competitive costs, resulting in increased use of biological pest control.

Introduction

Since the beginning of this century, the mass production of natural enemies has been considered as a means of improving biological control programmes, especially those based on inundative and seasonal inoculative releases. For general information on mass production and quality control of insects and other arthropods, we refer the reader to Morrison and King (1977), King and Morrison (1984), Singh (1984), Singh and Moore (1985), van Lenteren (1986a) and various chapters in this book. For mass production related to commercially produced natural enemies, we refer the reader to van Lenteren (1986b), van Lenteren and Woets (1988), Nicoli *et al.* (1994) and Bolckmans (1999). We shall not discuss the question of how to obtain a good stock colony to start a mass production, because this issue is addressed in Chapters 1, 6 and 7. In this chapter, we shall briefly summarize developments in the mass rearing of natural enemies for commercial biological control during the 20th century.

Mass production of beneficials is a 'skilful and highly defined processing of an entomophagous species through insectary procedures which results in economical production of millions of beneficial insects' (Finney and Fisher, 1964). This is true for most mass-rearing programmes, but there are important exceptions where mass production seems to be a fairly simple process.

© CAB International 2003. *Quality Control and Production of Biological Control Agents: Theory and Testing Procedures* (ed. J.C. van Lenteren)

The first step in a mass-rearing programme is a trial to rear the natural enemy on a natural host (the pest organism) in an economical way. Most natural enemies are reared in this way. However, several natural enemies are not mass-reared on their natural host because it is either too expensive or undesirable due to the risk of infection with the pest organism or concurrent infection with other pests or diseases when natural enemies are released on their natural substrate. In these cases a search is made for an opportunity to rear the natural enemy on an alternative host (and often an alternative host plant).

A subsequent step in making mass rearing more economical is to change from a natural host medium (host plant) to an artificial medium for rearing the host. Rearing insects on artificial diets was developed earlier this century and considerable progress has been made recently. Rearing on artificial diets is considerably cheaper as less expensively climatized space is needed, but artificial rearing may create serious quality problems, which will be discussed later in this chapter. Singh (1984) summarizes the historical development, recent advances and future prospects for insect diets as follows:

1. Some 750 species, mainly phytophagous insects, can be reared successfully on (semi-) artificial diets.
2. Only about two dozen species have been successfully reared for several generations on completely artificial diets.
3. Large-scale mass rearing on artificial media has been developed for fewer than 20 species of insects.
4. Quality control is essential, as there can be dietary effects on all critical performance traits of the mass-reared insect and also on the natural enemy produced on a host that was mass-reared on an artificial medium.
5. Suitable bioassays are important for answering the question 'what is the ultimate effect of the diet on the reared insect?'

A final step when trying to minimize rearing costs is the search for ways to rear the natural enemy on an artificial diet. This has been attained for several ecto- and endoparasitoids (e.g. *Trichogramma*) and a few predators (e.g. *Chrysoperla*). The technology for rearing natural enemies on diets is, however, far less developed than that for rearing pest species (Chapter 9; Grenier *et al.*, 1994).

The rapid development of commercial biological control based on mass-produced natural enemies can be illustrated well with data from Europe. About 150 species of natural enemies have been imported and released into Europe during the 20th century to control about 55 mite and insect pest species. Until 1970 this mainly concerned inoculative (classical) biological control. After 1970 many developments took place in greenhouses and annual field crops, and commercial biological control programmes for *c.* 50 pest species were developed by importing more than 60 species of natural enemies. In addition, more than 40 endemic species of natural enemies were employed in commercial biological control. For all these species, fine-tuned mass-production systems had to be developed. Our experience with the development of new biological control programmes has shown that dogmatism is useless when selecting natural enemies. This contrasts with the approach of earlier biocontrol workers (see, for example, DeBach, 1964). We have, for example, had excellent control results by releasing endemic natural enemies against exotic pests and vice versa: all combinations are worth trying (Table 12.1).

The most important species of natural enemies that are mass-reared in Europe are given in Table 11.2 of Chapter 11, and an overview of all available natural enemies is presented in Table 11.1 of the same chapter. Although on-farm production of natural enemies is possible, most growers purchase them from commercial suppliers. Many of the mass-production companies are, understandably, reluctant to provide information on many aspects of mass production. Our experience is that most of the natural enemies produced for biocontrol in protected cultivation are reared on their natural hosts (the pests) and host plants. Rearing on purely artificial media (without organic additives) is very rare, primarily because this technology is insufficiently developed for mass production and because this way of production may lead to poor performance of natural enemies when exposed to

Table 12.1. Commercial biological control of endemic or exotic arthropods with endemic or exotic natural enemies in Europe (the numbers reflect the number of combinations in which a certain natural enemy is used for control of a certain pest; situation in 1999).

Different combinations of natural-enemy use in Europe	
Use of endemic natural enemies for the control of endemic pests:	
example: *Chrysoperla carnea* for control of endemic aphid species	61
Use of endemic natural enemies for the control of exotic pests:	
example: *Diglyphus isaea* for control of exotic *Lyriomyza* species	40
Use of exotic natural enemies for the control of endemic pests:	
example: *Harmonia axyridis* for control of endemic aphid species	44
Use of exotic natural enemies for the control of exotic pests:	
example: *Encarsia formosa* for control of exotic whitefly species	47

their target hosts (for details, see van Lenteren, 1986a, and Chapters 1 and 2). Rearing conditions should be as similar as possible to the conditions under which the natural enemies will have to function in the field or greenhouse. Two examples of mass-production schemes, one for the predator *Orius* and the other for the parasitoid *Encarsia*, are presented in Figs 12.1 and 12.2.

Storage of Natural Enemies

It is necessary to have storage methods and facilities available to meet the requirements for good planning for a mass-production unit and because of the difficulty of accurately predicting demand from clients (both delivery dates and quantities). This is relatively simple for microbial biocontrol agents, such as fungi, viruses and bacteria, because they can often be stored in a resting stage for months or even years. Many predators and parasitoids can only be stored for a short time. This usually involves placing the natural enemies as immatures at temperatures between 4 and 15°C. Normally, storage only lasts several weeks, but even then reduction in fitness is the rule (Posthuma-Doodeman *et al.*, 1996). Storage of parasitoids at a low temperature (8°C) for 2 and 16 days, respectively, gives similar percent-

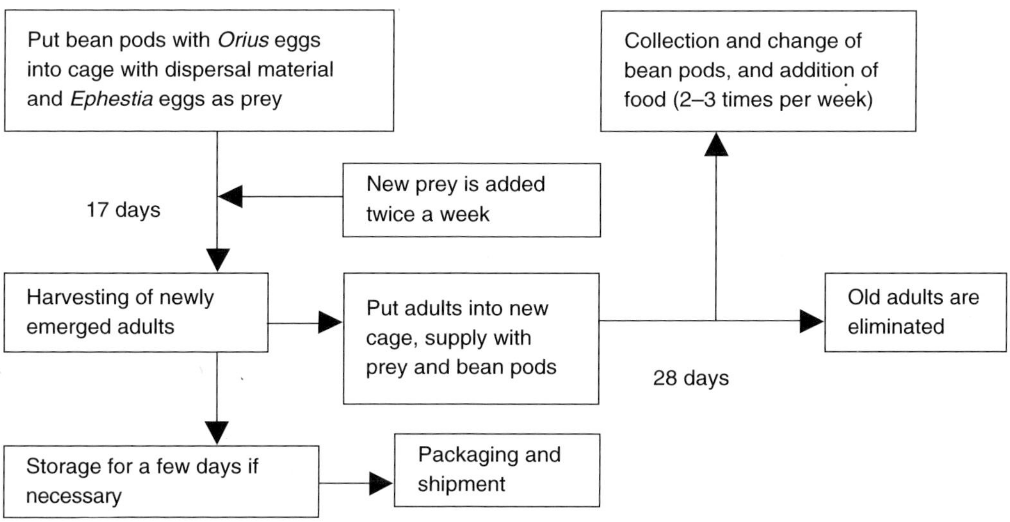

Fig. 12.1. Production scheme for the thrips predator *Orius* sp.

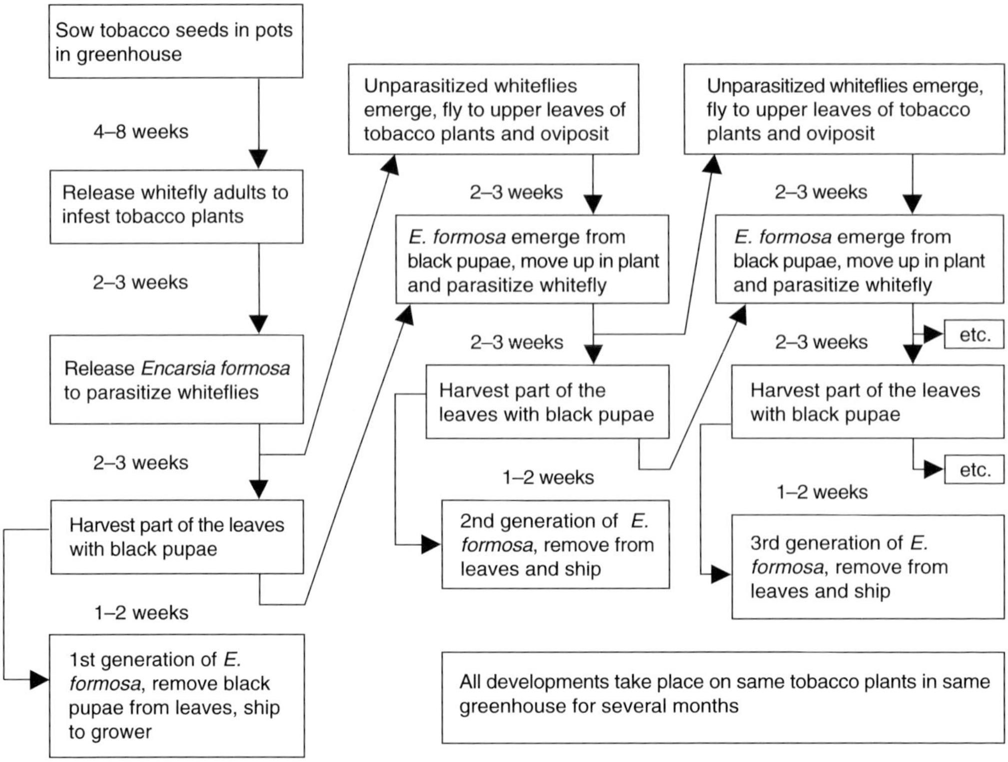

Fig. 12.2. 'Continuous' production scheme for the whitefly parasitoid *Encarsia formosa*.

ages of emergence, but the ability to fly is much lower for the parasitoids that were stored for 16 days (Fig. 12.3). This test was done with the short-range flight cylinder as described in Chapter 19. Storage during the adult stage leads to even higher and faster reduction in fitness than with storage of immatures. The pupal stage seems to be most suitable for short-term storage.

Data on long-term storage of natural enemies or their hosts are limited. Host material (e.g. eggs of *Sitotroga cerealella* and *Grapholita lineatum*) stored for long periods (in the case of *Grapholita* for up to 5 years) in liquid nitrogen could still be used for production of *Trichogramma* and *Trissolcus simoni*, respectively (Gennadiev and Khilistovskii, 1980). Eggs of *Ephestia kuehniella* can be sterilized by ultraviolet (UV) radiation or freezing, and then be stored at low temperature for several months without losing their value as alternative food for mass production of predators such as *Chrysoperla* and *Orius*.

The parasitoid *Diglyphus isaea* can be stored at a low temperature for at least 2 months, during which time mortality does not increase and fecundity remains the same (Burgio and Nicoli, 1994). Hagvar and Hofsvang (1991) reported that some species of Aphidiidae (e.g. *Aphidius matricariae*) can be stored at low temperatures for several weeks.

The possibility of storing beneficials in the diapausing stage has been studied, but most of this work has not yet led to practical application, because unacceptably high mortality occurred during the artificially induced diapause. There are, however, some positive exceptions. Diapausing adults of the predator *Chrysoperla carnea* can be stored at a low temperature for about 30 weeks while maintaining an acceptable level of survival and reproduction activity (Tauber *et al.*, 1993). Also the predator *Orius insidiosus* maintains good longevity and reproduction rate after storage in diapause for up to 8 weeks

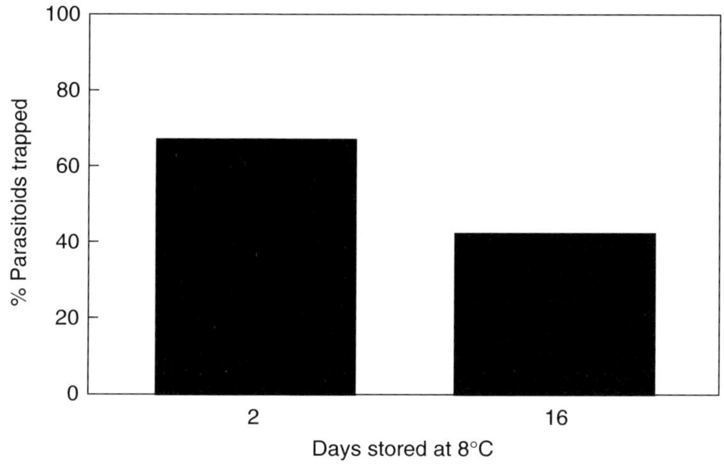

Fig. 12.3. Percentage *Encarsia formosa* females capable of flying (= reaching trap in short-range flight test) when stored for 2 and 16 days at 8°C.

(Ruberson *et al.*, 1998). The predator *Aphidoletes aphidimyza* can survive periods of 3–8 months when stored at 10°C (Tiitanen, 1988). Long-term storage of the diapausing stage of the parasitoid *Trichogramma*, has been successful for periods of up to a year and is now commercially exploited (J. Frandon, Biotop, Antibes, France, 1996, personal communication).

Long-term storage capability is very desirable for production companies, because:

- continuous production of the same quantity of beneficial insects is often economically more attractive than seasonal production of very large numbers;
- storage facilities enable them to build up reserve supplies of entomophages to compensate for periods of low production or periods of unexpected high demand;
- storage makes rearing possible at the best period of the year, e.g. at a period when host plants can be grown under optimal conditions.

Collection and Shipment of Natural Enemies

After production, the beneficials should be delivered to the growers as soon as possible. If delivery is looked after by the producer and occurs within 48 h of harvesting the organisms, no special shipment procedures are normally needed for parasitoids and non-cannibalistic predators other than protection against excessive heat, cold or rough handling. When transport takes several days, climatized containers should be used and it may be necessary to add food (e.g. honey in the case of parasitoids and pollen/prey for predators). A way to overcome problems with long times for transport of predators is for young stages to be packaged with food so that further development takes place during transport. Packaging of predators demands special attention when cannibalism is a common phenomenon. Many of the commercially available predators are generalists and exhibit cannibalism when kept at high densities, even if food is available in the containers for shipment. To reduce the risk of cannibalism, it is common to provide hiding-places for the natural enemy by using paper, buckwheat, vermiculite or wheat bran in the container (see Table 12.2). In the early days of mass production, the biological control agents were often collected and shipped on the host plant on which they were reared. With the internationalization of biocontrol, shipment on or in inert media became a necessity. Ingenious collection and shipping procedures have been developed.

Table 12.2. Survey of collection, shipment and release methods for biological control agents.

Natural enemy	Stage at which collected	Special handling	Counting	Packing	Shipping	Mode of introduction	Quality control
Amblyseius spp.	All stages		Estimate	In wheat bran	Plastic container	Sprinkling on plants	Yes
Aphidius spp.	Pupa	Removal from leaf	Weight		Plastic container	Small numbers in sheltered locations	Yes
Aphidoletes aphidimyza	Pupa		Weight	Vermiculite	Plastic container	Small numbers in sheltered locations	Yes
Chrysoperla carnea	Egg	Removal from leaf	Volumetr.	Buckwheat	Plastic container	Sprinkling on plants	Yes
Cryptolaemus montrouzieri	Adult		Counting	Paper strips	Plastic container	Small numbers on plants	
Dacnusa siberica	Adult		Counting		Plastic container	Tapped from container	Yes
Diglyphus isaea	Adult		Counting		Plastic container	Tapped from container	Yes
Encarsia formosa	Pupa on leaves	Removal from leaf	Volumetr.	Glued on cardboard	Paper box	Cards hung in plants	Yes
Eretmocerus mundus	Pupa on leaves	Removal from leaf	Volumetr.	Glued on cardboard	Paper box	Cards hung In plants	Yes
Harmonia axyridis	Larvae		Counting	Popcorn	Plastic container	Tapped from container on plants	
Leptomastix dactylopii	Adult		Counting		Plastic container	Tapped from container	Yes
Lysiphlebus testaceipes	Pupa	Removal from leaf	Counting		Plastic container	Small numbers in sheltered locations	
Macrolophus caluginosus	Nymphs and adult		Counting	Buckwheat + vermic.	Plastic container	Small numbers on plants	Yes
Neoseiulus spp.	All stages		Estimate	In wheat bran	Plastic container	Sprinkling on plants	Yes
Orius spp.	Nymphs and adult		Volumetr.	Buckwheat + vermic.	Plastic container	Small numbers on plants	Yes
Phytoseiulus persimilis	All moving stages		Volumetr.	In wheat bran	Plastic container	Sprinkling on plants	Yes
Trichogramma spp.	Pupa		Volumetr.	Glued on card	Paper box	Cards hung in plants or by sprinkling on plants	Yes

Poor shipping conditions frequently led to natural enemies arriving either dead or in poor condition. Difficulties in shipping can be considerable in countries where crops with the same target pest are not concentrated together and where distances are large. Most transport is still by truck, although an increasing quantity is sent by aircraft. Intercontinental transport problems are caused less by containerization than by the sometimes excessively long handling time at customs, which leads to high mortality or decrease in fitness. The logistics of shipments remains one of the main problems for the commercialization of biological control. Examples of the different techniques for collecting, counting, packaging and shipping of natural enemies are given in Table 12.2.

Release of Natural Enemies

Developmental stage at which organism is released

Entomophagous insects can be brought into greenhouses or the field in different stages of their development (see also Table 12.2):

- eggs (e.g. *Chrysoperla*);
- larvae or nymphs (e.g. *Chrysoperla, Phytoseiulus, Amblyseius, Orius*);
- pupae or mummies (e.g. *Aphidius, Trichogramma, Encarsia*);
- adults (e.g. *Dacnusa, Diglyphus, Orius, Phytoseiulus*);
- all stages together (e.g. *Phytoseiulus, Amblyseius*).

The stage in which the beneficials are introduced depends mainly on the ease of transport and manipulation in the field, but it is, of course, also important to release the natural enemy at a stage when it is most active at killing the pest. Usually the stage that is least vulnerable to mechanical handling is chosen and therefore a non-mobile stage, often the egg or pupa, is most suited for transport and release. In situations where it is difficult, but essential, to distinguish the natural enemy from the pest, the only solution is to introduce adults. Adult releases for parasitoids are advised only when younger natural-enemy stages cannot be distinguished or separated from the pest insect: the handling and releasing of delicate adult parasitoids is very difficult and often a large reduction of fertility is observed compared with the fertility of parasitoids when released as immatures. When the natural enemy is released in one of the developmental stages that do not prey upon or parasitize the host, the timing should be such that the active stage emerges at the right moment of pest population development. For some natural enemies the stage of release depends on pest development: when pest density is low, release of first-instar *C. carnea* suffices; when the infestation with the pest organisms is already relatively high, it is better to release second-instar larvae, which have a much higher predation capacity.

Methods of introduction

Beneficials are introduced into the field in many ways (Table 12.2). Eggs and pupae are either distributed over the field on their normal substrate (leaves of the host plant, e.g. *Chrysoperla* and *Encarsia*) or glued on paper/cardboard cards (e.g. *Encarsia, Trichogramma*). These stages of the natural enemies can also be collected and put into containers, which are then brought into the field (e.g. *Trichogramma*).

The mobile stages of natural enemies, larvae or nymphs and adults, can be put into the field in containers from which they emerge (e.g. many adult parasitoids and predators) or the grower can distribute natural enemies in these stages over the crop, for example by 'sprinkling' them on to the plant. In this case, the use of dispersal material (e.g. buckwheat, vermiculite) is often necessary in order to obtain a homogeneous distribution of small natural enemies. When natural substrates (e.g. buckwheat or wheat bran) are used as dispersal materials, they must be free of pesticides.

Instead of introducing the predator or parasitoid by itself, one can also introduce a whole 'production unit', e.g. 'banker-plants' containing the host insect and its natural enemy can be brought into a crop. When the introduced host population is almost exterminated, the natural enemies invade the surrounding crop (van Steenis, 1995).

The moment of introduction

In many cases, the natural enemies are released when the pest organism has been observed, although it is not unusual to apply 'blind releases' when sampling of the pest is difficult (e.g. whiteflies) or when pest populations develop very quickly, such as those of aphids and thrips. When pest generations are not yet overlapping early in the growing season, proper timing of the release(s) is essential so that the beneficials are available when the preferred host stages are present.

Determining the dosage, the distribution and the frequency of the releases is a very difficult problem, which is encountered in both inundative- and seasonal inoculative-release programmes. Release ratios are not critical in inundative-release programmes as long as it is possible to release a (super)abundance of natural enemies. This, however, may be limited by the cost of mass production. In seasonal inoculative programmes release ratios are more critical: if too few beneficials are released, effective control will be obtained after the pest has caused economic damage; if too many are released, there is a risk of exterminating the pest and thus eventually also the natural enemy. This is a practical problem in small tunnels and greenhouses. In the latter situation resurgence of the pest is likely and a serious threat. In these seasonal innoculative-release programmes, the release ratios are usually determined by trial and error, but the first simulation programmes are appearing for a more scientific estimate of release rates (number of releases, spacing between release points and timing of releases; see, for example van Roermund (1995) for seasonal inoculative releases and Sueverkropp (1997) for inundative releases).

Conclusions

Mass production of natural enemies has undergone a very fast development during the past three decades: the numbers produced have greatly increased, the spectrum of species available has widened dramatically and mass-production methods have clearly evolved. Developments in the area of mass production, quality control, storage, shipment and release of natural enemies have decreased production costs and led to better product quality, but much more can be done. Innovations in long-term storage (e.g. through diapause), shipment and release methods may lead to a further increase in natural-enemy quality, with a concurrent reduction in costs of biological control, thereby making it easier and more economical to apply.

References

Bolckmans, K.J.F. (1999) Commercial aspects of biological pest control. In: Albajes, R., Gullino, M.L., van Lenteren, J.C. and Elad, Y. (eds) *Integrated Pest and Disease Management in Greenhouse Crops*. Kluwer Academic Publishers, Dordrecht, pp. 310–318.

Burgio, C. and Nicoli, G. (1994) Cold storage of *Diglyphus isaea*. In: Nicoli, G., Benuzzi, M. and Leppla, N.C. (eds) *Proceedings 7th Global IOBC Workshop 'Quality Control of Mass Reared Arthropods', 13–16 September 1993, Rimini, Italy*, pp. 171–178.

DeBach, P. (ed.) (1964) *Biological Control of Insect Pests and Weeds*. Chapman & Hall, London, 844 pp.

Finney, G.L. and Fisher, T.W. (1964) Culture of entomophagous insects and their hosts. In: DeBach, P. (ed.) *Biological Control of Insect Pests and Weeds*. Chapman & Hall, London, pp. 329–355.

Gennadiev, V.G. and Khilistovskii, E.D. (1980) Long-term cold storage of host eggs and reproduction in them of egg-parasites of insect pests. *Zhurnal Obshchei Biologii* 41, 314–319.

Grenier, F., Greany, P. and Cohen, A.C. (1994) Potential for mass release of insect parasitoids and predators through development of artificial culture techniques. In: Rosen, D., Bennett, F.D. and Capinera, J.L. (eds) *Pest Management in the Subtropics – a Florida Perspective*. Intercept, Andover, UK, pp. 181–205.

Hagvar, E.B. and Hofsvang, T. (1991) Aphid parasitoids (Hymenoptera: Aphidiidae): biology, host selection and use in biological control. *Biocontrol News and Information* 12, 13–41.

King, E.G. and Morrison, R.K. (1984) Some systems for production of eight entomophagous arthropods. In: King, E.G. and Leppla, N.C. (eds) *Advances and Challenges in Insect Rearing*. USDA/ARS, New Orleans, pp. 206–222.

Morrison, R.K. and King, E.G. (1977) Mass production of natural enemies. In: Ridgway, R.L. and Vinson, S.B. (eds) *Biological Control by Augmentation of Natural Enemies*. Plenum, New York, pp. 183–217.

Nicoli, G., Benuzzi, M. and Leppla, N.C. (eds) (1994) *Proceedings 7th Global IOBC Workshop 'Quality Controy of Mass Reared Arthropods', 13–16 September 1993, Rimini, Italy*, 240 pp.

Posthuma-Doodeman, C.J.A.M., van Lenteren, J.C., Sebestyen, I. and Ilovai, Z. (1996) Short-range flight test for quality control of *Encarsia formosa*. *Proceedings Experimental and Applied Entomology* 7, 153–158.

Ruberson, J.R., Kring, T.J. and Elkassabany, N. (1998) Overwintering and diapause syndrome of predatory Heteroptera. In: Coll, M. and Ruberson, J.R. (eds) *Predatory Heteroptera: Their Ecology and Use In Biological Control*. Thomas Say Publications in Entomology, ESA, Lanham, Maryland, pp. 49–69.

Singh, P. (1984) Insect diets: historical developments, recent advances, and future prospects. In: King, E.G. and Leppla, N.C. (eds) *Advances and Challenges in Insect Rearing*. USDA/ARS, New Orleans, pp. 32–44.

Singh, P. and Moore, R.F. (eds) (1985) *Handbook of Insect Rearing*. Elsevier, Amsterdam, Vol. 1, 488 pp., Vol. 2, 514 pp.

Sueverkropp, B.P. (1997) Host-finding behaviour of *Trichogramma brassicae* in maize. PhD thesis, Agricultural University, Wageningen.

Tauber, M.J., Tauber, C.A. and Gardescu, S. (1993) Prolonged storage of *Chrysoperla carnea* (Neuroptera: Chrysopidae). *Environmental Entomology* 22, 843–848.

Tiitanen, K. (1988) Utilization of diapause in mass production of *Aphidoletes aphidimyza* (Rond.) (Dipt., Cecidomyiidae). *Annales Agriculturae Fenniae* 27, 339–343.

van Lenteren, J.C. (1986a) Evaluation, mass production, quality control and release of entomophagous insects. In: Franz, J.M. (ed.) *Biological Plant and Health Protection*. Fischer, Stuttgart, pp. 31–56.

van Lenteren, J.C. (1986b) Parasitoids in the greenhouse: successes with seasonal inoculative release systems. In: Waage, J.K. and Greathead, D.J. (eds) *Insect Parasitoids*. Academic Press, London, pp. 341–374.

van Lenteren, J.C. and Woets, J. (1988) Biological and integrated control in greenhouses. *Annual Review of Entomology* 33, 239–269.

van Roermund, H.J.W. (1995) Understanding biological control of greenhouse whitefly with the parasitoid *Encarsia formosa*: from individual behaviour to population dynamics. PhD thesis, Agricultural University, Wageningen.

van Steenis, M. (1995) Evaluation and application of parasitoids for biological control of *Aphis gossypii* in glasshouse cucumber crops. PhD thesis, Agricultural University, Wageningen.

13 Regulation of Import and Release of Mass-produced Natural Enemies: a Risk-assessment Approach

J.C. van Lenteren,[1] D. Babendreier,[2] F. Bigler,[2] G. Burgio,[3] H.M.T. Hokkanen,[4] S. Kuske,[2] A.J.M. Loomans,[1] I. Menzler-Hokkanen,[4] P.C.J. van Rijn,[5] M.B. Thomas[5] and M.G. Tommasini[6]

[1] *Laboratory of Entomology, Wageningen University, PO Box 8031, 6700 EH Wageningen, The Netherlands;* [2] *Swiss Federal Research Station for Agroecology and Agriculture, Zurich, Switzerland;* [3] *Department of Agroenvironmental Sciences and Technologies, University of Bologna, Italy;* [4] *Department of Applied Biology, University of Helsinki, Finland;* [5] *CABI Bioscience, Silwood Park, Ascot, Berkshire, UK;* [6] *CRPV (Centro Ricerche Produzioni Vegetali), Via Vicinale Monticino 1969, 47020 – Diegaro di Cesena (FC), Italy.*

Abstract

In the past 30 years, many exotic natural enemies have been imported, mass-reared and released as biological control agents. Negative effects of these releases have not been reported yet. The current popularity of biological control may, however, result in problems, as an increasing number of activities will be executed by persons not trained in the identification, evaluation and release of biocontrol agents. Therefore, protocols for risk assessment are being developed within the European Union (EU)-financed project 'Evaluating Environmental Risks of Biological Control Introductions into Europe' (ERBIC) as a basis for regulation of the import and release of exotic natural enemies. This chapter presents a summary of the situation concerning regulations for the import and release of natural enemies, a general framework for risk-assessment procedures for biological control agents, and a more detailed framework on the methodology for risk assessment of natural enemies. In the methodology for risk assessment, information on the potential of an agent to establish, its abilities to disperse and its direct and indirect effects, including host-specificity testing, is integrated.

Introduction

Biological control of insects by the introduction and permanent establishment of exotic natural enemies has been practised for over 100 years (classical or inoculative biological control; Greathead, 1995) and augmentative releases of beneficial insects for the control of pests in greenhouse and field situations date back to the 1920s (inundative and seasonal inoculative biological control; van Lenteren and Woets, 1988). In this period, hundreds of

© CAB International 2003. *Quality Control and Production of Biological Control Agents: Theory and Testing Procedures* (ed. J.C. van Lenteren)

species of exotic natural enemies have been imported, mass-reared and released, resulting in successful control of many species of pests (e.g. Greathead, 1995; Gurr and Wratten, 2000; van Lenteren, 2000). As far as we know, few problems have occurred concerning negative effects of these releases (e.g. Lynch and Thomas, 2000; Lockwood *et al.*, 2001; Lynch *et al.*, 2001). The introduction of natural enemies has largely been an empirical activity and has depended on the knowledge and insight of the biological control practitioner. Recently, a more scientific approach of evaluation of natural enemies before introduction is advocated and applied, with the aims: (i) to obtain insight into what constitutes an efficient natural enemy; (ii) to reduce research costs; and (iii) to limit risks of imports (van Lenteren, 1993).

As each organism that is introduced into a new region, be it for augmentative releases or for permanent biological control, may become established, extreme care should be exerted during the evaluation phase. The more than 5000 introductions of exotic natural enemies that have taken place during the past 100 years did not lead to environmental problems when the procedure of selection, importation and release was carefully applied (Greathead, 1995). In general, biological control workers prefer to import specialist natural enemies, attacking only one or a few related phytophagous species. The few documented problems caused by introduced organisms concern attempts to control pests other than insects, such as vertebrates and snails. The introduction of a generalist vertebrate predator (mongoose, *Herpestes* sp.) for rodent control in the 19th century resulted in economic and ecological problems, as this opportunist predator attacked all prey it could easily handle and contributed to the extinction of some endemic species. The introduction of other large generalist predators such as the giant toad (*Bufo marinus*) and predatory snails, had unfortunate side-effects (Greathead, 1995), and the liberation of such organisms has never been recommended by researchers of biological control (e.g. van Lenteren, 2000, 2001). To a large extent, only those organisms are released that experience has shown to have restricted host ranges and not to exhibit switches in host ranges (Aeschlimann, 1995).

The use of biological control has increased considerably during the past decades, as it offers a sustainable, economical and environmentally attractive alternative to chemical pest control. Arthropod biological practitioners are, however, confronted with criticism from environmentalists because they are fearful that the released natural enemies may attack: (i) beneficial non-target organisms, such as pollinators or other natural enemies; (ii) rare/endangered insects, such as butterflies; and (iii) other non-target organisms. This results in a bizarre conflict situation: on the one hand, biological control offers a very powerful option for reducing the many potentially negative environmental and human health effects of chemical pest control, while, on the other hand – though not found until now – the arthropod natural enemies may influence ecosystems in unwanted ways. The very positive effects of releases of natural enemies are in strong contrast with the effects of other types of introductions. Many intended introductions of animals and plants and many more unintended introductions of pest organisms have led to disasters worldwide (for examples, see Hokkanen and Lynch, 1995; van Lenteren, 1995).

The potential negative effects of arthropod biological control might be prevented when, as in weed biological control programmes (e.g. Wapshere, 1974; Blossey, 1995; Lonsdale *et al.*, 2001), not only is the effect on the target species determined, but also the effect on indigenous non-target species (van Lenteren, 1986, 1995; Blossey, 1995). Up until now, indigenous non-target species testing as part of a prerelease evaluation programme is rarely applied (van Lenteren and Woets, 1988; Waage, 1997), but there are a few examples where non-target species testing has been applied properly (e.g. Barratt *et al.*, 1999; Neuenschwander and Markham, 2001). Another way to reduce the risks of release of exotic natural enemies would be to first evaluate native natural enemies. Although this seems a logical approach, the experience of the last four decades shows that many exotic natural enemies were imported and released without preceding testing of native biocontrol agents (e.g. van Lenteren, 2000).

Because of the lack of in-depth behavioural and ecological studies concerning the risks of mass releases of exotic natural enemies, in 1998 we initiated a 4-year research project on 'Evaluating Environmental Risks of Biological Control Introductions into Europe' (ERBIC), which is funded by the European Union (EU) 4th Framework Programme (Lynch *et al.*, 2001). Within the EU-ERBIC project, and in collaboration with an Organization for Economic Cooperation and Development (OECD) working group, guidelines are being developed for harmonized information requirements for the import and release of invertebrate biological control agents used in augmentative biological control (van Lenteren *et al.*, 2003).

This chapter presents a summary of the situation concerning regulations for the import and release of natural enemies and a framework for risk-assessment procedures for biological control agents.

Current Situation Concerning the Import and Release of Exotic Natural Enemies

The potential risks of releases of exotic natural enemies have only recently received attention outside the biological control world, and an increasing number of countries now apply risk-assessment procedures before a new natural enemy can be imported or released. Currently, about 25 countries are using some form of regulation concerning the import of exotic biocontrol organisms, and the implementation of regulation is being discussed by many other countries. Some procedures (e.g. those of Australia, New Zealand and Hawaii; see articles in Lockwood *et al.*, 2001) are so strict that the import and release of exotic natural enemies are extremely difficult. Other countries have no regulations at all, so any species can be imported and released. There is a general trend, however, towards more stringent regulatory requirements (e.g. Barratt *et al.*, 1999). Governmental plant-protection divisions usually have the responsibility for permitting the introduction of agents, but many do not have the knowledge to make informed decisions. Consequently, there is a tendency for permits to be denied or delayed or to over-regulate imports, so that research costs for studies before introductions form an insurmountable threshold for biological control companies (Ravensberg, 1994). Based on work done in the 1970s for parasitoids of whiteflies in the 1970s and in 1980s for parasitoids of leafminers (which in both cases were imported from North America), it is estimated that costs associated with assessment of risks will be in the order of US$200,000 per natural enemy (J.C. van Lenteren, unpublished data).

Most European countries do not demand registration of macroorganisms such as mites, insects and nematodes. In Switzerland, Austria, Sweden and Hungary, it is, however, necessary to register these kinds of natural enemies. Switzerland has no specific administrative procedure and registration is handled on a case-by-case basis. Austria and Sweden apply regulations and, in Hungary, official registration is required but not yet strictly enforced. In Sweden it is no longer economically feasible for natural enemies to be used for control of minor pests or on limited acreage because of the high costs of registration fees (W. Ravensberg, Berkel en Rodenrijs, 2002, personal communication).

Other European countries are discussing the need for registration of macroorganisms. In the EU macroorganisms are still exempt from evaluation under the new pesticide legislation (Directive 91/414/EEC). Use of non-indigenous microorganisms is covered by the EU registration procedure, where more questions are asked about likely environmental impacts than for indigenous organisms. For macroorganisms European countries have very different criteria to allow importation and releases (ranging from no criteria to rather strict criteria, including information on possible environmental impacts (Ravensberg, 1994)). In the UK, Germany and Denmark, existing legislation applies to the import of alien organisms. In the UK, the release of non-indigenous organisms is prohibited under the Wildlife and Countryside Act and further supported by the Plant Health Order for pest species. Non-indigenous natural enemies have recently been included in the ruling, and an import licence is needed for these organisms through the Department of the Environment; the procedure for granting licences is still under review. For Germany also, procedures are under review; officially, non-indigenous natural enemies cannot be introduced. Denmark enforced a new Act on the Protection of the Environment, and releases of alien organisms (including biological control agents) are no longer permitted. Some European countries (e.g. Belgium) do not have any regulation to prevent the import and release of exotic arthropods. Harmonization within the EU is under discussion, and there are efforts to include the Food and Agriculture Organization (FAO) Code of Conduct for the Importation and Release of Biological Control Agents into the EU (Klingauf, 1995), but in our opinion the FAO Code of Conduct and the guidelines developed by the European Plant Protection Organization (EPPO) based on the FAO Code of Conduct are insufficiently specific (FAO, 1996; EPPO, 1998, and in preparation).

In the USA, the Plant Protection and Quarantine division of the Animal and Plant Health Inspection Service of the US Department of Agriculture demands an environmental assessment before a permit of introduction is released. More than 30 of these environmental assessments have been designed in the past 10 years. Examples are those for *Encarsia* and *Eretmocerus* spp. (see Royer, 1995). Environmental assessments have been issued for five microorganisms and 30 macroorganisms to be used as biological control agents for the control of nine species of weeds and 24 species of insects. When the release of an exotic biological control agent is expected not to result in negative effects, a finding of no significant impact (FONSI) is issued, based on an extensive review of the literature of the pest, pest-control methods and the natural enemies and after consultation with experts. The FONSI is supported by: (i) findings about the limited host range of the organism to be introduced; (ii) information on no negative effect on other natural enemies; (iii) data on potential effects on endangered and non-target species (although often limited); and (iv) evidence of no significant negative environmental impact.

A different procedure is followed by the International Institute of Biological Control (previously IIBC, currently CABI Bioscience) in the UK and associated countries. For new imports and releases of exotic organisms, this institute voluntarily prepares a dossier according to the FAO code of conduct. A dossier contains information on the pest and natural enemy, and an assessment of potential risks: (i) to non-target organisms; (ii) to human and animal health; (iii) to those handling the natural enemies; and (iv) of contaminants, and procedures for eliminating contaminants (for an example of such a dossier, see Cross and Noyes, 1995).

For a more detailed overview of existing regulations, including references, we refer the reader to OECD (2003).

A General Framework for Regulation of the Import and Release of Biological Control Agents

Available codes of conduct and guidelines produced by various organizations and countries (e.g. FAO, EPPO, North American Plant Protection Organization, CAB International, Austria, Australia, the Czech Republic, Japan, Hungary, Norway, Sweden, Switzerland and New Zealand) were studied. However, most guidelines (with the exception of those from New Zealand (Barratt *et al.*, 1999) and Australia (Paton, 1992)) are not very specific concerning criteria and methodology, so participants in the EU-ERBIC project decided to develop more specific guidelines, including methodology and criteria (van Lenteren *et al.*, 2003). Regulation procedures for biological control agents will – like those for chemical pesticides – be characterized by questions concerning four issues:

1. Characterization and identification of biocontrol agent (classical methods or molecular techniques, voucher specimens to be deposited, DNA fingerprinting in the case of taxonomic problems).
2. Health risks (for arthropod natural enemies these will be much easier to determine than for chemical agents).

3. Environmental risks (this is the most difficult part for invertebrate biocontrol agents).
4. Efficacy (efficacy of a biocontrol agent is defined as the ability to cause a significant reduction in the number of pest organisms, direct and indirect crop damage or yield loss. The efficacy of biocontrol agents can be highly variable if proper mass-rearing (Chapter 12) and quality control methods (Chapters 1, 2 and 19) are not applied. Efficacy is treated differently from cases with chemical control; as biocontrol agents often form part of an integrated pest management (IPM) programme, it is often not necessary to reach 90–100% control by the biocontrol agent alone, as long as the total IPM programme results in sufficient reduction of the pest or disease).

The environmental-risk assessment is the most critical and difficult part of the risk assessment procedure in biological control. Environmental assessments related to the release of exotic natural enemies are expected to be based on two items (OECD, 2003): (i) identification of potential hazards posed to the environment, based on collation of information and data from experiments and observations; and (ii) a summary of the risks and benefits of the release of the exotic natural enemy compared with alternative control methods. Postrelease reporting of any adverse effects on non-targets will be used to adjust environmental-risk assessments and to decide about future releases in other areas. Most biocontrol projects include postrelease studies to verify and monitor the establishment of a natural enemy (e.g. Cullen, 1997), but usually only the impact on the target species is studied. Barratt *et al.* (1999) propose to include non-target species in such follow-up studies, because only then can the predictive value of prerelease risk evaluations be estimated and testing be enhanced. Based on postrelease studies, Barratt *et al.* (1999) were able to compare the laboratory and field host ranges of two related parasitoids with large differences in their host ranges, and could conclude that laboratory-measured host ranges (i.e. host-specificity testing, see below) were indeed indicative of field host ranges.

Below, we concentrate our discussion on the identification of potential hazards posed to the environment as a result of the establishment, dispersal, host range and direct and indirect effects of release of the exotic natural-enemy species. Information about these issues will form the main point of the environmental risk assessment. The issue of quality control is very relevant for host risk-assessment testing: all this work should be done with natural enemies that are of good quality – otherwise risk will be underestimated.

Evaluation of the Ecological Factors Determining the Environmental Impact of an Introduced Agent

Establishment

The potential of an exotic natural enemy to establish will determine the extent of other tests/information needed for the environmental risk assessment. Conclusions can be drawn about the potential for the establishment of the natural enemy, based on information from the literature on: (i) abiotic factors (does the climate between area of origin and area of release match?); (ii) biotic factors (availability of non-target species suitable for reproduction, temporal and/or spatial matching of non-target organisms and biocontrol agent, diapause capabilities, winter survival); and (iii) combined biotic and abiotic factors (are other resources for survival and reproduction available?). Literature may suffice, but it may also be necessary to carry out laboratory and semi-field tests to prove non-establishment in the target area. If information indicates a very low probability that an agent can establish, the environmental assessment can be less extensive than in the case of a high potential for establishment.

Dispersal

It is important to determine the potential for dispersal of the biological control agent in order to answer the question of what the probability is of temporal and spatial encounter between the biological control agent and non-target species. This would be based on the mechanism of dispersal, life-span of the organism and local climate and habitat conditions in the area of release. If the agent does not disperse actively or passively for more than 10 m per season, no further information or studies are needed. If the agent does not establish but does disperse, dispersal experiments can be done in the target area. An inventory of non-target species should be made over time, space and habitat. Transect studies on dispersal speed (distance over time) and numbers dispersing (numbers over time) under normal climate and habitat conditions should be made. An alternative approach could be to count the number of hosts attacked, instead of the number of natural enemies dispersed. The attack of non-target hosts in various habitats should be checked, but also target insect on target host plant should be offered in these habitats. In this way, the presence of the biological control agent can be observed, and only then can one conclude that there is a very low risk for non-targets. If the agent can establish, similar experiments should be done in the country of origin to estimate dispersal capabilities. Any information on the possibility for secondary dispersal, e.g. mechanical or with crop, should be provided.

Host range

Before a specific testing scheme for host selection is designed, the following points need to be thought over (in this chapter, host is considered synonymous with prey).

1. The rearing of the host plant, host and natural-enemy species previous to testing should be described in a detailed way, among other reasons to be able to trace the effects of conditioning and learning. Learning behaviour by a natural enemy and the presence of semiochemicals from the host or host plant or their interaction may influence host acceptance patterns (Chapters 3 and 4; Vet and Dicke, 1992).

2. Test conditions should be described in a detailed way, as they may strongly influence host acceptance – for example, as a result of conditioning and learning (Chapters 3 and 4). Further, the host plant and host used in testing should be specified.

3. During testing, the target and non-target hosts should be offered in a natural host distribution pattern, on the natural host plant or part of that or on an alternative host plant that is not repellent for the natural enemy (van Dijken *et al.*, 1986; Sands, 1988; Follett *et al.*, 2000).

The choice of non-target species is difficult but critical. A procedure similar to the phylogenetic centrifugal method used for the evaluation of weed biocontrol agents is proposed, starting with non-target host species from the same genus and then progressing to those from the same tribe, subfamily, etc. (Wapshere, 1974; Lonsdale *et al.*, 2001). If none of the non-target species from the same genus is attacked, one can stop testing non-targets that are related to the target. If several species within the same genus of the target are attacked, then it would be appropriate to test non-targets from the same tribe, and so on.

In addition to testing of non-targets related to the target, several categories of other non-target species should be tested, such as: (i) non-related non-targets that occur in the same habitat of the target and are prone to attack; (ii) non-related non-targets that occur in other habitats that are explored by the natural enemy; and (iii) threatened, economic, aesthetic (symbolic) species even when not closely related to the target host. Available knowledge about host spectrum and habitats that are explored by the natural enemy can help in narrowing down the non-targets to be tested.

In a number of cases, host-specificity data from mono- or slightly oligophagous species found in the literature were confirmed when exposed to new non-target host species (e.g. Cameron and Walker, 1997). In contrast, other natural-enemy species that were considered to be monophagous or that had a rather restricted host range were found to attack a number of other host species in the area of release (e.g. Brower, 1991; Barratt *et al.*, 1997). Even more surprising was the finding that a polyphagous natural enemy at one place appeared to perform as a monophagous natural enemy after introduction in another region (Salerno, 2002). Conclusions about host specificity can, therefore, seldom be made solely on data collected in the native area of the biological control agent.

The special position of polyphagous predators should be considered when designing tests. Extreme care is needed in this case, and a wider host range has to be tested than with parasitoids because more intraguild predation is expected, as well as effects on higher trophic levels. Very specific (micro)habitat demands of the predator may limit its degree of polyphagy and may make it less risky.

During the EU-ERBIC project, we developed a sequential test for host-range testing, which is described in Table 13.1 and summa-

Table 13.1. Sequential test to determine the host range of a natural enemy.

Step 1. Petri-dish non-choice black-box test
The aim of the test is to answer the question: does the biological control agent attack the non-target organism in the appropriate stage? The non-target species are selected according to their: (i) phylogenetic relationship with the target; (ii) occurrence in the same microhabitat and proneness to attack; and (iii) status as endangered species (Lonsdale *et al.*, 2001). If none of the non-targets is attacked and the pest (control) species is attacked, one can stop testing and no direct effects on non-target species in the field are expected. If non-target species were attacked, go to step 2

Step 2. Petri-dish non-choice behavioural test
The aim of the test is to answer the question: does the biological control agent attack the non-target organism consistently? Check encounter and attack rate over time for non-target species to determine possible increase in acceptance due to increasing oviposition/predation pressure. If the non-target is not attacked at all and the pest (control) species is attacked, one can stop testing for that species and no direct effects on that non-target species in the field is expected. If the non-target is only attacked at the end of the observation period, then the risk of direct effects on that species is relatively small. If the non-target host is attacked a constant fixed percentage of times, then the risk might be considerable. For non-target species that are attacked, go to step 3

Step 3. Petri-dish choice test
The aim of the test is to answer the question: does the biological control agent attack the non-target when the target species is present? The choice test is with the target and non-target host. Check encounter and attack rate over time for non-target and target to determine host preference, eventual shifts in preference and possible increasing attack pressure on hosts that are not usually attacked due to the preferred host no longer being available. If there is no or low attack of non-target hosts and no shift in host preference over time, there is a low risk for direct effects on the non-target. If the non-target is easily attacked either from the start onwards or later during the observation, go to step 4

Step 4. Large-cage choice test
The aim of the test is to answer the question: does the biological control agent attack the non-target when the target species is present in a semi-natural situation? Present multiple host plants with various non-target and target hosts to the biological control agent in a large cage. Offer target and non-target hosts in as natural a situation as possible and on their natural host plants. Determine encounter and attack rates over time. For interpretation of results, see step 3. Non-target species that are easily attacked on their host plants pose a very high risk for non-target effects

Step 5. Field test
The aim of the test is to answer the question: does the biological control agent attack the non-target when the target species is present in a natural situation? This test can only be done if the biological control agent cannot establish in the target area (e.g. agents from tropical areas to be used in greenhouses in temperate climates). Release the natural enemy in the non-target habitat and determine the attack of the non-target species. If the target species is easily attacked and no or low attack of non-target occurs, there is a low risk for direct effects on the non-target. Non-target species that are easily attacked on their host plants in their habitat pose a very high risk for non-target effects

rized in Fig. 13.1 (for a more detailed description, see van Lenteren *et al.*, 2003).

Direct effects of released organism on other organisms in ecosystem

Direct effects of releases may be extinction or reduction in numbers of native non-target species. Certain categories of insects, such as hymenopteran and dipteran parasitoids, are extremely specific in host use and attack only one or a few species within an insect genus. Biocontrol scientists have therefore concluded that the introduction of these specific agents would not endanger the native fauna, contrary to the introduction of generalist vertebrate predators (e.g. Harris, 1990). Polyphagous insect predators may, however, attack non-target organisms and be harmful. Howarth (1985, 1991) presents circumstantial evidence that natural-enemy introductions may have led to a reduction of non-target species, but others (e.g. Funasaki *et al.*, 1988) believe these charges to be unjustified.

It is not an easy task to show that, in an ecosystem with a rare plant-eating insect species that is attacked by many different natural-enemy species, the introduction of a new natural enemy further reduces the density of the herbivore or merely replaces some other mortality factor. The extinction of pest or non-target organisms as a result of biological control is, however, extremely unlikely. Pests have seldom, if ever, been exterminated in the more than 100 years of insect biological control. Rather, a low population level of both pest and natural enemy has developed, as in natural ecosystems. In nature, it is the rule, rather than the exception, to find extremely low densities of

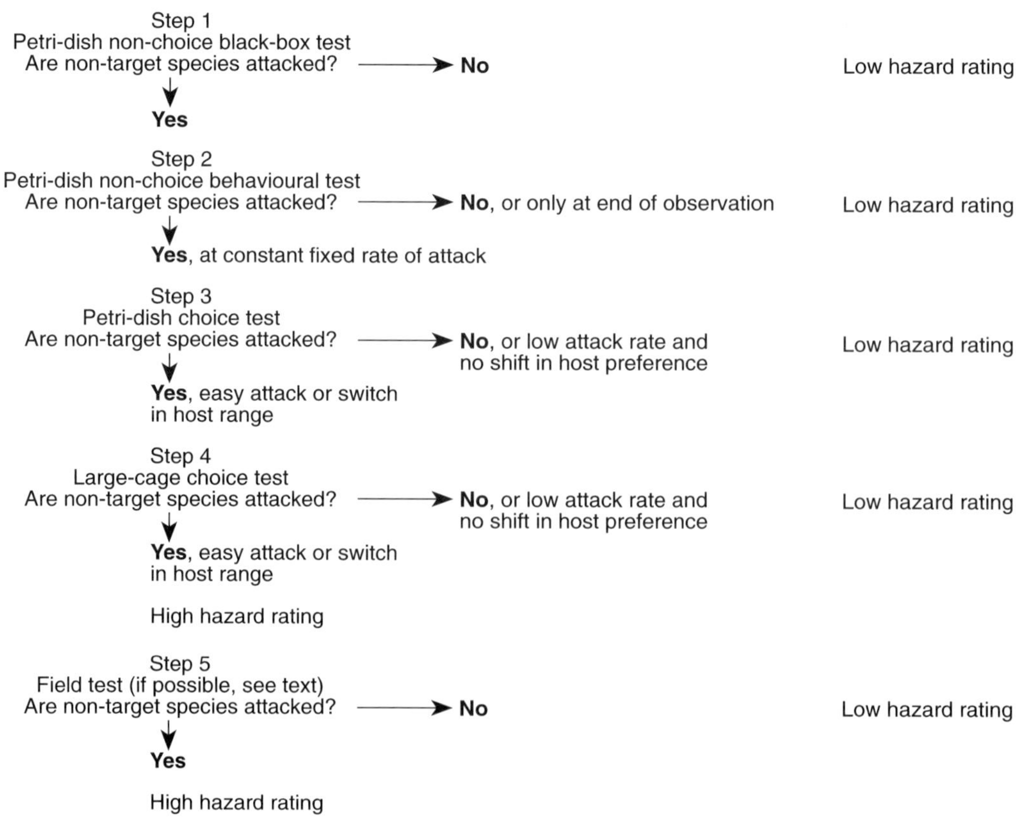

Fig. 13.1. Sequential host-specificity testing scheme.

both herbivores and their natural enemies, and these natural enemies are a substantial component of biodiversity. The search behaviour of natural enemies generally leads to the decision to leave host patches before parasitizing or eating all hosts or prey. Further, hosts have mechanisms of escaping their natural enemies in space and time, which reduce the chances of the host becoming extinct. Ecologists have long recognized the role that predators, parasitoids and pathogens play in reducing populations of plant-feeding organisms (herbivores – in agroecosystems, often pest insects), thereby 'keeping the world green' (e.g. Crawley, 1992).

In this section on direct effects, first testing for host specificity – the focal point of any environmental risk assessment – was described, followed by other direct effects, such as intraguild predation, competition and displacement, hybridization, effects on plants and risk of vectoring diseases below.

Intraguild predation

The occurrence of intraguild predation, which is the killing and eating of species that otherwise use similar resources (Rosenheim *et al.*, 1995; Brodeur and Rosenheim, 2001), can be checked in the literature. If intraguild predation effects are indicated for the specific or for related natural-enemy species or can be concluded from the biology of the natural enemy, then intraguild predation should be investigated case by case. Qualitative and quantitative effects should be listed and resulting effects estimated (positive, neutral or negative). Finally, a conclusion concerning risk should be drawn.

Competition and displacement

Check the literature if competition and displacement effects are indicated for the specific or related natural-enemy species or conclude from the biology of the natural enemy if effects are expected. Then investigate case by case. List qualitative and quantitative effects and estimate effects (positive, neutral or negative). Finally, draw a conclusion concerning risk.

Hybridization

Estimate or determine the likelihood of hybridization between the natural enemy and indigenous strains or biotypes of the same or very closely related natural-enemy species, discuss potential effects and draw a conclusion concerning risk.

Effects on plants

Effects of the natural enemy on plants should be provided if the agent is potentially a facultative herbivore. Check the literature if effects on target crop and non-target plants are indicated for the specific or related natural-enemy species or conclude from the biology of the natural enemy if effects are expected. Then investigate case by case. List qualitative and quantitative effects and draw a conclusion concerning risk.

Risk of vectoring diseases

Check the literature if vectoring of viruses or microorganisms that can negatively affect non-target organisms is indicated for the specific or related natural-enemy species (Chapter 10). Draw a conclusion concerning risk.

Indirect effects of released organism

An indirect effect of releases may be a reduction in the numbers of endemic natural enemies as a result of: (i) a strong reduction of their prey or host that is the target pest, which is attacked by the introduced natural enemy; and/or (ii) competition with endemic natural enemies for other hosts or prey of the introduced natural enemy. Further, the habitat may be modified as a result of indirect interactions. It has been suggested that, in some cases, releases of polyphagous predators has not only led to a decimation of pest caterpillars but also to a reduction of non-target caterpillars, resulting in a decline in native predacious wasps and native bird populations (Simberloff, 1992). Myriad indirect effects are possible. Species can interact through shared prey or hosts or shared predators, parasitoids or pathogens. Natural-history information on species involved in such indirect effects can be used (or collected) for prediction of such indirect effects. A case-study on host-range changes by Viggiani and Gerling (1994) showed that the acceptance of endemic hosts that are phylogenetically related to the original hosts of a newly introduced natural enemy or the acceptance of an accidentally imported pest by a native natural enemy of a related native pest species may occur.

For the environmental risk analysis, any known indirect effects or potential indirect effects on individual species and/or ecosystem should be reported. Indirect effects via target organisms (e.g. lower numbers of native natural enemies as a result of reduction of the target pest) are generally accepted and not considered negative, but indirect effects via non-target organisms on population and community level are usually considered negative. The problem is that each direct effect on a non-target is expected to result in a multitude of (small to large) indirect effects, and these can be positive, neutral or negative. Existing information on these effects is very limited, and estimating indirect effects is difficult. If the exotic biological control agent is expected to attack non-target species in high numbers, the direct and indirect effects will be considered too serious and establishment too risky.

Risk-assessment Methodology for Natural Enemies

The evaluation of risks related to releases of natural enemies demands the integration of many aspects of their biology, as well as information on ecological interactions. In the risk-assessment methodology proposed below, such an integration of information is presented. Usually, in a full risk-assessment process, three aspects are distinguished: (i) the risk-identification and evaluation procedure concerning the release of a natural enemy; (ii) a risk-management plan dealing with risk reduction and risk mitigation; and (iii) a risk/benefit analysis of the proposed release of a certain natural enemy, which should include a comparative performance of current and alternative pest-management methods, particularly based on environmental aspects.

Risk identification and evaluation

Normally, for a risk evaluation, one will identify the hazards and determine the probabilities that hazards will materialize. When more hazards are expected to occur, consider worst-case scenarios with accumulation of risks (e.g. attack of other natural enemies, attack of non-target and threatened species and ecosystem effects of the newly introduced natural enemy). Here the system proposed by Hickson *et al.* (2000) for environmental risk management in New Zealand is used as a starting-point for the development of a risk evaluation for biological control agents. In this system, five groups of risks are considered related to the release of exotic biological control agents: establishment, dispersal, host specificity, other direct effects and indirect non-target effects.

In order to assess risks, first the likelihood and the magnitude of adverse effects are estimated according to the matrix of magnitude × likelihood (Table 4 in Hickson *et al.*, 2000). As many of the descriptions in Hickson's table could not be used to estimate effect for the five groups of risks given above for biological control agents, a new list of descriptions for likelihood and magnitude was made; based on these descriptions, the EU-ERBIC project has developed a system where the calculation of risk is done in a numerical way and has presented case-studies where this system is illustrated (van Lenteren *et al.*, 2003). This system results in risk indices for the species under evaluation.

Interpretation of risk indices should be done with great care by biocontrol experts having experience with the natural-enemy species under evaluation. Host specificity, as earlier stated, is the crucial element in the whole evaluation process. Risk indices should not be seen as absolute values, from which a definitive conclusion can be drawn concerning import and release. We propose to use them within certain risk categories (low, intermediate, high risk) that will result in a proposal to release the agent (low-risk index), not to release the agent (high-risk index) or to come up with additional information (intermediate-risk index). Further, it should be clearly stated for which region a particular risk assessment was made (continent, part of a continent, ecoarea, country, part of a country, etc.), because risk indices will vary according to the region for which they were made. It would be best to determine indices for ecoareas, because these are rather well defined and meaningful biological units (ecoarea: an area with similar fauna, flora and climate and hence similar concerns about the introduction of biological control agents (FAO, 1999)).

Risk management

The next step of the risk-assessment process is to discuss risk management, including risk mitigation and risk reduction. For an example of information needed for risk management, we refer the reader to Cross and Noyes (1995).

Risk/benefit analysis

The final step in making a justified environmental risk analysis for a new biological control agent is to conduct a risk/benefit analysis, which should include a comparative performance of pest-management methods, particularly based on environmental aspects. The environmental benefits of the use of the proposed biocontrol agent should be compared with the environmental effects of currently used and other alternative control methods.

Discussion

Biological control has been practised for more than 25 centuries, and 'modern' forms of biological control, where exotic natural enemies are imported and released, for more than a century. This activity has resulted in long-term economic and environmentally benign solutions to severe pest, disease and weed problems. In contrast to chemical control, the biological control of insects and mites has not resulted in negative environmental or health effects.

The recent literature on introductions of natural enemies for insect control has not featured the role of biological control in extinctions. This is an important conclusion, as thousands of intended introductions have been made worldwide, and apparently the biological control scientists have correctly identified which natural enemies can be safely exported. In present-day biological control, governments that rigidly regulate the introduction of biological control agents, such as those of Australia and New Zealand, usually require that candidate agents undergo host-range testing to ensure that they will not become pests or threaten desirable species. This results in a general preference for highly specific natural enemies, as was already regarded as common sense among most biocontrol workers. Another effect of regulation is that first native natural enemies are evaluated as potential biocontrol agents, a development that we strongly support.

The topic of the implementation of a registration procedure for natural enemies is currently hotly debated by the biocontrol industry and regulators. The biocontrol industry foresees lengthy, cumbersome procedures leading to high costs and thus, in some cases, the impossibility of marketing an interesting natural enemy because of excessive costs. Regulators within ministries of environment and agriculture want to prevent unnecessary and risky releases of exotic organisms. The history of arthropod biocontrol shows that very few mistakes have been made up to now. This is a point in favour of the biocontrol industry and is in strong contrast to the problems that have been created by the accidental importation of pests and diseases on infested plant material by others. The current work by the EU-ERBIC project, in collaboration with an OECD working group (OECD, 2003), will hopefully result in a light and harmonized registration procedure that is not prohibitive for the biocontrol industry and will result in the preselection of safe natural enemies. The registration procedure also includes the specification of efficacy of the natural enemy. This efficacy will largely depend on good mass-production and quality control methods. The purpose of this registration procedure would be to keep biocontrol a respected, reliable and sustainable control method and to prevent the import and release of really unsafe natural enemies.

Acknowledgements

This study has been carried out with financial support from the Commission of the European Communities, Agriculture and Fisheries (FAIR) specific RTD programme CT97-3489, 'Evaluating Environmental Risks of Biological Control Introductions into Europe'. It does not reflect its views and in no way anticipates the Commission's future policy in this area. F. Bin, E. Conti, R. Romani and J. Salerno (University of Perugia, Italy) are thanked for contributing to the formulation of the ideas on risk assessment.

References

Aeschlimann, J.P. (1995) Lessons from post-release investigations in classical biological control: the case of *Microctonus aethiopoides* Loan (Hym., Braconidae) introduced into Australia and New Zealand for the biological control of *Sitona discoideus* Gyllenhal (Col., Curculionidae). In: Hokkanen, H.M.T. and Lynch, J.M. (eds) *Biological Control: Benefits and Risks*. Cambridge University Press, Cambridge, pp. 75–83.

Barratt, B.I.P., Evans, A.A., Ferguson C.M., Barker, G.M., McNeill, M.R. and Phillips, C.B. (1997) Laboratory nontarget host range of the introduced parasitoids *Microctonus aethiopoides* and *Microctonus hyperodae* (Hymenoptera: Braconidae) compared with field parasitism in New Zealand. *Environmental Entomology* 26, 694–702.

Barratt, B.I.P., Ferguson, C.M., McNeil, M.R. and Goldson, S.L. (1999) Parasitoid host specificity testing to predict field host range. In: Whithers, T.M., Barton Browne, L. and Stanley, J.N. (eds) *Host Specificity Testing in Australia: Towards Improved Assays for Biological Control.* CRC for Tropical Pest Management, Brisbane, Australia, pp. 70–83.

Blossey, B. (1995) Host specificity screening of insect biological control agents as part of an environmental risk assessment. In: Hokkanen, H.M.T. and Lynch, J.M. (eds) *Biological Control: Benefits and Risks*. Cambridge University Press, Cambridge, pp. 84–89.

Brodeur, J. and Rosenheim, J.A. (2001) Intraguild interactions in aphid parasitoids. *Entomologia Experimentalis et Applicata* 97, 93–108.

Brower, J.H. (1991) Potential host range and performance of a reportedly monophagous parasitoid, *Pteromalus cerealellae* (Hymenoptera: Pteromalidae). *Entomological News* 102, 231–235.

Cameron, P.J. and Walker, G.P. (1997) Host specificity of *Cotesia rubecula* and *Cotesia plutellae*, parasitoids of white butterfly and diamondback moth. In: *Proceedings of the 50th New Zealand Plant Protection Conference, Lincoln University, Canterbury*, pp. 236–241.

Crawley, M.J. (ed.) (1992) *Natural Enemies: the Population Biology of Predators, Parasites and Diseases.* Blackwell, Oxford.

Cross, A.E. and Noyes, J.S. (1995) Dossier on *Anagyrus kamali* Moursi, biological control agent for the pink mealybug, *Maconellicoccus hirsutus*, in Trinidad and Tobago. International Institute of Biological Control, Silwood, Ascot, 16 pp. + 1 Annex.

Cullen, J.M. (1997) Biological control and impacts on non-target species. In: *Proceedings of the 50th New Zealand Plant Protection Conference, Lincoln University, Canterbury*, pp. 195–201.

EPPO (1998) *Guideline for the Safety and Efficacy of Biological Control: First Introduction of Exotic Biological Control Agents for Research Under Contained Conditions.* EPPO, Rome, 4 pp.

FAO (1996) *Code of Conduct on the Import and Release of Biological Control Agents.* FAO, Rome, 21 pp.

FAO (1999) *Glossary of Phytosanitary Terms.* International Standards for Phytosanitary Measures Publication No. 5, Secretariat of the International Plant Protection Convention, Food and Agriculture Organization of the United Nations, Rome, 62 pp.

Follett, P.A., Duan, J., Messing, R.H. and Jones, V.P. (2000) Parasitoid drift after biological control introductions: re-examining Pandora's box. *American Entomologist* 46, 82–94.

Funasaki, G.Y., Li, P.Y., Nakahara, L.M., Beardsley, J.W. and Ota, A.K. (1988) A review of biological control introductions in Hawaii: 1890 to 1985. *Proceedings Hawaiian Entomological Society* 28, 293–305.

Greathead, D.J. (1995) Benefits and risks of classical biological control. In: Hokkanen, H.M.T. and Lynch, J.M. (eds) *Biological Control: Benefits and Risks*. Cambridge University Press, Cambridge, pp. 53–63.

Gurr, G. and Wratten, S. (eds) (2000) *Measures of Success in Biological Control*. Kluwer Academic Publishers, Dordrecht, 448 pp.

Harris, P. (1990) Environmental impact of introduced biological control agents. In: Mackauer, M., Ehler, L.E. and Roland, J. (eds) *Critical Issues in Biological Control*. Intercept, Andover, pp. 289–300.

Hickson, R., Moeed, A. and Hannah, D. (2000) HSNO, ERMA and risk management. *New Zealand Science Review* 57, 72–77.

Hokkanen, H.M.T. and Lynch, J.M. (eds) (1995) *Biological Control: Benefits and Risks*. Cambridge University Press, Cambridge.

Howarth, F.G. (1985) Impacts of alien land arthropods and molluscs on native plants and animals in Hawaii. In: Stone, C.P. and Scott, J.M. (eds) *Hawaii's Terrestrial Ecosystems: Preservation and Management*. University of Hawaii, Honolulu, pp. 149–178.

Howarth, F.G. (1991) Environmental impacts of classical biological control. *Annual Review of Entomology* 36, 485–509.

Klingauf, F.A.J. (1995) Registration requirements of biological control agents in Germany and in the European Union. In: Hokkanen, H.M.T. and Lynch, J.M. (eds) *Biological Control: Benefits and Risks*. Cambridge University Press, Cambridge, pp. 283–290.

Lockwood, J.A., Howarth, F.G. and Purcell, M.F. (eds) (2001) *Balancing Nature: Assessing the Impact of Importing Non-native Biological Control Agents*. Thomas Say Publications in Entomology: Proceedings, Entomological Society of America, Lanham, Maryland, 130 pp.

Lonsdale, W.M., Briese, D.T. and Cullen, J.M. (2001) Risk analysis and weed biological control. In: Wajnberg, E., Scott, J.C. and Quimby, P.C. (eds) *Evaluating Indirect Ecological Effects of Biological Control*. CAB International, Wallingford, pp. 185–210.

Lynch, L.D. and Thomas, M.B. (2000) Nontarget effects in the biocontrol of insects with insects, nematodes and microbial agents: the evidence. *Biocontrol News and Information* 21, 117–130.

Lynch, L.D., Hokkanen, H.M.T., Babendreier, D., Bigler, F., Burgio, G., Gao, Z.-H., Kuske, S., Loomans, A.J.M., Menzler-Hokkanen, I., Thomas, M.B., Tommasini, M.G., Waage, J., van Lenteren, J.C. and Zeng, Q.-Q. (2001) Indirect effects in the biological control of arthropods with arthropods. In: Wajnberg, E., Scott, J.C. and Quimby, P.C. (eds) *Evaluating Indirect Ecological Effects of Biological Control*. CAB International, Wallingford, pp. 99–125.

Neuenschwander, P. and Markham, R. (2001) Biological control in Africa and its possible effects on biodiversity. In: Wajnberg, E., Scott, J.C. and Quimby, P.C. (eds) *Evaluating Indirect Ecological Effects of Biological Control*. CAB International, Wallingford, pp. 127–146.

OECD (2003) *Guidance for Registration Requirements for Invertebrates as Biological Control Agents (IBCAs)*. OECD, Paris (in preparation).

Paton, R. (1992) Legislation and its administration in the approval of agents for biological control in Australia, In: Delfosse, E.S. and Scott, R.R. (eds) *Proceedings of the Eighth International Symposium on Biological Control of Weeds*. DSIR/CSIRO, Lincoln, Canterbury, pp. 653–658.

Ravensberg, W. (1994) Biological control of pests: current trends and future prospects. In: *Proceedings Brighton Crop Protection Conference, Pests and Diseases*. British Crop Protection Council, Farnham, pp. 591–600.

Rosenheim, J.A., Kaya, H.K., Ehler, L.E., Marois, J.J. and Jafee, B.A. (1995) Intraguild predation among biological control agents: theory and evidence. *Biological Control* 5, 303–335.

Royer, M. (1995) *Environmental Assessment for Field Releases of Certain* Encarsia *Species (Hymenoptera: Aphelinidae) for Biological Control of Armored Scale Insects and Whitefly*. APHIS, USDA, 15 pp. + 9 Appendices.

Salerno, G., Colazza, S. and Bin, F. (2002) *Nezara viridula* (Heteroptera: Pentatomidae) parasitism by the tachinid fly *Trichopoda pennipes* ten years after its accidental introduction into Italy from the New World. *BioControl* 47, 617–624.

Sands, D.P.A. (1988) Guidelines for testing host specificity of agents for biological control of arthropod pests. In: Zalucki, M.P., Drew, R.A.I. and White, G.G. (eds) *Proceedings 6th Australian Applied Entomological Conference, Brisbane*, pp. 556–560.

Simberloff, D. (1992) Conservation of pristine habitats and unintended effects of biological control. In: Kaufmann, W.C. and Nichols, R.J. (eds) *Selection Criteria and Ecological Consequences of Importing Natural Enemies.* Thomas Say Publications in Entomology: Proceedings, Entomological Society of America, Lanham, Maryland, pp. 103–117.

van Dijken, M.J., Kole, M., van Lenteren, J.C. and Brand, A.M. (1986) Host-preference studies with *Trichogramma evanescens* Westwood (Hym., Trichogrammatidae) for *Mamestra brassicae, Pieris brassicae* and *Pieris rapae. Journal of Applied Entomology* 101, 64–85.

van Lenteren, J.C. (1986) Evaluation, mass production, quality control and release of entomophagous insects. In: Franz, J.M. (ed.) *Biological Plant and Health Protection.* Progress in Zoology Series 32, Fischer, Stuttgart, pp. 31–56.

van Lenteren, J.C. (1993) Parasites and predators play a paramount role in pest management. In: Lumsden, R.D. and Vaughn, J.L. (eds) *Pest Management: Biologically Based Technologies.* American Chemical Society, Washington, DC, pp. 68–81.

van Lenteren, J.C. (1995) Frequency and consequences of insect invasions. In: Hokkanen, H.M.T. and Lynch, J.M. (eds) *Biological Control: Benefits and Risks.* Cambridge University Press, Cambridge, pp. 30–43.

van Lenteren, J.C. (2000) Measures of success in biological control of arthropods by augmentation of natural enemies. In: Gurr, G. and Wratten, S. (eds) *Measures of Success in Biological Control.* Kluwer Academic Publishers, Dordrecht, pp. 77–103.

van Lenteren, J.C. (2001) Harvesting safely from biodiversity: natural enemies as sustainable and environmentally friendly solutions for pest control. In: Lockwood, J.A., Howarth, F.G. and Purcell, M.F. (eds) *Balancing Nature: Assessing the Impact of Importing Non-native Biological Control Agents.* Thomas Say Publications in Entomology: Proceedings, Entomological Society of America, Lanham, Maryland, pp. 15–30.

van Lenteren, J.C. and Woets, J. (1988) Biological and integrated pest control in greenhouses. *Annual Review of Entomology* 33, 239–269.

van Lenteren, J.C., Babendreier, D., Bigler, F., Burgio, G., Hokkanen, H.M.T., Kuske, S., Loomans, A.J.M., Menzler-Hokkanen,I., van Rijn, P.C.J., Thomas, M.B. and Tommasini, M.G. (2003) Environmental risk assessment of exotic natural enemies used in inundative biological control. *BioControl* 48 1–36.

Vet, L.E.M. and Dicke, M. (1992) Ecology of infochemical use by natural enemies in a tritrophic context. *Annual Review of Entomology* 37, 141–172.

Viggiani, G. and Gerling, D. (1994) Host-range increase of indigenous and introduced parasitoids. In: Narang, K., Bartlett, A.C. and Faust, R.M. (eds) *Applications of Genetics to Arthropods.* CRC Press, Fort Lauderdale, pp. 147–157.

Waage, J.K. (1997) Global developments in biological control and the implications for Europe. In: Smith, I.M. (ed.) *EPPO/CABI Workshop on Safety and Efficacy of Biological Control in Europe.* Blackwell Science, Oxford, pp. 5–13.

Wapshere, A.J. (1974) A strategy for evaluating the safety of organisms for biological weed control. *Annals of Applied Biology* 77, 201–211.

14 Quality Assurance in North America: Merging Customer and Producer Needs

C.S. Glenister,[1] A. Hale[2] and A. Luczynski[3]

[1]*IPM Laboratories, Inc., 980 Main Street, Locke, NY 13092-0300, USA;* [2]*The Bug Factory, 1636 East Island Highway, Nanoose Bay, British Columbia, Canada V9P 9A5;* [3]*Biobugs Consulting Ltd, 16279 30B Ave., Surrey, British Columbia, Canada V4P 2X7*

Abstract

The history of a unified North American industry effort towards quality assurance began with the formation of the Association of Natural Bio-control Producers in 1990. Plans were first all-encompassing: to assess all steps in the product process from the quality of inputs into production through to the quality of the product on customer receipt, including the information supplied. Efforts eventually focused on the product received by the customer. North American producers joined with customers, researchers and government representatives to create a subcommittee of the American Society for Testing and Materials (ASTM) for development of quality assurance of beneficial organisms. By the year 2001, the ASTM subcommittee had created drafts for 16 species of beneficial organisms and ratified three standards.

The history of North American quality-assurance work is summarized and a case-study is presented on the quality of the biological control products that the growers received, focusing on *Encarsia formosa* and *Phytoseiulus persimilis.*

Introduction

Individual insectaries, distributors, customers and public employees have long had their own methods for evaluating quality of beneficial organisms. Many times, the methods have been unique to each entity. Two organizations in North America have influenced the direction of the standardization process: the Association of Natural Bio-control Producers (ANBP), representing producers, and the British Columbia (BC) Hot House Grower's Association (BCHHGA), representing users. Developments in these two sectors will be discussed below. The first unified effort concerning quality control by the North American insectaries began with the formation of the ANBP in 1990. The mission statement of the organization now stands as: ANBP's mission is to address key issues of the biological control industry through advocacy, education and quality assurance. Its four main programme areas are quality standards, education, research and regulation. The Association had more than 120 members in the year 2000, including 46 producers or suppliers.

© CAB International 2003. *Quality Control and Production of Biological Control Agents: Theory and Testing Procedures* (ed. J.C. van Lenteren)

History of the Development of ANBP Quality Assurance: the Producer Approach

The original ANBP idea was that quality assessments needed to start with assessing inputs into production and end with the customer. Eventually it was decided that the market-place was the most important and effective 'police force' for addressing quality issues. Thus, the producers' job was to create the tools whereby the market could assess quality.

Certification discussions, 1990–1992

At first, the Quality Control Committee (QCC) investigated the possibility of creating a certification process. A twofold process of certifying insectaries for acceptable in-house quality control and of quality evaluations by independent laboratories was envisioned. A survey was sent to producer members, which asked detailed questions on rearing protocols. Some insectaries shared this information quickly, while others held this information as proprietary. Eventually, members decided that in-house rearing should remain proprietary and that product should be evaluated at the end of production, not during production (L.A. Merrill, 1995, unpublished). In 1992, the QCC had received guideline submissions for 'live product arrival to distributor or supplier' from two producers. Their information included: product examination on arrival, identification, short-term storage, long-term storage where applicable, shipping and random-sampling information.

Certification was one thing, but the procedure for determining that a product meets or fails certification standards created a quagmire of issues that caused these efforts rapidly to come to a halt. The QCC avoided further controversy by turning to proactive steps to build ethics into the framework of how products were represented for sale. They proceeded to avoid action in regard to procedures that by their very nature involved punitive measures, leaving that to the discretion of the market-place.

Product profiles, 1991

The next step of the QCC was to address a malpractice issue: some beneficial insects were being sold for purposes that were unsupportable. For example, *Trichogramma* species were being sold for controlling gypsy moths. The QCC began work to ensure that factual information was available to the consumer to support the appropriate use of biological controls. Producers were requested to list necessary factual information on a product profile to help the market use the products properly. The intent was that the profiles be concise, easy-to-read documents geared to the distributor and end-user.

The word 'label' was avoided in order to avoid confusion with the regulatory requirements of the Environmental Protection Agency (EPA) for pesticide labels. EPA pesticide-label requirements would represent an onerous burden on small businesses. Required safety data cost millions of dollars and there are annual federal and state fees required for registration of these labels. The information guidelines were called 'product profiles'. During 1992, the committee received 'product profile' submissions from five producers, which included: common name, origin, scientific name, environmental needs, biology, hosts, quantity, release instructions, producer name, compatibility with pesticides and warranty and disclaimer (optional). Eventually, 22 product profiles were submitted by at least 12 producers (L.A. Merrill and C.S. Glenister, 20 July 1995, unpublished).

Quality assurance paper, 1993–1998

An intense effort by biological control producers, practitioners and researchers to record the industry status of development and needs took place during a conference in 1993. The presentations and the book resulting from that meeting (Ridgway *et al.*, 1998) covered concepts, practices and needs in quality assurance and regulation, as well as biological control in greenhouses, field crops, ornamentals, vegetables and dairies. The chapter on quality assurance (Penn *et al.*,

1998), entitled 'Quality assurance by the commercial producer of arthropod natural enemies', was an intense effort led by Penn and Scriven (two commercial producers) and assisted by Ridgway and Inscoe (two employees of the US Department of Agriculture Agricultural Research Service (USDA ARS) to incorporate the concepts of Deming's total quality management into documentation on what quality assurance is. Two major advances for the North American biological control industry resulted:

1. Following the Deming method of total quality management, the customer is an integral part of quality assurance. This is well summarized in the abstract from 'Quality assurance by the commercial producer of arthropod natural enemies' (Penn *et al.*, 1998):

> The primary goal of the commercial insectary is to satisfy the customer. However, providing an abundant and consistent supply of high-quality natural enemies that can be effectively used in integrated pest management programs is not always sufficient to accomplish that goal. Such an environmentally sensitive product, composed primarily of living organisms, must be produced, delivered, and applied with special care to achieve the desired outcome. Of utmost importance is the customers' ability to recognise the results arising from the application of these natural enemies and to compare competitive products. Therefore, a reliable quality assurance program, consisting of comprehensive quality control and extensive customer involvement, is highly desirable. Specific examples to illustrate various aspects of a quality assurance program are taken from experiences with some representative arthropods: predacious insects, *Chrysoperla* spp.; a predacious mite, *Phytoseiulus persimilis* Athias-Henriot; and egg parasitoids, *Trichogramma* spp. In the insectary, a quality control program is normally made up of 3 major components: production control, process control, and product control. Standards (i.e. predetermined specifications) are established to measure variables throughout the system, such as temperature, humidity, content or quality of diet, fecundity, rate of growth, survival, vigour, sex ratio, and numbers. Methods of storing, packaging, and transport are included as a part of the transition between quality control and customer involvement. Customer involvement is an integral part of a quality assurance program. An effective educational program is necessary to ensure that the customer's expectations of quality are compatible with reality and that they know how to evaluate the product on arrival and how to use it most effectively. Customer feedback is important to refine information flow and guide product improvement, thereby leading to product optimization and repeat sales.

2. Accurate, concise definitions for quality-assurance terms were published by Penn *et al.* (1998):

Quality: The suitability of a product for its intended use and its degree of excellence in comparison with standards.

Standard: A specification for use as a basis for comparison in evaluating quantity, value or quality.

Quality management: A system for monitoring and directing desired outcome.

Total quality management: A programme or philosophy that includes the customer as a major component of a dynamic system in which processes and products are continuously examined for ways of improvement. A fundamental aspect of total quality management is identification of the desired characteristics of a product, followed by achievement of these characteristics through minimizing variance at defined critical control points.

Quality assurance: The process by which confidence in the quality of a product is developed. Quality assurance goes beyond quality control in that it involves both the procedures whereby the producer makes certain that the quality of the product is maintained and the processes through which the customer, the ultimate user, gains confidence in that product. For this reason, customer involvement and customer education are critical components of quality assurance.

Quality control: A system for verifying and maintaining a desired level of quality in a product or process by careful planning, continued inspection and corrective action where required. Three major interrelated elements of quality control are production control, process control and product control.

Production control: Management of the consistency, reliability and timeliness of production output through monitoring items such as materials, equipment, environments, schedules and personnel.

Process control: The means of assuring the performance of production processes through sampling of all life stages of the arthropods in production, using indicators to predict quality and minimize variance from predetermined standards.

Product control: Assurance of conformity of the product to acceptable standards of quality through monitoring procedures applied at the end of production. Such procedures substantially increase the probability that the product will be effective in performing its intended function but cannot predict performance because of the highly variable conditions to which the product may be exposed after the product is shipped.

Product-control guidelines: Suggested standards and processes for product control procedures that might be adopted by commercial insectaries, trade associations and other groups to provide consistency in product control.

Customer involvement: Selection of product, application of the product, evaluation of the results and feedback, with encouragement of continual informational flow.

Product profile: A document supplied by a producer, often with a product shipment, to provide the customer with information on the nature of the product being shipped and methods of storing and using the product properly. Details such as identity, quantity, origin, life stages shipped, sex ratio and expiration date are usually included. Alternatively, similar information of a more generic nature may be developed by an organization of producers.

Product handling and evaluation: Environmental conditions and actions designated by the customer, including assessments of quantity and quality of the product prior to application.

Application: The release of the natural enemies or otherwise putting them to use.

Performance evaluation: Assessments of the effects of natural-enemy application upon the target pest population.

Quality testing performed at California Polytechnic Institute, 1994–1995

With the objective of creating a laboratory that could independently test beneficial organisms, a pilot project was established to create a working link between the biological control producers and the university, funded by the National Biological Control Institute (NBCI) and ANBP (G. Scriven, unpublished, 1994, 'Quality assurance, production, delivery, and release of *Phytoseiulus persimilis*'). Under the guidance of Dr J. Wheatley, the following natural enemies were tested: *Trichogramma brassicae, Cryptolaemus montrouzieri, Phytoseiulus persimilis, Aphytis melinus* and *Chrysoperla rufilabris*. Existing International Organization for Biological Control (IOBC) Guidelines (Chapter 19) were used for *Trichogramma* and *P. persimilis* and the IOBC *Chrysoperla carnea* guideline was used for *C. rufilabris*. There were no IOBC guidelines for *Cryptolaemus* and *Aphytis*, so here only numbers, shipping mortality and sex ratio were tested. In addition, random sampling and statistics were studied for *Trichogramma* and *P. persimilis*. Most of the samples met the guidelines. However, the idea of establishing a certification laboratory with student labour was quietly abandoned, because it was concluded that certification testing needed to be done by experienced and skilled laboratory personnel.

The project resulted in substantive questions and proposals:

1. It needs to be clarified what we are trying to measure. The IOBC protocols are an internal activity for the producer to use. Product assessment and any certification procedure would be activities done by agencies that were independent of the producer and, as such, should be receiving product as if they were an end-user. Protocols need to be established for product at the end-user.

2. A tiered system of quality analysis was proposed:

Tier I: Test if numbers, mortality, species conforms to package information.
Tier II: Test the sex ratio and minimum specified longevity.
Tier III: Test fecundity and flight.

3. It was proposed that fecundity should be tested on groups of 50 or more unsexed individuals housed individually with abundant prey/hosts and a population reproductive index (PRI) generated that could replace the time-consuming sexing, longevity and fecundity testing. PRI would be computed as the total number of eggs laid, divided by the numbers of housed individuals. PRI represents the relative ability of the population to reproduce itself.
4. Inexpensive chemical techniques are needed for the identification of *Trichogramma* species.
5. Methods of selecting a random sample to go with each protocol are needed.
6. In any case where adults were used for reproduction prior to being sold, that should be stated on the package.

Consideration of International Organization for Standardization (ISO) certification, 1995

During 1995, some ANBP members considered working within the ISO framework for certification. Two points quickly became clear: (i) the process was very expensive; and (ii) good product standards would be necessary for this certification to have any relationship to product quality. Further, the ISO system appeared to certify process, not product quality.

Quality assurance definitions adopted, 1995

The ANBP Board voted unanimously in 1995 to adopt the quality assurance definitions in the draft of what eventually became Penn *et al.* (1998) (see above).

Standard package information adopted, 1996

Since product profiles were too cumbersome to attach to every package, the ANBP discussed and unanimously adopted (ANBP annual meeting in Sacramento, California, 28 January 1996) the following standard for minimum package information to appear on the package:

Package labels show at the minimum:

- the name(s) of the beneficial species in the package;
- the number of beneficial individuals in the package and the packing date;
- the level of purity of the predators and parasitoids in the package. This statement of purity does not refer to the prey or carrier in the package.

The package contents are as claimed on the label:

- the species in the package are as claimed on the label;
- the number of live individuals of the stated species is at least 100% of the quantity stated on the package;
- the purity of the beneficial species at the time of packing is as stated on the label.

The consensus of the producers was that:

- an individual company's guarantee covers the condition of beneficials on arrival at the initial destination;
- distributors need standards of their own.

American Society for Testing and Materials (ASTM) subcommittee formed, 1998

After discussions at the ANBP meetings in 1997 and 1998, the group unanimously agreed to formalize an ASTM standards development activity as Subcommittee E35.30 within ASTM Committee E35 on Pesticides. The group suggested that the main committee broaden its name to include biological pest-control agents. They approved the subcommittee title 'Subcommittee E35.30 on Natural Multi-cellular (Metazoan) Biological Control Organisms' and its scope:

> The development of standard definitions, classifications, appropriate test methods, and recommended practices for quality, handling, distribution and use of natural multi-cellular biological control organisms. The activities will

be coordinated with related committees in ASTM and with other professional and government organizations.

ASTM can be characterized shortly as follows:

- Organized in 1898.
- Provides a management system and the administrative framework for the development of voluntary, consensus standards and promotion of related knowledge.
- Open membership, one vote per 'voting interest'.
- Producers, users, academia and government represented.
- 34,000 members, 100 countries represented, 131 technical committees, 2200 subcommittees, thousands of task groups.
- Cost to participate in ASTM: $75 per year administrative fee, travel to meetings, time.

ASTM process for development of quality-assurance standards

The ASTM process for the development of quality-assurance standards is as summarized here. Task groups are formed for each biological control organism or product. Each task group has a leader, who coordinates standard development, writes drafts, distributes drafts to collaborators (who review drafts and verify test methods) and submits standards for ballot. The following steps are usually followed.

Appointment of task-group leader (writer)

The task-group leader volunteers to write a standard on a biological control organism. Generally the organism is selected for one of two reasons: (i) there is a pressing need for standardization of that particular organism for communication between producer and distributor or end-user; or (ii) the organism has not yet been described by IOBC guidelines, creating a void in the industry in terms of product specifications. The task-group leader is an individual with extensive experience with the biological control organism being tested and may be a producer, supplier, end-user, scientist, regulator or consultant.

Appointment of task-group members (collaborators)

Additional task-group members consist of volunteers, who review draft standards, test described methods and give constructive feedback to the task-group leader to incorporate into the next draft standard. These collaborators must also have extensive knowledge and familiarity with the biological control organism being tested.

Preparation of draft standard

The first step in preparing a draft standard is selecting the appropriate format or type of standard. The ASTM document can be prepared as a classification, guide, practice, specification, terminology or test method. The majority of standards in preparation for biological control organisms are written as standard specifications with embedded test methods. Although ASTM is rigid in its requirements for style and form, the organization allows flexibility in content and standards are developed on a voluntary basis within working groups.

Circulation of draft standard for peer review and revision

The task-group leader completes draft standards, which are circulated to task-group members and any other interested stakeholders for comment. The task-group leader incorporates suggested revisions into a new draft standard, which is recirculated. This process of revision and circulation for review can be repeated for many cycles.

Acceptance of revised standard by task-group members and submission for ballot

Task-group members accept revised standard and recommend submission to balloting process. The completed draft standard is submitted to ASTM headquarters, where it is prepared for ballot.

Voting by E35.30 subcommittee members

When a draft standard is submitted to ASTM, a formal ballot is prepared and circulated to all voting members of subcommittee E35.30 by fax or electronically on ASTM's website at www.astm.org. Votes are recorded as affirmative, negative or abstaining. Generally, abstaining votes are from subcommittee members unfamiliar with the product being tested. A single negative vote will stop the process, although negative votes must be accompanied by a written explanation or they will be discounted. When a standard has had a negative vote, there are several options. The standard may be withdrawn, the standard may be revised and resubmitted or the negative vote can be addressed and the standard resubmitted as is.

Membership accepts or rejects standard

If a standard is accepted, it is published by ASTM. If the standard is rejected, it is either withdrawn or revised and resubmitted for ballot.

Publication

Once accepted, the new standard is published in the *Annual Book of ASTM Standards*. An ASTM standard is subject to revision at any time by the responsible technical committee and must be reviewed every 5 years and, if not revised, either reapproved or withdrawn. In addition to the *Annual Book of ASTM Standards*, the ASTM organization offers a variety of publishing options. For example, the biocontrol community could request that standards be published as a compilation or handbook to be used as a reference by producers, distributors and end-users of biological control organisms. Each ASTM standard is available as a separate copy from ASTM. General and purchasing information is available on ASTM's website at www.astm.org.

Quality Control by the British Columbia Greenhouse Vegetable Industry: a Customer Approach

The greenhouse vegetable industry in British Columbia, Canada, is a young but rapidly growing sector of agriculture. The industry exceeded 500 acres in the year 2000 and is valued at over $200 million. Its produce label is recognized internationally for its quality and commitment to the biological control of arthropod pests. The reliance on biological control programmes has its foundation in customers' demands for reduced pesticide products and government policies on pesticide registration and produce exportation. It reflects, however, largely on BC growers' initiative, organizational abilities and compliance in the adoption of new ideas. This part of the chapter describes the process that led the BC growers to become skilful and committed users of biological control programmes and involved participants in the process of quality assurance for biological control agents.

The majority of BC greenhouse vegetable operators are members of BCHHGA and their produce is graded and marketed through BC Hot House Foods Inc. (BCHHF). BCHHGA represents the industry on regulatory issues and manages the BC Greenhouse Vegetable Industry Development Trust Fund (the Trust). The Trust was established in 1990 in partnership with the BC government, with the government and individual operators contributing to the fund. Its goal was to support research addressing the most relevant needs of the industry and to provide a structure for the process of identifying, addressing and implementing research results. The Trust has been managed by the Greenhouse Vegetable Research Council (BCGVRC), which includes administrators, technical advisers and growers. Each year approximately 25 projects are funded by the Trust. In the past 10 years approximately $4.6 million were spent on research, 60% of which were allocated to projects addressing disease and insect- and mite-pest problems. Development or expansion of biological control programmes for insect and mite pests on greenhouse vegetables absorbed the majority of the 60% expenditure. The ongoing process of identifying pest problems and implementing biological control solutions gradually improved growers' skills and made them committed users of the biological control programmes. Over the years, it significantly reduced reliance on pesticides for the control of insect and mite pests.

The British Columbia Hot House Growers' Quality Control Project

In 1995/96, BC greenhouse growers were increasingly concerned about the quality and the accuracy of packaging of supplied biologicals. In response to growers' concerns, BCGVRC funded a 3-year (1997–1999) 'quality control' project designed to improve the accuracy of *P. persimilis* (Athios-Henriot) and *Encarsia formosa* Gahan packaging by providing ongoing and constructive feedback on the quality of delivered products to participating suppliers. Results from the 3-year study have shown that the accuracy of shipments of *P. persimilis* and *E. formosa* varied significantly among the four participating producers.

In 1997, 38% of examined containers of *P. persimilis* were below the stated content whereas, in 1998 and 1999, only 11% and 13% were below the stated content, respectively. Some of the participating suppliers had consistently maintained over 95% shipping accuracy (Fig. 14.1; Luczynski and Caddick, 2000). The range of product packages below the stated quantity for four participating producers decreased from 0–57% in 1997 to 3–15% and 4–25% in 1998 and 1999, respectively. Some of the participating suppliers had consistently maintained over 95% of their packages at or above the stated quantity. During the course of the project, only six containers of *P. persimilis* out of 340 examined had excessive mortality, which indicates that packing below the stated content was the main cause of lower-than-expected counts.

In 1997, 73% of shipments of *E. formosa* were below the stated content and the shortage was reduced to 31.6% by 1998 and to 26.1% by 1999 (Fig. 14.1). Percentage of emergence was added to the quantification of *E. formosa* as an indirect measure of insect quality. This was done to alert producers about possible problems in rearing, storage or transport of wasps. Lower-than-stated content on the shipment of *E. formosa* always coincided with a low (45–65%) percentage of emergence. This suggests that causes and effects of poor emergence should be examined for inclusion in future quality standards.

The constructive feedback provided by the project proved very productive and accuracy in quantities received matching those claimed by suppliers of the two beneficial species improved significantly across the BC greenhouse vegetable industry. A quality consultant for the BCHHGA joined the ASTM subcommittee, became the technical contact and writer of the standard on *P. persimilis* and actively participated in drafting several other standards. Results from this project increased growers' confidence in the use of biologicals and increased their demand for better quality assurance of supplied products. They also stressed the importance of product standardization and customer involvement in the assessment of product quality.

Merging of Producer and Customer Approaches Under ASTM

The format of the ASTM standardization process presented all biological control stakeholders with a unique opportunity to work together in addressing quality assurance issues. The format employed by the ASTM requires involvement of a balanced group of producers, distributors and users of biological control agents, academics and government regulatory agencies in the process. In 1999, the BC quality control project broadened its objectives to include participation in the development of standards for biological control agents through ASTM. Quantification methods used during the quality control project have been incorporated into two of the ASTM standards balloted in autumn 2001. Standards are in process for 14 species and three standards have been balloted and ratified (Table 14.1). The drafts that have been submitted for ballot are:

- Standard specification for *Phytoseiulus persimilis* (Athias-Henriot) (Acarina: Phytoseiidae).
- Standard specification for *Encarsia formosa* Gahan (*Hymenoptera: Aphelinidae*).
- Standard specification for information included with packaging of multicellular biological control organisms.

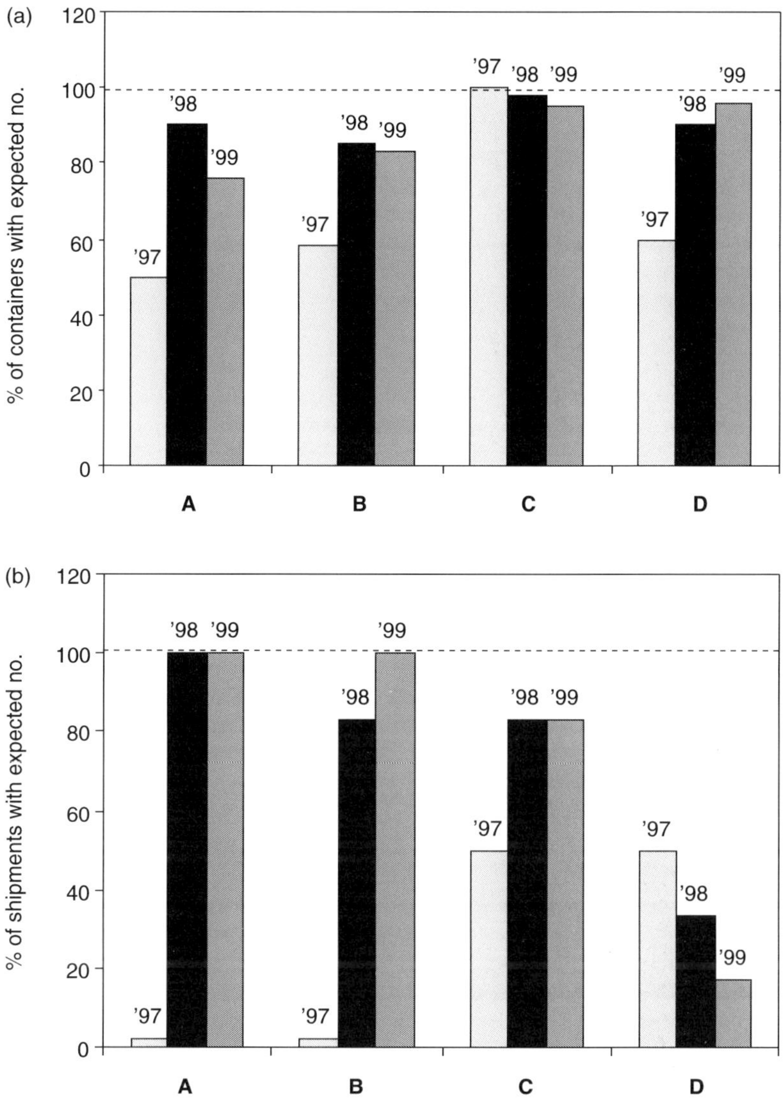

Fig. 14.1. Percentage of containers with (a) *P. persimilis* and shipments with (b) *E. formosa* with quantities stated on the label for 1997, 1998 and 1999 for four participating suppliers (A, B, C, D). (The numbers of *P. persimilis* containers examined in 1997, 1998 and 1999 were as follows: A: 30, 30, 25; B: 50, 27, 18; C: 26, 30, 20; and D: 30, 30, 24, respectively. The numbers of *E. formosa* shipments examined in '97, '98 and '99 were as follows: A: 4, 6, 6; B: 3, 14, 5; C: 4, 6, 6; and D: 4, 6, 6, respectively.)

Standardization will accelerate now because the first standards will serve as templates for others. For example, the *P. persimilis* standard is likely to serve as a template for standards for other predatory mite species. The subcommittee has agreed that face-to-face meetings twice per year were necessary to keep up momentum and communication. Given that the greatest organizational hurdles have been accomplished, many more of the standards drafts should go to ballot within the next year. The next hurdle will be to increase the scope of the standards to include vigour and ability to perform.

On the customer side, a new Canada-wide project has been initiated by the BCGVRC

Table 14.1. ASTM standards in progress.

ASTM E35.30 task groups		
Species	Activity start	Status November 2001
Goniozus legneri	22 April 1999	3rd draft
Feltiella acarisuga	15 May 1999	3rd draft
Podisus maculiventris	22 April 1999	3rd draft
Chrysoperla rufilabris	4 Oct. 1998	2nd draft
Galendromus occidentalis	22 April 1999	1st draft
Trichogramma brassicae	12 June 1998	3rd draft
Trichogramma pretiosum	4 Oct. 1998	With above
Phytoseiulus persimilis	12 June 1998	Ratified
Cryptolaemus montrouzieri	12 June 1998	5th draft – to ballot
Encarsia formosa	4 Oct. 1998	Ratified
Eretmocerus eremicus	22 April 1999	5th draft
Aphidoletes aphidimyza	22 April 1999	2nd draft
Entomophagous nematodes, label specification	30 Aug. 1999	1st draft
Entomophagous nematodes, counting method	30 Aug. 1999	1st draft
Hypoaspis miles	30 Oct. 1999	1st draft
Dicyphus hesperus	30 Oct. 1999	
Delphastus catalinae	30 Oct. 1999	
Information specification	31 Mar. 2000	Ratified
Terminology specification	31 Mar. 2001	1st draft – to ballot

where users, producers and scientists will form a partnership for the development of simple assays ('user-friendly tests') to assess shipment accuracy and to test performance of selected biological control agents. For the first time, customers will monitor the quality of the biocontrol agents at their final destination. This will maintain constructive dialogue on quality assurance between users and suppliers of biocontrol agents and will lead to the improvement of the quality of the product and efficacy of biocontrol programmes. The ASTM subcommittee has expressed interest in including these user tests in standards as well.

Acknowledgements

Lee Anne Merrill and Sinthya Penn initiated and compiled most of the product-profile project. Glenn Scriven reformatted the profiles to their present form and entered them on the internet. Norman Leppla provided ideas, inspiration and interpretation from the inception of the ANBP and its quality assurance efforts up to the present. The NBCI contributed three annual grants during 1999, 2000 and 2001 for the start-up phase of the new ASTM subcommittee E35.30.

References

Luczynski, A. and Caddick, G. (2000) *Assessing Accuracy of Shipments of* P. persimilis *and* E. formosa *and Field Performance of* E. formosa. Technical Report 99-09, British Columbia Greenhouse Vegetable Research Council, 28 pp.

Penn, S.L., Ridgway, R.L., Scriven, G.T. and Inscoe, M.N. (1998) Quality assurance by the commercial producer of arthropod natural enemies. In: Ridgway, R.L., Hoffman, M.P., Inscoe, M.N. and Glenister, C.S. (eds) *Mass-reared Natural Enemies: Application, Regulation and Needs.* Proceedings of Thomas Say Publications in Entomology, Entomological Society of America, Lanham, Maryland, pp. 202–227.

Ridgway, R.L., Hoffman, M.P., Inscoe, M.N. and Glenister, C.S. (eds) (1998) *Mass-reared Natural Enemies: Application, Regulation and Needs.* Proceedings of Thomas Say Publications in Entomology, Entomological Society of America, Lanham, Maryland.

15 State of Affairs and Future Directions of Product Quality Assurance in Europe

K.J.F. Bolckmans
Koppert Biological Systems, PO Box 155, 2650 AD Berkel and Rodenrijs, The Netherlands

Abstract

Beneficial arthropods will only become a mainstream pest-management tool when good-quality organisms are constantly available and when sufficient knowledge and technical support are provided for farmers for the proper use of these natural enemies. These requirements demand great challenges from the producers of natural enemies in all phases of production and transport of biocontrol agents. In this chapter, the history of quality control of biocontrol agents in Europe is sketched and the future of guideline development is discussed. Furthermore, it is argued that the biocontrol industry should move from quality control to quality assurance, because quality assurance looks at product quality from a proactive viewpoint, while quality control is a retrospective activity that focuses only on the quality of a product at the end of the production chain. Aspects of quality assurance during production and transport are described. Finally, the issue of certification of product quality is discussed.

Introduction

Almost 90 different natural enemies are on the market for greenhouse biocontrol. About 25 of these natural enemies are reared in very large numbers by commercial producers for use in greenhouses. Natural enemies are produced by about 85 companies worldwide, of which 25 are based in Europe. Chapter 1 and Bolckmans (1999) give an overview of the current biological control industry.

Beneficial arthropods will only continue to become a mainstream pest-management tool when a consistent supply of good-quality organisms is available and on the condition that sufficient knowledge and technical support are provided for the growers for the proper use of these natural enemies. These requirements present extreme challenges to commercial insectaries, who must generate very large numbers of delicate and short-living organisms of consistent quantity and quality to meet a market demand that can fluctuate enormously and unpredictably (Penn *et al.*, 1998).

This chapter describes the current state of affairs and future directions for product quality control by European natural-enemy producers.

© CAB International 2003. *Quality Control and Production of Biological Control Agents: Theory and Testing Procedures* (ed. J.C. van Lenteren)

Increasing Importance of Product Quality

Food-safety issues and the increased awareness of the consumer of the importance of safe and healthy food have put pressure on farmers and growers to certify their produce. A group of European supermarkets has developed a certification scheme, which is currently known as the EUREPGAP guidelines. Also European legislation puts the final responsibility for food safety at the farmer level. As a result, farmers increasingly request their suppliers to certify the quality of their products. They not only want to be assured that the supplied products contain the stated numbers and that the organisms are sufficiently fit but also that no potential pests or diseases are transferred on to their crop through releasing natural enemies.

Growers tend to organize themselves more and more in grower groups and cooperatives, and the average size of greenhouse operations has grown substantially over the past decade. These larger growers or grower groups increasingly use their buying power to assure optimal quality of all their supplied products and services. In case of defective quality, litigation is another instrument growers increasingly use to assure product quality.

Another development is that, from time to time, suggestions have been made by governments or others that efforts should be put into regulation of the quality of natural enemies. It is clear that it is a general legal requirement that the numbers of organisms supplied should equal or exceed the number stated on the label and invoice (Penn *et al.*, 1998), but other characteristics to be put into a framework of regulation are still under discussion. Several countries already apply a system for regulating the import and release of natural enemies (see Chapter 13 for a description of the state of affairs).

When supplying precious products with a limited shelf-life in such a highly competitive market, product quality could become an important competitive advantage.

History

The first publication about quality control in commercial mass-reared natural enemies appeared in the 1980s (van Lenteren, 1986). In 1980 the International Organization for Biological and Integrated Control of Noxious Animals and Plants (IOBC) created the global working group 'Quality Control of Mass-reared Arthropods'. This working group focused largely on quality issues and techniques for large governmental facilities, where arthropods are used for region- or nationwide eradication programmes, using sterile-insect techniques. The fifth workshop of this global IOBC working group in Wageningen, The Netherlands, in 1991 led to a sequence of workshops funded by the European Union and in collaboration with IOBC (see Chapter 1 for details), during which European commercial producers and scientists together developed a set of practical quality control guidelines and standards for 20 different natural enemies.

The resulting guidelines quickly became known as the 'IOBC guidelines' (van Lenteren, 1994). They include both quality control method descriptions (further called 'guidelines') and prescribed minimum quality standards for various biological parameters (further called 'standards'). The IOBC guidelines specify an easy method for checking the quantity of natural enemies per unit and testing qualitative characteristics, such as emergence rate, mortality, sex ratio, longevity, fecundity, adult size, predation/parasitization rate and flight activity in the laboratory.

After the grant from the European Union expired, the International Biocontrol Manufacturers Association (IBMA), an association of mainly European-based commercial producers, organized a workshop in Rotterdam, The Netherlands, in September 2000 to further the work of the IOBC working group. In 1990 the Association of Natural Biocontrol Producers (ANBP), an association of mainly North American-based commercial producers, started to develop their own set of quality control guidelines, which are largely based on the IOBC guidelines (see

Chapter 14). In order to ensure the acceptance of these guidelines by their members, end-users and scientists, ANBP chose to develop the guidelines using the process provided by the American Society for Testing and Materials (ASTM). ASTM is a not-for-profit organization that provides a forum for the development and publication of voluntary consensus standards for materials, products, systems and services.

In October 2001, the IBMA and ANBP organized their first joint meeting in Washington, DC, USA, and discussed the future alignment of their efforts in developing quality control methods and standards.

Future of Guideline Development

During the Washington 2001 meeting, it became clear that it is very important to keep the current efforts of both IBMA and ANBP as much aligned as possible, both in generally accepted generic testing methods and quality standards and in the layout and communication of the guidelines. Furthermore, it was felt that the link with the scientific community is very important for acceptance of the guidelines by scientists and regulators, for the quality of the developed methods and for developing new methods. It was felt that this would be best achieved through reinforcing close links with the IOBC through the global IOBC working group 'Arthropod Mass-rearing and Quality Control' (AMRQC).

From 1992 onwards, most quality control methods have been developed through close collaboration between industry and academic researchers. This collaboration allows the development of scientifically sound methodologies, while evaluating and validating them under practical circumstances by commercial producers. This collaboration has also ensured that the developed methodologies are now broadly accepted by both the scientific world and the whole biocontrol industry. Currently, it is mostly the producers of natural enemies that are refining and optimizing the quality control guidelines and developing new guidelines in the framework of the IBMA.

Future work will be focused on the following:

- Developing testing guidelines and standards for additional natural enemies. For all the major natural enemies for greenhouse use, quality control guidelines are now in place. However, for several natural enemies, quality control guidelines still need to be developed and/or validated. This includes beneficials that are typically used in outdoor crops, such as citrus, olives, almonds and maize, or beneficials that are used for the biological control of filth flies, urban pests and storage pests.
- Optimizing and validating existing guidelines and standards through ring testing. Repeatability of the test methods is very important. The test methods need to be sufficiently robust to ensure that performing the tests on the same batch of natural enemies by different experienced quality control technicians yields the same results. Also, the agreed quality standards need to be realistic. Therefore, proposals for quality standards always need to be validated through rigorous ring testing by multiple producers.
- Optimizing sample numbers and sampling sizes to ensure statistically sound tests. There are currently no clearly agreed standards for the number of samples to be taken from a batch. Also, the number of subsamples and sample sizes require critical analysis of their statistical soundness. Because most methods are very laborious, a very pragmatic approach is currently used. The optimal number of samples required for quantity control will largely depend on the standard deviation of the packaging process that is used.
- Developing less labour-intensive methods. The methods that are currently used are rather labour-intensive. Developing less labour-intensive methods could reduce the costs involved in quality control, allow the performance of more tests, enable more samples and larger sample sizes to be taken and make counting results available more quickly. Recent

advances in the fields of machine vision techniques, digital-image analysis, rapid counting and sorting techniques and quantitative immunological and molecular techniques could open up possibilities for the development of new, less laborious methods for quality and quantity control. In many cases, the results of quality control tests become available when the products have already been sold. This is especially the case with products that are sold as pupae (e.g. *Encarsia formosa*, *Eretmocerus eremicus*, *Aphidius colemani*, *Aphidius ervi*). Instead of performing emergence tests, it might be possible with some products to assess the viability of the pupae at an earlier stage using biochemically based techniques.

- Developing simple testing methods and standards for end-users. Most quality control guidelines are too tedious to be performed by growers. Therefore, simple indicative tests need to be developed to allow growers to quickly assess the quality of the products upon receipt. However, it currently remains unclear whether the results of such approximate tests have sufficient power and accuracy to allow for formal complaints and product replacement or, worse, for liability claims. The IOBC working group focused on developing guidelines and quality standards at the producers' level. It is still unclear whether these standards are also valid upon arrival of the product at the end-users' facilities, especially for parameters such as emergence rate, fecundity, longevity, parasitism/predation rate and flight propensity. If the present methods cannot be used, different quality standards need to be developed for products after shipping.
- Developing testing methods relating laboratory tests to field performance. It is not clear to what degree the currently used laboratory quality control tests give sufficiently reliable information about performance of the natural enemies in the greenhouse or field. This subject is discussed in Chapters 16 and 17 by Steinberg and Cain, and Luck and Forster, respectively.
- Quality standards: minimum or average? At this moment, there is no agreement within the industry as to whether quantitative product specifications are minimum quantities or average quantities. This issue has led to numerous debates. It is well known that, for many products, there is a rather important variability in the number of natural enemies per bottle due to the packaging techniques used. It is my opinion that 95% of the bottles of one batch should contain, as a minimum, the number of natural enemies that is stated on the bottle. For the remaining 5% of the bottles, underpacking of more than 10% should not be acceptable. But it is important to realize that putting an exact number of natural enemies in a bottle is in practice rather difficult. Automation of the packing process based on volumetric, weighing or 'counting-and-diluting' techniques often leads to even higher variability than manually pootering individual insects into a container. More research and development needs to be done on mechanizing this very important stage of the production process in order to get a better grip on the packaged quantities and therefore product quality. Packing techniques with a small standard deviation also require a smaller number of samples to obtain a reliable estimate of the average number of insects per bottle.
- Harmonizing the developed guidelines. The format of the tests, the timing and frequency of the different tests and the specified test conditions could use some further harmonization.

From Quality Control to Quality Assurance

The main focus of the activities of commercial producers and scientists in the IOBC working group has been on product quality control. Quality control is a retrospective activity that focuses on the quality of a product at the end of the production chain. Quality control does not contribute to product quality unless action is taken when products are detected that do not meet predefined product specifications. Quality assurance, on the other hand, looks at product quality from a proactive viewpoint. The fundamental idea is that, if the production process is well organized and of high quality, the result of the production process should also be of high quality. Already in the 1980s, Chambers and Ashley (1984) and Leppla and Fisher (1989) introduced the notion of total quality management for the production of natural enemies. They made a distinction between product control, production control and process control. Product control rejects faulty products, production control assures consistency of production output for a given production process and process control tells how well a production process is performing (see Chapter 2 for details).

To date these notions are still not widely recognized by commercial producers of natural enemies. The establishment of a total quality-management programme requires substantial extra investments, which are limited for most commercial producers. However, a more proactive approach to product quality not only requires investment in total quality-management techniques but also requires a more profound knowledge of those factors that influence product quality during the production and logistic chain.

What is Optimal Quality?

Many researchers and also end-users have the feeling that mass-reared natural enemies should have the same quality as the original collected field population. This is not only an illusion: it is an unnecessary and expensive goal to pursue (van Lenteren, 1991). Van Lenteren and Tommasini (1999) state:

> Rather than pursuing a scientific approach on the development of quality control, we would like to follow a more pragmatic way. The aim of releases of mass produced natural enemies is to control a pest. In this context the aim of quality control should be to determine whether a natural enemy is still in a condition to properly control the pest. Formulated in this way we do not deal with terms like maximal or optimal quality, but something like *acceptable quality*.

In the case of seasonal inoculative biological control, it is mainly the descendants of the introduced natural enemies that will control the target pest. In the case of inundative releases, the released individuals should be able to reduce the pest population. The characteristics to be measured should be limited in number but directly linked to field performance and determining whether the released individuals are still of acceptable quality. However, the currently applied quality control techniques are still labour-intensive and therefore expensive. The development of accurate but less labour-intensive techniques would allow more regular testing of product quality than currently performed by many companies.

Quality Assurance in Production

Many authors have discussed changes in genetic variability and inbreeding during mass rearing. Van Lenteren and Tommasini (1999) give a recent overview. Although this subject is still poorly understood by scientists and producers, the recent evolution of molecular techniques makes possible a more profound insight into the influence of mass-rearing conditions on the genetic variability over the coming years. However, producers should already implement in a pragmatic way guidelines for the minimum size of founder populations, composition of founder populations (collect from a wide area with different climatic and ecological characteristics), regular strain renewal or alternatively gene infusion and maintaining mother cultures under variable laboratory environments. Valuable information can be found in Joslyn (1984) and Chapters 1, 6 and 7.

Mass-rearing natural enemies in an economic way resulting in biocontrol agents that are affordable by the grower will inevitably always be an 'unnatural activity'. However, the negative effects of unfavourable rearing conditions, such as food availability, overcrowding, unfavourable climatic conditions, etc., on the fitness of the natural enemies and therefore on product quality can be managed through the establishment of well-specified rearing protocols, rigorous production control and quality measuring throughout the production cycle. Preventing these problems requires a good deal of knowledge, experience with mass rearing and 'green fingers'.

Probably the least understood and least studied factor that influences product quality is the effect of microbial contamination by entomopathogenic fungi, bacteria, viruses or microsporidia (Shapiro, 1984; Sikorowski, 1984). Except for predatory mites, very little is known about the importance of microbial diseases for parasitic hymenopterous wasps, dipterous gall-midges, hemipterous bugs and coccinellid beetles. An overview of current knowledge is given by Bjørnson and Schütte in Chapter 10. As the knowledge in this field increases, additional standard testing guidelines to check for the presence of entomopathogens could be developed.

A concern that has emerged more recently is the risk of plant-disease transmission, particularly viruses, by natural enemies or their hosts that might accidentally or intentionally be included in the product. More specifically, the fear of Pepino mosaic virus (PepMV) by European tomato growers has led several producers of natural enemies to investigate the potential of vectoring plant viruses by natural enemies, such as *Macrolophus caliginosus*. We believe that strict hygienic production procedures, adapted rearing facilities, choice of plants that are not hosts for such dangerous plant viruses, routine checks for plant viruses and occasional inspections by local plant-protection services are the best way to ensure a clean product. As a general rule, plant material should be avoided as much as possible in the final product. If it is necessary to include some plant material to assure optimal product quality and survival, it should be free from plant diseases and the plant material should, when the natural enemies are released, not come in direct contact with the crop.

All biocontrol products should in principle be pure and free from contamination by potential pests. However, the presence of a limited number of the target host, such as whiteflies in the case of whitefly parasitoids or spider mites in the case of spider-mite predators, can often not be avoided and should also not necessarily be a problem, on the condition that they do not provide a risk for transmission of plant diseases or creating a pest.

Taxonomy is a crucial element in the quality assurance of beneficial arthropods. Insectaries should have established procedures to verify the identity of the natural enemies produced. Although a shortage of taxonomists places severe limitations on authoritative identification of some hard-to-identify organisms, such as *Trichogramma* spp., every reasonable effort should be made to use available resources to obtain proper identification (Penn *et al.*, 1998).

Harvesting and Formulating

Harvesting and formulation techniques can have a large impact on product quality due to their technical nature and the intensive mechanical handling of the mass-reared natural enemies. All producers have developed their own, proprietary techniques for harvesting, purifying, counting and formulating natural enemies. The nature of the applied technique and the diligence with which it is performed will determine the effect on product quality. Bottling techniques not only can cause mechanical damage but also may lead to great variability in numbers of natural enemies per bottle. The composition of the carrier material (type of material, granule size and moisture content) and the availability of food or carbohydrates (honey) and water can greatly influence the survival of the natural enemies during storage and transportation. The design of release cards for parasitized pupae (*E. formosa*, *E. eremicus*) and the type of glue and the dosing technique that are used to apply the pupae to the cards all influence the quality of the final product. Depending on the product and duration of transport, products are typically overpacked between 5 and 15% in order to compensate for mortality during the logistic chain to the end-user.

Quality Assurance in Logistics

Storage

Many producers of natural enemies operate on a national or even international scale. Most natural enemies need to be kept cool during the logistic chain. Optimal storage temperatures and maximum storage times differ depending on the species. Therefore rooms at different temperatures are needed both at the producers' facilities and at the facilities of local subsidiaries or distributors. Strict written protocols and regular internal audits have to ensure that products are stored under the prescribed conditions and that maximum storage times are respected throughout the logistic chain. The performance of storage facilities and the stability of storage temperatures can be monitored with dedicated climate computers or even with simple data-loggers.

The developmental stage of the natural enemy not only influences the production cost and the possibility of mechanizing parts of the production and/or packaging process but also influences shelf-life and survival during storage and transportation. Whitefly parasites and certain aphid parasites, for example, can be stored much longer in the pupal stage and are less sensitive to mechanical damage during the harvesting and formulation process than when stored as nymphs or adults.

Packaging for transportation

During transportation from the producer to the local subsidiary or distributor and from there to the end-user, the products have to be kept cool. Large-volume transportation is done by truck or air cargo under climate-controlled conditions. Transportation to the end-user is often done with delivery vans, which need to be climate-controlled as well, especially in areas with extreme climatic conditions. During the logistic chain, the products are packed in insulated, polystyrene boxes, together with ice-packs. Many producers have worked on the optimization of such packaging techniques. Since ice-packs have a limited lifespan, even when transport takes place under climate-controlled conditions, transport has to be fast to ensure the arrival of an optimal product. Delays through slow customs-clearance procedures at international borders and airports may be very detrimental for the quality of the products. Fortunately, nowadays the customs of most countries are used to weekly importation of natural enemies. Special attention has to be paid to air exchange in packages when large volumes of certain natural enemies are transported. If very large volumes of *Amblyseius cucumeris* or bumble bees are packed and transported in a way that does not allow sufficient air exchange, carbon dioxide can build up to very high levels, resulting in mortality of the organisms.

Minimum label requirements

Labelling of biological control products is currently not regulated in most countries. Also, in the biocontrol industry, there is currently no formal agreement on minimum label requirements for natural enemies, although it is always a point of great attention at meetings of IBMA and ANBP. I propose the following minimum label requirements: scientific name of the natural enemy, target hosts, number of individuals per unit (bottle, card, sachet, etc.), use-by date, batch number for tracking and tracing in case of complaints, optimal storage temperature and method of release. Because of the limited 'shelf-life' of natural enemies, products need to be packed and prepared just before shipping. For international exporting companies, last-minute labelling of relatively small numbers of products with different labels in different languages for different countries becomes extremely complicated and therefore expensive to organize. Therefore, several companies have developed universal pictograms to provide the above-mentioned information for the customer. The customer can find further information and explanation of the pictograms in his own language in a booklet of user guidelines or on a poster that is supplied at least once a year to the customer.

With natural enemies, it is not sensible to mention release rates on the label since the release rate, number of releases and release frequency will depend on the crop, time of the year, region, country, pest pressure and pest-management strategy. This specific information is usually provided in the booklet of user guidelines and through personal advice from an experienced technical adviser or scout, who bases his/her advice on observations in the crop.

Quality assurance by distributors and end-users

Experience has taught that, for natural enemies as for everything else, the chain is only as strong as its weakest link. Natural enemies need to be kept under the correct storage conditions until the very last moment. Several companies have developed information leaflets or booklets and even training for distributors and end-users to ensure optimal handling of their products. Implementation of use-by dates prevents distributors from storing products for too long. Distributors should be trained to perform quality control tests. Results from quality tests performed by distributors or at the facilities of local subsidiaries yield valuable information about the influence of transportation of natural enemies and therefore allow appropriate measures to be taken to improve shipping conditions, packaging and shipping times. Another well-known technique is to include temperature data-loggers in shipments of natural enemies. These data-loggers give information that can be used to improve shipment. Furthermore, the use of tools such as data-loggers, temperature-sensitive stickers, tip indicators and tilt indicators in or on transportation boxes ensures that producers are able to check whether transport companies are performing shipment under the agreed conditions.

Suppliers typically advise the end-users to release the natural enemies immediately or as soon as possible after delivery. Growers usually do not have the facilities to store the products correctly. Further storage at the grower's facility only unnecessarily prolongs the time in transit of the natural enemies. Finally, the release method may also influence the product quality. Release cards with parasitized pupae, for example, need to be handled with care during introduction in the greenhouse to avoid damaging the pupae. Bottles with natural enemies should not be left in direct sunlight while the grower is releasing natural enemies.

Tracking and tracing techniques

Implementation of tracking and tracing techniques, such as batch codes, allows the history of a product to be traced back in case of complaints or quality testing by end-users or distributors. The use of batch codes makes it possible to look for common sources of quality problems and to determine whether the cause lies at the production/packaging level or in the logistic chain when several complaints from different sources reach the producer.

Certification

Consumers and supermarkets nowadays demand farmers to certify their produce (see earlier in this chapter). As a result, farmers increasingly request their suppliers to certify the quality of their products, including products from the biocontrol industry. Voluntary quality-certification programmes, such as the International Organization for Standardization (ISO) 9000 or European Federation for Quality Management (EPQM), offer a framework to clearly establish and describe all processes within a company in written standard operating procedures and protocols. Certification might reduce the chance for errors and mistakes through strict procedures, but, without adequate production installations, reliable production techniques and experienced and skilled personnel, it will not necessarily guarantee high-quality products. On the other hand, excellent installations, techniques and people without a streamlined organization and procedures will not result in a reliable output of high-quality products.

In 2002, the Israeli-based company Bio Bee Biological Systems (Sde Eliyahu, Israel) became the first producer of natural enemies to obtain an ISO 9000 certificate.

Another method of certification is through establishment of a voluntary industry quality label, which is audited through independent external auditors. There are currently no concrete plans among the European producers to develop such an industry quality label, although the topic has come up at several IBMA meetings and been intensively debated.

Conclusions

Optimal product quality is in the interest of the producers of natural enemies. Growers quickly recognize a poor-quality product, which is very often a reason for changing to another supplier. Suppliers of structurally poor-quality products are not able to remain in business for a long time, but they often cause harm for the method of biological control. Producers of natural enemies need to be able to supply products of constant and consistent quality to be successful in the market.

Regulation often leads to stronger rules than originally intended. Over-regulation of biological control risks severely damaging the creative dynamics of this valuable alternative to chemical pesticides. Therefore industry needs to impose a certain level of self-regulation. By signing the IBMA Charter of Principles, members of the IBMA commit themselves to certain production ethics. The next step might be to establish an industry quality label that is audited by an independent auditing company.

Assurance of consistent product quality can be achieved through a combination of professional production systems under suitable rearing conditions, experienced and skilled production and research and development professionals, implementation of total quality management, development of accurate and consistent packing techniques, implementation of rigorous quality control, development of labour-friendly quality control methods, use of suitable packaging for transportation, short transportation times through a reliable cooled chain, implementation of batch codes for tracking and tracing, use of use-by dates and education and training of distributors and end-users.

References

Bolckmans, K.J.F. (1999) Commercial aspects of biological pest control. In: Albajes, R., Gullino, M.L., van Lenteren, J.C. and Elad, Y. (eds) *Integrated Pest and Disease Management in Greenhouse Crops*. Kluwer Academic Publishers, Dordrecht, pp. 310–338.

Chambers, D.L. and Ashley, T.R. (1984) Putting the control in quality control in insect rearing. In: King, E.G. and Leppla, N.C. (eds) *Advances and Challenges in Insect Rearing*. US Department of Agriculture, Agricultural Research Service, Southern Region, New Orleans, Louisiana, pp. 256–260.

Joslyn, D.J. (1984) Maintenance of genetic variability in reared insects. In: King, E.G. and Leppla, N.C. (eds) *Advances and Challenges in Insect Rearing*. US Department of Agriculture, Agricultural Research Service, Southern Region, New Orleans, Louisiana, pp. 20–29.

Leppla, N.C. and Fisher, W.R. (1989) Total quality control in insect mass production for insect pest management. *Journal of Applied Entomology* 108, 452–461.

Penn, S.L., Ridgway, R.L., Scriven, G.T. and Inscoe, M.N. (1998) Quality assurance by the commercial producer of arthropod natural enemies. In: Ridgway, R.L., Hoffman, M.P., Inscoe, M.N. and Glenister, C.S. (eds) *Mass-reared Natural Enemies: Application, Regulation and Needs*. Proceedings of Thomas Say Publications in Entomology, Entomological Society of America, Lanham, Maryland, pp. 202–227.

Shapiro, M. (1984) Micro-organisms as contaminants and pathogens in insect rearing. In: King, E.G. and Leppla, N.C. (eds) *Advances and Challenges in Insect Rearing*. US Department of Agriculture, Agricultural Research Service, Southern Region, New Orleans, Louisiana, pp. 130–142.

Sikorowski, P.P. (1984) Microbial contamination in insectaries. In: King, E.G. and Leppla, N.C. (eds) *Advances and Challenges in Insect Rearing*. US Department of Agriculture, Agricultural Research Service, Southern Region, New Orleans, Louisiana, pp. 143–153.

van Lenteren, J.C. (1986) Parasitoids in the greenhouse: successes with seasonal inoculative release systems. In: Waage, J.K. and Greathead, D.J. (eds) *Insect Parasitoids*. Academic Press, London, pp. 341–374.

van Lenteren, J.C. (1991) Quality control of natural enemies: hope or illusion? In: Bigler F. (ed.) *Proceedings 5th Workshop Global IOBC Working Group 'Quality Control of Mass Reared Organisms', Wageningen, 25–26 March 1991*, pp. 1–14.

van Lenteren, J.C. (1994) Quality control guidelines for 21 natural enemies. *Sting, Newsletter on Biological Control in Greenhouses, Wageningen* 14, 3–24.

van Lenteren, J.C. and Tommasini, M.G. (1999) Mass production, storage, shipment and quality control of natural enemies. In: Albajes, R., Gullino, M.L., van Lenteren, J.C. and Elad, Y. (eds) *Integrated Pest and Disease Management in Greenhouse Crops*. Kluwer Academic Publishers, Dordrecht, The Netherlands, pp. 276–294.

16 The Relationship between Results from Laboratory Product-control Tests and Large-cage Tests Where Dispersal of Natural Enemies is Possible: a Case-study with *Phytoseiulus persimilis*

S. Steinberg and H. Cain
Bio-Bee Biological Systems Sde Eliyahu, Bet Shean Valley, 10810 Israel

Abstract

Most of the quality control guidelines used until now relate to criteria that are measured under laboratory conditions. In this chapter, the development of a dispersal test of predatory mites in large cages is described. The aim of this test is to evaluate dispersal and reproduction capacities of natural enemies under semi-natural conditions.

Introduction

The term 'product control' for a natural enemy was defined by Penn *et al.* (1998) as:

> Assurance of the conformity of the product to acceptable standards of quality through monitoring procedures applied at the end of production. Such procedures substantially increase the probability that the product will be effective in performing its intended function but cannot predict performance because of the highly variable conditions to which the product may be exposed after the product is shipped.

This definition reflects the idea that because the aim of a natural-enemy release is pest suppression, the adopted product-control standards should be relevant to the intended use of the natural enemy. Thus terms such as 'maximal' or 'optimal' quality may need to be replaced by 'acceptable quality' at the gate of the producer, bearing in mind that different criteria are appropriate for different types of biological control programmes (van Lenteren, 1991).

Van Lenteren and Tommasini (1999) stress that, if commercial producers of natural enemies are going to apply quality control on a regular basis, then the characteristics of product control to be measured should be few in number but directly linked to the field performance. They foresee that, in the future, testing flight and/or field performance of natural enemies will become a necessity in order to show the relevance of the laboratory product-control measurements. Hence, the latter will only be adequate when a good

© CAB International 2003.

correlation is found between laboratory results and flight tests or field performance. Van Lenteren and Tommasini (1999) mention several attributes of field populations of natural enemies that might change once the latter are introduced into the laboratory under mass-rearing conditions. Searching efficiency and dispersal characteristics (e.g. flight and/or walking behaviour) are among the key attributes that might be influenced by the mass-rearing conditions and might therefore negatively affect the field performance of the natural enemy.

A list of criteria for product control of the commercially available predatory mite *Phytoseiulus persimilis* Athias-Henriot was initially included in the protocols developed by the International Organization for Biological Control (IOBC) global working group 'Quality Control of Mass-reared Arthropods' (van Lenteren and Steinberg, 1991). The updated measure of fecundity for *P. persimilis* determined in a laboratory test should be a minimum of ten eggs per female during a 5-day test (van Lenteren, 1998). Penn *et al.* (1998) criticize the idea of laboratory tests, stating that, although it can be helpful in determining predator fitness, testing individual predatory mites is time-consuming and does not necessarily provide adequate information on the performance of the predators in the field. For instance, *P. persimilis* mites applied directly to 'hot spots' where prey is readily available do not require the same attributes as predatory mites distributed in a crop where prey is more dispersed. Thus product-control standards for a natural enemy must allow flexibility according to the conditions of its use and the expected results (Penn *et al.*, 1998).

Here we report on the initiative taken by a commercial natural-enemy producer to quantify the relationship between the earlier-developed laboratory fecundity test and a newly developed greenhouse cage test, where dispersal of the natural enemy is possible. *P. persimilis* was chosen as a model because of the key role it plays in many integrated pest management (IPM)/biocontrol programmes and hence its high economic value in commercial augmentative biocontrol worldwide. The greenhouse cage test is described in detail in this chapter. The details of the laboratory fecundity test for *P. persimilis* can be found in the last version of the product-control guidelines established by the IOBC/European Community (EC) Concerted Action Group on Quality Control (van Lenteren, 1998) and in Chapter 19.

Materials and Methods

Plants

Cucumber plants were grown in a 'perlight' medium in buckets under greenhouse conditions. The plants reached up to 15 true leaves and were trellised in a normal procedure. They were kept completely separate from one another, both to avoid migration of the mites from one plant (= replicate) to another (see below) and also to provide enough space for the person to check the leaves and move freely between the plants. Tanglefoot glue was smeared on the trellising cords between the plants as another measure to ensure that no mites could migrate from one plant to another.

Infestation by spider mites

A piece of brown bean leaf was taken from the greenhouse mass-production unit, checked thoroughly for absence of predatory mites and counted for presence of *c.* 25 mobile stages of the two-spotted spider mite *Tetranychus urticae* Koch. The infested leaf was placed on a cucumber leaf at the top or at the lower part of the plant (according to the treatment, see below). Spider-mite infestation was allowed to develop for a period of 48 h, at the end of which a colony consisting of a few dozens of mobile stages as well as eggs of spider mites was covering *c.* 50% of the leaf underside.

Introduction of the predatory mites

Forty-eight hours after initial infestation, five females of *P. persimilis*, randomly chosen from a 7-day-old storage container (after harvest), were placed in a small plastic vial and introduced on a leaf located eight to 14 leaves away from the spider-mite-infested spot.

Parameters measured

Two treatments were tested initially, each in five replicates. In one treatment, the spider-mite-infested leaf was located at the lower part of the plant and *P. persimilis* was introduced on the top leaves of the plant. The second treatment was arranged the other way round, i.e. the infested leaf at the top of the plant and the predators introduced on the lower leaves of the plant. Twenty-four hours after introduction of the predatory mites, their presence on the spider-mite-infested leaf (= target leaf) was recorded twice a day, in the morning and in the afternoon, using a magnifying glass. This procedure lasted for 7 days. The day on which the first predatory mite was observed on the target leaf was registered. At the end of the 7-day period, the target leaves were detached and examined under a stereomicroscope in order to count all *P. persimilis* stages, i.e. the number of adults, young stages and eggs.

Results and Discussion: Searching-capacity Tests

Tables 16.1 and 16.2 show clearly that the initial tendency of *P. persimilis* is to move upward. When introduced on the lower leaves of the plant, it took the predatory mites 24 h to reach the target leaf located ten to 14 leaves above. This trend was recorded in four out of five replicates (Table 16.1). After 1 week, all developmental stages of the predator were present on the target leaf (Table 16.1). When introduced on the top leaves of the plant, *P. persimilis* reached the spider-mite-infested leaf located eight to 11 leaves downwards in only two out of five replicates 3 and 5 days following the initial introduction (Table 16.2).

For the sake of developing a standard searching-capacity bioassay, we suggest locating the spider-mite-infested leaf in the upper part of the plant and introducing *P. persimilis* in the lower part of the plant. Further, for practical releases in the crop, it is recommended that the predatory mites are introduced on the lower leaves of the plant, especially in vertically trellised crops, such as cucumber, tomato, sweet pepper, etc., so that they can quickly find their way to the prey located above.

Results and Discussion: Testing Effect of Storage

After development of the test, the searching-ability bioassay was performed to study whether the predatory mites differ in their searching ability as a result of storage. Adults of *P. persimilis* stored at 8–10°C for 5 and 18 days were compared simultaneously, using the greenhouse cage test. The results show no difference between the searching capacity of predatory mites exposed to short or long storage periods. In four out of five replicates, the 5-day-stored adults found

Table 16.1. Searching ability of *Phytoseiulus persimilis* (*P.p*) when placed on the lower leaves of a cucumber plant with spider mites at the top of the plant.

			Situation of *P.p* population on the infested leaf after 7 days		
Rep. no.	Leaf distance*	Day *P.p* first found on the infested leaf	Adults	Young stages	Eggs
1	11	2	1	5	13
2	10	1	1	–	4
3	14	1	5	1	16
4	11	1	1	3	2
5	14	1	4	9	25

*Between the spider-mite-infested leaf and the leaf on which *P. persimilis* was introduced.

Table 16.2. Searching ability of *Phytoseiulus persimilis* (*P.p*) when placed on the top leaves of a cucumber plant with spider mites at the lower part of the plant.

Rep. no.	Leaf distance*	Day *P.p* first found on the infested leaf	Situation of *P.p* population on the infested leaf after 7 days		
			Adults	Young stages	Eggs
1	11	–	–	–	–
2	11	–	–	–	–
3	11	3	2	–	2
4	8	–	–	–	–
5	9	5	1	1	1

*Between the spider-mite-infested leaf and the leaf on which *P. persimilis* was introduced.

their way to the prey within the first 24 h (Table 16.3). In all six replicates, the 18-day-stored adults found the target leaf within 24 h (Table 16.4). Establishment of the predators on the target leaf was recorded in four replicates for the 5-day-stored adults and in all the six replicates for the 18-day-stored adults (Tables 16.3 and 16.4).

Results and Discussion: Comparing Laboratory Performance and Searching Capacity

The first attempts to compare the performance of *P. persimilis* in a laboratory fecundity test with its performance in a greenhouse-cage searching-ability test were

Table 16.3. Searching ability of 5-day-old *Phytoseiulus persimilis* (*P.p*). The spider-mite-infested leaf is located above the predator's release site.

Rep. no.	Leaf distance*	Day *P.p* first found on the infested leaf	Situation of *P.p* population on the infested leaf after 7 days		
			Adults	Young stages	Eggs
1	10	1	4	8	–
2	9	1	6	–	–
3	10	–	–	–	–
4	9	1	2	–	–
5	10	1	1	–	–

*Between the spider-mite-infested leaf and the leaf on which *P. persimilis* was introduced.

Table 16.4. Searching ability of 18-day-old *Phytoseiulus persimilis* (*P.p*). The spider-mite-infested leaf is located above the predator's release site.

Rep. no.	Leaf distance*	Day *P.p* first found on the infested leaf	Situation of *P.p* population on the infested leaf after 7 days		
			Adults	Young stages	Eggs
1	11	1	3	–	–
2	10	1	1	–	9
3	9	1	3	–	–
4	10	1	3	5	–
5	9	1	4	–	–
6	8	1	2	–	1

*Between the spider-mite-infested leaf and the leaf on which *P. persimilis* was introduced.

carried out in June 2001 and January 2002. This comparison was executed by examining simultaneously, in the fecundity test and the searching-ability test, individual predators sampled from the same batch. Results are presented in Table 16.5. Good laboratory performers of June 2001 showed a mean fecundity of 21.8 ± 1.7 eggs per female per 5 days, which is more than two times higher than the product-control minimum criterion of ten eggs per female per 5 days, and a survival rate of 85%, which is higher than the minimum criterion of 80% survivors at the end of the 5-day fecundity test. Simultaneously, other female predators from the same batch have demonstrated a 73% response in the searching-ability test (response = predators found on the target leaf within 24 h following their release, and predator population established on the target leaf at the end of the trial (5 days after its start)). Both 'fat' and 'flat' (see section on *Phytoseiulus* in Chapter 10) female predators from the batch of 9 January 2002 showed poor fecundity and high mortality in the laboratory, and a searching ability rate of response fluctuating between 37.5% and 57%, respectively (Table 16.5). Another series of 'fat' females of *P. persimilis*, which was tested on 15 January 2002, yielded *c.* 11 eggs per female per 5 days and 83% survival, but a low response in the searching-ability test (37.5%, Table 16.5).

Data are still insufficient to correlate between bad/good performers in a laboratory fecundity test and their performance in the greenhouse cage test (see discussion of disease-infected predatory mites and extreme dispersal behaviour in Chapter 10). Such testing will be carried out in the near future because of the problematic situation with diseased predatory mites that may be found in mass-rearing situations.

Conclusions

The greenhouse-cage searching-ability test, where dispersal of *P. persimilis* is possible, is indeed halfway to field performance, which is defined by van Lenteren and Tommasini (1999) as: 'Capacity to locate and consume or parasitise prey/host in crop under field conditions.' Therefore, the kind and strength of the correlation that will be found between the laboratory product-control test and the greenhouse-cage test will apparently shed more light on the relevance of a laboratory quality test as a key product-control criterion that is practised by many natural-enemy producers around the world.

As to the greenhouse-cage bioassay itself, based on the limited data generated up till now, we propose that 80% responding predatory mites should be adopted as a minimum criterion. We suggest a minimum of 30 replicates per treatment for statistically sound testing. This proposal has to be verified by ring testing, which will be conducted in the near future.

Table 16.5. Comparison between laboratory fecundity test and a greenhouse-cage searching-ability test for the predatory mite *Phytoseiulus persimilis*.

Date/batch	Fecundity ± SE*	Survival (%)	Response (%)**
June 2001	21.8 ± 1.71 (n = 17)	85	73 (n = 11)
9 January 2002 ('fat' females)	6.41 ± 1.21 (n = 22)	77	37.5 (n = 8)
9 January 2002 ('flat' females)	Very low (n = 2)	47	57 (n = 7)
15 January 2002 ('fat' females)***	10.96 ± 1.33 (n = 26)	83	37.5 (n = 8)

*Eggs per female per 5 days.
**In the searching-ability test.
***Sampled from the same batch of 9 January 2002.

References

Penn, S.L., Ridgway, R.L., Scriven, G.T. and Inscoe, M.N. (1998) Quality assurance by the commercial producer of arthropod natural enemies. In: Ridgway, R.L., Hoffmann, M.P., Inscoe, M.N. and Glenister, C.S. (eds) *Mass-reared Natural Enemies: Application, Regulation and Needs.* Proceedings of Thomas Say Publications in Entomology, Entomological Society of America, Lanham, Maryland, pp. 202–230.

van Lenteren, J.C. (1991) Quality control of natural enemies: hope or illusion. In: Bigler, F. (ed.) *Proceedings 5th Global IOBC Workshop 'Quality Control of Mass Reared Arthropods', Wageningen, The Netherlands, 25–28 March 1991*, pp. 1–14.

van Lenteren, J.C. (1998) *Sting 18, Newsletter on Biological Control in Greenhouses.* Wageningen Agricultural University, Wageningen, 32 pp.

van Lenteren, J.C. and Steinberg, S. (1991) A preliminary list of criteria for quality control of beneficial arthropods used commercially in greenhouse crops. In: Bigler, F. (ed.) *Proceedings 5th Global IOBC Workshop 'Quality Control of Mass Reared Arthropods', Wageningen, The Netherlands, 25–28 March 1991*, pp. 195–199.

van Lenteren, J.C and Tommasini, M.G. (1999) Mass production, storage, shipment and quality control of natural enemies. In: Albajes, R., Gullino, M.L., van Lenteren, J.C. and Elad, Y. (eds) *Integrated Pest Management in Greenhouse Crops.* Kluwer Publishers, Dordrecht, pp. 276–294.

17 Quality of Augmentative Biological Control Agents: a Historical Perspective and Lessons Learned from Evaluating *Trichogramma*

R.F. Luck and L.D. Forster
Department of Entomology, University of California, Riverside, CA 92521, USA

Abstract

Augmentative biological control involves one or more releases of a natural enemy in an attempt to suppress and maintain a pest population at subeconomic densities. The notion of releasing parasitoids augmentatively for pest suppression was initially proposed in the late 1800s. However, its first sustained use involved the suppression of the citrophilous mealybug, *Pseudococcus calceolariae* Fernald, a pest of citrus in southern California, which began sometime between 1913 and 1917. The biological control agent, the coccinellid *Cryptolaemus montrouzieri* Mulsant, initially introduced as a classical biological control agent, was unable to survive in sufficient numbers to affect control without augmentation. This coccinellid is still being used in citrus to suppress mealybug pests and it is still commercially available. The initial success of this tactic led to an expansion in its use against other pests, beginning with the most widely used augmentative biological control agents, *Trichogramma* species. Their use began in the late 1920s, when S.E. Flanders developed a mass-production system for them. In this chapter, we first summarize this historical origin and then illustrate the role of fundamental research and its interaction with theory in improving augmentative biological control's predictability, using *Trichogramma* species (Hymenoptera: Trichogrammatidae) as examples. Furthermore, we extend the notion of the quality of a biological control agent by defining it in terms of the attributes that make an agent successful against a particular pest under field conditions. We provide several examples in which *Trichogramma* species are assessed by using behavioural observations under laboratory conditions, employing a technique first used by European researchers. However, we frame this evaluation in the context of a parasitoid's reproductive success, defined in terms of its offspring's reproductive prospects – that is, the offspring's characteristics that allow them to maximize their reproduction in the field on the targeted pest. It is this field reproduction that provides biological control. We suggest that this behavioural approach provides a means of evaluating the likely prospects for augmentative biological control by a particular agent and the potential to manipulate the agent and its interaction with its host to enhance its success within an economic framework.

© CAB International 2003. *Quality Control and Production of Biological Control Agents: Theory and Testing Procedures* (ed. J.C. van Lenteren)

Introduction

Augmentative biological control utilizes one to several releases of a natural enemy to suppress a pest during the course of a season or a crop's production cycle. Permanent establishment with consistent pest suppression in the absence of augmentation is not its aim. Frequently, augmentative releases are an outgrowth of an unsuccessful or partially successful effort to establish a natural enemy permanently, i.e. a classical biological control programme (Smith and Armitage, 1931; Flanders, 1949). Under such circumstances, augmentative releases are meant to supplement an established complex of endemic and/or exotic natural-enemy populations during critical periods when the natural-enemy complex is incapable of suppressing the pest consistently on its own. Augmentative biological control is one tactic in a pest-management strategy that seeks sustainability in the management of a pest complex (e.g. Rabb *et al.*, 1976; Flint and van den Bosch, 1981; Haney *et al.*, 1992; Trumble and Morse, 1993; Luck *et al.*, 1997).

The notion of periodically releasing natural enemies was first suggested by F. Enock (1895) at a meeting of the London Entomological and Natural History Society. He suggested the possibility of 'farming' *Trichogramma*. Flanders (1949) also credits Felix Gillet, the Horticulture Commissioner of California, with a similar notion. In an 1882 meeting in El Dorado, California, the Horticultural Commissioner stated that: 'it is surprising [given all the money spent to fight noxious insects that we] have never tried to raise ichneumon flies by the million and let them loose wherever there are any insect pests to destroy'. Also, Decaux (1899) employed natural-enemy releases as part of an integrated-control tactic for fruit pests in France. Finally, Kot (1964, p. 278) cites Radeckij as initiating experiments in 1911 on rearing and introducing *Trichogramma evanescens* Westwood for the control of *Cydia pomonella* (L.) (Lepidoptera: Tortiricidae). Radeckij collected the parasitoid from Astrakhan province in Turkistan and introduced it into Turkistani apple orchards.

The first extensive and sustained augmentative biological control effort against a pest or group of pests, however, involved the mass production of a ladybird, *Cryptolaemus montrouzieri* Mulsant (Coccinellidae: Coleoptera), to suppress a complex of mealybugs (Homoptera: Pseudococcidae) in southern California citrus. In this chapter, we first summarize this historical origin and then illustrate the role of fundamental research and its interaction with theory in improving augmentative biological control's predictability, using *Trichogramma* species (Hymenoptera: Trichogrammatidae) as examples.

The Early Evolution of Augmentative Biological Control

An early introduction of an exotic pest and its augmentative control

The first sustained augmentative biological control project began before 1917 and involved the citrophilous mealybug, *Pseudococcus calceolariae* Fernald (= *gahani* Green), a pest of citrus in southern California. It had appeared in California's citrus groves around 1913 and spread throughout the coastal areas of southern California, causing substantial damage to citrus (Smith and Armitrage, 1920; Quayle, 1938). The available control method, hydrogen cyanide fumigation, which was used to control soft scales (Homoptera: Coccidae) and armoured scales (Homoptera: Diaspididae), was ineffective in controlling the mealybug (Quayle, 1938). Earlier, foreign exploration for natural enemies of mealybugs in general had discovered a ladybird, the mealybug destroyer, in Australia. It had been imported and released in California's citrus groves in 1892, and again in 1909, for suppression of a variety of mealybugs, including the citrus mealybug, *Planococcus citri* (Russo), the obscure or Baker's mealybug, *Pseudococcus obscurus* Essig (= *maritimus* (Ehr.)), and the long-tailed mealybug, *Pseudococcus longispinus* (Targioni-Tozzetti), all of which were present in California prior to the citrophilous mealybug's discovery. However, *C. montrouzieri* was unable to survive the cool winter temperatures in southern California's coastal valleys and, although it was able to survive along the immediate southern California coast, its numbers were often insufficient to effect mealybug suppression (Bartlett, 1974).

With the discovery and spread of the citrophilous mealybug in 1913, the citrus industry was faced with a serious problem that threatened the crop's viability. In response to the problem, the citrus industry promoted the establishment of insectaries to mass-produce the beetle (Smith and Armitage, 1920, 1931; Quayle, 1938). The techniques for doing so had been worked out by the early 1920s and, for the next decade, the beetle was the principal means of suppressing this pest. In 1930, at the height of the campaign against this pest, 16 insectaries had been established and were producing 20 million beetles annually for release in infested groves. Meanwhile, foreign exploration for natural enemies to suppress this pest continued. Success was finally achieved when two parasitoids, *Coccophagus gurneyi* Compere (Hymenoptera: Aphelinidae) and *Tetracneumus pretiosus* Timberlake (Hymenoptera: Encyrtidae), were discovered in Australia in the late 1920s. Their introduction suppressed the citrophilous mealybug to subeconomic densities (Quayle, 1938) and it remains a rare insect to this day in California's citrus groves.

Commercial insectaries in California

However, an effective and permanent biological control agent has yet to be discovered for one of the mealybug pests, the citrus mealybug. Hence, augmentative releases are still used in some southern California citrus groves. During the 70 years in which this beetle has been mass-reared, the rearing methods have improved and become more efficient. Even though the number of insectaries producing the beetle has declined, the number of beetles produced exceeds the 1930 levels of 20 million that were being produced at the height of the citrophilous mealybug campaign. In 1946, just prior to the introduction of synthetic organic pesticides in citrus (Ebeling, 1950), seven insectaries were producing 40 million ladybirds annually (Flanders, 1949). By 1963, only three insectaries still produced this predator, but they produced 30 million beetles annually (Fisher, 1963). Although modified and made more efficient, the same production technique that had been developed by Smith and Armitage (1920, 1931) in the 1910s was still being used in 1963 (Fisher, 1963) and it continues to be used to this day. Adult mealybug destroyers can still be purchased from any one of nine suppliers (Cranshaw *et al.*, 1996; Hunter, 1997). Thus, from its initial conceptualization and development in the late 1910s, augmentative biological control has been an integral part of citrus pest management in California, and it still remains so (Flanders, 1951; DeBach and White, 1960; Fisher, 1963; Lorbeer, 1971; Graebner *et al.*, 1984; Morino and Luck, 1992; Forster and Luck, 1997; Luck *et al.*, 1997), although the percentage of growers utilizing this tactic has waxed and waned.

The success of this pest-suppression tactic in southern California citrus has led to the development of commercial insectaries selling natural enemies (Dietrick, 1981; Hunter, 1997) or insectaries sustained by citrus protective districts, e.g. the Fillmore Citrus Protective District, a grower cooperative (Smith and Armitage, 1931; Lorbeer, 1971; Graebner *et al.*, 1984). Historically, many of the insectaries had also been organized by regional governmental organizations within southern California, i.e. by the County Agricultural Commissioner, which provided the land, supervised the building, maintained the facilities and produced and distributed the natural enemies (Smith and Armitage, 1931). The cost of the facilities and the natural-enemy production was borne by the citrus industry in that county or district via an initial assessment and by subsequent levies on each grower's annual production of citrus fruit (Smith and Armitage, 1931). With the advent of synthetic organic pesticides in the 1950s, however, the county-based insectaries disappeared and only the private and grower-owned insectaries remained (Graebner *et al.*, 1984). More recently, these grower-owned insectaries have also become associated with licensed pest-control advisers, who provide grower clients with information and recommendations about their pest-management options (Lorbeer, 1971; Dietrick, 1981).

One consequence of augmentative biological control's initial success in citrus was its subsequent development and spread to other citrus pests and their natural enemies and to other commodities in a variety of locations and contexts (e.g. Flanders, 1930, 1942; Voegelé *et al.*, 1975; Stinner, 1976; Hassan, 1981; Ables and Ridgway, 1981; Ridgway and Vinson, 1976; King *et al.*, 1985; Bigler, 1986; van Lenteren and Woets, 1988; Ridgway *et al.*, 1998). Pesticide resistance and the environmental impacts of pesticides, along with increasing public concern and growing regulations against pesticide use, have been major factors stimulating the development of alternative pest-control tactics. Augmentative biological control has been an important alternative pest-suppression tactic.

Augmentative biological control consists of three elements: (i) the mass production of an augmentative biological control agent(s) and its economics; (ii) the agent's release and impact on a target's population density in the field – that is, the mechanics of release, along with the ecology and population dynamics of the agent and its host or prey; and (iii) the suppression of a pest complex as part of a sustainable pest-management programme at a specific geographical location, while producing a crop economically.

Fundamental to an augmentative biological control programme is the development of a means of producing the natural enemy inexpensively and in large numbers, while maintaining their quality at a level for them to be effective. Traditionally, this entails developing a means of producing: (i) the host plant or plant product; (ii) the host insect; and (iii) the natural enemy or enemies (Flanders, 1949; Fisher, 1963). Historically, the development of a production system has involved serendipity, practical experience and tenacity, coupled with a great deal of 'art' or 'intuition' gained from practical experience, along with experimentation to verify the hunches and intuition that have been gained from observation and experience (e.g. Flanders, 1930, 1949; Finney and Fisher, 1964).

Economics of citrus pest management

That an integrated pest management (IPM) approach involving augmentative biological control has practical value to the management of California's citrus pests can be seen in the response by a group of citrus growers who formed the Fillmore Citrus Protective District in 1922 (Graebner *et al.*, 1984). This district was organized to spread the cost of pest management among its grower members and thereby reduce each grower's individual pest-management costs while increasing pest control's reliability through the use of natural enemies. Thus, it is the cost-effectiveness of the whole IPM programme that is the important criterion, not the cost of controlling each individual pest species or of producing each natural-enemy species (Graebner *et al.*, 1984). For example, in the decade between 1971 and 1980, the average Protective District grower spent \$71.88 ha^{-1} annually for pest control. This contrasted with the average pest-control costs of \$362.50 ha^{-1} for a non-member – that is, an orange grower in the same Ventura County location of California but who does not belong to the district. In the last year of the decade, 1980, a member grower paid \$32.50 ha^{-1} in assessment fees and spent an additional \$60 ha^{-1} for the control of various pests, such as weeds, ants and brown-rot control (a fruit rot, *Phytophthora* spp.) (Weppler, 1998). Thus, the \$92.50 expended by a grower member for pest control during the last year of the decade was still substantially less than the decade average (\$362.50 ha^{-1}) paid by the average, non-member grower during the entire decade. This was the case even though producing the soft-scale parasitoid, *Metaphycus helvolus* (Compere) (Hymenoptera: Encyrtidae), to control black scale, *Saissetia oleae* (Olivier) (Homoptera: Coccidae), is more expensive than a single pesticide application for this pest, as judged by the parasitoid's market price as charged by the commercial insectaries (Cranshaw *et al.*, 1996). However, it is the disruptive nature of these pesticide applications for black- and California red-scale suppression that causes the overall increase in pest-management costs. The application of broad-spectrum pesticides disrupts the natural enemies, which then requires additional pesticide use, i.e. the pesticide treadmill. Thus, it is the total pest-management cost for economic suppression that is the criterion, not the cost of producing a particular natural enemy or the cost of suppressing a particular pest.

Many pest-management researchers and practitioners will tend to dismiss the Fillmore Citrus Protective District results as unique to this specific crop and locality and assume that it has little applicability to other situations or circumstances. While the particular solutions discussed above are specific to certain citrus regions in southern California, 'such thinking misses the point. The specific solution is not as important as the process ... What is important are the factors and process by which these people recognised, approached, confronted, and solved their problems' (Graebner *et al.*, 1984).

Lessons from Augmentative Biological Control with *Trichogramma*

History of use and mass production of *Trichogramma*

Trichogramma species (Hymenoptera: Trichogrammatidae), which have become the most widely released augmentative biological control agent worldwide (Li, 1994; Smith, 1996; Chapter 1), are the quintessential augmentative biological control agents. They are easily reared in large numbers at low cost and they parasitize a variety of lepidopteran pests. Smith (1996) reported that *Trichogramma* species have been used against pests in some nine commodities on 32 million ha, but this might be an overestimate (see Chapter 1). The largest users have been the People's Republic of China (Huffaker, 1977; Li, 1994) and the former USSR (Beglyarov and Smetnik, 1977). The best examples documenting the effectiveness of these parasitoids as augmentative biological control agents, however, are those by Voegelé and his colleagues in France (Voegelé *et al.*, 1975); by Hassan and his colleagues in Germany (Hassan, 1981); and by Bigler and his colleagues in Switzerland (Bigler, 1986). These programmes involve the suppression of the one to two generations of the European cornborer, *Ostrinia nubialis* Hübner, occurring in maize in northern Europe.

The first use of *Trichogramma* of which we are aware arose from an attempt to release and establish two exotic species from Austria for the control of the exotic brown-tail moth, *Nygmia phaerorrhoea* (Donovan) (= *Euproctis chrysorrhoea* L.) (Lepidoptera: Lymantridae), in the north-eastern USA during the early 1900s (Howard and Fiske, 1911, pp. 256–260) (but see Decaux, 1899). An endemic American *Trichogramma* species, *T. minutum* Riley (= *T. pretiosum* Riley (see Pinto, 1998)), was also collected from brown-tail moth egg masses in the north-eastern USA. Both the American and the European species were reared on brown-tail moth egg masses and the parasitized eggs were stored at cool temperatures during the winter to synchronize their emergence with the presence of the moth's egg masses in the field. In 1908/9, large numbers of the European species were reared and released but, as expected from laboratory observations, these releases were unsuccessful. *Trichogramma* had difficulty in penetrating the chorion of the moth eggs or reaching the lower layers of the multilayered, setae-covered egg mass.

It was the development of a mass-production system for *Trichogramma* by Flanders (1930), however, that finally spurred the use of these parasitoids as augmentative biological control agents. His development of a production system for this wasp was stimulated when codling-moth eggs were detected as being heavily parasitized by a *Trichogramma* species in 1926 in a southern California walnut grove. This level of parasitization was thought to have arisen from the presence of eggs of a migrating butterfly, the painted lady, *Vanessa cardui* L. (Lepidoptera: Nymphalidae), which laid its eggs on herbaceous species in spring, especially in disturbed habitats (Scott, 1986). Flanders assumed that the availability of these butterfly eggs early in the season allowed *Trichogramma* to parasitize and build up its density on them and then move on to codling-moth eggs. Thus, Flanders reasoned, if these parasitoids could be reared in sufficient numbers early in the season and released to coincide with the codling moth's period of oviposition during the first generation, the moth might be suppressed to subeconomic densities (Flanders, 1930).

The *Trichogramma* species that Flanders (1934) encountered on *Vanessa cardui* was probably *T. pretiosum*, a species usually associated with hosts in herbaceous and brushy habitats in western North America, whereas that parasitizing codling-moth eggs in the west was probably *T. platneri* Nagarkatti, a species common on the eggs of arboreal Lepidoptera (Pinto, 1998). *T. platneri* is a member of the *T. minutum* complex. *T. minutum* Riley is the eastern arboreal counterpart of *T. platneri*, the western member of the complex. They cannot be distinguished from each other morphologically but they are reproductively incompatible, even though they readily interbreed (Stouthamer *et al.*, 2000). These two species are reared by a number of insectaries. However, it is impossible to determine which species comprises a culture – much less whether the culture consists of a single species. If *T. minutum* is released augmentatively in western North America and encounters *T. platneri* in the field, the two species will interbreed but no female offspring will be produced from such matings. Thus, if the eastern species is released within the western species' distribution or vice versa, the population densities of both species will be severely depressed because of the absence of female offspring (Stouthamer *et al.*, 2000). This emphasizes the importance of understanding the systematic infrastructure of biological control agents, as has been repeatedly emphasized by a number of biological control workers (e.g. Rosen, 1986; Gordh and Beardsley, 1999; Stouthamer *et al.*, 2000).

After Flanders (1934) tested several hosts on which to mass-rear the wasp, including the Mediterranean flour moth, *Anagasta* (*Ephestia*) *kuehniella* (Zeller) (Lepidoptera: Pyralidae), the potato tuber moth, *Phthorimaea operculella* (Zeller) and the Angoumois grain moth, *Sitotroga cerealella* (Oliver) (Lepidoptera: Gelechiidae), he chose *S. cerealella* eggs reared on wheat kernels for mass-producing *Trichogramma*. The total production per unit weight of grain reached its maximum much more quickly with wheat than with maize kernels. However, he maintained his small cultures on maize because they required less handling of equipment to maintain the small colony. Thus, the rearing system he employed depended on his rearing objective, a part of which sought to minimize rearing and maintenance costs. He eliminated *A. kuehniella* eggs as a host for *Trichogramma* because it was much more susceptible to larval parasitism and its webbing habits caused problems in handling the culture (Flanders, 1930). Better sanitary methods and rearing techniques have now minimized these latter factors as problems and *A. kuehniella* eggs are also used for mass-producing *Trichogramma* (e.g. Voegelé *et al.*, 1975; Bigler, 1986). The eggs of these two moths are the principal hosts used to mass-rear *Trichogramma* species except in the People's Republic of China (Smith 1996). Eggs of the giant silkworms, *Saamia cynthia* (Drury) (Lepidoptera: Saturnidae) and *Antherea perniyi* (Gnérin-Mádneville) (Lepidoptera: Saturnidae), and the rice-grain moth, *Corcyra cephalonica* (Lepidoptera: Pyralidae), are the principal hosts used in the People's Republic of China (Huffaker, 1977).

Trichogramma size, quality and reproductive success

It has been known for some time that the size of the host used by *Trichogramma* or the number of *Trichogramma* emerging from a host influences the size of the emerging wasps. For example, Howard and Fiske (1911) noted that the size of emerging wasps was related to the number of wasps with which they emerged; in this case the host was the browntail moth. Flanders (1930), in his monograph on the mass-rearing of *Trichogramma*, mentioned that the host egg used for rearing this wasp influenced the size, longevity and fecundity of the emerging adult *Trichogramma*. Similarly, Salt (1940) noted that host size influenced the size of the emerging adult: the smaller the host, the smaller the wasp. However, it was Klomp and Teerink (1962, 1967) who showed, in an elegant set of experiments, that the wasp measured a host's size and laid a characteristic number of eggs in that host. Larger hosts received more eggs than smaller hosts. Schmidt and Smith (1985, 1987) confirmed Klomp and Teerink's (1962, 1967) findings by showing that the wasp measured a host egg during its initial transit across the egg. When Klomp and Teerink (1962) moved their *Trichogramma* female from the initial host to either a larger or a smaller host following the wasp's initial transit across the first host, the female laid a clutch of eggs in the second host characteristic of that normally laid in the host from which she was moved. Thus, she had measured the host's size before she was moved. That there is a characteristic number of wasps emerging from an egg of a given host species can be seen clearly in field data, such as those of Oatman and Platner (1971, 1978) and their collaborators (Oatman *et al.*, 1983).

To understand the role of nutrition in relation to host quality and the quality of the resulting *Trichogramma* offspring, Barrett and Schmidt (1991) investigated the amount of amino acids present in several hosts used by *T. minutum* and related them to the amount of amino acids present in the emerging wasp. They found that *Sitotroga* eggs, which typically issue a single *Trichogramma* offspring, contained an average of 2.2 μg of total amino acids and produced a wasp that contained an average of 1.0 μg of amino acids. In contrast, a *Manduca sexta* (L.) (Lepidoptera: Sphingidae) egg contained 94 μg of amino acids on average, after that present in the chorion had been subtracted. Using the amino acid content of a *Trichogramma* emerging from an *S. cerealella* egg as the minimum required for producing a wasp, they suggested that 80–90 *T. minutum* should be produced from an *M. sexta* egg. However, in their laboratory experiments, Schmidt and Smith (1987) only obtained ten to 20 *T. minutum* wasps from an *M. sexta* egg, depending on the type of experiment they were conducting. They found that the number of *Trichogramma* eggs allocated to a host depended on the degree of host exposure – that is, the amount of the egg's surface that was exposed to the wasp during its initial transit across the host. If the egg touched its neighbours on all sides, it had less exposed surface and the wasp encountered the sharp junction between the host and its adjoining neighbour sooner than if it was a solitary host. The wasp thus laid fewer eggs in the host that touched its neighbours. However, with the solitary host, the wasp covered more distance before it encountered the junction between the round, solitary host and the substrate on which it was laid. The experimental findings of Schmidt and Smith (1987) with solitary *M. sexta* eggs match the results obtained by Oatman and Platner (1971, 1978) for this same, normally solitary host during a 3-year field study in southern California tomato fields. Oatman and Platner (1971, 1978) reported an average of 21.1, 26.7 and 26.3 *T. pretiosum* per *M. sexta* egg, respectively, for each of the years in their study. Thus, the number of eggs a *Trichogramma* allocates to this host (20–30 eggs (Oatman and Platner, 1971, 1978)) is nowhere near the number that this host can support, i.e. 80–90, based on the nutritional requirements of a *Trichogramma* emerging from a *Sitotroga* egg (Barrett and Schmidt, 1991).

Is bigger better?

So why should *Trichogramma* measure host size and regulate the number of eggs it lays in a host? It has been shown that wasp size, up to a point, influences a female's success in encountering hosts in the field: the larger the female parasitoid, the more likely she will encounter a host and, therefore, reproduce (Fig. 17.1; Kazmer and Luck, 1995; Bennett and Hoffmann, 1998). Kazmer and Luck (1995) conducted a set of field experiments to determine whether the ability of *T. pretiosum* females to find hosts in a southern California tomato field was related to their size. They estimated the size distribution of searching females by sampling a host population in a southern California tomato field. During a single day, they collected parasitized hosts by randomly sampling tomato plants in their experimental plots and collecting from them the eggs of the cabbage looper, *Trichoplusia ni* (Hübner) (Lepidoptera: Noctuidae), the tomato fruit worm, *Helicoverpa* (= *Heliothis*) *zea* (Broddie) (Lepidoptera: Noctuidae), and

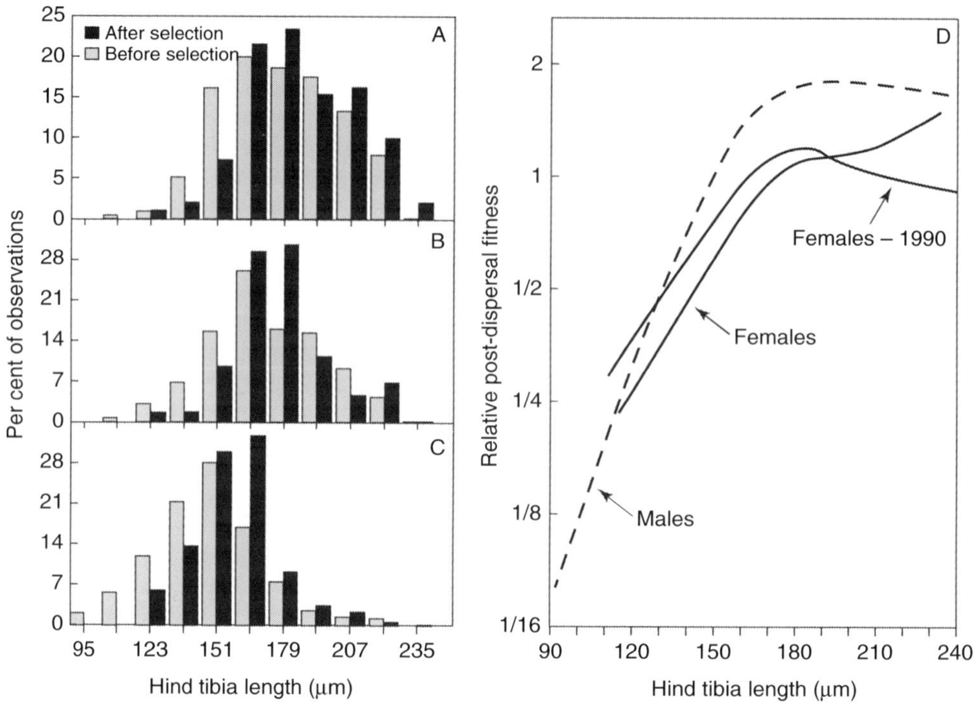

Fig. 17.1. The effect of wasp size on their ability to find hosts and mates in the field. (a)–(c) show the frequency distribution of *Trichogramma pretiosum* Riley emerging from field-collected eggs, i.e. 'before selection' for dispersal ability (light grey bars) and sizes of adults dispersing to egg cards placed in the field after wasp emergence, i.e. 'after selection' (dark grey bars). Hind tibia lengths were used as an index of wasp size: (a) data for females in 1988, (b) for females in 1990, and (c) for males in 1990. (d) Shows the data expressed as the relative success of wasps in locating hosts or mates as a function of wasp size. (Redrawn from Kazmer and Luck, 1995.)

the tomato hornworm, *M. sexta*. These are the species of moths whose eggs are used by *T. pretiosum* in this cropping system (Oatman and Platner, 1971, 1978).

After collecting them in the field, the eggs were returned to the laboratory and individually isolated in gelatin capsules for wasp emergence. Each emerged wasp was then mounted on a microscope slide and its hind tibia length was measured as an index of its size. The frequency distribution of wasp sizes estimated the size distribution of wasps emerging and searching for hosts in the tomato field. To obtain an estimate of the size of wasps finding hosts in the field, these workers glued 15–35 irradiated *T. ni* eggs on a card. When the sampled generation was due to emerge in the field (6–8 days after the initial sample), 400 egg cards with *T. ni* hosts were attached to tomato plants in the field and the cards were inspected daily for ovipositing wasps. *Trichogramma* that appeared on these cards were collected as adults and returned to the laboratory and their hind tibia lengths were measured. The size distribution of female wasps emerging from the field-collected eggs was then compared with that of wasps appearing at the egg cards in each of 2 years. If a female's size was unrelated to her ability to locate hosts in the field, we would expect the size distribution of females emerging from field-collected eggs and those encountering hosts in the field to be the same. They were not. The size of females appearing at the egg cards was significantly larger than that of females emerging from the field-collected eggs, suggesting that the smaller females were less able to find hosts in the field than their larger counterparts (Fig. 17.1A, B). Thus, it appeared that the larger wasps had an advantage in finding more hosts, which probably translates to the production of more offspring.

Assuming that the behaviour involved in allocating offspring number to a host is heritable, the results of this field experiment imply that the larger females left more progeny in future generations – that is, on average they parasitized more hosts in the field. Kazmer and Luck (1995) tested this idea by using these size distributions to compare the reproductive potential, i.e. fitness, of small versus large females (i.e. their ability to encounter hosts). A female's relative fitness increased with female size until a threshold value for hind tibia length of 170 μm, after which it levelled off (Fig. 17.1D). This implies that females with a hind tibia length larger than 170 μm do not have a reproductive advantage over those that have a hind tibia 170 μm long. A similar comparison was made for males emerging from field hosts with those appearing at small cages containing virgin female *T. pretiosum* in the field. Again, large males were more successful at encountering virgin females in the field up to a point (Fig. 17.1C). Several other studies with different *Trichogramma* species (Bennett and Hoffmann, 1998) or with different parasitoid species have shown similar results (Visser, 1994; West *et al.*, 1996).

In general, the hind tibia lengths of female *T. pretiosum* emerging from hosts parasitized in the field, i.e. those hosts listed by Pinto (1998), suggest that the size range of females emerging from *S. cerealella* and *E. kuehniella*, i.e. 120–140 μm, lie in the bottom quartile of the size range for *T. pretiosum* females emerging from their field hosts (Bai *et al.*, 1992; Fig. 17.2). Furthermore, if the size of the emerging offspring were unimportant,

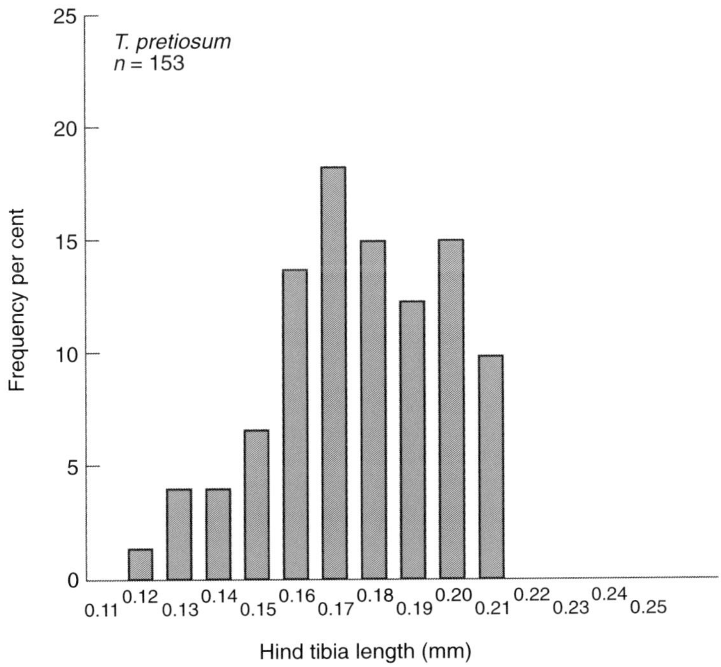

Fig. 17.2. The size distribution of *Trichogramma pretiosum* Riley emerging from the eggs of various lepidopteran species, including *Argraulis vanillae* (L.) (Nymphalidae); *Vanessa* sp. (Nymphalidae); *Helicoverpa* (=*Heliothis*) *zea* (Boddie) (Noctuidae); *Spodoptera exigua* (Hübner) (Noctuidae); *Trichoplusia ni* (Hübner) (Noctuidae); *Manduca sexta* (L.) (Sphingidae); Hesperidae; unknown Lepidoptera. (Redrawn from Bai *et al.*, 1992.)

we would not expect to see the right-handed skew in wasp size that occurs in these results (Fig. 17.2). The number of eggs potentially available in a female's ovaries (Bai *et al.*, 1992; Smith, 1996; Honda and Luck, 2001), along with her walking speed (Honda and Luck, 2001), is known to increase with wasp size. These results suggest that the average per capita efficiency of *Trichogramma* in an augmentative biological control programme could be enhanced two- to threefold if larger wasps were used. However, the cost of producing them must be less than two to three times that of producing wasps from *S. cerealella* or *E. kuhniella* eggs (e.g. Bai *et al.*, 1992). Marston and Ertle (1973) estimated that the cost per wasp of producing larger *Trichogramma* in the desired size range is about 40 times that of producing them from *S. cerealella* eggs. While these costs can be reduced substantially with recent improvements in mass production and implementation, the production costs for the larger wasps still seems to preclude their economic production.

Can small size be compensated by release of larger numbers of parasitoids?

Perhaps, in some cases, releasing higher rates of smaller wasps can compensate for the lower encounter rates manifested by the smaller wasps (Stinner *et al.*, 1974). However, the use of smaller *Trichogramma* can have consequences for the success of a biological control programme other than encounter rates. As has been widely documented, lifetime fecundity and the potential egg load, i.e. the number of mature eggs present in the ovaries, also increase with wasp size (e.g. Waage and Ng, 1984; Bai *et al.*, 1992; Bourchier *et al.*, 1993). This can be important in the biological control of pests that lay their eggs in masses or clusters (Honda and Luck, 2001). For example, in California, *T. platneri* is a frequent constituent of the natural-enemy complex associated with two sporadic avocado pests, the avocado leaf-roller, *Amorbia cuneana* Walshingham (Lepidoptera: Tortricidae), and the omnivorous looper, *Sabulodes aegrotata* (Guneeè) (Lepidoptera: Geometridae) (Oatman *et al.*, 1983). In evaluating *T. platneri*'s potential as an augmentative biological control agent for these pests, Honda and Luck (2001) found that, for *T. pretiosum* reared on *S. cerealella* or *E. kuehniella*, several females must encounter an egg mass or egg cluster if a substantial number of the eggs in these masses or clusters are to be parasitized. An *A. cuneana* female lays about 30 eggs per egg mass, whereas an *S. aegrotata* female lays about 14 eggs per egg cluster. *T. platneri* reared on *S. cerealella* (i.e. a small wasp) lays an average of 1.5 eggs per *A. cuneana* egg and parasitizes 16 of the 30 eggs in an average *A. cuneana* egg mass, whereas wasps reared on *T. ni* eggs (a large wasp) parasitize 26 of the 30 eggs in an average *A. cuneana* egg mass. Similarly, a *T. platneri* female reared on *S. cerealella* (a small wasp) lays five eggs per *S. aegrotata* egg and parasitizes one or two of the 14 eggs in an average *S. aegrotata* egg cluster, whereas wasps reared on *T. ni* eggs (a large wasp) parasitize five of the 14 eggs in average egg cluster. Thus, large wasps are more effective in parasitizing the eggs of both moth species. But note that, even with large wasps, only one-third of the eggs in an average *S. aegrotata* egg cluster are parasitized.

This disparity is even more pronounced when we compare a small versus large wasp's investment (time) in parasitizing an *S. aegrotata* versus an *A. cuneana* egg: *S. aegrotata* eggs are more difficult for even large *T. platneri* to parasitize. It takes a large wasp about 54 min to drill, drum and lay five eggs, the normal clutch size allocated to an *S. aegrotata* egg; but only 28% of the first attempts to parasitize an egg are successful. Another 50% of the parasitoids are successful on their second attempt, adding another 21 min to the time each wasp invests in parasitizing a host egg. The remaining 21% are successful on their third attempt, but that adds another 25 min to a wasp's investment. If we take the weighted average of the time a large wasp invests in allocating each of its five offspring to an *S. aegrotata* egg, it takes about 15 min per wasp egg. In contrast, a large *T. platneri* parasitizing an *A. cuneana* egg mass is always successful on its first attempt and it invests 5 min for each of the

1.5 eggs it allocates to an *A. cuneana* egg. The difference in the time that a small *T. platneri* invests for each egg it lays in these two host species is even greater. A small wasp can expect to invest 20 min for each of the five eggs it lays in an *S. aegrotata* egg, whereas it still invests the same 5 min for each of the 1.5 eggs it lays in an *A. cuneana* egg.

Effect of host quality on acceptance and oviposition by parasitoid

This disparity in the value of the two hosts is also reflected in the percentage of a wasp's egg load that is expended on each host species (Honda and Luck, 2001). A large *T. platneri* retains 39% of its egg load when it finishes parasitizing an *S. aegrotata* egg cluster, whereas it retains 24% of its egg load when it finishes parasitizing an *A. cuneana* egg mass. This difference in egg-load retention is even more pronounced in the smaller *T. platneri*. The small wasp retains 77% of its egg load if it is exposed to an *S. aegrotata* egg cluster, whereas it retains 24% of its egg load if it is exposed to an *A. cuneana* egg mass. This occurs even though the size of a wasp offspring arising from an average clutch allocated to an egg of either host species is the same and the female offspring show the same relationship between their size and their egg loads. Thus, from the parental wasp's perspective, there appears to be little difference in reproductive value among the wasp offspring produced from these two hosts other than the time that she must invest in obtaining them, i.e. in parasitizing each host species. Honda and Luck (2001) interpreted these results as suggesting that parasitizing *S. aegrotata* eggs was less valuable per unit time than parasitizing *A. cuneana* eggs. Thus, after allocating a few clutches to an *S. aegrotata* egg cluster, a female *T. platneri* leaves the host egg cluster because it is more likely to obtain better returns for her searching time if she seeks hosts other than *S. aegrotata* (see Stephens and Krebs, 1986). Honda and Luck (2001) also concluded that it is unlikely that augmentative releases of *T. platneri* would suppress *S. aegrotata* to subeconomic densities on its own.

Honda and Luck (2001) viewed the pattern of egg retention as an index reflecting the reticence that a *Trichogramma* species manifested in exploiting a host species. It is related to the reproductive value that a female obtains in investing its time in exploiting a particular host. This reticence was revealed using a modification of the direct observational techniques first outlined by van Dijken *et al.* (1986) to evaluate three strains of *Trichogramma* species for use as augmentative agents against several pests of cabbage in The Netherlands. This evaluation technique was also used by van Bergeijk *et al.* (1989) to determine why *Trichogramma brassicae* Bezdenko (= *maidis* Pentureau et Voegele) reared on *A.* (= *E.*) *kuehniella* for more than a few generations became less effective in suppressing the Swiss populations of European cornborer in 1980. Bigler (1986) had developed a mass-release programme against this borer, but in one year, 1980, the programme was ineffective (see Chapter 19 for more details). By changing the rearing method for *T. brassicae*, the quality of the parasitoid was dramatically improved. Subsequently, a parasitoid line was established that was continuously reared on the European cornborer under semi-natural conditions in a greenhouse. This line supplied the wasps that were mass-produced on *A. kuehniella* eggs for the five generations required to obtain sufficient numbers for release (van Bergeijk *et al.*, 1989). Van Bergeijk *et al.* (1989) used behavioural observations to determine why the control failure occurred. They found that the number of encounters, drills and parasitizations of *Ostrinia nubilalis* Hübner (Lepidoptera: Pyralidae) eggs or egg masses declined with the number of generations that *T. brassicae* had been reared on *A. kuehniella* eggs. *T. brassicae* reared on *O. nubilalis* readily accepted *A. kuehniella* eggs, whereas only 30% of those reared on *A. kuehniella* eggs for 12 or more generations accepted *O. nubilalis* eggs. Moreover, of those that accepted, i.e. oviposited in, this host, only 35% of them yielded wasp offspring. Van Bergeijk *et al.* (1989) interpreted their results as arising from two processes. First, because food for the wasp offspring was limited in *A.*

kuehniella eggs (see Barrett and Schmidt, 1991), *T. brassicae* produced on this host were smaller and less vigorous. Thus, the smaller wasps were likely to have had more difficulty parasitizing *O. nubilalis* egg masses. Secondly, the increasing rejection rate of *O. nubilalis* eggs by *T. brassicae* also had a genetic basis. They pointed out that the rejection rate of *O. nubilalis* eggs increased with the number of generations that *T. brassicae* had been reared on *A. kuehniella* eggs. Presumably, the size of the wasps in the first generation that emerged from *A. kuehniella* eggs was similar to that of wasps emerging after 12 generations on this host. If this was indeed the case, then wasp size *per se* could be eliminated as the only explanation for the decrease in acceptance of *O. nubilalis* eggs. This, coupled with the failure of some parasitized hosts to issue wasp offspring, suggested that both phenotypic (i.e. deterioration) and genetic changes (i.e. a failure to recognize and develop in *O. nubilalis* eggs) were responsible for the change in acceptance and suitability of *O. nubilalis* eggs to *T. brassicae* reared on *A. kuehniella* eggs.

Summary

While trial and error, serendipity, experience and economics play major roles in developing an augmentative biological control programme, laboratory and field experiments must still form the foundation for such programmes (e.g. Taylor and Stern, 1971; Parker and Pinnell, 1974; van Dijken *et al.*, 1986). They provide the framework in which to evaluate the likely prospects for a programme's success or for modifying elements in a programme that would enhance its success. Traditionally, augmentative biological control has been viewed as simply a matter of releasing sufficient numbers of a biological control agent, such as *Trichogramma*, to ensure that a high proportion of the potential host population is encountered. This assumes that the only factor determining the success of an augmentative biological control programme is the rate at which hosts are encountered. Growing evidence suggests otherwise. The degree of biological control that can be expected is much more likely to involve the particular attributes manifested by these hosts in terms of the quality of the parasitoid offspring arising from them. The particular host attributes assessed by a parasitoid as part of her 'decision' to allocate offspring are specific to and vary with the host species, parasitoid species, parasitoid state, host condition and circumstance. Thus, host attributes such as egg size, stage of a developing embryo within the host egg, chorion hardness or multilayered egg masses, i.e. those attributes associated with host acceptance by *Trichogramma* species, are but examples. Each parasitoid–host interaction is likely to differ and each interaction needs to be assessed.

The experimental approach advocated here involves behavioural observations, but observations that are tied to the size and sex of the resulting offspring and the degree to which the parental female is willing to exploit a particular type of host resource. This latter aspect measures the reticence manifested by a parental female in exploiting a host resource in terms of the proportion of her egg load she is willing to lay in that host type (e.g. species, size, age, stage of development, etc.). A parasitoid will often exploit a few marginal hosts if she has had little success in encountering hosts, but such hosts will seldom be heavily exploited. Thus, the reticence in host use allows host preferences to be assessed and ranked. This then determines the attributes of those hosts that are most likely to be exploited in the field, i.e. whether augmentative biological control is likely to succeed. This ranking of host characteristics is also the basis for assessing the status of the host (pest) population in the field.

The behavioural approach we advocate is rooted in a parental female's reproductive success – that is, in terms of her offspring's reproductive prospects (= a parental female's reproductive success). Given a parental female with a particular set of behavioural attributes (e.g. how readily she accepts hosts of a given size or how willing she is to invest time in exploiting a specific host stage), her reproductive success is defined in terms of the success that her offspring manifest in reproducing when compared with the prospects of those produced by a parental female with differing attributes (e.g. in accepting larger or smaller hosts) (Dawkins, 1982, p. 179). Thus, a wasp's reproductive fitness can be viewed as having three components: clutch size (= number of eggs laid per host), her lifetime production of clutches and her influence on the fitness of her offspring arising from these clutches (e.g. offspring size and sex). In solitary parasitoids, host acceptance often involves the choice of host size on which to lay female versus male offspring (Charnov, 1982; Godfray, 1994; van Alphen and Jervis, 1996). This component is important to augmentative biological control with solitary, idiobiont parasitoids because offspring usually develop on or in a single host. A female thus influences the reproductive prospects of her offspring through her choice of a host's developmental stage and size, which is often strongly correlated with the sex of the offspring she allocates to it (Charnov, 1982; Waage and Ng, 1984; Luck and Podoler, 1985; Godfray, 1994; Kazmer and Luck, 1995; van Alphen and Jervis, 1996; Luck *et al.*, 2000).

The theoretical underpinning of this behavioural approach rests on the assumption that, evolutionarily, those lineages that have successfully reproduced in the past are likely to be those that manifest host-selection behaviours that maximize their offspring's reproductive prospects under current circumstances. Moreover, these are likely to be the behaviours that are present in a population of potential biological control agents. These host-acceptance behaviours then set the limits within which these agents can be expected to exploit a particular host resource. They set the conditions that influence a female's choice of hosts, in terms of both the degree to which she will exploit them, as in the case of *T. platneri* parasitizing *S. aegrotata* eggs, and the number and sex of offspring she will allocate to each of them (Waage and Ng, 1984; Luck *et al.*, 2000). This later, via offspring size and associated egg load, influences their likely encounter rates with hosts after they emerge, as for example with *T. pretiosum* parasitizing the eggs of *T. ni*, *H. zea* and *M. sexta* in the tomato fields of southern California (Kazmer and Luck, 1995). Offspring inheriting the behaviours associated with these choices, i.e. host acceptance or clutch size in relation to host size, are those most likely to leave offspring that will themselves reproduce in subsequent generations. These behavioural 'choices' form the basis for the behavioural assessment utilized by van Dijken *et al.* (1986) to evaluate host recognition and host acceptance (e.g. van Bergeijk *et al.*, 1989). It is the approach we have advocated here but with the addition of an evolutionary perspective. Thus, a natural concordance exists between a parasitoid's reproductive fitness and biological control – namely, how readily a parental female oviposits in a host directly translates to the prospects for biological control, both augmentative and classical – and it arises historically from the reproductive success that the lineage has had when choosing such hosts.

References

Ables, J.R. and Ridgway, R.L. (1981) Augmentation of entomophagous arthropods to control pest insects and mites. In: Papavizas, G.C. (ed.) *Beltsville Symposia in Agriculture Research. (5) Biological Control in Crop Production*. Allenheld, Osmun, London, pp. 273–303.

Bai, B., Luck, R.F., Forster, L., Stephens, B. and Janssen, J.A.M. (1992) The effect of host size on quality attributes of the egg parasitoid, *Trichogramma pretiosum*. *Entomologia. Experimentalis et Applicata* 64, 37–48.

Barrett, M. and Schmidt, J.M. (1991) A comparison between the amino acid composition of an egg parasitoid and some of its hosts. *Entomologia. Experimentalis et Applicata* 59, 29–41.

Bartlett, B.R. (1974) Introduction into California of cold-tolerant biotypes of the mealybug predator, *Cryptolaemus montrouzieri*, and laboratory procedures for test natural enemies for cold-hardiness. *Environmental Entomology* 3, 553–556.

Beglyarov, G.A. and Smetnik, A.I. (1977) Seasonal colonization of entomophages in the USSR. In: Ridgway, R.L. and Vinson, S.B. (eds) *Biological Control by Augmentation of Natural Enemies*. Plenum Press, New York, pp. 283–328.

Bennett, D.M. and Hoffmann, A.A. (1998) Effects of size and fluctuating asymmetry on field fitness of the parasitoid *Trichogramma carverae* (Hymenoptera: Trichogrammatidae). *Journal of Animal Ecology* 67, 580–591.

Bigler, F. (1986) Mass production of *Trichogramma maidis* Pint. et Voeg. and its field application against *Ostrinia nubilalis* Hbn. in Switzerland. *Journal of Applied Entomology* 101, 23–29.

Bourchier, R.S., Smith, S.M. and Song, S.J. (1993) Host acceptance and parasitoid size as predictors of parasitoid quality for mass-reared *Trichogramma minutum*. *Biological Control* 3, 135–139.

Charnov, E.L. (1982) *The Theory of Sex Allocation*. Princeton University Press, Princeton, New Jersey, 355 pp.

Cranshaw, W., Sclar, D.C. and Cooper, D. (1996) A review of 1994 pricing and marketing by suppliers of organisms for biological control of arthropods in the United States. *Biological Control* 6, 291–296.

Dawkins, R. (1982) *The Extended Phenotype*. W.H. Freeman, San Francisco, 307 pp.

DeBach, P. and White, E.B. (1960) *Commercial Mass Production of the California Red Scale Parasite,* Aphytis lingnanensis. Bulletin 770, California Agricultural Experiment Station, 58 pp.

Decaux, F. (1899) Destruction rationnelle des insectes qui attaquent les arbres fruitiers par l'emploi simultané des insectices, des insectes auxiliaires, et par las prection et l'élevage de leurs ennemis naturels les parasites. *Journal de la Société Nationale d'Horticulture de France* 22, 158–184.

Dietrick, E.J. (1981) Commercial production of entomophagous insects and their successful use in agriculture. In: Papavizas, G.C. (ed.) *Beltsville Symposia in Agriculture Research. (5) Biological Control in Crop Production*. Allenheld, Osmun, London, pp. 151–160.

Ebeling, W. (1950) *Subtropical Entomology*. Lithigtoe Process, San Francisco, 747 pp.

Enock, F. (1895) Remarks on *Trichogramma evanescens* Westw. recorded by the Secretary of the South London Entomological and Natural History Society. *The Entomologist* 28, 283.

Finney, G.L. and Fisher, T.W. (1964) Culture of entomophagous insects and their hosts. In: DeBach, P. (ed.) *Biological Control of Insect Pests and Weeds*. Chapman & Hall, London, pp. 328–355

Fisher, T.W. (1963) *Mass Culture of* Cryptolaemus *and* Leptomastix, *Natural Enemies of Citrus Mealybug*. Bulletin 797, California Agricultural Experiment Station, 37 pp.

Flanders, S.E. (1930) Mass production of egg parasites of the genus *Trichogramma*. *Hilgardia* 4, 465–501.

Flanders, S.E. (1934) Storage production. *Journal of Economic Entomology* 27, 1197.

Flanders, S.E. (1942) Propagation of black scale on potato sprouts. *Journal of Economic Entomology* 35, 687–689.

Flanders, S.E. (1949) Culture of entomophagous insects. *The Canadian Entomologist* 81, 257–274.

Flanders, S.E. (1951) Mass culture of California red scale and its golden chalcid parasites. *Hilgardia* 21, 1–42.

Flint, M.L. and van den Bosch, R. (1981) *Introduction to Integrated Pest Management*. Plenum Press, New York, 240 pp.

Forster, L.D. and Luck, R.F. (1997) The role of natural enemies of California red scale in an IPM program. In: *Proceedings of the International Society of Citriculture VIII International Citrus Congress, 12–17 May 1996, Sun City, South Africa,* Vol. 1, pp. 504–507.

Godfray, H.C.J. (1994) *Parasitoids: Behavior and Evolutionary Ecology*. Princeton University Press, Princeton, New Jersey, 473 pp.

Gordh, G. and Beardsley, J.W. (1999) Taxonomy and biological control. In: Bellows, T.S. and Fisher, T.W. (eds) *Handbook of Biological Control*. Academic Press, New York, pp. 45–55

Graebner, L., Moreno, D.S. and Baritelle, L.L. (1984) The Fillmore Protective District: a success story in integrated pest management. *Bulletin of the Entomological Society of America* 30, 27–33.

Haney, P.B., Morse, J.G., Luck, R.F., Griffths, H., Grafton-Cardwell, E.E. and O'Connell, N.V. (1992) *Reducing Insecticide Use and Energy Costs in Citrus Pest Management*. Publication 15, Division of Agriculture and Natural Resources, Statewide Integrated Pest Management, University of California, Berkeley, 62 pp.

Hassan, S.A. (1981) Mass-production and utilization of *Trichogramma*. 2. Four years of successful biological control of the European corn borer. *Mededelingen van de Faculteit Landbouwwetenshappen, Rijksuniversiteit Gent* 46, 417–428.

Honda, J.Y. and Luck, R.F. (2001) Interactions between host attribute and wasp size: a laboratory evaluation of *Trichogramma platneri* as an augmentative biological control agent for two avocado pests. *Entomologia. Experimentalis et Applicata* 100, 1–13.

Howard, L.O. and Fiske, W.F. (1911) The importation into the United States of the parasites of the gipsy moth and the brown-tail moth. *United States Department of Agriculture Bureau of Entomology Bulletin* 90, 312 pp.

Huffaker, C.B. (1977) Augmentation of natural enemies in the People' Republic of China. In: Ridgway, R.L. and Vinson, S.B. (eds) *Biological Control by Augmentation of Natural Enemies*. Plenum Press, New York, pp. 329–339.

Hunter, C.D. (1997) *Suppliers of Beneficial Organisms in North America*. PM 94–03, Department of Pesticide Regulation, California Environmental Protection Agency, Sacramento, California, 34 pp.

Kazmer, D.J. and Luck, R.F. (1995) Field tests of the size fitness hypothesis in the egg parasitoid *Trichogramma pretiosum*. *Ecology* 76, 412–425.

King, E.G., Bull, D.L., Bouse, L.F. and Phillips, J.R. (1985) Biological control of bollworm and tobacco budworm in cotton by augmentative releases of *Trichogramma*. *The Southwestern Entomologist, Supplement* 8, 198 pp.

Klomp, H. and Teerink, B.J. (1962) Host selection and the number of eggs per oviposition in the egg parasitoid, *Trichogramma embryophagum*. *Nature* 195, 1020–1021.

Klomp, H. and Teerink, B.J. (1967) The significance of oviposition rates in the egg parasite, *Trichogramma embryophagum* Htg. *Archives Néederlandaises de Zoologie* 17, 350–375.

Kot, J. (1964) Experiments in the biology and ecology of species of the genus *Trichogramma* Westw. and their use in plant protection. *Ekologia Polska Series A* 12, 243–303.

Li, L.-Y. (1994) Worldwide use of *Trichogramma* for biological control on different crops: a survey. In: Wijnberg, E. and Hassen, S.A. (eds) *Biological Control with Egg Parasitoids*. CAB International, Wallingford, UK, pp. 37–53.

Lorbeer, H. (1971) Integrated biological control in Fillmore citrus groves. *California Citrograph* 56 (6), 199–201.

Luck, R.F. and Podoler, H. (1985) Competitive exclusion of *Aphytis lingnanensis* by *A. melinus*: potential role of host size. *Ecology* 66, 904–913.

Luck, R.F., Forster, L.D. and Morse, J.G. (1997) An ecologically based IPM program for citrus in California's San Joaquin Valley using augmentative biological control. In: *Proceedings of the International Society of Citriculture, VIII International Citrus Congress, 12–17 May 1996, Sun City, South Africa*, Vol. 1, pp. 504–507.

Luck, R.F., Jenssen, J.A.M., Pinto, J.D. and Oatman, E.R. (2000) Precise sex allocation and sex ratio shifts by the parasitoid *Trichogramma pretiosum*. *Behaviour Ecology and Sociobiology* 49, 311–321.

Marston, M. and Ertle, L.R. (1973) Host influence on the bionomics of *Trichogramma minutum*. *Annals of the Entomological Society of America* 66, 1155–1162.

Morino, D.S. and Luck, R.F. (1992) Augmentative releases of *Aphytis melinus* (Hymenoptera: Aphelinidae) to suppress California red scale (Homoptera: Diaspididae) in southern California lemon orchards. *Journal of Economic Entomology* 85, 1112–1119.

Oatman, E.R. and Platner, G.R. (1971) Biological control of the tomato fruitworm, cabbage looper, and hornworms on processing tomatoes in southern California, using mass releases of *Trichogramma pretiosum*. *Journal of Economic Entomology* 64, 501–506.

Oatman, E.R. and Platner, G.R. (1978) Effects of mass releases of *Trichogramma pretiosum* against lepidopterous pests on processing tomatoes in southern California, with notes on host egg population trends. *Journal of Economic Entomology* 71, 896–900.

Oatman, E.R., Platner, G.R., Wyman, J.A., van Steenwyk, R.A., Johnson, M.W. and Browning, H.W. (1983) Parasitization of lepidopterous pests on fresh market tomatoes in southern California. *Journal of Economic Entomology* 76, 452–455.

Parker, F.D. and Pinnell., R.E. (1974) Effectiveness of *Trichogramma* spp. in parasitizing eggs of *Pieris rapae* and *Trichoplusia ni* in the laboratory. *Environmental Entomology* 3, 935–938.

Pinto, J.D. (1998) Systematics of the North American species of *Trichogramma* Westwood (Hymenoptera: Trichogrammatidae). *Memoirs of the Entomological Society of Washington* 22, 1–287.

Quayle, H.J. (1938) *Insects of Citrus and Other Subtropical Fruits*. Comstock Publishing, Ithaca, New York, 583 pp.

Rabb, R.L., Stinner, R.E. and van den Bosch, R. (1976) Conservation and augmentation of natural enemies. In: Huffaker, C.B. and Messenger, P.S. (eds) *Theory and Practice of Biological Control*. Academic Press, New York, pp. 233–254.

Ridgway, R.L. and Vinson, S.B. (eds) (1976) *Biological Control by Augmentation of Natural Enemies*. Plenum Press, New York, 480 pp.

Ridgway, R.L., Hoffman, M.P., Inscoe, M.N. and Glennister, C.S. (eds) (1998) *Mass-reared Natural Enemies: Application, Regulation, and Needs*. Thomas Say Foundation in Entomology, Proceedings of the Entomological Society of America, 332 pp.

Rosen, D. (1986) The role of taxonomy in effective biological control programs. *Agriculture, Ecosystem and Environment* 15, 121–129.
Salt, G. (1940) Experimental studies in insect parasitism. VII. The effects of different hosts on the parasites *Trichogramma evanescens* Westw. (Hym. Chalcidoidea). *Proceedings of the Royal Entomological Society, London, Series A* 15, 81–124.
Schmidt, J.M. and Smith, J.J.B. (1985) Host volume measurement by the parasitoid wasp *Trichogramma minutum*: role of curvature and surface area. *Entomologia Applicata et Experimentalis* 39, 213–221.
Schmidt, J.M. and Smith, J.J.B. (1987) Measurement of host curvature by the parasitoid wasp *Trichogramma minutum* and its effect on host examination and progeny allocation. *Journal of Experimental Biology* 125, 271–285.
Scott, J.A. (1986) *The Butterflies of North America*. Stanford University Press, Stanford, California, 583 pp.
Smith, H.S. and Armitage, H.M. (1920) Biological control of mealybugs in California. *California Department of Agriculture Monthly Bulletin* 9, 104–158.
Smith, H.S. and Armitage, H.M. (1931) *The Biological Control of Mealybugs Attacking Citrus*. Bulletin 509, California Agricultural Experiment Station, Berkeley, 74 pp.
Smith, S.M. (1996) Biological control with *Trichogramma*: advances, successes, and potential of their use. *Annual Review of Entomology* 41, 375–406.
Stephens, D.W. and Krebs, J.R. (1986) *Foraging Theory*. Monographs in Behavior and Ecology, Princeton University Press, Princeton, New Jersey, 247 pp.
Stinner, R.E. (1976) Biological control with *Trichogramma*: advances, successes, and potential of their use. *Annual Review of Entomology* 22, 515–531.
Stinner, R.E., Ridgway, R.L. and Morrison, R.K. (1974) Longevity, fecundity, and searching ability of *Trichogramma pretiosum* reared by three methods. *Environmental Entomologist* 3, 558–560.
Stouthamer, R., Jochemsen, P., Platner, G.R. and Pinto, J.D. (2000) Crossing incompatibility between *Trichogramma minutum* and *T. platneri* (Hymenoptera: Trichogrammatidae): implications for biological control. *Environmental Entomologist* 29, 832–837.
Taylor, T.A. and Stern, V.M. (1971) Host preference studies with the egg parasite *Trichogramma semifumatum* (Hymenoptera: Trichogrammatidae). *Annals of the Entomological Society of America* 64, 1381–1390.
Trumble, J.T. and Morse, J.G. (1993) Economics of integrating the predacious mite, *Phytoseiulus persimilis* (Acari: Phytoseiidae), with pesticides in strawberries. *Journal of Economic Entomology* 86, 879–885.
van Alphen, J.J.M. and Jervis, M.A. (1996) Foraging behaviour. In: Jervis, M.A. and Kidd, N.A.C. (eds) *Insect Natural Enemies: Practical Approaches to Their Study and Evaluation*. Chapman & Hall, London, pp. 1–62.
van Bergeijk, K.E., Bigler, F., Kaashoek, N.K. and Pak, G.A. (1989) Changes in host acceptance and host suitability as an effect of rearing *Trichogramma maidis* on a factious host. *Entomologia Experimentalis et Applicata* 52, 229–238.
van Dijken, M.J., Kole, M., van Lenteren, J.C. and Brand, A.M. (1986) Host preference studies with *Trichogramma evanescens* Westw. (Hym. Trichogrammatidae) for *Mamestra brassicae, Pieris brassicae* and *Pieris rapae*. *Journal of Applied Entomology* 101, 64–85.
van Lenteren, J.C. and Woets, J. (1988) Biological and integrated control in greenhouses. *Annual Review of Entomology* 33, 239–269.
Visser, M.E. (1994) The importance of being large – the relationship between size and fitness in females of the parasitoid *Aphaereta minuta* (Hymenoptera: Braconidae). *Journal of Animal Ecology* 63, 963–978.
Voegelé, J., Stengel, M., Schubert, G., Daumal, J. and Pizzol, J. (1975) Premiers résultats sur l'introduction en Alsace sous forme de lâchers saisonniers de l'écotype moldave de *Trichogramma evanescens* Wests. contre la purale du maïs, *Ostrinia nubilalis* Hübner. *Annales de Zoologie, Écologie Animale* 7, 535–551.
Waage, J.K. and Ng, S.M. (1984) The reproductive strategy of a parasitic wasp. I. Optimal progeny and sex allocation in *Trichogramma evanescens*. *Journal of Animal Ecology* 53, 401–415.
Weppler, R.A. (1998) Studies on the rearing of *Metaphycus helvolus* (Compere) (Hymenoptera: Encyrtidae) for augmentative release against black scale, *Saissetia oleae* (Olivier) (Homoptera: Coccidae). Masters thesis, University of California, Riverside, California, 140 pp.
West, S.A., Flanagan, K.E. and Godfray, H.C.J. (1996) The relationship between parasitoid size and fitness in the field, a study of *Achrysocarides zwoelferi* (Hymenoptera: Eulophidae). *Journal of Animal Ecology* 65, 631–639.

18 Towards the Standardization of Quality Control of Fungal and Viral Biocontrol Agents

N.E. Jenkins[1] and D. Grzywacz[2]
[1]CABI Bioscience, Silwood Park, Ascot SL5 7TA, UK; [2]Natural Resources Institute, University of Greenwich, Chatham ME4 4TB, UK

Abstract

An essential feature of the production of all microbial control agents is an effective quality control system. Well-defined product specifications with accompanying quality control procedures help to maximize product performance, ensure product safety, standardize manufacturing costs and reduce the risks for supply failure, thus building user confidence. A production system that does not have a quality control system is one whose output is uncontrolled and a lack of thorough quality feedback can result in batches of product with variable concentrations of active agent. This results in products with variable performance, leading to control failures by users and serious loss of user confidence. Strict quality control procedures are not only essential for product consistency, but also for safety. Where quality control is inadequate, microbial contamination of the final product is inevitable. It must be recognized that quality control procedures can be more complex and technologically demanding than the production procedures themselves. It is largely on the effectiveness of these control procedures that the long-term success of fungal and viral products depends.

Introduction

In promoting the adoption of microbial control agents by farmers, the reliability of the product is a crucial issue in ensuring acceptance and sustained use. In this chapter, quality control of fungal and viral biocontrol agents is discussed; we refer the reader to Lisansky *et al.* (1993) for quality control of *Bacillus thuringiensis* products. The issue of erratic performance of fungal and viral biocontrol agents has been recognized as a significant factor in the limited successful commercialization of these agents (Lisansky, 1997). It has been widely perceived that fungal and viral control agents have not, to date, achieved a level of efficacy comparable with that of their chemical counterparts or with that of the leading bacterial agent, *B. thuringiensis*. Many of the products that have been placed on the market have been characterized as 'weak products with poor efficacy and questionable quality control' (Harris, 1997).

The implementation of an effective quality control system should be at the heart of any programme to develop a successful microbial control product (Shieh, 1989). However, microbial control agents are unlike chemical pesticides in their manufacturing requirements and characteristics. To be effective, microbial con-

© CAB International 2003. *Quality Control and Production of Biological Control Agents: Theory and Testing Procedures* (ed. J.C. van Lenteren)

trol agents based upon fungi and baculoviruses (BV) depend on the successful completion of a complex infection process by the pathogen. Monitoring the quality of microbial control agents, therefore, can be more complex than with chemical insecticides. A biocontrol product that has an infectious agent as the active ingredient is the result of a complex biotic process, whether manufactured using *in vitro* or *in vivo* technology. The quality-assessment procedure of the final product needs, therefore, to extend back to all levels of the production process, including any insects or cell lines used to multiply the agent.

Given the crucial role of quality control in fungal and viral production, it is a matter of concern that standardized protocols for the quality control of these products are not yet widely agreed or accepted, a situation similar to that in the production of natural enemies (Chapters 1, 2 and 19). Even where regulatory procedures for registration of fungal and viral control agents are enforced, no standardized guidelines for quality control procedures are available. Manufacturers of these microbial control products are therefore required to develop their own quality control procedures, which has resulted in a disparity in standards between manufacturers of similar products. Further, in those countries where registration of microbial control products is not enforced, manufacturers are not obliged to develop or conduct any quality control procedures.

There are reports that a number of fungal and viral products are failing to meet acceptable standards (Grzywacz, 1995; Kern and Vaagt, 1996). Unless this matter is addressed effectively, there is a serious danger in these countries that poor-quality products, with their inevitable failures, will erode farmer confidence in microbial control products and significantly retard the promotion of this promising technology.

Production and Quality Control of Fungal Pathogens and Antagonists

Production technology

Fungal products used in biological control cover a wide range of fungal genera and applications, including use as fungal antagonists, plant-growth enhancers and weed- and insect-control agents. Most of these fungi are produced *in vitro*. The scale of production ranges from small-scale agar-based production, such as is used for *Phlebiopsis gigantea*, a product registered in the UK for application to conifer stumps to protect them against *Fomes annosus* (J.E. Pratt, Midlothian, UK, 1999, personal communication), through to industrial-scale units, such as that built by Mycotech Corp., which is capable of producing tonnes of *Beauveria bassiana* product annually (Stephens, 1997).

Methods of production vary considerably. Many are solid-substrate fermentations based on cereal grains, such as rice (Alves and Pereira, 1989; Jenkins *et al.*, 1998), or, occasionally, non-nutritive substrates, such as clay granules (Guillon, 1997). Some encourage the production of conidia on the surface of static liquid culture (Kybal and Vlcek, 1976; Ferron, 1981), while others use deep-tank liquid fermentation to produce mycelial products, blastospores or submerged conidia (McCabe and Soper, 1985; Jackson and Bothast, 1990; Reinecke *et al.*, 1990; Jenkins and Prior, 1993). Arbuscular mycorrhizal fungi are produced on the roots of plants cultivated in glasshouses or growth chambers (J. Parat, Paris, France, 2000, personal communication). Whichever system is used and at whatever scale it is employed, there is a need for strict and standardized quality control procedures to ensure that the product is safe, viable and effective.

Culture maintenance

Once selected for development as microbial control agents (if not before), fungal isolates should be lodged with a recognized culture collection and formally identified by a recognized fungal taxonomist. Taxonomic procedures are becoming more and more complex and it is now generally accepted that some kind of molecular identification is needed in addition to the traditional morphological characteristics formally used to classify fungal species (Bridge and Arora, 1998).

A formal identification will place the fungal isolate within a taxonomic rank, initially at species level. Many fungal species are widely distributed in the environment, and therefore there is also a need to be able to identify the individual isolate that is under consideration. This can be undertaken through techniques, such as DNA fingerprinting (Caetano-Anolles *et al.*, 1991; Schlick *et al.*, 1994; Bridge *et al.*, 1997), which allow the identification of either single strains or groups of very closely related strains. This level of identification is particularly important as it provides a mechanism for tracking the progress and fate of the agent in the environment, a validation check for the purity and accuracy of the formulations and a standard reference that may be used to register or protect individual isolates (Schlick *et al.*, 1994; Edel, 1998; McClintock, 1999; Neal and Newton, 1999).

It has long been accepted that serial subculturing of fungal isolates on agar can result in the loss of certain characteristics of the original isolate (Thomas and Smith, 1994). This can be avoided by careful management of fungal material from the point of original isolation. Once a clean, uncontaminated culture has been obtained, subcultures from this should be made and stored, using as many long-term storage methods as possible. Two methods that are commonly used by large culture collections are lyophilization (freeze-drying), cryopreservation (using liquid nitrogen) and storage at −80°C. All of these techniques require the purchase of relatively expensive equipment but can be contracted out to culture-collection laboratories with experience in their use. Lyophilization or freeze-drying can be very useful as a large number of freeze-dried ampoules can be prepared and stored long-term (> 10 years) at room temperature (Smith and Kolkowski, 1996). Liquid nitrogen or cryopreservation is also an effective method of long-term storage, but maintenance of cultures preserved using this technique requires a regular supply of liquid nitrogen and highly specialized preservation tanks. Cultures preserved using this method are therefore maintained at the site of preservation and must be requested from storage each time that a source of inoculum is required. Storage at −80°C requires a specialized freezer, but may be the most economical option for small companies interested in maintaining their own culture collection. Techniques requiring less sophisticated equipment can also be effective and can be used as a backup to the above techniques. These include silica-gel storage, storage under mineral oil, soil storage and water storage (Smith and Kolkowski, 1996). Some of these techniques are not suitable for all types of fungi. The most appropriate method/methods should be selected for each organism to be preserved. In preparation for mass production, a large stock of fungal material should be prepared and stored using a suitable long-term storage method. Fresh starter material for each production run should then be prepared, using material direct from the stock.

Recent studies on long-term preservation techniques have uncovered some worrying effects on both phenotypic characteristics and genetic stability of some of the fungal isolates studied (Ryan, 1999). Detailed studies on a number of mitosporic fungi have shown that both lyophilization and cryopreservation can cause detectable molecular polymorphisms. This was particularly marked if generalized protocols (as opposed to one that has been optimized for the particular fungus in question) were used for preservation (Ryan, 1999). This highlights the need for careful consideration before placing valuable isolates in long-term storage. In response to this, Ryan *et al.* (2000) have developed a key for use in the selection of fungal preservation techniques.

One safeguard against the potential loss of valuable characteristics of some mycopesticides (particularly mycoinsecticides) is the regular passaging of the fungal isolate through the host or closely related species. However, this process should be carefully controlled to ensure that the population used for passaging is not harbouring latent infections, which may result in cross-contamination with or isolation of the wrong pathogen.

Production process and record keeping

Production processes, downstream processing (harvest and postharvest) and extraction methods vary widely from product to product, so it is not possible to implement a single set of standardized procedures to cover all. However, within any given production process, standardized procedures for process monitoring and recording should be applied to ensure consistent production of reliable, high-quality products.

In countries where product registration is enforced, a complete description of the process used for production of the agent is generally required (McClintock, 1999; Neal and Newton, 1999). In-process monitoring and quality control should be an integral part of this description. Since culture conditions are known to affect a number of aspects of pathogen performance (Burges, 1998b), including virulence (Lane *et al.*, 1991b), storage longevity (Trinci *et al.*, 1990; Lane *et al.*, 1991a; Hong *et al.*, 2002) and durability, careful monitoring and recording of the fermentation process is critical to ensuring a product of consistent efficacy.

Records of all the following parameters should be kept for each production batch and maintained on file so that the information gathered can be back-checked according to batch number as and when required.

Temperature

Temperature has a direct effect on yield; hence fermentations are generally run at or close to the optimum temperature for the growth of the fungal isolate. Since large-scale fermentations generate heat, most large-scale fungal production systems have some degree of inbuilt temperature monitoring and regulation. Temperature should be monitored and recorded over the whole fermentation period.

Moisture content

In solid-substrate fermentations, moisture content should be adjusted to give optimum production (Jenkins *et al.*, 1998) and checked postinoculation to ensure that it does not vary significantly from batch to batch.

pH

Entomopathogenic fungi tend to be tolerant of pH and will grow over a wide pH range (Ibrahim *et al.*, 1993), whereas fungal antagonists have a narrower pH tolerance (Jackson *et al.*, 1991). The degree to which pH is monitored in process will depend largely on the effect that variations in pH have on yield. pH tends to be monitored closely in liquid fermentations but, because of the difficulty in controlling pH in solid-substrate fermentations, it is often disregarded. None the less, records of pH can be a useful reference. In fermentations started with non-sterile media, low pH can be an advantage as it inhibits unwanted bacterial growth (Bartlett and Jaronski, 1988) and, where this is used as a technique in reducing contamination, pH should be carefully recorded and monitored.

Contamination monitoring

All microbial fermentations carry the potential risk of contamination. In *in vitro* systems, the unintentional parallel culture of contaminants with the desired fungal product will inevitably lead to a lower yield of product from the fermentation, due to competition for nutrients and, in some cases, the inhibitory effect of secondary metabolites. If the contaminating organisms are included in the final product, this can lead to dilution of the active ingredient in the packaged product, particularly if contamination has not been closely monitored and quantified. With respect to safety, the presence of contaminants increases the risk of production of potentially harmful organisms. Where production is *in vivo*, however, the presence of other microbes is inevitable; indeed, in the production of arbuscular mycorrhizal fungi, the presence of some associated microorganisms is important for both fungus and plant. In such a system, only microorganisms that are potentially harmful or may be detrimental to the production process are considered to be contaminants.

For all production systems, each stage of the production process must be monitored for the entrance of contaminants to enable the early detection of problems and the identification of the source of contamination should it occur. Careful monitoring at each stage will help in reducing as far as possible the lost production time required to eradicate the problem. Furthermore, results from contamination checks throughout the course of production can be used to show that harmful pathogens have not entered the system (see section on safety). A simple method for contamination monitoring is to take samples of the medium/substrate (pre- and postinoculation) and the inoculum itself and to plate each sample on generalist agar media, such as nutrient agar and Sabouraud dextrose agar (SDA). These check plates should be incubated (at approximately 25°C) for 2–3 days and observed for the appearance of bacterial and fungal contaminants, respectively. Should contaminants appear on any of the plates, the source may be identified by a process of elimination so that remedial action can be taken. If contaminants have entered the system, the batch should be discarded. Clearing contaminated batches soon after discovery reduces the risk of introducing high levels of contamination into the production environment, which may persist to affect subsequent batches.

Safety

The intrinsic safety of any microbial control agent can be evaluated by carrying out the appropriate mammalian and ecotoxicological safety tests. These tests tend to be expensive, but are a requirement of registration of microbial agents in countries where registration of biological agents is enforced (McClintock, 1999; Neal and Newton, 1999). Use of microbial agents for biological control cannot simply be assumed to be safe and some degree of testing should be employed in all cases.

Safety issues also arise in respect of microbial contamination of the final product. Standards for biological purity need to be set; these should include a thorough assessment of the production process to establish that human pathogens are unlikely to enter the system under normal operating procedures. In addition to the standard contamination-monitoring checks described in the previous section, these precautions should be sufficient evidence that harmful microorganisms are unlikely to be present in the product.

Levels of contamination in unformulated products can be quantified using a simple dilution series plated on to a generalist agar medium (Jenkins *et al.*, 1998). In order to set an acceptable level of contaminating organisms, baseline levels of contaminants entering the system during downstream processing and during formulation should be established. Again, in-process contamination-monitoring records can be used as evidence to show that contaminants have not multiplied during the fermentation process.

The final aspect of product safety relates to the genetic and phenotypic stability of the organism being produced. Provided that master cultures have been carefully maintained, there should be no cause for concern. Phenotypic stability should be monitored during routine laboratory manipulation of the isolate, so laboratory personnel should become familiar with the typical characteristics of the isolate and should report any changes observed outside the normal growing characteristics. If doubts arise about fungal efficacy, checks using baseline data from molecular characterization should be carried out to ensure that genetic change is not detectable, although the sensitivity of many of these techniques to detect small genetic changes can be limited. Any changes in phenotypic characteristics indicate a need to draw on fresh material from long-term storage.

Efficacy

Efficacy is the most critical factor in ensuring product performance and long-term acceptance. If a product has high viability and virulence/pathogenicity, the chances of field efficacy are maximized. However, understanding the ecology and the effect of environmental factors on the fungus is essential to ensure reliable field performance of even the highest-quality product.

Viability is easily measured in fungi; most laboratories working with these agents have an established method for assessing product viability. For conidial products, a relatively quick, and probably the most accurate, method is to determine the percentage germination of the conidia. Using a dilute suspension of conidia plated on to a nutritive agar medium, viable and non-viable spores can be counted under a microscope following incubation for a predetermined time and temperature (Goettel, 1987; McClatchie *et al.*, 1994; Hong *et al.*, 1999). Other variations, such as liquid media (Milner *et al.*, 1997) or fluorescence staining (Firstencel *et al.*, 1990; Jimenez and Gillespie, 1990) can also be used. However, the use of colony-forming unit (cfu) determination from dilution plating is not recommended for viability testing as it is prone to too many inaccuracies (Jenkins and Grzywacz, 2000).

For microbial inoculants, efficacy is particularly difficult to measure under standardized conditions, as these products tend to confer long-term benefits on the crop as a whole. These effects are not easily measured over the short term and raise some challenges to manufacturers of these products.

For fungal pathogens, virulence is generally verified with a standard bioassay using the target organism (plant, insect or fungus). The dose/dosage used should be low to enable differences in virulence/pathogenicity to be detected, and bioassays should always be run against a standard batch of product as the quality of the test organism can vary over time. For many products, it may not be necessary to conduct bioassays on every batch. However, a series of bioassays should be carried out initially to establish that batches do not vary significantly from each other provided production conditions are standardized. It should then be sufficient to carry out periodic checks on random batches, provided production conditions do not change. Any changes in the production process would necessitate a further series of assays to identify any resultant changes in efficacy of the product.

However, given that bioassays are notoriously variable and their utility in detecting small changes in virulence/pathogenicity is limited (Robertson *et al.*, 1995), the determination of acceptable limits for batch-to-batch variation is difficult. Variability in bioassays occurs as a result of a number of factors, including the natural variability in the host organism and hence its susceptibility to the fungal pathogen, variation in the dose of the pathogen acquired and the conditions under which the bioassay is conducted. Steps can be taken to reduce variability; these include maintenance of the host organism (fungus, plant or insect) under controlled/standardized conditions, health monitoring of the host population (particularly in respect of plants and insects), the development of a dosing system that as far as possible ensures that the dose applied is completely acquired and standardization of the bioassay incubation conditions (temperature, humidity, etc.). These measures, in addition to the use of a standard and a control against which to compare the test sample, can reduce the effect of variation in the host population and bioassay procedure. Thus, the performance of the test sample should be considered in relation to that of the standard – using a ratio, for example – rather than in absolute terms. All bioassay data should be archived to build up an accurate picture of the variation in assays over time.

Storage

Shelf-life is a critical limiting factor in the use and acceptance of microbial biocontrol products. If microbial control is to compete with chemical-control products, shelf-life should be as long as possible and, preferably, should not require refrigeration or special treatment (Jones and Burges, 1998). A number of factors affect the shelf-life of a fungal product. Nutrition and culture conditions during fermentation have already been highlighted in earlier sections; in addition, methods of downstream processing and harvesting, product moisture content, formulation and packaging all play an important role in successful long-term storage (Hong *et al.*, 2000).

Significant progress has been made over recent years in the understanding of conidial longevity (Hong *et al.*, 1997, 1998, 1999; Burges, 1998b). These studies have shown that conidia of *Metarhizium anisopliae* var. *acridum* (= *Metarhizium flavoviride*) behave in a similar way to seeds when stored under dry conditions. Applying equations and experimental protocols used for estimating seed longevity (Ellis, 1988; Roberts and Ellis, 1989; Dickie *et al.*, 1990), it is possible to define constants for conidia of individual species of mitosporic fungi and thus predict the longevity of fungal products in storage (Hong *et al.*, 1997, 1998). When packaged in hermetically sealed, foil-lined sachets at low moisture content ($< 5\%$), conidia of *M. anisopliae* var. *acridum* can be stored for up to a year at 30°C and > 6 years under refrigeration. Further, it is possible to predict shelf-life under more practical fluctuating temperature conditions, using a modification of the predictive model (Hong *et al.*, 1999). Recently, these models have been applied to data for conidia of *B. bassiana* (Hong *et al.*, 2002), showing that, although conidia of different fungal species can vary in their absolute longevity (*M. anisopliae* var. *acridum* has greater conidial longevity than *B. bassiana*), the relative effect of moisture content and temperature on the lifespan of conidia is similar between species. Thus, it is possible to harness these models to give clear estimates of the likely shelf-life of a dry conidial product when stored under a range of conditions. For other product types (blastospores, dried mycelium, aqueous formulations, etc.) no such models exist, but data obtained during optimization of the harvesting or drying process can and should be used to clearly indicate the expected shelf-life of the product.

Most important with regard to shelf-life is that each batch of product should perform similarly to the next during long-term storage. To ensure that this is the case, nutritional conditions, method of downstream processing, final moisture content of the product and packaging should be standardized. Provided these parameters are consistent for all batches, storage temperature becomes the primary factor in determining likely shelf-life (Hong *et al.*, 2000). As much information as possible on the effects of storage temperature on shelf-life should be given on the packet. Clear labelling can help to take the mystery out of the use of mycopesticides and help to increase user confidence.

Specifications

Specifications over and above the standards laid out above include information such as the amount of active ingredient (ai) per gram or millilitre of product (e.g. conidia g^{-1} or ml^{-1} formulation). Further specifications might include particle-size spectrum, which is critical for many application techniques, and other physical properties related to the formulation, such as suspensibility, wettability, flowability and emulsion stability.

Recommended minimum quality control for fungi

Careful culture maintenance and preservation

- Accurate identification and characterization of the fungal strain by a recognized taxonomist.
- Use of established long-term storage methods for the isolate obtained from the original isolation.
- Use of single-spore (ss) isolation to reduce genetic drift.
- For production, use of stock culture material bred from the ss isolation and kept in long-term storage.

In-process contamination monitoring

- Take subsamples at critical stages throughout the course of production for the detection of contaminants entering the system.

Total contaminating-organism count (assuming prior screens for human pathogens are clear)

- Measured by dilution plating on to generalist media. The acceptable level of contamination will be product-specific, but should always be backed up with results of in-process contamination monitoring to show that contaminants have not entered during the production process and have not multiplied with the production fungus. Under ideal conditions, a typical limit for an unformulated fungal product could be as low as 1×10^6 cfu g^{-1} in pure conidia powder containing 5×10^{10} conidia g^{-1}.

However, the issue of detecting potential human pathogens in the final product requires further investigation.

Viability

- Greater than 80% viable using a direct counting method, e.g. counting germinated spores under the microscope (up to advised product expiry date when kept at recommended temperature).

Virulence/potency

- This aspect of quality control requires further study before defined procedures and limits can be set. In principle, a standard bioassay appropriate to the agent in question should be developed and the performance of product batches compared with a fungal standard at appropriate intervals. Product performance should be compared with that of the standard and a tolerance limit set beyond which batches are retested to confirm loss in efficacy and discarded on second failure. The difficulties of setting up such a system for microbial inoculants have been discussed in the section on efficacy above.

Other checks appropriate to specification

- Moisture content.
- Quantity of active ingredient per gram product.
- Particle-size distribution as established by the producer and indicated on the label (e.g. for stable ultra-low-volume (ULV) formulations, all particles should be < 100 μm, 99.9% should be < 60 μm and 80% < 10 μm).

Production and Quality Control of Baculoviruses

Production technology

The mass production of insect viruses, primarily BV, such as nucleopolyhedroviruses (NPV) and granuloviruses (GV), for use as insecticides is currently exclusively by *in vivo* propagation, although significant efforts are under way to make mass production in cell culture both practicable and cost-effective (Reid and Weiss, 2000).

A virus is produced *in vivo* by infecting larval stages of a susceptible host insect and then rearing the insects to allow the virus to multiply and complete its replication, before harvesting the infected insects in order to extract the propagated virus. The physical facilities and systems employed for *in vivo* production range from simple low-technology methods, involving the use of wild-collected larvae being reared *in situ* or in simple rearing units (Moscardi, 1999), to automated mass-production facilities (Guillon, 1997). High-technology facilities use insect larvae specially reared in near-sterile conditions involving capital equipment and a level of sophistication comparable to anything elsewhere in the biotechnology industry.

Low-technology, field propagation of BV has been regarded as a viable option for developing countries (Prior, 1989). Commercial production of *Anticarsia gemmatalis* NPV is carried out in Brazil using this approach (Moscardi, 1999). However, production of BV is most commonly a laboratory-based production system, using insects that are reared in a specialized facility or collected from the field, or a mixture of the two. To inoculate the larvae, these are put on a substrate of either fresh natural food or artificial diet that has been sprayed with BV or has BV incorporated into it. The larvae are then reared until virus propagation has been completed or the larvae die of infection. In some systems larvae are harvested alive just prior to death (a step that may reduce productivity but also reduces bacterial contamination), though more usually they are harvested when dead (Grzywacz *et al.*, 1997). The harvested larvae are generally frozen prior to processing. Larvae are then processed by macerating in water and filtered to remove integument, mandibles and other hard parts that might block sprayers. Most current formulations used in developing countries are simple suspensions of unpurified BV, although commercial formulations can be quite sophisticated and may contain synergists, phagostimulants, structuring agents, spreaders, stickers and stabilizing agents (Burges, 1998). Additives such as wetting agents and gustatory stimulants – for example, molasses or crude sugar – may be introduced to tank mixes prior to spraying to improve efficacy and persistence (Burges and Jones, 1998).

The process of *in vivo* BV production is conceptually very simple and can require minimal equipment. In practice, maintaining a sustained production of virus is demanding and it is only with well-developed production procedures and a rigorous standard of process quality control that effective production can be both attained and maintained.

Culture maintenance

Quality control in BV production starts with the raw materials, which, for *in vivo* production, include not only the pathogen but also the live host insects. The virus used for inocula should be a properly identified strain that has been characterized by DNA restriction analysis. The inoculum should be purified using gradient centrifugation (Hunter Fujita *et al.*, 1998) to ensure that it is virtually free of contaminating insect pathogens, which could infect the hosts used for production and interfere with viral replication, reducing BV productivity. One practice that should always be avoided is to use recycled BV from production as inoculum for new batches without characterization or purification. This is a common practice in some low-technology NPV-production systems and can be a major cause of problems.

The insect cultures used for BV production should be free of parasites and monitored for the presence of latent virus infections. For this reason, the preferred hosts for production are disease-free laboratory insects reared under controlled conditions and on a standardized artificial diet. The use of laboratory-reared insects ensures that larvae can be infected at the right developmental stage for optimum productivity. Several studies have shown that optimizing the larval age/weight at infection is valuable in maximizing BV productivity in individual larvae (Cherry *et al.*, 1997; Grzywacz *et al.*, 1998).

Ideally, detailed records of egg laying, hatching rates, pupal size and larval/pupal mortality rates should be maintained, as well as rates of pupal deformation (Moore *et al.*, 1985). These data should be plotted on graphs and scrutinized at regular intervals, as deviations from norms are often the best early indicator of insect health problems. It can also be very useful to carry out regular weekly microscopic examinations on any 'sickly' insects and a few randomly selected ones to check for possible infection, using a standard guide to entomopathogens, such as that of Poinar and Thomas (1984).

Some insects appear to adapt very readily to culture and can survive for many (> 50) generations without the need to introduce new genetic stock; *Spodoptera littoralis* is a good example of this (McKinley *et al.*, 1984). Such cultures are relatively easy to maintain and are a valuable asset, as the introduction of new insects is the commonest route by which unwanted infections enter a culture. However, laboratory cultures of other species, e.g. *H. armigera*, have been observed to decline in vitality after a number of generations and require regular introductions of new insects. Where insects are to be introduced into established cultures, rigorous checking of the new insects is important. The new stock should first be held in a separate quarantine unit. Precautions should include egg washing, rearing in pairs or small groups and keeping eggs separate to minimize the spread of infection. Combined with close examinations of larvae for signs of ill health and microscopic examination of females for infection (after they have finished egg laying), these precautions can be used to eliminate contaminated breeding lines (Chapter 10).

The design of the insect production facility needs to be carefully planned to minimize the danger of the colony becoming infected by either the production BV isolate or any wild pathogens. Examples of designs for appropriate facilities are available in the literature (Bell *et al.*, 1981; Hunter-Fujita *et al.*, 1998) and published protocols and methodologies for mass-rearing host insects are available (e.g. Singh and Moore, 1985).

The use of field-collected insects does have significant potential drawbacks. It can be difficult for untrained collectors to distinguish larvae of susceptible and non-susceptible species. Larvae from the field can be carrying pathogens that may compete with the selected BV and reduce the yield of BV. Further, microsporidia and *Bacillus* spp. can be difficult to distinguish from NPV occlusion bodies (OB) unless staff are well trained. However, with proper procedures, the use of field-collected larvae can still be a viable option for cost-effective mass production of BV for some species of insect host (Moscardi, 1999).

Production process and record keeping

The production process for BV is essentially simple as the virus reproduces itself readily and efficiently if ingested by a susceptible host. Ensuring that this host has appropriate conditions of temperature and humidity, with an adequate food supply, is all that is required to achieve virus propagation. However, full and adequate records of all quality control data need to be kept and reviewed regularly. There is a tendency for staff involved in any day-to-day production to become accustomed to a routine and not to notice when small systematic deviations from norms occur. These may be crucial signs that a problem is building up and it is important that these are investigated early. Records should be scrutinized regularly by a separate reviewer who is not involved directly in day-to-day production.

Host contamination

By creating a favourable environment for BV multiplication, the production system is vulnerable to contamination by unwanted entomopathogens. Especially problematic are chronic pathogens, such as some GV, microsporidia and cypoviruses (CPV), whose initial occurrence may cause little obvious problem but whose uncontrolled spread can significantly reduce BV production. Fungi that colonize unused diet or faeces may also become a problem, but this can usually be controlled with effective hygiene and the use of fungicide in the diet (Singh, 1977).

Nutrition

For choice of food the options lie between an artificial diet, usually an agar-based formula, or fresh plant material. The latter is cheap and can provide a balanced diet for species of insects that are nutritionally demanding. However, fresh material has drawbacks in that nutritional quality may vary according to season or growth stage and field-harvested material may be contaminated with unwanted entomopathogens. For this reason, specially grown plant material is preferable and systems for growing sprouting seedlings, such as wheat or barley, can ensure a regular controllable supply for species that require artificial diet to be supplemented by fresh plant material.

However, most NPV production uses agar-based artificial diets (Singh, 1977). Not only do these ensure that nutritional quality is controlled, but also the cooking or autoclaving stages in diet preparation sterilize the diet effectively to prevent unwanted infection. Diets are usually made in batches and cold-stored for a time prior to use; this should be systematically monitored and batches of diet should be disposed of as they reach expiry (*c.* 7 days). Records of ingredients and supplies should also be kept so that any problems can be checked back to determine if faulty ingredients are to blame. All ingredients should be stored properly – in refrigerators or freezers, as appropriate – and batches replaced at predetermined intervals. Cereal-based diet ingredients should be purchased in bulk and subsamples tested before use on insects (e.g. a suitable storage species) to check for the presence of insecticide residues, or they should be obtained from an organic supplier.

The objective in mass insect rearing is to produce large numbers of larvae of predictable size for virus infection continuously in a controlled manner. This requires precise control of the conditions in which eggs are incubated and larvae reared. Therefore, conditions (temperature, light cycles and humidity) should be closely monitored and clear records kept and regularly checked. Records of production parameters, such as insect weights at infection and harvesting, should also be kept and reviewed on a monthly basis. Even so, large-scale production of larvae with synchronized predictable growth is difficult to achieve in practice (Shieh, 1989). Ensuring an adequate, sustained supply of appropriate larvae for infection is probably the most demanding part of any BV production process and is where serious problems are most common.

Yields

The numbers of OB in BV products should be counted by light microscopy and bioassayed regularly, either at weekly intervals or on every batch produced, to check for yield. These counts should be made when insects are harvested so that any production problems are identified quickly. Microscope counts of NPV suspensions (Wigley, 1980; Hunter-Fujita *et al.*, 1998) are preferred over indirect or non-visual methods, such as the enzyme-linked immunosorbent assay, protein assay or use of particle counters (Jones, 2000). Counting is best done on the technical product – that is, the processed BV suspension prior to final formulation. These protocols for regular counting can be time-consuming but they are at the core of effective quality control, and failure to implement an effective system may lead to production problems later.

DNA profiling

When a new stock of inoculum is made (see section on culture maintenance), DNA profiling of BV produced should be carried out to monitor the strain consistency, as genetic diversity is common in BV (Gettig and McCarthy, 1982). Routine DNA profiling of selected production batches can also serve to detect any viral contamination by unwanted NPV, GV or even CPV. Samples for DNA should be processed within a week or two at most and not saved up for later analysis; otherwise contamination by a foreign virus may be detected too late to prevent it spreading throughout the production system. When this occurs, a major shut-down and disinfection of the facility is often the only solution.

Harvesting

Processing of NPV for formulation as microbial control agents may be relatively simple, involving no more than the preparation of crude suspensions simply by macerating the infected insect cadavers to release the BV. Insect debris, such as skin and mandibles, is then filtered out to prevent blockages in sprayers. This is often as far as processing goes for many products. More advanced processing protocols include settling or the use of centrifugation to concentrate the NPV and remove lipid and insect-derived cellular debris. This procedure also removes some of the by-products that provide contaminant microflora with substrates on which to multiply. However, simple centrifugation in water does not remove many microbial contaminants, as bacterial spores, etc., tend to pellet with NPV OB (Grzywacz *et al.*, 1997). Bacteriostatic agents and pH buffers can be added to stabilize formulations by reducing the multiplication of bacterial contaminants.

Safety

BV are obligate parasites that can replicate only in arthropods or arthropod cell lines and so are inherently safe for humans. Many BV are also highly host-specific; therefore the main safety issue arises from possible contamination of production systems by human or veterinary pathogenic bacteria and fungi. Studies have shown that a variety of species may contaminate BV production systems and multiply either in host insects during life or on the cadavers or faeces (Podgwaite *et al.*, 1983).

The setting of arbitrary standards for non-pathogenic contaminant bacteria is problematic. Early BV commercial products, such as ELCAR, were established with the vigorous limit of 1×10^7 cfu g^{-1} non-pathogenic microbial contaminants (Burges, 1981). It can be argued that this was unnecessarily demanding, contributing nothing significant to improving real safety while burdening the product with an excessively expensive quality control system (H.D. Burges, personal communication).

In BV-infected larvae harvested dead under high ambient temperatures (26°C), bacteria multiply rapidly after larval death and contaminant bacteria may reach 2×10^8 cfu per larva. In larvae harvested alive or a few hours after death, levels of 2×10^7 cfu can be achieved. As a reasonable target, a total non-pathogen microbial contaminant level of 1×10^8 cfu ml^{-1} for a product with an NPV content of 1×10^9 OB ml^{-1} for liquid formulations or 5×10^8 g^{-1} for dry powders would seem both attainable and defensible. If further research confirms the essentially benign nature of *in vivo*-produced BV contaminant microbes, then an even more relaxed standard of 1×10^9 cfu ml^{-1} might be considered acceptable. Current Environmental Protection Agency (EPA) guidelines specify that human or non-target animal pathogens should be absent and that if the production system can support such pathogens then batches must be tested for their presence.

Efficacy

Bioassays are, with microscopy, an essential tool for quality control in BV production. *In vivo* assays are central to quality control on biological pesticides as they are the only means by which the potency of a preparation can be established or monitored and unacceptable batches identified (Hughes and Woods, 1986). Bioassay techniques must be appropriate to the product in question and several recent publications provide guidance to standard methods both for assays and analysis of results (e.g. Evans and Shapiro, 1997; Hunter Fujita *et al.*, 1998; Jones, 2000). The choice of assay depends upon many factors, but assays with neonate larvae are simple and allow many comparisons to be carried out with relative ease (Jones, 2000). The use of neonates may also reduce some of the variability that may occur when older larvae are used. In all assays, the use of laboratory-reared insects from a standard culture will produce more consistent results than wild-collected larvae and the latter should be avoided where possible.

However, there is an inherent, widely recognized problem: that the variability of bioassays is very great and that the results for a host/BV system may vary by up to three orders of magnitude (Robertson *et al.*, 1995). This high variability of course creates problems if it is the intention to identify batches of BV product whose activity is too low. To some extent, the variability problem can be addressed by including a standard virus preparation of known OB concentration, and the results recorded in terms of potency ratio against this standard. The standard should be a reference sample that has been characterized using DNA analysis. Assay variability also improves markedly with operator experience, so that standardizing product monitoring should involve using specific staff who have experience with the assay system (Evans and Shapiro, 1997).

In seeking to set acceptable limits for product potency based upon bioassays, it is probably not productive to set strict a priori limits for acceptable potency. One basis for determining an acceptable potency standard is for quality control limits to be developed during the product-development cycle, as every system will require its own criteria (H.D. Burges, personal communication). In the absence of such data, a guide might be that batches of product found to have a potency of less than 0.5 should be assayed again. These variability problems are not restricted to BV and fungi; similar variation is seen in assays of *B. thuringiensis* and even chemical insecticides (Robertson *et al.*, 1995). The most appropriate stages in the BV production process for efficacy bioassays are on the BV after any purification/processing but before formulation – which is normally called the technical product (TP) – or on the final formulated product (FP). Bioassays on the TP are normally preferred as the results then allow the formulators to blend different batches of TP, which may have different activity, and to dilute the TP appropriately to give an FP of specified standard activity.

Specifications

The specifications for BV products should include the identity and amount of active ingredient. The amount should be expressed both as minimum % w/w TP and as OB g^{-1} dry powder or OB ml^{-1} for liquid formulations, or OB equivalents for formulated products containing synergists or enhancing factors. The % of inert ingredients should also be specified. For liquid formulations, viscosity at a specified temperature is normally included in the specifications. The particle size is also important and either a particle-size spectrum or the results of other particle-size estimations, such as a wet sieve test, should be included. Most importantly, specifications should include recommended storage conditions and estimated shelf-life, at both an ambient temperature, e.g. 20°C, and under recommended storage conditions.

Storage

The shelf-life of BV products has been a significant problem for consumer acceptance. It has been proposed that 18 months is a minimum target shelf-life for BV products (Ignoffo and Couch, 1981). However, most products currently available need to be either refrigerated, deep-frozen, dried or lyophilized to achieve shelf-lives comparable to those of chemical products. In the OB of BV, the infective virions are encased within a protein crystal matrix that protects the virus. Under the right conditions, BV can remain active for many years. Some factors, such as ultraviolet (UV) light, are well known to cause the inactivation of BV even in suspension (Ignoffo and Garcia, 1992) and appropriate UV-opaque packaging should be used to prevent this. The OB is also pH-sensitive and, while very stable between pH 4.0 and 9.0, activity is reported to deteriorate outside this range (Ignoffo and Garcia, 1966). Thus, products need to be checked for pH and buffered if necessary. Where formulations are prepared as dried powders, then the use of moisture-excluding packaging is important as BV powders are often hygroscopic and the absorption of water can seriously reduce BV viability.

The metabolic activity of microbial contaminants in unpurified NPV products can also deter potential users through the production of pungent and very unpleasant odours associated with bacterial fermentation. This needs to be controlled both through more effective clean-up of harvested NPV and through maintenance of effective handling and storage conditions throughout the supply chain, from production to end-user.

Recommended minimum quality control for BV

DNA restriction profiles

- Characterization of the isolate, using at least three commonly used restriction enzymes, e.g. *Eco*R1, *Hin*dIII, *Pst*1, *Bam*H1

OB microscope counts (NPV)

- On the technical product (after processing prior to final formulation) or on formulated product (this may not be practicable with some powder formulations due to presence of carrier).

Total bacterial count

- Dilution plating on generalist agar media for formulated product. Ideally, viable counts should normally not exceed 1×10^8 cfu ml^{-1} for formulated products with an activity of 1×10^9 OB ml^{-1} or 5×10^8 cfu g^{-1} for formulated dry products.

Virulence (pathogenicity)

- Standardized bioassay (median lethal dose (LD_{50} or LC_{50}) on final product, using defined instar of a laboratory-reared strain of the target pest. Results should confirm that the potency ratio is within defined limits specified for the product and determined during product development or > 0.5 of a standard virus preparation.

Conclusions

Quality control criteria are similar for fungi and viruses in that products composed of either organism must be safe, virulent to the target pest, uncontaminated with human pathogens and of consistent quality. However, the issues surrounding the maintenance of high-quality products based on these organisms can be quite different. Fungi can be produced *in vitro*, which avoids the complications of maintenance of healthy host cultures, the risks of latent infections in the host and the inevitable bacterial contamination from saprophytic bacteria that are associated with *in vivo* production. Viability can be easily assessed in fungi, but can be checked in virus preparations only by bioassay. Hence, the methodologies for achieving high-quality products are quite different for these two types of microorganism.

In countries where registration of microbial control products is not yet enforced, the biocontrol industry is unregulated and products tend to be of poor quality. Reasons for not establishing adequate quality control include increased cost (materials and labour), lack of information on standard procedures and the absence of an agreed set of acceptable standards. In countries where regulatory procedures are enforced for microbial control products, registration requires that products are safe and have been adequately tested against non-target organisms. Apart from this, there are no set requirements for quality assurance and manufacturers are left to develop their own procedures. This results in a disparity in the degree to which different manufacturers control product quality.

Given that poor-quality products can result in loss of user confidence in microbial biocontrol products in general, it is long overdue for a set of standard quality control guidelines to be produced to protect the industry as a whole. These quality standards and methods for their implementation need to be both realistic and economically appropriate. It is therefore essential that they be drawn up through close collaboration between manufacturers of biocontrol products, researchers and the appropriate regulatory authorities. Only in this way can regulations be developed that are neither too onerous for producers nor inappropriate for the special nature of infectious agents.

Acknowledgements

Sections from this chapter have been previously published as a paper in the journal *Biocontrol Science and Technology*; http://www.tandf.co.uk (Jenkins and Grzywacz, 2000).

References

Alves, S.B. and Pereira, R.M. (1989) Production of *Metarhizium anisopliae* (Metsch) Sorok and *Beauveria bassiana* (Bals.) Vuill. in plastic trays. *Ecossistema* 14, 188–192.

Anon. (1993) Production: the key technology for biopesticides. *IMPACT AgBioIndustry* February, 7–13.

Bartlett, M.C. and Jaronski, S.T. (1988) Mass production of entomogenous fungi for biological control of insects. In: Burge, M.N. (ed.) *Fungi in Biological Control Systems*, pp. 61–85.

Bell, R.A., Owens, C.D., Shapiro, M. and Tardif, J.R. (1981) Mass rearing and virus production. In: Doane, C.C. and McManus, M.L. (eds) *The Gypsy Moth: Research Towards Integrated Pest Management*. Technical Bulletin 1584, Forest Service, US Department of Agriculture, Washington, DC, pp. 599–655.

Bridge, P.D. and Arora, D.K. (1998) PCR for species definition. In: Bridge, P.D., Arora, D.K., Reddy, C.A. and Elander, R.P. (eds) *Applications of PCR in Mycology*. CAB International, Wallingford, pp. 63–84.

Bridge, P.D., Pearce, D., Rutherford, M. and Rivero, A. (1997) VNTR derived oligonucleotides as PCR primers for population studies in filamentous fungi. *Letters in Applied Microbiology* 24, 426–430.

Burges, H.D. (1981) Safety, safety testing and quality control of microbial pesticides. In: Burges, H.D. (ed.) *Microbial Control of Pests and Plant Diseases 1970–1980*. Academic Press, London, pp. 737–767.

Burges, H.D. (ed.) (1998a) *Formulation of Microbial Biopesticides*. Kluwer Academic Publishers, Dordrecht.

Burges, H.D. (1998b) Formulation of mycoinsecticides. In: Burges, H.D. (ed.) *Formulation of Microbial Biopesticides*. Kluwer Academic Publishers, Dordrecht, pp. 131–185.

Burges, H.D. and Jones, K.A. (1998) Formulation of bacteria viruses and protozoa to control insects. In: Burges, H.D. (ed.) *Formulation of Microbial Biopesticides*. Kluwer Academic Publishers, Dordrecht, pp. 33–127.

Caetano-Anolles, G., Bassam, B.J. and Gresshoff, P.M. (1991) DNA amplification fingerprinting using very short arbitrary oligonucleotide primers. *Biotechnology* 9, 553–557.

Cherry, A.J., Parnell, M., Grzywacz, D., Brown, M. and Jones, K.A. (1997) The optimization of *in vivo* nuclear polyhedrosis virus production of *Spodoptera exempta* (Walker) and *Spodoptera exigua* (Hubner). *Journal of Invertebrate Pathology* 70, 50–58.

Cherry, A.J., Jenkins, N.E., Heviefo, G., Bateman, R. and Lomer, C.J. (1999) Operational and economic analysis of a west African pilot-scale production plant for aerial conidia of *Metarhizium* spp. for use as a mycoinsecticide against locusts and grasshoppers. *Biocontrol Science and Technology* 9, 35–51.

Dickie, J.B., Ellis, R.H., Kraak, H.L. and Ryder, K. (1990) Temperature and seed storage longevity. *Annals of Botany* 65, 197–204.

Edel, V. (1998) PCR in mycology: an overview. In: Bridge, P.D., Arora, D.K., Reddy, C.A. and Elander, R.P. (eds) *Applications of PCR in Mycology*. CAB International, Wallingford, pp. 1–20.

Ellis, R.H. (1988) The viability equation, seed viability nomographs, and practical advice on seed storage. *Seed Science and Technology* 16, 29–50.

Evans, H.F. (1981) Quantitative assessment of the relationship between dosage and response of the nuclear polyhedrosis virus of *Mamestra brassicae*. *Journal of Invertebrate Pathology* 8, 279–283.

Evans, H.F. and Shapiro, M. (1997) Viruses. In: Lacey, L.A. (ed.) *A Manual of Techniques in Insect Pathology*. Academic Press, San Diego, pp. 17–54.

Ferron, P. (1981) Pest control by the fungi *Beauveria* and *Metarhizium*. In: Burges, H.D. (ed.) *Microbial Control of Pests and Diseases 1970–1980*. Academic Press, London pp. 465–482.

Firstencel, H., Butt, T.M. and Carruthers, R.I. (1990) A fluorescence microscopy method for determining the viability of entomophthoralean fungal spores. *Journal of Invertebrate Pathology* 55, 258–264.

Gettig, R.R. and McCarthy, W.I. (1982) Genotypic variation among wild isolates of *Heliothis* spp. nuclear polyhedrosis virus. *Virology* 117, 245–252.

Goettel, M.S. (1987) Conidial viability of the mosquito pathogenic hyphomycete, *Tolypocladium cylindrosporum*, following prolonged storage at −20°C. *Journal of Invertebrate Pathology* 50, 327–329.

Grzywacz, D. (1995) Homeopathic viruses for control of the podborer *Helicoverpa armigera*. *IPM and IRM Newsletter for Legume Crops in Asia* 2, 4–5.

Grzywacz, D., McKinley, D., Jones, K.A. and Moawad, G. (1997) Microbial contamination in *Spodoptera littoralis* nuclear polyhedrosis virus produced in insects in Egypt. *Journal of Invertebrate Pathology* 69, 151–156.

Grzywacz, D., Jones, K.A., Moawad, G. and Cherry, A. (1998) The *in vivo* production of *Spodoptera littoralis* nuclear polyhedrosis virus. *Journal of Virological Methods* 71, 115–122.

Guillon, M. (1997) Production of biopesticides: scale up and quality assurance. In: *Microbial Insecticides: Novelty or Necessity?* Proceeding Monograph Series No. 68, British Crop Protection Council, pp. 151–162.

Harris, J.G. (1997) Microbial insecticides – an industry perspective. In: *Microbial Insecticides: Novelty or Necessity?* Proceeding Monograph Series No. 68, British Crop Protection Council, pp. 41–50.

Hong, T.D., Ellis, R.H. and Moore, D. (1997) Development of a model to predict the effect of temperature and moisture on fungal spore longevity. *Annals of Botany* 79, 121–128.

Hong, T.D., Jenkins, N.E., Ellis, R.H. and Moore, D. (1998) Limits to the negative logarithmic relationship between moisture content and longevity in conidia of *Metarhizium flavoviride*. *Annals of Botany* 81, 625–630.

Hong, T.D., Jenkins, N.E. and Ellis, R.H. (1999) Fluctuating temperature and the longevity of conidia of *Metarhizium flavoviride* in storage. *Biocontrol Science and Technology* 9, 165–176.

Hong, T.D., Jenkins, N.E. and Ellis, R.H. (2000) Effect of duration of development and drying regime on the longevity of conidia of *Metarhizium flavoviride*. *Mycological Research* 104, 662–665.

Hong, T.D., Gunn, J., Ellis, R.H., Jenkins, N.E. and Moore, D. (2002) Relative humidity, temperature, and the equilibrium moisture content of conidia of *Beauveria bassiana* (Balsamo) Vuillemin: a quantitative approach. *Journal of Stored Products Research* 38, 33–41.

Hughes, P.R. and Woods, H.A. (1986) *In vivo* and *in vitro* bioassay methods for baculoviruses. In: Granados, R.R. and Federichi, B.A. (eds) *The Biology of Baculoviruses*. Vol. II. *Practical Application for Insect Control*. CRC Press, Boca Raton, Florida, pp. 2–30.

Hunter Fujita, F.R., Entwistle, P.F., Evans, H. and Crook, N.E. (1998) *Insect Viruses and Pest Management*. Wiley and Sons, Chichester, 620 pp.

Ibrahim, Y.B., Lim, T.K., Tang, M.K. and Teng, H.M. (1993) Influence of temperature, pH and selected growth media on germination, growth and sporulation of *Aschersonia placenta* and *Hypocrella raciborskii*. *Biocontrol Science and Technology* 3, 55–61.

Ignoffo, C.M. and Couch, T.L. (1981) The nuclear polyhedrosis virus of *Heliothis* species as a microbial insecticide. In: Burges, H.D. (ed.) *Microbial Control of Pests and Plant Diseases 1970–1981*. Academic Press, London, pp. 330–362.
Ignoffo, C.M. and Garcia, C. (1966) The relation of pH to the activity of inclusion bodies of a *Heliothis* nuclear polyhedrosis. *Journal of Invertebrate Pathology* 8, 426–427.
Ignoffo, C.M. and Garcia, C. (1992) Combinations of environmental factors and simulated sunlight affecting inclusion bodies of the *Heliothis* (Lepidoptera: Noctuidae) nucleopolyhedrosis virus. *Environmental Entomology* 21, 210–213.
Jackson, A.M., Whipps, J.M. and Lynch, J.M. (1991) Effects of temperature, pH and water potential on growth of four fungi with disease biocontrol potential. *World Journal of Microbiology and Biotechnology* 7, 494–501.
Jackson, M.A. and Bothast, R.J. (1990) Carbon concentration and carbon-to-nitrogen ratio influence submerged-culture conidiation by the potential bioherbicide *Colletotrichum truncatum* NRRL 13737. *Applied and Environmental Microbiology* 56, 3435–3438.
Jenkins, N.E. and Grzywacz, D. (2000) Quality control of fungal and viral biocontrol – assurance of product performance. *Biocontrol Science and Technology* 10, 753–777.
Jenkins, N.E. and Prior, C. (1993) Growth and formation of true conidia by *Metarhizium flavoviride* in a simple liquid medium. *Mycological Research* 97, 1489–1494.
Jenkins, N.E., Heviefo, G., Langewald, J., Cherry, A.J. and Lomer, C.J. (1998) Development of mass production technology for aerial conidia for use as mycopesticides. *Biocontrol News and Information* 19, 21N–31N.
Jimenez, J. and Gillespie, A.T. (1990) Use of the optical brightener Tinopal BOPT for the rapid determination of conidial viabilities in entomogenous Deuteromycetes. *Mycological Research* 94, 279–283.
Jones, K.A. (1998) Spray application criteria. In: Burges, H.D. (ed.) *Formulation of Microbial Biopesticides*. Kluwer Academic Publishers, Dordrecht, pp. 367–375.
Jones, K.A. (2000) Bioassays of entomopathogenic viruses. In: Navon, A. and Ascher, K.R.S. (eds) *Bioassays of Entomopathogenic Microbes and Nematodes*. CAB International, Wallingford, pp. 95–139.
Jones, K.A. and Burges, H.D. (1998) Technology of formulation and application. In: Burges, H.D. (ed.) *Formulation of Microbial Biopesticides*. Kluwer Academic Publishers, Dordrecht, pp. 7–30.
Kern, M. and Vaagt, G. (1996) Pesticide quality in developing coutries. *Pesticide Outlook* October, 7–10.
Kybal, J. and Vlcek, V. (1976) A simple device for stationary cultivation of microorganisms. *Biotechnology and Bioengineering* 18, 1713–1718.
Lane, B.S., Trinci, A.P.J. and Gillespie, A.T. (1991a) Endogenous reserves and survival of blastospores of *Beauveria bassiana* harvested from carbon- and nitrogen-limited batch cultures. *Mycological Research* 7, 821–828.
Lane, B.S., Trinci, A.P.J. and Gillespie, A.T. (1991b) Influence of cultural conditions on the virulence of conidia and blastospores of *Beauveria bassiana* to the green leafhopper, *Nephotettix virescens*. *Mycological Research* 7, 829–833.
Lisansky, S. (1997) Microbial biopesticides. In: *Microbial Insecticides: Novelty or Necessity?* Proceeding Monograph Series No. 68, British Crop Protection Council, pp. 3–10.
Lisansky, S., Quinlan, R. and Tassoni, G. (1993) Bacillus thuringiensis *Production Handbook: Laboratory Methods, Manufacturing, Formulation, Quality Control, Registration*. CPL Scientific, Newbury, 124 pp.
McCabe, D. and Soper, R.S. (1985) Preparation of an entomopathogenic fungal insect control agent. *United States Patent*, No. 4,530,834.
McClatchie, G.V., Moore, D., Bateman, R.P. and Prior, C. (1994) Effects of temperature on the viability of the conidia of *Metarhizium flavoviride* in oil formulations. *Mycological Research* 98, 749–756.
McClintock, J.T. (1999) The federal registration process and requirements for the United States. In: Hall, F.R. and Menn, J.J. (eds) *Methods in Biotechnology*, Vol. 5, *Biopesticides: Use and Delivery*. Humana Press, Totowa, New Jersey, pp. 453–471.
McKinley, D.J., Smith, S. and Jones, K.A. (1984) *The Laboratory Culture and Biology of* Spodoptera littoralis *Boisduval*. Report of the Tropical Development and Research Institute L67, 13 pp.
McKinley, D.J., Moawad, G., Jones, K.A., Grzywacz, G. and Turner, C. (1989) The development of nuclear polyhedrosis virus for the control of *Spodoptera littoralis* in cotton. In: Green, M.B. and de Lyon, D.J.B. (eds) *Pest Management in Cotton*. Society of Chemical Industry, Ellis Horwood, Chichester, pp. 93–100.
Milner, R.J., Staples, J.A. and Lutton, G.G. (1997) The effect of humidity on germination and infection of termites by the hyphomycete, *Metarhizium anisopliae*. *Journal of Invertebrate Pathology* 69, 64–69.

Moore, R.F., Odell, T.M. and Calkins (1985) Quality assessment in laboratory reared insects. In: Singh, P. and Moore, R.F. (eds) *Handbook of Insect Rearing* Vol. 1. Elsevier, Amsterdam, pp. 488.

Moscardi, F. (1999) Assessment of the application of baculoviruses for control of Lepidoptera. *Annual Review of Entomology* 44, 257–289.

Neal, M. and Newton, P. (1999) Registration/regulatory requirements in Europe. In: Hall, F.R. and Menn, J.J. (eds) *Methods in Biotechnology*, Vol. 5, *Biopesticides: Use and Delivery*. Humana Press, Totowa, New Jersey, pp. 453–471.

OECD (1996) *Data Requirements for Registration of Biopesticides in OECD Member Countries: Survey Results.* Environment Monograph No. 106, Organization for Economic Cooperation and Development, Paris.

Podgwaite, J.D., Bruen, R.B. and Shapiro, M. (1983) Micro-organisms associated with production lots of the nuclear polyhedrosis virus of the gypsy moth *Lymantria dispar*. *Entomophaga* 28, 9–16.

Poinar, G.O., Jr and Thomas, G.M. (1984) *Laboratory Guide to Insect Pathogens and Parasites*. Plenum Press, New York, 392 pp.

Prior, C. (1989) Biological pesticides for low external-input agriculture. *Biocontrol News and Information* 10, 17–22.

Reid, S. and Weiss, S. (2000) Baculovirus production *in vitro* – recent developments. In: Gazzoni, D.L. (ed.) *Abstracts of XXI International Congress of Entomology, Igassu Falls, Brazil, 20–26th August 2000*, Book 1. Embrapa, Sojo, abstract 1998, p. 504.

Reinecke, P., Andersch, W., Stenzel, K. and Hartwig, J. (1990) Bio 1020, a new microbial insecticide for use in horticultural crops. In: *Proceedings of Brighton Crop Protection Conference – Pests and Diseases, Brighton, 1990*, pp. 49–53.

Roberts, E.H. and Ellis, R.H. (1989) Water and seed survival. *Annals of Botany* 63, 39–52.

Robertson, J.L., Preisler, H.K., Ng, S.S., Hickle, L.A. and Gelernter, W.D. (1995) Natural variation: a complicating factor in bioassays with chemical and microbial pesticides. *Journal of Economic Entomology* 88, 1–10.

Ryan, M.J. (1999) The effect of preservation regime on the physiology and genetic stability of economically important fungi. PhD thesis, University of Kent.

Ryan, M.J., Smith, D. and Jeffreys, P. (2000) A decision based key to determine the most appropriate protocol for the preservation of fungi. *World Journal of Microbiology and Biotechnology* 16, 183–186.

Schlick, A., Kuhls, K., Meyer, W., Lieckfeldt, E., Borner, T. and Messner, K. (1994) Fingerprinting reveals gamma-ray induced mutations in fungal DNA: implications for identification of patent strains of *Trichoderma harzianum*. *Current Genetics* 26, 74–78.

Shieh, T.R. (1989) Industrial production of viral pesticides. *Advances in Virus Research* 36, 315–343.

Singh, P. (1977) *Artificial Diets for Insects, Mites and Spiders*. IFI/Plenum, New York, 96 pp.

Singh, P. and Moore, R.F. (1985) *Handbook of Insect Rearing*, Vol. 2. Elsevier, Amsterdam, 514 pp.

Smith, D. and Kolkowski, J. (1996) Fungi. In: Hunter-Cevera, J.C. and Belt, A. (eds) *Preservation and Maintenance of Cultures Used in Biotechnology and Industry*. Academic Press, USA, pp. 101–132.

Stephens, D. (1997) Fungus factory – Mycotech produces biopesticides to control whitefly, thrips, and aphids. *Farm Chemicals* September, 30.

Thomas, V. and Smith, D. (1994) Cryogenic light microscopy and the development of long-term cryopreservation techniques for fungi. *Outlook on Agriculture* 23, 163–167.

Trinci, A.P.J., Lane, B.S. and Humphreys, A.M. (1990) Optimisation of cultural conditions for the production and longevity of entomopathogenic fungi. In: *Proceedings of Vth International Colloquium on Invertebrate Pathology and Microbial Control*. Society of Invertebarate Pathology, Adelaide, Australia, pp. 116–120.

Wigley, P.J. (1980) Counting micro-organisms. In: Kalmakoff, J. and Longworth, J.F. (eds) *Microbial Control of Insect Pests*. Bulletin 228, New Zealand Department of Science and Industrial Research, Wellington, New Zealand, pp. 9–35.

19 Guidelines for Quality Control of Commercially Produced Natural Enemies

J.C. van Lenteren,[1] A. Hale,[2] J.N. Klapwijk,[3] J. van Schelt[3] and S. Steinberg[4]

[1]*Laboratory of Entomology, Wageningen University, PO Box 8031, 6700 EH Wageningen, The Netherlands;* [2]*The Bug Factory, 1636 East Island Highway, Nanoose Bay, British Columbia, Canada V9P 9A5;* [3]*Koppert Biological Systems, PO Box 155, 2650 AD Berkel and Rodenrijs, The Netherlands;* [4]*Bio-Bee Biological Systems Sde Eliyahu, Bet Shean Valley, 10810 Israel*

Abstract

Lack of quality control procedures during the mass production of natural enemies may lead to failures in biological pest control. An example is presented of deteriorating efficiency of a natural enemy before quality control was applied and the measures that were taken to restore quality. Although some producers of beneficial insects and mites have applied quality control in one form or another for more than 30 years, it was only during the past 10 years that collaboration between researchers and the biocontrol industry resulted in the development of the 30 harmonized quality control guidelines that are presented in this chapter. Most guidelines have been ring-tested by commercial producers; some are still in development. For each natural enemy the test conditions and the quality control criteria are given. Finally, future developments in quality control are discussed, such as the need to develop flight and performance tests, and the wish of farmers to obtain a set of simple tests for quality control of natural enemies once they have arrived at the farm.

Introduction

The literature on quality control of mass-produced arthropods presents several examples of poorly functioning organisms when quality control guidelines are not applied or are neglected (e.g. Chapter 2; Calkins and Ashley, 1989). Cases where inferior natural enemies resulted in failure of biological control are well known among the biocontrol community, but are seldom published. The following text, concerning a failure in biocontrol and the way this was solved by applying quality control, comes from Bigler (1994):

> In Switzerland, *Trichogramma brassicae* has been mass-produced since 1975 and applied commercially against the European corn borer, *Ostrinia nubilalis*, in maize since 1978. A significant loss in field efficacy was observed in 1980 [Fig. 19.1]. By changing the mass-production system and the colony

© CAB International 2003. *Quality Control and Production of Biological Control Agents: Theory and Testing Procedures* (ed. J.C. van Lenteren)

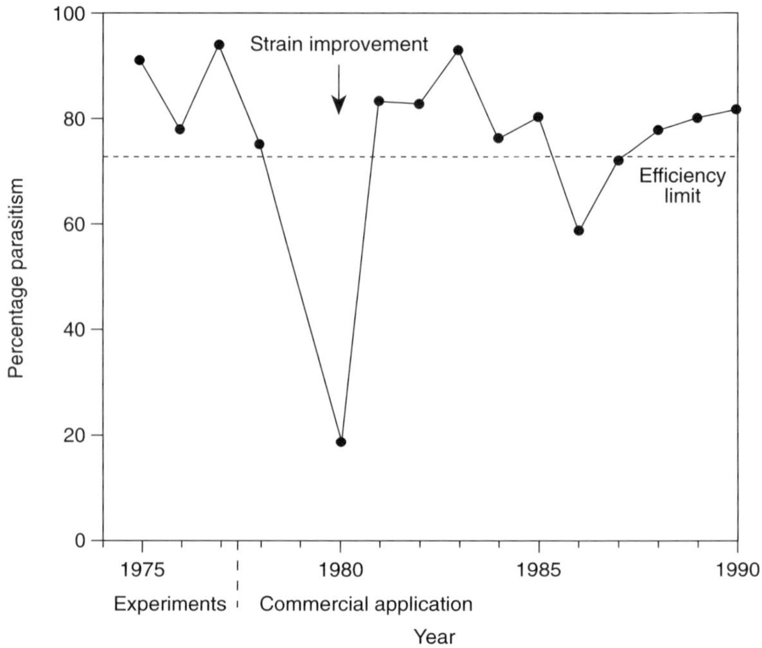

Fig. 19.1. Percentage parasitism of *Ostrinia nubilalis* eggs by *Trichogramma brassicae* in maize from 1975 to 1990 in Switzerland (after Bigler, 1994).

maintenance, it was possible to improve the performance of the strain and achieve the efficiency limit of at least 75% parasitisation in the field.

A thorough analysis of the production system and the performance requirements of *T. brassicae* under the maize growing conditions in Switzerland led to the discovery of important traits which are crucial for a high efficacy. Since attributes like locomotory activity, host acceptance, host suitability and temperature tolerance were negatively affected by the former rearing system, we developed a new production unit. At the same time, risk evaluations of other deteriorations in the strain were performed and methods for measuring single traits and the field performance were developed.

In recent years, the production system of *T. brassicae* was changed from a short period production with a high daily output to a long period production with a low daily output. Improvements of the long-term storage of the parasitoids (diapause) have prolonged the mass-production period from 2 to 9 months per year . . .

The total production system consists of three subunits, namely the European corn borer rearing unit as natural host of the wasps, the *Ephestia kuehniella* unit as factitious (mass-rearing) host and the *T. brassicae* unit. Each of these subunits has its own quality control system . . . Quality control procedures in the host units concentrate on: the rearing diet quality (origin, storage, grind, mixture), egg hatch, larval and pupal weight per unit rearing diet, egg production, egg sterilisation efficiency, temperature and humidity regulation, ventilation control to prevent health hazards by insect scales and sanitation procedures to avoid insect diseases.

Eggs of the European corn borer are used to produce the sting stock of *T. brassicae*. This population is permanently reared under semi-field conditions, i.e. in a field insectary in summer and in a greenhouse with fluctuating temperatures in winter; corn borer egg masses are attached to corn plants and the adults of the parental generation emerge 2 to 4 m away from the plants so that the egg masses must be reached by flight. A portion of the sting stock is regularly used to develop the strain on eggs of *E. kuehniella*. From our experience we know that the performance decreases with an increasing number of generations reared on the factitious host. Therefore, we recommend production of no more than six to seven generations before releasing the parasitoids for biological control purposes . . .

> Since the sting stock is reared under near-natural conditions we do not expect deterioration. However, once a year we assess in the laboratory the [quality control] parameters and, in addition, we measure the field efficacy (percentage parasitism). Since we know from our previous experiments that a change of quality attributes does not occur (or is not measurable) within the first generations on *E. kuehniella* eggs, we quantify the parameters only once a year. The sixth generation (F6) is normally sold to the farmers . . . A few rapid tests (parasitism, emergence, and sex ratio) are made on each batch when the parasitised eggs are shipped immediately . . . The final user (the farmer) is in general not able to do any performance tests. Therefore, government institutes, with financial support of the *Trichogramma* producers, accomplish the tests.

Bigler (1994) concludes that:

> Quality control in *Trichogramma* mass-rearings is one of the measures used to avoid failures in biological control with these parasitoids. The extremely artificial rearing conditions, compared to the habitat where they are released, call for the establishment of sophisticated quality control concepts . . . The importance of single performance attributes has to be established and related to field performance. The methods must be quick, simple and reliable. A single trait will never predict the overall performance accurately and therefore, the best combination of a set of laboratory methods must be developed. Whereas performance of the parasitoids in the field is the best indication of a good rearing system, low field efficacy does not tell us the causes. Regular performance control, carried out in the laboratory, will either indicate deterioration of performance and initiate corrections, or make us confident to produce wasps that are within the quality specifications.

This work by Bigler's group has initiated the development of the *Trichogramma* guideline, as presented in this chapter. There have been various attempts to simplify the quality control tests for *Trichogramma* (see, for example, Chapter 1, section 'State of affairs concerning application of quality control worldwide,' and the work of Greenberg (1991) and Silva *et al.* (2000)), but without result. On the other hand, the guideline developed by Bigler's group has been used successfully by Hassan and Wen Qing Zhang (2001) to measure variability in the quality of *Trichogramma* sold by several commercial suppliers in Germany.

Currently Used Quality Control Guidelines

The history of mass production and quality control of natural enemies is described in Chapters 1, 11 and 12. The current situation concerning quality control in North America and Europe is presented in Chapters 14 and 15, respectively. Recent developments and collaboration between natural-enemy producers and scientists in North America, Europe, Australia, New Zealand and South Africa have resulted in harmonization of quality control guidelines. Although some producers have applied quality control in one form or another for more than 30 years, the real catalyst for designing guidelines was a European Community (EC) Concerted Action that made it possible to organize a series of meetings with representatives from industry and academia (van Lenteren, 1998; van Lenteren and Tommasini, 1999). During five EC-funded workshops in the 1990s, and later at meetings under the umbrella of the International Biocontrol Manufacturers Association (IBMA) and the Association of Natural Bio-control Producers (ANBP), quality control guidelines were written for the 30 species of natural enemies that are most often used in commercial biological control (Table 19.1), and these have been tested and adapted by commercial producers of biological control agents. These guidelines cover features that are relatively easy to determine in the laboratory (e.g. emergence, sex ratio, lifespan, fecundity, adult size and predation/parasitism rate). Recently, IBMA and ANBP have taken the initiative to update and further develop quality control guidelines.

The guidelines described in this chapter refer to product-control procedures, not to production or process control. They were designed to be as uniform as possible so that they can be used in a standardized manner by many producers, and elements of the tests

Table 19.1. Natural enemies for which quality-control guidelines have been developed.

Amblyseius (*Neoseiulus*) *degenerans* Berlese (Acarina: Phytoseiidae)	
Anthocoris nemoralis (Fabricius) (Hemiptera: Anthocoridae)	Provisional test
Aphelinus abdominalis Dalman (Hymenoptera: Aphelinidae)	Provisional test
Aphidius colemani Viereck (Hymenoptera: Braconidae)	
Aphidius ervi (Haliday) (Hymenoptera: Braconidae)	
Aphidoletes aphidimyza (Rondani) (Diptera: Cecidomyiidae)	
Aphytis lingnanensis Compere and *Aphytis melinus* DeBach (Hymenoptera: Aphelinidae)	
Chrysoperla carnea Steph. (Neuroptera: Chrysopidae)	Provisional test
Cryptolaemus montrouzieri Mulsant (Coleoptera: Coccinellidae)	Provisional test
Dacnusa sibirica Telenga (Hymenoptera: Braconidae)	
Dicyphus hesperus Wagner (Hemiptera: Miridae)	
Diglyphus isaea (Walker) (Hymenoptera: Eulophidae)	
Encarsia formosa Gahan (Hymenoptera: Aphelinidae)	
Eretmocerus eremicus (Rose) (Hymenoptera: Aphelinidae)	
Eretmocerus mundus Mercet (Hymenoptera: Aphelinidae)	Provisional test
Hypoaspis miles Berlese (Acarina: Laelapidae)	Provisional test
Leptomastix dactylopii Howard (Hymenoptera: Encyrtidae)	
Macrolophus caliginosus Wagner (Hemiptera: Miridae)	
Neoseiulus californicus McGregor (Acarina: Phytoseiidae)	
Neoseiulus cucumeris (Oudemans) (Acarina: Phytoseiidae)	
Orius spp. (*O. aldibipennis, O. insidiosus, O. laevigatus, O. majusculus*) (*Hemiptera: Anthocoridae*)	
Phytoseiulus persimilis Athias-Henriot (Acarina: Phytoseiidae)	
Podisus maculiventris Say (Hemiptera: Pentatomidae)	Provisional test
Trichogramma brassicae Bezd. (= *Trichogramma maidis*) (Hymenoptera: Trichogrammatidae)	
Trichogramma cacoeciae Marchal (Hymenoptera: Trichogrammatidae)	
Trichogramma dendrolimi Matsumura (Hymenoptera: Trichogrammatidae)	

can be used by distributors, pest-management advisory personnel and farmers. The standard elements of the quality control guidelines are given in Table 19.2. The tests should preferably be carried out by the producer after all handling procedures just before shipment. It is expected that the user (farmer) will only perform a few aspects of the quality test, e.g. per cent emergence or number of live adults in the package. Some tests are to be carried out frequently by the producer, i.e. on a daily, weekly or batch-wise basis. Others will be done less frequently, i.e. on an annual or seasonal basis, or when rearing procedures are changed. In the near future, flight tests and field-performance tests are expected to be added to these guidelines. Such tests are needed to show the relevance of the laboratory measurements. Laboratory tests are only adequate when a good correlation has been established between the laboratory measurements, flight tests and field performance.

The quality control guidelines presented in this chapter are applied by a number of companies that mass-produce natural enemies in Europe and North America (Chapters 14 and 15) and are used by others to compare performance of the same species of natural enemy produced by different companies (e.g. O'Neil *et al.*, 1998; Luczynski and Caddick, 2000; Hassan and Wen Qing Zhang, 2001). Depending on the size of the company and the number of natural-enemy species they produce, they may apply from one to more than 20 tests. Understandably, very few data are made public by the companies, although extensive exchange of information of test results took place during the development of the quality control guidelines from 1991 to 1998. Nowadays, the biocontrol industry has developed a ring-testing system

Table 19.2. General quality-control criteria for mass-reared natural enemies.

Criteria already in use	
Quantity:	number of live natural-enemy organisms in container
Sex ratio:	minimum percentage females (male-biased ratio may indicate poor rearing conditions)
Emergence:	emergence rate to be specified for all organisms sold as eggs or pupae
Fecundity:	number of offspring produced during a certain period (for parasitoids, fecundity is also an indication of the host-kill rate)
Longevity:	minimum longevity in days
Parasitism:	number of hosts parasitized during a certain period
Predation:	number of prey eaten during a certain period
Adult size:	hind tibia length of adults, sometimes pupal size (size is often a good indication for longevity, fecundity and parasitization/predation capacity)
Criteria to be added in near future	
Flight:	short- or long-range flight capacity
Field performance:	capacity to locate and consume prey or parasitize hosts in crop under field conditions

Comments

- Quality control is done under standardized test conditions of temperature (usually 22 ± 2°C or 25 ± 2°C), relative humidity (usually 75 ± 10%) and light regime (usually 16L : 8D), that are specified for each test
- All numbers/ratios/sizes should be mentioned on the container or packaging material
- Fecundity, longevity and predation capacity tests can often be combined
- Expiration date for each shipment should be given on packaging material
- Guidelines should be usable for all product formulations

Original designers:	names of the persons who made the first design of the guideline
Coordinators:	names of the persons who collect new information for the guideline and will adapt the guideline when needed; (e-mail) addresses of coordinators can be found at: http://www.agrsci.dk/plb/iobc/sting/sting23.htm (who is who in biological control)
Updated guidelines:	will be available at www.AMRQC.org

L, hours light; D, hours dark.

for development guidelines for new species of natural enemies and adaptation of old guidelines.

Future Additions to Current Quality Control Guidelines

The producers of natural enemies work together with biological control researchers to develop flight tests and field-performance tests. The importance of these flight tests has been discussed by several authors (see introduction and Bigler (1994)), but testing of these aspects is still rare. Flight tests are supposed to be essential to determine quality if the natural enemy has been reared under conditions where flight was not needed to find hosts or prey, which is often the case under crowded mass-rearing conditions. Flight tests are also needed when the natural enemy is seriously manipulated during mass rearing and preparation for shipment (e.g. removal of pupae from leaves and gluing pupae to cardboard cards) and when storage periods are long (see Chapter 12). Correlation between values obtained at laboratory testing and field performance is important to be able to select a limited set of laboratory criteria that give meaningful information about performance after release.

Bigler (1994) provides information about laboratory testing and field performance. Also Silva *et al.* (2000) describe and use an interesting test that was initially developed by Greenberg (1991) to evaluate the searching and dispersal ability of parasitoids in a maze in the laboratory. Silva *et al.* (2000)

measured the performance of *Trichogramma* in a maze in the laboratory to predict its dispersal capacity in the field. Disappointingly, it appeared that the laboratory bioassay with the maze did not properly predict dispersal capacity of *Trichogramma* in the field. Many other dispersal tests are in development, but data obtained in the field are often rare. An interesting approach for a field-performance test has been described by van Schelt and Ravensberg (1990). Their goal was to compare the capacity to control *Ostrinia nubilalis* in maize by *Trichogramma maidis* that were either obtained from diapause storage or freshly reared. In the laboraratory, percentage emergence, sex ratio and fecundity were determined of diapause and freshly reared parasitoids. Vials with parasitoids of the same samples as the laboratory material were put at a central release point in a maize crop. From the release point, cards with sentinel *O. nubilalis* eggs were hung on maize plants in eight directions, with an interval of 1 m and up to 10 m away from the release point. Percentage parasitism was determined on these cards. The laboratory results showed no differences in emergence and fecundity between the diapause and fresh parasitoids, but the sex ratio of the diapause parasitoids was lower than that of fresh ones. The field tests showed that diapause and fresh parasitoids dispersed in all directions, but that percentage parasitism by fresh parasitoids was higher than that of diapause parasitoids (van Schelt and Ravensberg, 1990). The results obtained with one of the flight tests are described below to illustrate developments in this area.

A short-range flight test has been developed for *Encarsia formosa*, i.e. a test where the parasitoid has to fly a distance of 4–20 cm (Posthuma-Doodeman *et al.*, 1996). Such distances are similar to distances between leaves in a plant. We have experienced that some methods of producing or storing *E. formosa* can lead to defective individuals that are unable to fly even such short distances, and that was the reason for developing this test. This short-range flight test is run in a glass cylinder that has a glass cover with sticky material on the underside. A barrier of repellent material (e.g. Blistex lippomade), 4 cm in height, is applied to the vertical wall of the cylinder to prevents wasps from walking to the sticky material on the glass cover plate at the top (Fig. 19.2). Parasitoids are put on leaves or on cards on the bottom of the cylinder. The whole set-up consists of standardized parts, is easy to assemble and reusable and uses a small amount of space (400 cm^2) per glass cylinder. Counting of the trapped wasps can be done rapidly (2 min per cylinder) and without manipulation of the cylinder. The effects of parasitoid rearing, handling and storage conditions can be evaluated with this test. This test can also be used for concurrent measurement of immature mortality and parasitoid emergence pattern, elements that are included in the current quality control guideline.

The short-range flight test is suitable, among other things, for evaluating the effect of storage periods, temperature and handling procedures on the flight capability of

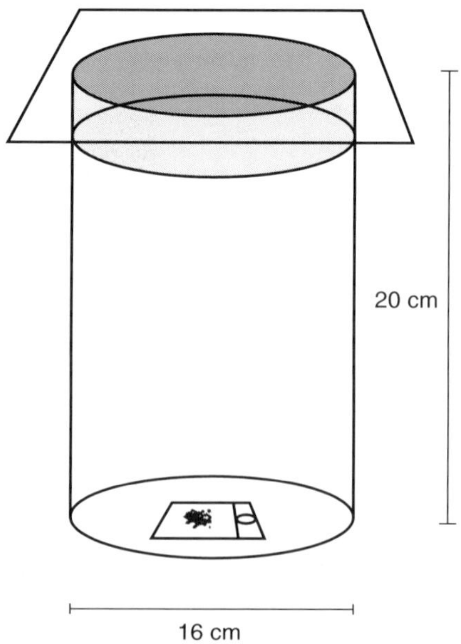

Fig. 19.2. Set-up of short-range flight test for *Encarsia formosa.* The glass cylinder of 16 cm diameter and 20 cm high is covered with a glass plate, which has a circle of glue on the underside. On the inside wall at the upper part of the cylinder, a 4 cm high strip of repellent material is applied so that the parasitoids cannot walk into the glue, but have to fly. Pupae on leaf or paper are put on the glass bottom of the cylinder.

E. formosa and is expected to be included in the standard testing procedure in the near future. This short-range flight test has already provided important additional information for the quality control measurements discussed above. Two examples are given here, one concerning storage of parasitoids and the other about the effect of temperature on the flight capacity of parasitoids. Storage of parasitoids at a low temperature (8°C) for 2 and 16 days, respectively, gives similar percentages of emergence, but the ability to fly is much lower for the parasitoids that were stored for 16 days (Fig. 19.3). The test also clearly demonstrates the effect of temperature on ability to fly: while many parasitoids reach the top of the cylinder at 25°C, very few do so at 18 and 15°C (Fig. 19.4).

A short-range flight test based on the one used for *E. formosa* has been developed for *Trichogramma* by Dutton and Bigler (1995) and is discussed by Prezotti and Parra (2002). Flight tests need further improvement for easy and reliable use.

Another point to be considered for future work is the strong wish of farmers in some countries to be able to test the quality of natural enemies after transport and before release in the field or greenhouse. Luczynski and Caddick (see Chapter 14; Luczynski and Caddick, 2000) used elements of the quality control guidelines developed for *Phytoseiulus* and *Encarsia* to compare the quality of different products when arriving at the greenhouse, just before release. Testing helped in this case to improve products, and results from this project increased growers' confidence in the use of biological control and increased their demand for better quality assurance of supplied products.

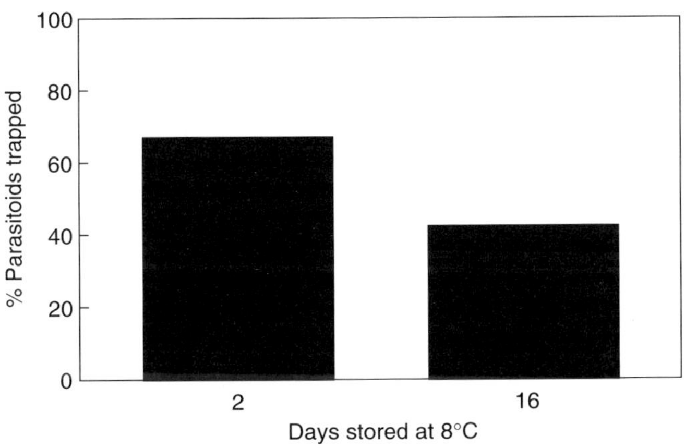

Fig. 19.3. Percentage *Encarsia formosa* females capable of flying (= percentage of parasitoids trapped in glue of glass plate) when stored for 2 and 16 days at 8°C.

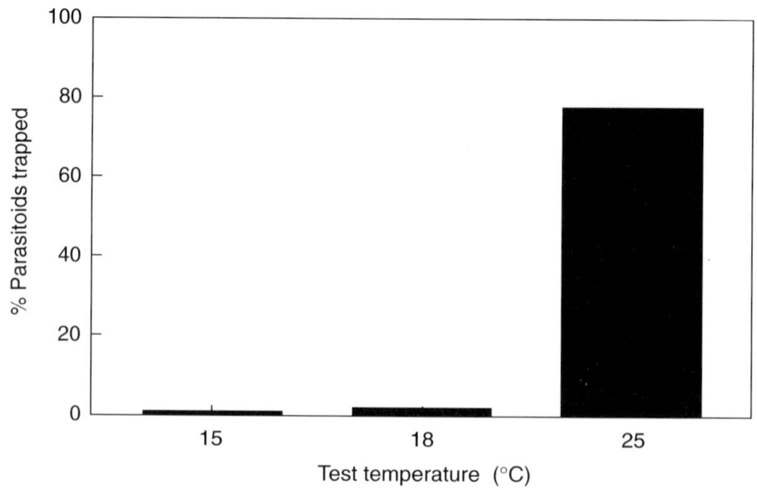

Fig. 19.4. Percentage *Encarsia formosa* females capable of flying (= percentage of parasitoids trapped in glue of glass plate) at three temperatures.

Description of Quality Control Tests in Alphabetical Order

Anthocoris nemoralis (Fabricius) (Hemiptera: Anthocoridae) Provisional test

Test conditions

Temperature:	25 ± 1°C
Relative humidity (RH):	70 ± 5%
Light regime:	16 hours light (L) : 8 hours dark (D)

Quality control criteria

Quantity	The number of living adults/nymphs as specified on the container. Batch-wise or weekly test.
Sex ratio	⩾ 45% females; $n = 100$; a seasonal test.
Fecundity	⩾ 30 eggs per female per 28 days; $n = 30$ pairs; an annual test.

Description of testing methods

Counting method	Take a container with *Anthocoris* and leave it for about 20 min at 8°C. Then sieve the material of the container to separate vermiculite, buckwheat husk and insects; thereafter all the individuals can be counted using an aspirator.
Fecundity	Take about 100 *Anthocoris* adults from a bottle ready for shipment or from a rearing cage with adults that emerged less than 24 h ago. Maintain these individuals for 4 days in a cage, feeding them with *Ephestia kuehniella* eggs and supplying a French bean pod. Then determine their sex under stereomicroscope. Subsequently, put a pair of predators in a ventilated transparent container of *c.* 75 ml (3–4 cm of diameter) that contains a French bean pod. Provide new *E. kuehniella* eggs *ad libitum*. Every 2–3 days the bean pod is substituted with a fresh one, eggs are counted and new prey is added. The total number of eggs is calculated for a 28-day oviposition period. Exclude

data from females that have died accidentally or got lost during the test period.

Designers and coordinators: M. Mosti and J. Vermeulen (provisional test).

Aphelinus abdominalis Dalman (Hymenoptera: Aphelinidae) Provisional test

Test conditions

Temperature:	22 ± 2°C
RH:	75 ± 5%
Light regime:	16L : 8D

Quality control criteria (for adults or mummies)

Quantity	The number of live adults and / or mummies as specified on the label.
Adult mortality	< 10% per package; $n = 300$, based on sample from three containers; batch-wise or weekly
Emergence rate	⩾ the number of live adults that should emerge from the package as specified by the manufacturer. A minimum of three containers should be counted. Emergence rate ⩾ 75% within 12 days ($n = 300$). A weekly or batch-wise test.
Sex ratio	⩾ 40% female, $n = 300$, batch-wise or weekly.
Fecundity	⩾ 10 eggs per female during 2 days, $n = 30$; annual test.

Description of testing methods

Quantity and emergence	Specify the number that should emerge from the mummies. Put the mummies with the carrier material in a ventilated container. Run the test for a maximum of 12 days. For calculating the emergence rate, count the total number of mummies in the container and calculate the per cent emergence according to the formula: (no. of adult wasps / no. of mummies) × 100.
Sex ratio	Mix all the adult wasps from the emergence test. Take a sample of 300 adults and put them on transparent tape. Use a magnification of 35×. Count the number of female wasps. Females have a yellow abdomen with an ovipositor (small stripe over the middle of the abdomen). Males are usually smaller and have a slightly darker abdomen with a small point at the end. At the end of this small point a little thread can be seen (see Fig. 19.5).
Fecundity Leaf discs on agar Day 0	Put an ample amount of black mummies that are close to emergence in a container. Put some droplets of honey in the container or on the gauze. Put the container in a climate room temperature (T = 22°C).

Preparing the bio-assay. The bio-assay tray consists of a round plastic petri-dish tray with a lid that can be closed tightly (diameter 77 mm; height 31 mm; Bock, Art. Nr. 41113). A piece of gauze is incorporated into the lid for ventilation. Pour 1 cm of water agar (1%) into the tray and cool to 30°C. Just before it solidifies, an aubergine leaf disc or a few tomato leaf discs are put upside down on the agar. It is very important to use a fresh leaf with maximal turgor; otherwise the lifespan of the leaf will be too short for the test period (11 days). It is best to pick leaves early in the morning. Put 30 adult *Macrosiphum euphorbiae* on to the leaf, using a fine brush. Place the Petri dishes

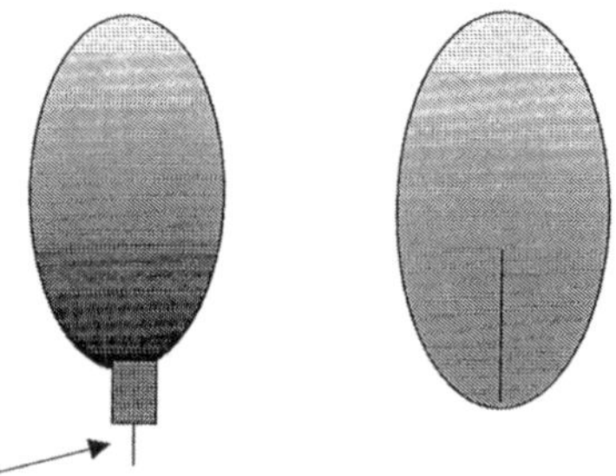

Fig. 19.5. Abdomen of adult male and female wasps.

	upside down on a ventilated tray, to simulate a more natural situation for the aphids and to prevent the leaf from becoming sticky with honeydew. Make 30 trays.
Day 1	Collect the emerged wasps in another container. Put some droplets of honey in the container or on the gauze. Place the container with the wasps in a climate room (T = 22°C).
Day 3	Remove the adult aphids from the dishes with a moist brush. Check the number of offspring (25–50 per tray, age of aphids 0–48 h). Place the container with emerged wasps in a cold room (8–12°C) for 5 min. Tap the wasps from the container on to a smooth white surface. Place small vials over the wasps and after they have walked in, close the vial. Select 30 females by checking them under a binocular microscope (see Fig. 19.5). Tap the vial to release individual females on to the Petri dishes in the cold room. Place the dishes upside down at 22°C.
Day 5	Remove the wasps from the dishes after 48 h.
Days 6–12	Check the quality of the leaves. If the quality is poor, remove the aphids to a new Petri dish with a fresh leaf.
Day 13	Count the number of black mummies per dish.

Designers and coordinators: J. van Schelt and G. Burgio (provisional test).

Aphidius colemani Viereck (Hymenoptera: Braconidae)

Test conditions

Temperature:	25 ± 2°C
RH:	75 ± 5%
Light regime:	16L : 8D

Quality control criteria for mummies

Quantity and emergence	⩾ the number of live adults that should emerge from the package as specified by the manufacturer. A minimum of three containers should be counted. Emergence rate ⩾ 45% (n = 500). A weekly or batch-wise test.
Sex ratio	⩾ 45% females, n = 150, a seasonal test.
Fecundity	⩾ 60 mummies per female on the first day when tested on *Aphis gossypii*; n = 30, an annual test.

⩾ 35 mummies per female on the first day when tested on *Myzus persicae*; $n = 30$, an annual test.

Description of testing methods

Quantity and emergence

Specify the number that should emerge from the mummies. Put the mummies with the carrier material in a container (height 15 cm, diameter 9 cm) with a cork in the bottom. The lid should have one or more holes with gauze for ventilation. Put some droplets of honey on the outer side of the gauze. By removing the cork, mummies and carrier material can be transferred to a new container every day. The container with emerged adults can be frozen and subsequently counted. Continue until no more wasps emerge. Run the test for a maximum of 7 days.

An alternative method for collecting the emerged adults: put the mummies with the carrier material in a ventilated container (15 cm height, 9 cm diameter) with a lid at its bottom. An inverted funnel is glued to the upper part of the container. A glass collecting tube is fitted, by means of a cork, to the neck of the funnel. A standard light source, e.g. fluorescent tube, is placed *c.* 20 cm above the collecting apparatus. The whole system, except for the collecting tube, is covered by a dark cloth to force the emerging wasps towards the collecting tube. Change the collecting tube every day for 7 days and count the total number of adult parasitoids caught in the tube. Add to this the number of wasps that remained at the bottom of the apparatus.

For calculating the emergence rate, count the total number of mummies in the container and figure the per cent emergence according to the formula: (no. of adult wasps/no. of mummies) × 100.

Sex ratio

Mix all the adult wasps from the emergence test. Take a sample of 100 adults and count the number of female wasps. Females are distinguished from males by their pointed abdomen (ovipositor). The length of the female abdomen is almost equal to wing length. The male abdomen is more rounded at the end and is always shorter than the wings. The females should amount to more than 45% of the total.

Fecundity

This test can be done either with leaf discs on agar or on whole plants.

Leaf discs on agar
Day 1

Preparing the bio-assay. The bio-assay tray consists of a round plastic Petri-dish tray with a lid that can be closed tightly (diameter 77 mm; height 31 mm; Bock, Art. Nr. 41113). A piece of gauze is incorporated into the lid for ventilation. Pour 1 cm of water agar (1%) into the tray and cool to 30°C. Just before it solidifies, a cucumber leaf disc (when tested on *A. gossypii*) or a sweet-pepper leaf disc (when tested on *M. persicae*) is put upside down on the agar. It is very important to use a fresh leaf with maximal turgor; otherwise the lifespan of the leaf will be too short for the test period (11 days). It is best to pick leaves early in the morning. Put 30 adult *A. gossypii* or 30 *M. persicae* on to the leaf using a fine brush. Place the Petri dishes upside down on a ventilated tray, to simulate a more natural situation for the aphids and to prevent the leaf from becoming sticky with honeydew. Remove the adult aphids after 1 day. By doing this, between 100 and 150 young aphids (first and second nymphal stage) per tray can be used for testing. Prepare 30 trays.

	Put an ample amount of *Aphidius* sp. mummies that are close to emergence in a container. Put some droplets of honey in the container or on the gauze. Put the container in a climate room (25°C).
Day 2	Remove the adult aphids from the dishes with a moist brush. Check the number of offspring (> 100 per tray). Place the container with emerged wasps in a cold room (8–12°C) for 5 min. Tap the wasps from the container on to a smooth white surface. Place small vials over the wasps and after they have walked in, close the vial. Select 30 females by checking them under a stereoscopic microscope. Tap the vial to release individual females on to the Petri dishes in the cold room. Place the dishes upside down at 25°C for 24 h.
Day 3	Remove the wasps from the dishes after 24 h.
Days 4–10	Check the quality of the leaves. If the quality is poor, remove the aphids to a new Petri dish with a fresh leaf.
Day 11	Count the number of mummies per dish.
Comment:	It is possible that *M. persicae* needs 2 days to produce enough offspring. Adjust the scheme accordingly.
Whole plants	
Day 1	Use potted cucumber plants (30×) pruned to bear two/three leaves, or small sweet-pepper plants. Place a plastic ventilated cylinder over the plant. Use a layer of vermiculite to seal the cylinder at the underside. Put 30 adult *A. gossypii* or 30 *M. persicae* on to each plant, using a fine brush.
Days 2–12	Follow the same protocol as described above for the system with leaf discs on agar.
Comment:	It is possible that *M. persicae* needs 2 days to produce enough offspring. Adjust the scheme accordingly.

Design: J. van Schelt and S. Steinberg.
Coordinators: J. van Schelt and S. Steinberg.

Aphidius ervi (Haliday) (Hymenoptera: Braconidae)

Test conditions

Temperature:	22 ± 2°C
RH:	75 ± 5%
Light regime:	16L : 8D

Quality control criteria for adults

Quantity	⩾ the number of live adults as specified on the container.
Adult mortality	⩽ 8% of the number of adults present in the container, based on three containers sampled and $n = 500$ or more; a weekly or batch-wise test.

Quality control criteria for mummies

Quantity	⩾ the number of live adults that should emerge from the package as specified by the manufacturer.
Emergence	Emergence rate ⩾ 75% ($n = 250$). A weekly or batch-wise test.
Sex ratio	⩾ 45% females, $n = 150$, a seasonal test.
Fecundity	⩾ 35 mummies/female in 2 h when tested on *M. euphorbiae*, an annual test.

Description of testing methods

Quantity – adults: Count the number of dead wasps. Put the containers in the freezer for a minimum of 2 h. Count the number of wasps.

Quantity and emergence – mummies: Specify the number that should emerge from the mummies. Put the mummies with the carrier material in a container (height 15 cm, diameter 9 cm) with a cork in the bottom. The lid should have one or more holes with gauze for ventilation. Put some droplets of honey on the outer side of the gauze. By removing the cork, mummies and carrier material can be transferred to a new container every day. The container with emerged adults can be frozen and subsequently counted. Continue until no more wasps emerge. Run the test for a maximum of 8 days.

An alternative method for collecting the emerged adults: put the mummies with the carrier material in a ventilated container (15 cm height, 9 cm diameter) with a lid at its bottom. An inverted funnel is glued to the upper part of the container. A glass collecting tube is fitted, by mean of a cork, to the 'neck' of the funnel. A standard light source, e.g. fluorescent tube, is placed *c.* 20 cm above the collecting apparatus. The whole system, except for the collecting tube, is covered by a dark cloth to force the emerging wasps towards the collecting tube. Change the collecting tube every day. Within 7 days, count the total number of adult parasitoids caught in the tube. Add to this the number of wasps that remained at the bottom of the apparatus.

For calculating the emergence rate, count the total number of mummies in the container and figure the per cent emergence according to the formula: (no. of adult wasps/no. of mummies) × 100.

Sex ratio: Mix all the adult wasps from the emergence test. Take a sample of 100 adults and count the number of female wasps. Females are distinguished from males by their pointed abdomen (ovipositor). The length of the female abdomen is almost equal to wing length. The male abdomen is more rounded at the end and is always shorter than the wings. The females should amount to more than 45% of the total.

Fecundity: This test is done with leaf discs on agar.

Day 1: Preparing the bio-assay. The bio-assay tray consists of a round plastic Petri-dish tray with a lid that can be closed tightly (diameter 77 mm; height 31 mm; Bock, Art. Nr. 41113). A piece of gauze is incorporated into the lid for ventilation. Pour 1 cm of water agar (1%) into the tray and cool to 30°C. Just before it solidifies, an aubergine leaf disc is put upside down on the agar. It is very important to use a fresh leaf with maximal turgor; otherwise the lifespan of the leaf will be too short for the test period (11 days). It is best to pick leaves early in the morning. Put 10 adult *M. euphorbiae* on to the leaf, using a fine brush. Place the Petri dishes upside down on a ventilated tray, to simulate a more natural situation for the aphids and to prevent the leaf from becoming sticky with honeydew. Remove the adult aphids after 1 day. By doing this, between 30 and 40 young aphids (first and second nymphal stages) per tray can be used for testing. Prepare 60 trays.

Put an ample amount of *A. ervi* mummies that are close to emergence in a container. Put some droplets of honey in the container or on the gauze. Put the container in a climate room (22°C).

Day 2: Remove the adult aphids from the dishes with a moist brush. Check the number of offspring (> 30 per tray). Place the container with

	emerged wasps in a cold room (8–12°C) for 5 min. Tap the wasps from the container on to a smooth white surface. Place small vials over the wasps and, after they have walked in, close the vial. Select 30 females by checking them under a stereoscopic microscope. Tap the vial to release individual females on to the Petri dishes in the cold room. Place the dishes upside down at 22°C for 30 min. Bring the dishes to the cold room. Tap the wasps into a new Petri dish after 5 min. Place the dishes upside down at 22°C for 90 min. Remove the wasps from the dishes.
Days 3–10	Check the quality of the leaves. If the quality is poor, remove the aphids to a new Petri dish with a fresh leaf.
Day 11	Count the number of mummies per dish.

Design: J. van Schelt.
Coordinators: J. van Schelt and J. Vermeulen.

Aphidoletes aphidimyza **(Rondani) (Diptera: Cecidomyiidae)**

Test conditions

Temperature:	22 ± 2°C
RH:	75 ± 5%
Light regime:	16L : 8D

Quality control criteria

Quantity	Number of adult insects as specified on the label; a weekly test.
Emergence rate	> 70% emergence within 7 days; $n = 150$; a weekly test.
Sex ratio	⩾ 45% females; $n = 150$; a weekly test.
Fecundity	> 40 eggs per female within 3 days; $n = 25$; annual test.
Flight activity	Simple test, see below; annual test.

Description of testing methods

Quantity	This test is made for a standard product of 1000 mummies per 0.1 l of vermiculite (22 g). Estimating the number of *Aphidoletes aphidimyza* pupae per bottle: Weigh the content of the bottle. Mix carefully and take three samples of 1 g from the material. Count the number of pupae per sample. Note that pupae may be lumped together. Calculate the total number of pupae per bottle.
Number of midges and emergence rate	Put half of the original material back in the bottle. Put the bottle with open cap in a bucket. Put white paper on the bottom of the bucket. Close the bucket carefully and place it at 22°C. After 6–7 days, most midges will emerge. Place the bucket in a freezer for at least 4 h. After this period the midges can be counted. Multiply the count by 2. Dissect 150 pupae to assess per cent emergence.
Sex ratio	Take 150 midges at random from the emergence test and sex them. Males have long hairy antennae, females bear short antennae without hairs (see Fig. 19.6).
Fecundity	Estimate the number of days required till emergence of the pupae. This can be done by dissecting the pupae and checking the development of eyes, legs and wings:

- no legs, no eyes — 7 days
- legs, white eyes — 4–5 days

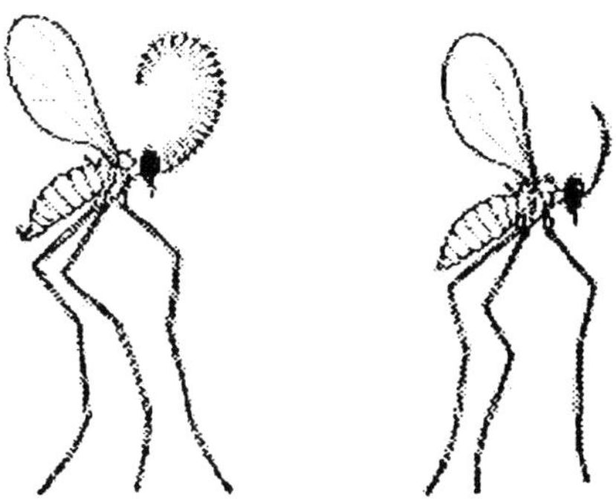

Fig. 19.6. Adult male (left) and female (right) of *Aphidoletes aphidimyza*.

- reticulation of the eyes clearly visible — 2–3 days
- wing formation: dark coloured — 1 day

Place an open bottle in a large cage (> 40 cm × 40 cm × 40 cm). The midges can easily crawl through a layer of 12 cm of vermiculite. To ensure proper mating bring in some cobwebs from spiders. Cobwebs can be sampled with a metal ring (diameter 20 cm). The ring should be placed horizontally in the corner at the rear end of the cage. The midges will emerge during dawn and night. Mating will take place during the first night after emergence (van Lenteren *et al.*, 2002).

Put a plant with aphids in the cage (e.g. wheat with *Rhopalosiphum padi*). This will serve as a source of carbohydrates and will stimulate egg production.

Day 0 — On the day you expect the emergence of the pupae, prepare 25 trays with *A. gossypii* (around 100) on cucumber (see testing method *A. colemani*, day 1). Sweet pepper with *Myzus* sp. can also be used. Aphids may be of variable age. Midges will emerge in the large cage in the evening and night.

Day 1 (morning) — Place the cage in a cold room (5–10°C) for 10 min. Female midges (determine visually) can be gently tapped into the 25 trays. Place the trays upside down in the climate room. To ensure proper ventilation, place the trays on a piece of gauze.

Day 4 — Count the number of eggs in the trays. Eggs are oval and orange-coloured. Also inspect the sides of the tray and the lid. Some eggs may be hatched; small larvae can hide under the aphids.

Calculate the number of eggs per female. If many zeros (> 5) are found and midges are still alive at the end of the test, something has gone wrong in the mating and the test should be repeated.

Flight test — Put 250 pupae in a tray. Place the tray in cylinder with a diameter of approximately 25 cm. Make a small ring of grease (any kind of grease will do, as long as it does not melt) at 5 cm from the bottom. Put the cylinder in a large cage or leave it open in the climate room. When the majority of the midges have emerged count the number on the bottom of the cylinder (non-fliers) and count the number of white skins in the tray (= total emerged). Calculate the percentage of fliers.

Design: J. van Schelt.
Coordinators: J. van Schelt and B. Spencer.

Aphytis spp. (_A. lingnanensis_ (Compere), _A. melinus_ (De Bàch)) (Hymenoptera: Aphelinidae)

Test conditions

Temperature:	25 ± 2°C
RH:	60 ± 10%
Light regime:	16L : 8D
Rearing host:	*Aspidiotus nerii*

Quality control criteria

Quantity	The number of live adults as specified on the container, $n = 2$; weekly or batch-wise test.
Sex ratio	70 ± 10% females; 100 adults from bulk material; monthly.
Survival in transport	⩾ 80%; 5 × units; 5 days, $T = 17 \pm 2$°C; monthly.
Fecundity and longevity	⩾ 20 offspring per 5 days per female; 80% of females should live at least 5 days; $n = 30$; seasonal test.

Description of testing methods

Sex ratio	Examination of 100 adults with binocular microscope. Presence of ovipositor on ventral surface to indicate females.
Survival in transport	Approximately 1000 adults are confined to a capsule which also contains sufficient honey in the lid. After 5 days, the capsules are checked for mortality. Mortality after day 5 is < 20%; at least monthly test or batch-wise if batches were exposed to special treatments (e.g. storage procedures, long-range shipments).
Fecunditiy and longevity	30 females (age 24 h) are confined individually in a plastic capsule with a male and a drop of honey. The capsule is affixed to the surface of a squash infested with *A. nerii*. Each female should have access to 100 scales. The capsule and *Aphytis* are removed at day 5. A parasitism assessment of the scale is performed at day 14. Minimum fecundity is 20 pupae per female; mortality after day 5 is ⩽ 20%.

Comments

Recommend replacement of culture with field-collected strain every 2 years.

Initial design: D.F. Papacek and A.C. Dove.
Coordinators: D.F. Papacek and A.C. Dove.

Chrysoperla carnea Steph. (Neuroptera: Chrysopidae) Provisional test

Test conditions

Temperature:	25 ± 1°C
RH:	70 ± 5%
Light regime:	16L : 8D

Quality control criteria

WHEN SHIPPED AS EGGS

Quantity	The number of eggs as specified on the package; a weekly test.
Hatching rate	⩾ 65% within 5 days; $n = 200$; eggs must be isolated to prevent cannibalism after emergence; a weekly test.
Predator quality	⩾ 65% of newly hatched larvae has to develop to second-instar larvae within 4 days using aphids as prey; to be conducted once a year or when the rearing system is changed.

Predator efficiency/ searching capability	⩾ 25 *M. euphorbiae* or 50 *M. persicae* should be consumed within 4 days by a second-instar larva; $n = 10$; an annual test.

WHEN SHIPPED IN SECOND LARVAL STAGE

Quantity	Number of live predators as specified on the package; a weekly test.
Predator quality	⩾ 65% of second-instar larvae have to develop to third-instar larvae within 5 days using aphids as prey; to be conducted once a year or when the rearing system is changed.
Predator efficiency/ searching capability	⩾ 25 *M. euphorbiae* or 50 *M. persicae* should be consumed within 4 days by a second-instar larva; $n = 10$; an annual test.

Description of testing methods

Preparation of agar-leaf for testing predator quality and predation capacity	Use a round Petri dish (13.5 cm diameter and 2 cm height) that can be closed very tightly; a nylon mesh is incorporated into the lid to facilitate air exchange. Trays are filled with agar solution (1.5%) to a depth of 5 mm; just before the agar solidifies a leaf is placed with its upper surface on agar; then transfer 30 adult aphids with a toothpick on to the leaf; close the Petri-dish tray and turn it upside down. All aphid adults are removed 24 h later and the offspring are counted (offspring should be ⩾ 100). Three species of aphids can be used as prey items: *A. gossypii* on cucumber, *M. euphorbiae* on strawberry or potato, or *M. persicae* on sweet pepper.

FOR *CHRYSOPERLA* SHIPPED AS EGGS

Predator quality	Offer individual, freshly emerged larvae at least 50 prey items on a leaf on agar as described above; $n = 30$.
Predator efficiency	As described for larvae, see below.
Searching capacity	As described for larvae, see below.

FOR *CHRYSOPERLA* SHIPPED AS LARVAE

Predator quality	Offer individual, freshly emerged second-instar larvae about 100 aphids on a leaf on agar as described above; $n = 30$.
Predator efficiency	Prepare agar–leaf set-up as described above; introduce second-instar larva into Petri dish and count number of aphids consumed after 4 days.
Searching capability	This test is carried out on a plant (i.e. on sweet pepper in the 'ten-leaf stage' or on potato) put into a transparent plastic cylinder (20 cm diameter and 50 cm high) with nylon mesh on the top for air exchange and pressed on to a container filled with vermiculite to prevent escape of larvae. Take a leaf with the offspring of 30 adult aphids, count the aphids and cut the leaf in equal pieces, then distribute these pieces all over the plant to get an even distribution. Half a day later the aphids have moved on to the plant; then transfer a second *Chrysoperla* instar on each plant. Count the remaining aphids 4 days later.

Initial design: M.G. Tommasini and R. Mayer.
Coordinator: M. Mosti (provisional test).

Cryptolaemus montrouzieri Mulsant (Coleoptera: Coccinellidae) **Provisional test**

Test conditions

Temperature:	25 ± 2°C
RH:	70 ± 5%
Light regime:	16L : 8D

Quality control criteria

Quantity	Average number of live predators as specified on the container; $n = 3$ containers; a weekly or batch-wise test;
Sex ratio	⩾ 40% females; seasonal; $n = 100$; for identification of sex, examine anterior femur: orange for males and black for females.
Longevity	Minimum 30 days, reached by at least 80% of the females examined in the fecundity test; $n = 30$; a seasonal test.
Fecundity	⩾ 50 eggs per female per 10 days; $n = 30$; a seasonal test.

Description of testing methods

Quantity	Hold samples at 10°C for at least 1 h prior to testing to reduce activity of beetles. To accurately count total number of beetles, carefully remove any packing material, allowing the beetles to drop back into container. If beetles have been shipped in a large container, it is advisable to transfer the beetles to a snap-top vial or bottle for ease in handling. Gently tap a portion of the beetles on to the counting tray and immediately start counting beetles. Record the count, then brush the counted beetles into the pill-tray reservoir. Store the counted beetles in a second container with a snap-on lid. Repeat the process until all the beetles are counted. While counting the beetles, record the presence of any live arthropods other than adult *Cryptolaemus* beetles. Since the beetles start to warm during the counting process, the sample container must be repeatedly tapped on a hard surface to cause the beetles to drop to the bottom of the container to prevent escape.
Fecundity	For each batch to be tested, prepare 30 replicates. Each replicate consists of one female beetle in a small ventilated tube or cup containing a 2 cm piece of potato sprout infested with live citrus mealy-bug egg masses, nymphs and adults of a sufficient quantity to provide *ad libitum* diet and oviposition sites. After 48 h, transfer the female to a freshly prepared container. Count the eggs present in the old container by inspecting the infested sprout, using a dissecting microscope, and record time of first observation of *C. montrouzieri* eggs. Tease apart mealy-bug colony and egg masses with a blunt probe and count and record presence of *C. montrouzieri* eggs. It is necessary to distinguish between the eggs of *C. montrouzieri* and those of *Planococcus citri*. Eggs of *C. montrouzieri* are larger and pale yellow. Terminate the test if eggs have not been observed after 7 days of observation.

Designers and coordinators: A. Hale and S. Steinberg (provisional test).

Dacnusa sibirica Telenga (Hymenoptera: Braconidae)

Test conditions

Temperature:	22 ± 2°C
RH:	60 ± 5%
Light regime:	16L : 8D

Quality control criteria

Quantity	⩾ number of live adults specified on the label; a weekly or batch-wise test.
Adult mortality	⩽ 5% of the number of adults present in the container, based on three containers sampled and $n = 500$ or more; a weekly or batch-wise test.
Sex ratio	⩾ 45% females; $n = 100$; conducted four times per year.
Fecundity	⩾ 45 offspring per female within 3 days; $n = 15$; an annual test. Daily oviposition of single female wasps on brown beans (*Phaseolus vulgaris*) infested with sufficient *Liriomyza trifolii* and a source of carbohydrate (other hosts if *L. trifolii* is not available: *Liriomyza bryoniae* or *Chromatomyia syngenesiae*).

Description of testing methods

FECUNDITY TEST: *DACNUSA SIBIRICA* USING *CHROMATOMYIA* AS HOST

Plant	*Sonchus oleraceus* grown from seed for *c.* 8 weeks.
Host leaf-miner	*C. syngenesiae*, a minimum of 100 second-instar larvae per plant (*c.* 8 days after eggs laid).
Apparatus	60 cm Perspex cube cages.
Day 0	Place a single female *Dacnusa* into a cage containing three plants infested with leaf-miner on 3 consecutive days and leave for 72 h.
Day 3	Remove the females from the cages. Leave the plants *in situ* until the leaf-miner larvae have pupated (*c.* 5 days later).
Day 8±	Cut plants off, save leaf and stem tissue containing pupae and discard the rest. Store pupae in a ventilated plastic box (*c.* 30 cm × 30 cm × 15 cm is ideal), lined with several layers of absorbent paper, at RH of 80% and 22°C for 21 days.
Day 29	Freeze the box for 24 h and count *Dacnusa* adults.

FECUNDITY TEST: *DACNUSA SIBIRICA* USING *LYRIOMYZA* AS HOST

Plant	*P. vulgaris* grown from seed 2 weeks before starting the test.
Host leaf-miner	*L. bryoniae.*
Preparations	Two weeks before starting the test, sow *c.* 300 *P. vulgaris* seeds in small pots with soil. Ensure that the plants can grow under optimum conditions (sufficient light and nitrogen, no sciarid flies), as strong plants are required. After 2 weeks remove the tops of the plants, leaving only the spade leaves. About 500 1-day-old female *L. bryoniae* are then released on to the plants (the leaf-miners are provided with honey from the time they emerge from the pupae). Keep the plants in a greenhouse at about 22–25°C until the first larvae have reached second-instar (*c.* 7 days).
Day 0	Count the approximate number of larvae on the infested plants. Place as many plants in a cage as is necessary to provide about 150 larvae per parasite. Do not use plants with more than 30 larvae per

	leaf. Collect 15 female wasps from a bottle. Put one female in each cage.
Day 3	Put/fix a sleeve or ventilated box around the leaves to collect the larvae, which will pupate within about 5 days at 22°C. The RH must be 80%.
Days 5–8	Check that all the larvae have pupated. Before they hatch, collect all the pupae in a Petri dish and store them at 22°C and 80% RH.
Day 20 ±	Check that all the pupae have hatched and count the number of adults emerged after killing them in a freezer. Calculate the number of offspring produced per female during the 3-day test.

Design: J. Dale, P. Smytheman and R. Greatrex.
Coordinators: P. Smytheman, J. Dale and R. Greatrex.

Dicyphus hesperus **Wagner (Hemiptera: Miridae)**

Test conditions

Temperature:	22°C ± 2°C
RH:	75 ± 10%
Light regime:	16L : 8D

Quality control criteria

Quantity	⩾ the number of live adults and nymphs as specified on the label; weekly test.
Mortality	⩽ 5% of the number of live adults and nymphs as specified on the label; weekly test.
Sex ratio	⩾ 45% females; $n = 100$; seasonal test.
Fecundity	⩾ 7 eggs per female per 72 h; $n = 30$; annual test.

Description of testing methods

Quantity	Place packages in the freezer for at least 1 h. Count the insects.
Mortality	Count the dead insects left in the packages after live insects have been allowed to move to another container.
Sex ratio	Take a sample of 500 insects found in the quantity test and determine sex ratio.
Fecundity	Collect 30 females 7–10 days after their final moult from the mass rearing. Place them individually on a tobacco-leaf disc (5 cm diameter) with a midrib in the middle, placed upside down on a 4 mm layer of agar (1%). The container has to be ventilated with at least 2 cm between the agar and the lid. Feed them *E. kuehniella* eggs *ad libitum*. Remove the insects after 72 h and examine leaf discs for predator eggs under a stereomicroscope. Eggs will be embedded in leaf midrib and veins. The average number of eggs laid should be ⩾ 7 eggs per female per 72 hours.

Design: C. Castane.
Coordinators: J. Klapwijk and K. Jans.

***Diglyphus isaea* (Walker) (Hymenoptera: Eulophidae)**

Test conditions

Temperature:	25 ± 2°C
RH:	70 ± 10%
Light regime:	16L : 8D
Light intensity:	< 300 lx; direct lighting to be prevented – it affects the activity of females.

Quality control criteria

Quantity	⩾ number of live adults specified on the label; a weekly or batch-wise test.
Adult mortality	⩽ 8% of the number of adults present in the container, based on three containers sampled and n = 500 or more; a weekly or batch-wise test.
Sex ratio	Females ⩾ 45% of live adults; n = 100; a batch-wise test.
Fecundity	70% of females laying eggs within 1 week in a Petri-dish test; n = 30; an annual test.

Description of testing methods

Fecundity

Isolate individual females, chosen at random from the container, in Petri dishes of *c.* 12 cm diameter. Use Parafilm to seal the dish. For each replicate, staple a brown bean leaf (*P. vulgaris* L.) infested by *c.* 10 L_2–L_3 larvae of *L. trifolii* or *L. bryoniae* to a slightly moistened filter-paper and place it in a dish. Add some honey droplets to the lid of the dish. Alternatively, *S. oleraceus* L. leaves infested by *C. syngenesiae* Hardy can be used.

Replace the leaf daily for up to 7 days and check for presence of parasitoid eggs in the mines. Leaves can be stored prior to dissection for up to 24 h at 4–8°C. For each female the test is completed once the first parasitoid egg is found. At the end of 7 days, calculate the percentage of females that have laid eggs. Include the total females which died naturally. The number of females accidentally killed or lost should be omitted from the calculations, but must be recorded. (See Fig. 19.7.)

Design: G. Nicoli and S. Steinberg.
Coordinators: S. Steinberg, P. Smytheman and G. Burgio.

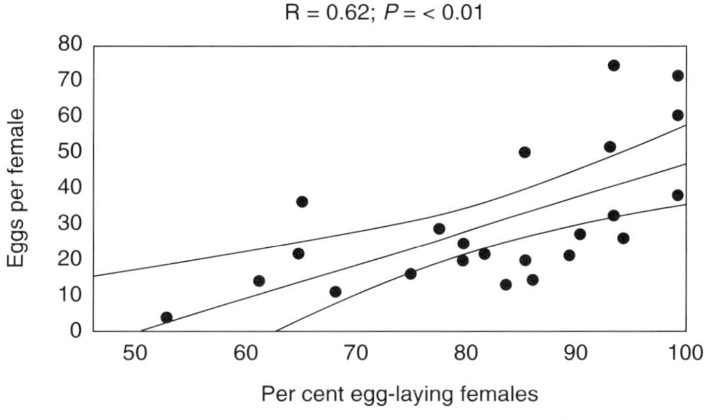

Fig. 19.7. Prediction of fecundity. The number of eggs laid per female in 7 days can be predicted from the relationship between this value and the percentage of egg-laying females (each replicate: n = 30 females). The fecundity data were obtained using the testing method described above (G. Nicoli, unpublished data).

Encarsia formosa **Gahan (Hymenoptera: Aphelinidae)**

Test conditions

Temperature:	22 ± 2°C
RH:	60–90%
Light regime:	16L : 8D

Quality control criteria

Emergence rate	⩾ the number of adults specified on the label that will emerge over a 2-week period; $n = 1000$; a weekly or batch-wise test.
Sex ratio	⩾ 98% females; $n = 500$; an annual test.
Fecundity	⩾ 7 eggs per female per day for days 2, 3 and 4 after emergence of the adult; $n = 30$ females; an annual test.

Description of testing methods

Emergence	Specify the number of adults that should emerge before conducting the test. Take three subsamples that make up 1000 or more full black pupae in total. Put the samples in a closed container for 2 weeks and then determine the number of emerged adults. This can be done by counting the number of emerged adult parasites or by comparing the number of empty pupae at the start and at the end of the test. A combination of both counting methods will give the most reliable results. The quantity of emerged adults should achieve the number specified on the label.
Sex ratio	Take a sample of 500 of the adults from the emergence test and count the number of male wasps. These are completely black and easily distinguished from the females, which have a yellow abdomen. The number of females should be ⩾ 98%.
Fecundity	
Day 1	Put an ample amount of black pupae that are close to emergence in a container. Remove all adult parasites the night before the day on which the test animals will be collected from the container.
Day 2	Collect 30 freshly emerged females at about 10 o'clock; put each into a small container with a droplet of honey until the following day. This is to feed them and to get them through the preoviposition period.
Day 3	The test is conducted on individual females in small round plastic Petri-dish-type trays (min. diameter 35 mm; height 15 mm), which can be closed very tightly. A nylon mesh is incorporated into the lid to facilitate air exchange. Trays are filled with agar solution (1%) to a depth of 10 mm. Just before the agar solidifies, a tobacco leaf disc is placed with its upper surface in contact with the agar. The leaf disc should contain at least 25 whitefly larvae (*Trialeurodes vaporariorum*) in the third and fourth instar. To ensure an optimum quality of leaf disc, pick leaves and prepare trays early in the morning and use leaves with a density of max. 3 larvae cm^{-2}. Prepare 30 trays in total and release one female per tray.
Day 4	Provide the female with a new supply of whitefly larvae by placing her in a new tray. Do this around 10 o'clock in the morning, again.
Day 5	Repeat day 4.
Day 6	Remove the parasites from the whitefly larvae. Keep all whitefly that were exposed to *E. formosa* in closed containers to prevent unwanted

parasitism after the test. Count all black pupae after 14 days. The average number of black pupae per female per day should be ⩾ 7. This test should be performed in the period August to October.

Design: J.C. van Lenteren and W.J. Ravensberg.
Coordinators: J. Klapwijk, P. Smytheman and A. Luczynski.

Eretmocerus eremicus **(Rose) (Hymenoptera: Aphelinidae)**

Test conditions

Temperature:	22°C ± 2°C
RH:	75 ± 10%
Light regime:	16L : 8D

Quality control criteria

Emergence	⩾ the number of adults as specified on the label which will emerge over a 2-week period; *n* = 1000; a weekly or batch-wise test.
Sex ratio	⩾ 45% females; *n* = 500; an annual test.
Fecundity	⩾ 45 eggs per female per 72 h for days 2–4 after emergence of the adult; *n* = 30 females; an annual test.

Description of testing methods

Emergence: Take at least three subsamples that make up 1000 or more yellow pupae in total. Put the samples in a closed, ventilated container for 2 weeks and then determine the number of emerged adults. This can be done by comparing the number of empty pupae at the start and at the end of the test. The quantity of emerged adults should achieve the number specified on the label.

Sex ratio: Take a sample of 500 of the adults from the emergence test and count the number of female wasps. These can be distinguished from the males by the shape of the antennae (see Fig. 19.8) and the brighter yellow colour. The number of females should be ⩾ 45%.

Fecundity

Day 1: Put an ample amount of yellow pupae that are close to emergence in a container. Remove all adult parasites the night before the day on which the test animals will be collected from the container.

Day 2: Collect 30 freshly emerged females and males in the morning; put them together in a container with a droplet of honey until the following day. This is to allow them to mate, to feed them and to get them through the preoviposition period.

Day 3: The test is conducted on individual females in small round plastic Petri-dish-type trays (min. diameter 50 mm; height 15 mm), which can be closed very tightly. A nylon mesh is incorporated into the lid to facilitate air exchange. Trays are filled with agar solution (1%) to a depth of 10 mm. Just before the agar solidifies, a tobacco leaf disc is placed with its upper surface in contact with the agar.

The leaf disc should contain at least 60 whitefly larvae (*T. vaporariorum*) in the second and third instar. To ensure an optimum quality of the leaf disc, pick leaves early in the morning and prepare trays immediately. Use leaves with a density of max. 3 larvae cm^{-2}. Prepare 30 trays in total and release one *E. eremicus* female per tray.

Day 4 Provide the female with a new supply of whitefly larvae by placing her in a new tray. Do this around the same time in the morning.

Day 5 Repeat day 4.

Day 6 Remove the parasites from the whitefly larvae. Keep all whitefly that were exposed to *E. eremicus* in closed containers to prevent unwanted parasitism after the test. Remove adult whitefly emerging from unparasitized pupae to keep the leaf disc in optimum condition.

Day 20 Count all yellow pupae. The average number of yellow pupae per female per day should be ⩾ 15.

Design: J.N. Klapwijk.
Coordinators: J.N. Klapwijk and Pete Smytheman.

Fig. 19.8. Antennae of *Eretmocerus eremicus*: male (left) and female (right).

Eretmocerus mundus **Mercet (Hymenoptera: Aphelinidae)** **Provisional test**

Test conditions

Temperature: 25 ± 2°C
RH: 65 ± 5%
Light regime: 16L : 8D

Quality control criteria (based on shipment of parasitized pupae)

Emergence See *Eretmocerus eremicus*.

Sex ratio ≥ 50% females, $n = 5$ bottles, a sample of 100 wasps from each bottle; batch-wise or seasonal test.

Fecundity ≥ 10 offspring per female, $n = 30$; seasonal or annual test.

Description of testing methods

Emergence See *Eretmocerus eremicus*

Sex ratio Take a sample of about 500 wasps from the emergence test and count the number of females. Sex can be easily determined by the antennae and the dark brown coloration on the male's dorsothorax.

Fecundity The test is performed on cotton plants infested with second-to-third-stage nymphs of *Bemisia tabaci*. The cotton should be about 20 cm high with two or three true leaves. Each plant should be infested with about 30 *B. tabaci* adults 12 days earlier in order to get enough second-to-third-stage nymphs, which are preferred by *E. mundus*. Place the plants in a well-ventilated transparent cage (~4 l volume, ~30 cm high, 15 cm wide, 15 cm long).

Allow newly emerged wasps (males and females) to mate and feed (honey should be supplied as food) for 24 h in a container. After 24 h, select randomly 30 *E. mundus* females from this container and place each female separately in one of the cages with the plants. After 12–14 days, take off the leaves and count the number of *E. mundus* pupae/larvae. An alternative option is to wait 10 more days and then count the empty pupae with circular emergence hole and dead pupae.

Coordinators: E. Chiel and S. Steinberg (provisional test).

***Hypoaspis miles* Berlese (Acarina: Laelapidae) Provisional test**

Test conditions

Temperature:	20°C ± 1°C
RH:	70 ± 5%

Quality control criteria

Quantity	The number of live predators as specified on the label (including larvae, nymphs and adults, excluding eggs). The test should be done batchwise, just before delivery.
Sex ratio	⩾ 50% of the adults present should be female. Sex ratio is said to be significantly influenced by the type of food supplied to the female mite and/or the density of mites at the mass rearing. Sex ratio is normally female biased (0.66–0.85)
Fecundity	Test is in development.

Description of testing methods

Quantity

H. miles is normally propagated and sold in a mixture of sphagnum and vermiculite. The mites appear as adult individuals as well as in the larval and the nymph stages and in a mixture with mould mites (*Tyrophagus putrescentiae*), which forms the feeding material. Alternatively, nematodes can be used as food.

The small 'Berlese technique' is used when counting the mites (see description under *Neoseiulus cucumeris*). First of all, a homogenized and representative sample is taken. The bottle should be kept at room temperature (18–25°C) for at least half an hour, because the mites have to be active enough to let themselves fall through the sieve. The number and the size of the sample(s) will influence the precision of the counting. A sample-size of 1 g (*c.* 5–6 ml) will normally result in a good count. Put the sample in a metal sieve, 6 cm diameter, height 2.5 cm, mesh size 450 μm and 48% open. The material must be spread evenly on the sieve. The sieve is then placed under a lamp of 150 W at a distance of 6 cm. To ensure that all mites are driven out and not burnt off, the heating must be started with a warming up time of 4 min. After 4 min the lamp is turned up to full power for about 17 min. Under the sieve the falling mites are caught on white tape with substantial adhesiveness. The number of mites can be counted directly on this tape. Note that, with a humidity of the material of *c.* 45–55%, this method works; whenever the humidity is higher it does not work any more and a new lamp–sieve distance and heating time will have to be determined.

Based on the numbers counted, a confidence interval can be calculated (i.e. it can be estimated within a certain confidence level – e.g. 95% – that the numbers of mites are caught inside a specific calculated interval). If, for example, a test is made on a 95% confidence interval and the lower value of the confidence interval is specified on

	the label, there is a probability of 97.5% that the specified number is correct. As mentioned above, the number of samples will affect the confidence interval of the mites. Similarly, the variation within the samples is crucial for how narrow the confidence interval will be.
Sex ratio	A sample of mites is examined and the numbers of males and females are counted. The male can be distinguished on its ventral shield which is not divided, the male has an oval genital opening in front of the ventral shield. The female has three ventral shields: a sternal, a genital and an anal shield. Different size of samples results in different certainty of the sex-ratio determination as shown from an example in our testing: a sample of 120 mites causes an uncertainty of *c.* ± 6.9%, 240 mites causes a certainty of *c.* ±4.8 %, 480 mites causes a certainty of *c.* ± 3.4 %.

Initial design: B. Larsen and J. Reitzel.
Coordinators: J. Vermeulen and S. Mulder (provisional test).

Leptomastix dactylopii Howard (Hymenoptera: Encyrtidae)

Test conditions

Temperature:	25 ± 2°C
RH:	70 ± 5%
Light regime:	16L : 8D

Quality control criteria

Quantity	⩾ the number of live adults as specified on the label; a weekly or batch-wise test.
Adult mortality	⩽ 10%, based on three containers sampled and n = 500 or more; a weekly or batch-wise test.
Sex ratio	⩾ 45% of the number specified on the label should be females; the sex ratio does not necessarily have to be 45% as long as there are enough females in the container; a 4-weekly test.
Fecundity	⩾ 40 offspring per female per 14 days; n = 30 females; an annual test

Description of testing methods

Fecundity	Place a single potato tuber with short sprouts and infested by an ample amount of L3 females of citrus mealy bug, *P. citri*, in a ventilated container. Introduce a single pair of *L. dactylopii* into the container. Leave the system as it is for 14 days. By the end of the 14 days, take out the pair of wasps from each container. Collect emerging adults of *L. dactylopii* from the 21st day up to the 31st day from the beginning of the experiment. Calculate the cumulative number of adults emerging during this period.

Initial design: S. Steinberg and M. Tommasini.
Coordinators: M. Mosti and M. Kole.

Macrolophus caliginosus Wagner (Hemiptera: Miridae)

Test conditions

Temperature:	22 ± 2°C
RH:	75 ± 10%
Light regime:	16L : 8D

Quality control criteria

Quantity	⩾ the number of live adults and nymphs as specified on the label; weekly test.
Mortality	⩽ 5% of the number of live adults and nymphs as specified on the label; weekly test.
Sex ratio	⩾ 45% females; $n = 100$; seasonal test.
Fecundity	⩾ 7 eggs per female per 72 h; $n = 30$; annual test.

Description of testing methods

Quantity	Place packages in the freezer for at least 1 h. Count the insects.
Mortality	Count the dead insects left in the packages after live insects have been allowed to move to another container.
Sex ratio	Take a sample of 500 insects found in the quantity test and determine sex ratio.
Fecundity	Collect 30 females 7–10 days after their final moult from the mass rearing. Place them individually on a tobacco-leaf disc (5 cm diameter) with a midrib in the middle, placed upside down on a 4 mm layer of agar (1%). The container has to be ventilated, with at least 2 cm between the agar and the lid. Feed them *E. kuehniella* eggs *ad libitum*. Remove the insects after 72 h and examine leaf discs for predator eggs under a stereomicroscope. Eggs will be embedded in leaf midrib and veins. The average number of eggs laid should be ⩾ 7 eggs per female per 72 h.

Design: C. Castane.
Coordinators: J. Klapwijk and K. Jans.

Neoseiulus californicus McGregor (Acarina: Phytoseiidae)

Test conditions

Temperature:	25 ± 1°C
RH:	75 ± 5%
Light regime:	16L : 8D

Quality control criteria

Quantity	Average number of living predators as specified on the container. Every week or batch-wise check.
Sex ratio	⩾ 60% females; $n = 100$; for identification of sex, mount the individuals on microscopic slides; once a year.
Fecundity	⩾ 7 eggs per female, during a period of 5 days ($n = 30$); a seasonal test.
Longevity	⩾ 5 days reached by 80% of the females examined in the fecundity test. $n = 30$; a seasonal test.

Description of testing methods

Quantity — Empty the contents of a bottle into a container, weigh the contents. Mix thoroughly with a spoon. Take three samples of 2 g (for instance, in a small cup). Empty the small cup in the upper sieve of a set of two sieves (upper one 315 μm, lower one 90 μm). Wash out the cup through the sieves. Run cold water through the sieves for a few minutes. All stages of the mite pass through the upper sieve and remain in the lower sieve. Trickle hot water on the sieve so that the mites are dead but not completely destroyed. Submerge the lower part in a shallow dish with a bit of water and detergent, and move the sieve gently. The mites in the sieve are evenly distributed now. Remove the sieve and wipe the underside of the sieve dry with a piece of tissue-paper. Place the sieve on top of a graph-paper circle covered by plastic. Count the predatory mites in the 2 g sample. Take the weighed mean of the three samples to calculate the amount of predatory mites in the bottle.

Fecundity — Prepare leaf discs of 2.7 cm diameter of brown beans (*P. vulgaris*) or sweet pepper (*Capsicum annuum*) infested with ten to 15 mobile stages of the two-spotted spider mite (*Tetranychus urticae*). An ample amount of spider-mite eggs should also be present. Place the discs on agar, their infested side facing upward, in small plastic containers (diameter 32 mm and 15 mm height). The container lid should fit tightly to prevent escape of mites and should incorporate proper ventilation. Take females of *N. californicus* at random from the bottle. Put a single adult female on each leaf disc (= one replicate) by allowing mites to walk on to a fine brush. Place the containers with leaf discs upside down, i.e. infested side facing downward, so as to simulate the true orientation of the predatory mites in the field. After 48 h, transfer the females to freshly prepared containers. Count the eggs and/or larvae present in the old container. Leave the female predator in the new container for another 72 h. Then count the number of eggs and/or larvae. Add up the two counts to get the total fecundity for 5 days. Calculate the mean and standard error for the 30 replicates. Exclude individuals that do not lay eggs throughout the test, but indicate their number.

Initial design: J. van Schelt and S. Steinberg.
Coordinators: S. Mulder and E. van Baal.

Neoseiulus cucumeris **Oudemans (Acarina: Phytoseiidae)**

Test conditions

Temperature:	22 ± 1°C
RH:	70 ± 5%
Light regime:	16L : 8D

Quality control criteria

Quantity — The number of live predators as specified on the label, excluding eggs, at the time of delivery (for both containers and controlled-release systems).

Sex ratio — ⩾ 50% females; $n = 100$; an annual test.

Fecundity ⩾ 7 eggs per female over a period of 7 days. Count from the second day of testing; $n = 30$ females; an annual test.

Description of testing methods

Quantity *Neoseiulus cucumeris* is normally sold as a mix of bran, bran mites (as a food source) and the predatory mites themselves. Both the ratio of the two mite species and the concentration can vary considerably, depending on the product and the producer.

Mites can be washed out of the material with (hot) water, though counting is not easy because of reflection and difficult identification of the mite species. A more accurate method is to use a 'Berlese technique', as described below. Mites are driven out of the material with the heat of a lamp. The advantage is that the mite species are clearly visible and dead mites will remain behind in the sieve. Make sure that a 'warming-up time' is allowed for. This gives the small bran mites the chance to walk downwards before getting burned. Full heat is needed to drive the predatory mites out of the sieve.

Use material from one container or four sachets, depending on the product. Empty the contents into a bucket and weigh the content. Mix thoroughly with a spoon to get a homogeneous mixture. According to the density, take the following samples:

Density of *N. cucumeris*	Sample size
1000 5 g^{-1}	0.5 g
500 5 g^{-1}	0.5 g
250 5 g^{-1}	1.0 g
100 5 g^{-1}	1.0 g

Put the material directly in a sieve of 6 cm diameter, 2.5 cm height, mesh width 333 μm, 42% open. Spread the material as evenly as possible. Place the sieve at a distance of 4 cm under a lamp of 150 W. The warming-up time should take 5 min. Full power for an extra 10 min (see Figs 19.9 and 19.10). Put a piece of black sticky tape under the sieve to trap the falling mites.

The number of mites can be counted directly with a grid or if the mites can still walk over the glue, they can be killed in the freezer (20 min). Use fibre light to prevent melting of the glue. The stickiness of the glue is very important. When the tape is not sticky many mites will walk off the tape.

An alternative for the black sticky tape is to use a black plate with a ring of pure detergent as a barrier. Mites must be killed in the freezer immediately after extracting them. The humidity of the material is also very important. Within the range of 16.5–19% there does not seem to be a difference in the counting. At a higher humidity of the material, there may be a different total heating time, because mites stay longer in the material.

Fecundity

Day 0 The test is conducted on individual female mites in small round plastic Petri-dish trays (32 mm diameter; 15 mm height), which can be closed tightly. A nylon mesh is incorporated into the lid for ventilation. Trays are filled with agar solution (1%) to a depth of 5 mm. Just

	before the agar solidifies a sweet-pepper leaf disc is placed with its upper surface in contact with the agar. Care should be taken to ensure that a leaf disc has a vein and some hairs for egg deposition. Good contact between the leaf disc and the agar solution is also necessary to prevent predatory mites from hiding. Thirty 'big' mated females of *N. cucumeris* are taken from the commercial product. An ample amount of killed, fresh *E. kuehniella* eggs is added as food every day. Place the trays upside down in the climate room to simulate the natural leaf position.
Day 1	Remove the eggs laid on the first day. Do not include them in the total number of eggs laid.
Day 2	Count the egg laying of the predatory mites and provide the females with new trays as described on day 0.
Day 3	Count the eggs while removing them.
Day 4	Repeat day 2.
Day 5	Repeat day 3.
Day 6	Repeat day 2.
Day 7	Repeat day 3.
Day 8	Repeat day 3. The average number of eggs per female should be ≥ 7. Do not include eggs that are laid on the first day of the test.

Initial design: J. van Schelt and F. Stepper.
Coordinators: S. Mulder and A. Hale.

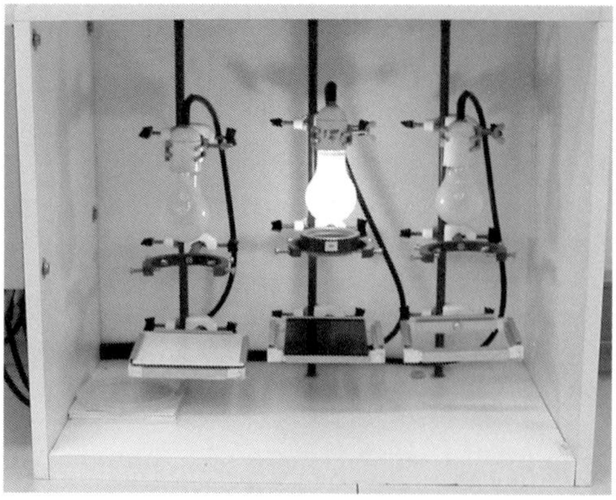

Fig. 19.9. Set-up of mite extractor.

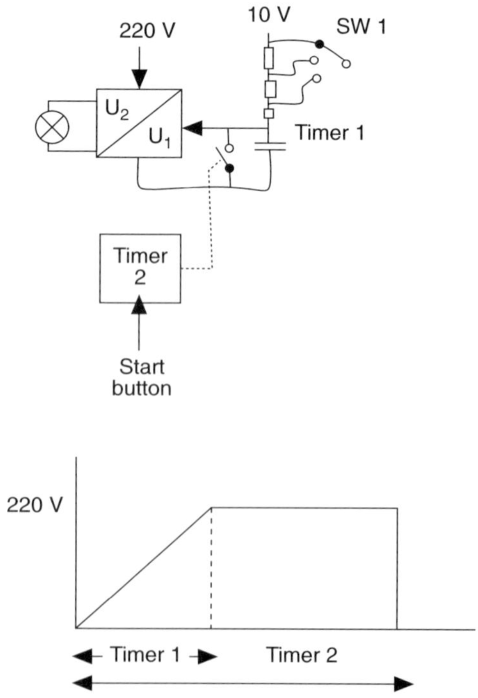

Fig. 19.10. Electric scheme of mite extractor.

***Neoseiulus* (*Amblyseius*) *degenerans* Berlese (Acarina: Phytoseiidae)**

Test conditions

Temperature:	22 ± 1°C
RH:	70 ± 5%
Light regime:	16L : 8D

Quality control criteria

Quantity	The number of mites as specified on the label, excluding eggs.
Sex ratio	⩾ 50% females; n = 100; an annual test.
Fecundity	⩾ 7 eggs per female over a period of 7 days; n = 30; an annual test

Description of testing methods

Quantity — Mix the contents of the package thoroughly by tumbling the product container. If the product container does not allow proper mixing use an alternative container. Weigh the contents of the package and take four random samples, each of 3 g. Spread each sample on a white paper and count the mites by gently stirring the material (vermiculite). First, count the live predators running away; the counted mites will be killed. Next, sieve the material, rinsing it with cold water. The upper sieve opening size is 440 μm, the lower is smaller (90 μm) so that the mites will be collected in it. Count the remaining mites (only live ones, check if they are still moving their legs) and add this number to the total number of mites counted earlier. Then calculate the total number of mites: (number of mites in 4 × 3 g/12) × the weight of the contents of the container.

A 'dry' method similar to the *Phytoseiulus persimilis* procedure can also be used; it has the advantage that one can distinguish between living and dead mites.

Sex ratio — Identification of sex is done by mounting individuals on microscopic slides. After mounting, gently push the body contents out of the body and check the ventral side with a magnification of 100×. Adding a droplet of a lactic acid solution will facilitate the procedure. Distinguish between females and males by checking the ventral and genital shields (see Fig. 19.11).

Fecundity — See testing method for *Neoseiulus cucumeris*. An ample amount of freshly hand-collected sweet-pepper pollen is used as food, however. The average number of eggs per female should be ⩾ 7. Do not include the eggs laid during the first day of the experiment.

Initial design: S. Steinberg and W.J. Ravensberg.
Coordinator: J. Vermeulen.

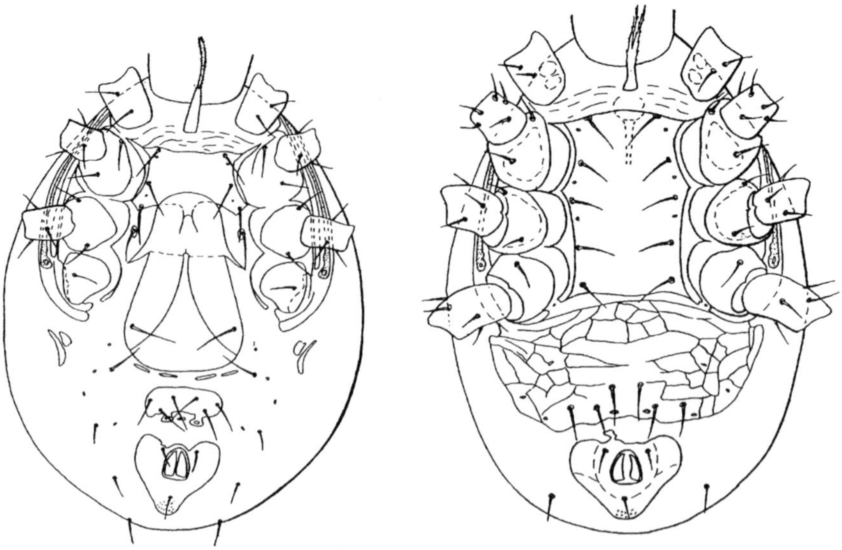

Fig. 19.11. The female (left) and male (right) genital shields of *Neoseiulus degenerans* (after Evans, 1953).

Orius spp. (O. aldibipennis, O. insidiosus, Orius laevigatus, O. majusculus) (Hemiptera: Anthocoridae)

Test conditions

Temperature:	22–25°C
RH:	70 ± 5%
Light regime:	16L : 8D

Quality control criteria

Quantity	The number of live adults/nymphs as specified on the container; species name(s) to be indicated on the label; a weekly test.
Sex ratio	⩾ 45% females; $n = 100$ (picked at random; to distinguish between male and female, see Fig. 19.12); a seasonal test.
Fecundity	⩾ 30 eggs per female per 14 days; $n = 30$ pairs; an annual test.

Description of testing methods

Counting method	After a short period (*c.* 20 min) at 8°C, the material in the container is sieved to separate vermiculite, buckwheat husk and insects; thereafter all the individuals can be counted using an aspirator.
Fecundity	Take about 100 *Orius* adults from a bottle ready for shipment or from a rearing cage with adults that emerged less than 24 h ago. Maintain these individuals for 2–3 days in a cage, feeding them with *E. kuehniella* eggs and supplying a French bean pod. Then determine their sex under stereomicroscope (see Fig. 19.12). Subsequently, put a pair of predators in a ventilated transparent container of *c.* 75 ml (3–4 cm of diameter), which contains a piece of French bean pod. The bean pod has to be cut between two seeds, to prevent egg laying on the inside of the bean pod. Provide new *E. kuehniella* eggs *ad libitum*. Every 2–3 days the bean pod is substituted with a fresh one; eggs are counted and new prey is added.

The total number of eggs is calculated for a 14-day oviposition period. Exclude data from females that have died accidentally or got lost during the test period.

Design: M.G. Tommasini and K. Bolckmans.
Coordinators: J. Vermeulen and M. Mosti.

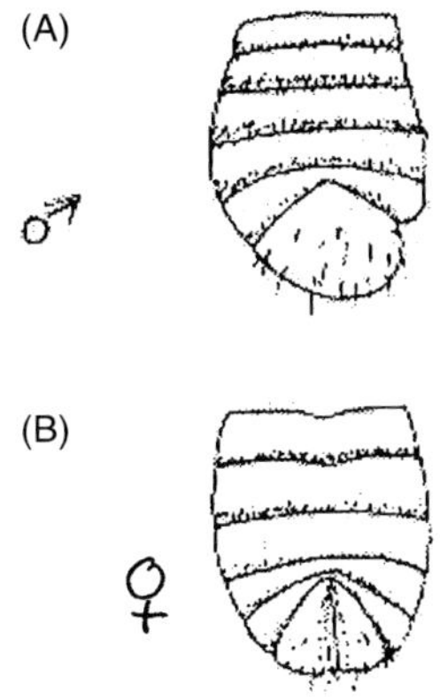

Fig. 19.12. Abdomen of *Orius*. Males have a curled, swollen abdomen (A), females have an elongated ovipositor (B).

Phytoseiulus persimilis Athias-Henriot (Acarina: Phytoseiidae)

Test conditions

Temperature:	22–25°C
RH:	70 ± 5%
Light regime:	16L : 8D

Quality control criteria

Quantity	Average number of live predators as specified on the container; $n = 3$ containers; a weekly or batch-wise test.
Sex ratio	⩾ 70% females; $n = 100$; once a year; for identification of sex, mount the individuals on microscopic slides.
Longevity	Minimum 5 days, reached by at least 80% of the females examined in the fecundity test; n= 30; a seasonal test.
Fecundity	⩾ 10 eggs per female per 5 days; $n = 30$; a seasonal test.

Description of testing methods

Quantity

Mix the contents of the package thoroughly by tumbling the product container. If the product container does not allow proper mixing, use an alternative container. Take a minimum of five samples per container, each sample consisting of 2% of carrier weight or volume for 200–500 ml containers or minimum 5% of carrier weight or volume for < 200 ml containers. Make sure that the material is remixed immediately before taking each sample.

Analyse the samples one by one. Spread the sample on a white sheet of paper under a warm bulb. Count the live predators running out of the material. Add to it the count of live predators remaining in the material. From these, estimate the total number of *P. persimilis* per container by dividing x (the mean number of adults per sample) by f (the fraction of the total contents of the vermiculite carrier in one sample) (x/f).

Fecundity

Prepare leaf discs of 2.5–3.5 cm diameter of brown beans (*P. vulgaris*) infested with an ample amount of the two-spotted spider mite (*T. urticae*, all developmental stages). Place the discs on agar, their infested side facing upward, in small plastic containers of the same diameter and of 2 cm height. The container lid should fit tightly to prevent escape of mites and should allow proper ventilation.

Take females of *P. persimilis* at random from the rearing unit. Put a single adult female on each leaf disc (= one replicate) by allowing mites to walk on to a fine brush. Place the containers with leaf discs upside down, i.e. infested side facing downward, so as to simulate the true orientation of the predatory mites in the field. After 48 h, transfer the female to a freshly prepared container. Count the eggs and/or larvae present in the old container. Leave the female predator in the new container for another 72 h. Then count the number of eggs and/or larvae. Add up the two counts to get the total fecundity for 5 days. Calculate the mean and standard error for the 30 replicates. Exclude individuals that do not lay eggs throughout the test, but indicate their number.

Preparation of agar substrate for fecundity tests, quantities for 6 containers:

- Boil 100 cm^3 of plain tap water.
- Add 1 g of Bacto agar when water temperature is 65°C.
- Boil solution again while stirring.
- Cool the agar solution to 40–45°C (keep on stirring while cooling).
- Pour agar solution in containers.
- Put the leaf discs on the agar when it is still liquid. Make sure the margins of the leaf discs are slightly dipped in the agar. This will prevent the disc from dehydrating prematurely.

Design: S. Steinberg and J. Dale.
Coordinators: S. Steinberg, J. Dale and A. Luczynski.

Podisus maculiventris Say (Hemiptera: Pentatomidae) **Provisional test**

Test conditions

Temperature:	22–25°C
RH:	70 ± 5%
Light regime:	16L : 8D

Quality control criteria

Quantity	Average number of live predators or viable eggs as specified on the container; $n = 3$ containers; a weekly or batch-wise test.
Eclosion rate	≥ 80% hatch; $n = 100$; weekly or batch-wise test.
Viability	≥ 90% active bugs, turgid, not desiccated, $n = 100$; weekly or batch-wise test.

Longevity	Minimum 30 days, reached by at least 80% of the females examined in the fecundity test; n=30; a seasonal test.
Fecundity	⩾ 50 eggs per female per 10 days; $n = 30$; a seasonal test.

Description of testing methods

Quantity	Nymphs: To accurately count the total number of insects, gently tap the bottom of container on a hard surface to remove all insects clinging to lid. Empty entire contents of bottle on to wire screen of appropriate size to separate vermiculite from insects and/or remove any other packing material. Tapping holding container on a hard surface will allow insects to fall to the bottom of the tray to prevent escape during counting procedure. Begin counting the number of living, active *Podisus* by hand-sorting one by one from one side of the counting tray into a second container. Record numbers of live insects. Eggs: Sieve packing material to separate eggs. Measure 10% of sample by weight or volume and count eggs while viewing with dissecting microscope. A fine paintbrush is useful for manipulating eggs. Multiply by 10 to estimate total number in container. Eggs can then be used to determine eclosion rate.
Eclosion rate	Prepare ten Petri dishes with one 2 cm piece of fresh bean pod in each dish to provide moisture. Take eggs of *P. maculiventris* at random from a shipping unit. Place ten eggs in each Petri dish (= one replicate) by transferring with a fine brush. Petri-dish covers must fit tightly or be taped to prevent escape of nymphal neonates and should allow proper ventilation. Check containers daily and record when first nymphs observed. After 48 h, record number of nymphs and non-viable eggs.
Viability	Pretest conditions: If containers of *P. maculiventris* have been chilled, warm to 22°C and hold for minimum of 1 h prior to testing viability. Acceptable *Podisus* are indicated by motility, grasping ability and/or lacking signs of desiccation or starvation, i.e. dorsoventrally flattened insects and dull colour patterns indicate poor quality.
Fecundity	For each replicate, confine a mating pair within a small, ventilated cage or large Petri dish. Provide insects with a water wick and *ad libitum* prey (yellow meal-worms, *Tenebrio molitor*). Check cages daily for 10 days and record eggs produced. Remove eggs daily after counting them. Add up the daily counts to get the total fecundity for 10 days. Calculate the mean and standard error for the 30 replicates. Exclude individuals that do not lay eggs throughout the test, but indicate their number.

Designers and coordinators: A. Hale and P. DeClercq (provisional test).

Trichogramma brassicae Bezd. (= _T. maidis_) (Hymenoptera: Trichogrammatidae)

Test conditions

Temperature:	23 ± 2°C
RH:	75 ± 10%
Light regime:	16L : 8D
Rearing hosts:	*E. kuehniella* *Sitotroga cerealella*

Species identification

The species is specified on the label and verified by the producer.[1]

Quality control criteria

Sex ratio	⩾ 50% females; 100 adults assessed on ten release units each or 5 × 100 adults of bulk material; at least weekly or batch-wise test if batches were exposed to special treatments (e.g. storage).
Number of females[2]	As indicated on label; determined as for sex ratio.
Fecundity and longevity	⩾ 40 offspring per 7 days per female; 80% of females should live at least 7 days; $n = 30$; monthly or batch-wise test.
Natural-host parasitism	⩾ 10 parasitized hosts per 4 h per female.

Description of testing methods

Fecundity and longevity	30 females (age 24 h) are confined individually in glass tubes; at least 200 factitious host eggs (< 24 h) are glued with water on to a small cardboard strip; a small droplet of honey and a droplet of water are added directly to the wall of the vial. Eggs of *E. kuehniella* (< 24 h old) are ultraviolet (UV)-irradiated and provided on day 1 and removed after day 7; fresh eggs of *S. cerealella* are provided on days 1, 3 and 5. The number of living adults is recorded after day 7. Egg cards are incubated and the number of black eggs is counted not earlier than day 10. Minimum fecundity after day 7 is 40 offspring per female; mortality after day 7 is < 20%; at least monthly test or batch-wise if batches were exposed to special treatments (e.g. storage procedures, long-range shipments).
Natural-host parasitism	30 females (age 24 h) are confined individually in tubes; two fresh egg masses of at least 20 eggs per egg mass of *O. nubilalis* (< 24 h old) are added for 4 h; honey and water are provided as described above; after separation of the egg masses from the females they are incubated for 3 days; the number of black eggs is counted; the mean number of black eggs is ⩾ 10 per female. The host-cluster acceptance rate (= females parasitizing at least one host egg) should be ⩾ 80%. This measure is important because parasitism drops drastically if a high proportion of females does not accept their hosts. This is especially true at low host densities and when hosts occur in batches. Often, parasitoids find only one egg mass during their lifetime and a high percentage acceptance is therefore crucial. This test is an indirect measure of the acceptance and suitability of the natural-host egg. The test should be performed two to four times per year depending on the rearing system (number of generations reared on the factitious hosts).

Comments

[1] Molecular techniques are available at Laboratory of Entomology, Wageningen University, The Netherlands. Test necessary once a year, sample size min. 30 individuals.

[2] The emergence period and pattern depend on the mixture of developmental stages released together and must be specified on the label.

Initial design: F. Bigler.
Coordinator: S.A. Hassan.

Trichogramma cacoeciae Marchal (Hymenoptera: Trichogrammatidae)

Test conditions

Temperature:	23 ± 2°C
RH:	75 ± 10%
Light regime:	16L : 8D
Rearing hosts:	*E. kuehniella* *S. cerealella*

Species identification

The species is specified on the label and verified by the producer.[1]

Quality control criteria

Sex-ratio	100% females; 100 adults assessed on ten release units each or 5 × 100 adults of bulk material; at least weekly or batch-wise test if batches were exposed to special treatments (e.g. storage).
Number of females[2]	As indicated on label, determined as for sex ratio.
Fecundity and longevity	⩾ 30 offspring per 7 days per female; 80% of females should live at least 7 days; $n = 30$; monthly or batch-wise test.
Natural-host parasitism	⩾ 5 parasitized hosts per 4 hours per female.

Description of testing methods

Fecundity and longevity	30 females (age 24 h) are confined individually in glass tubes; at least 200 factitious host eggs (< 24 h) are glued with water on a small cardboard strip; a small droplet of honey and a droplet of water are added directly to the wall of the vial. Eggs of *E. kuehniella* (< 24 h old) are UV-irradiated and provided on day 1 and removed after day 7; fresh eggs of *S. cerealella* are provided on days 1, 3 and 5. The number of living adults is recorded after day 7. Egg cards are incubated and the number of black eggs is counted not earlier than day 10. Minimum fecundity after day 7 is 30 eggs per female; mortality after day 7 is < 20%; at least monthly test or batch-wise if batches were exposed to special treatments (e.g. storage procedures, long-range shipments).
Natural-host parasitism	30 females (age 24 h) are confined individually in tubes; approx. 40 fresh eggs of natural host (*Cydia pomonella* or *Adoxophyes orana*, as available; < 24 h old) are added for 4 h; honey and water are provided as described above; after the separation of the eggs from the females, they are incubated for 3 days; the number of black eggs is counted; mean number of black eggs ⩾ 5 per female. This test is an indirect measure of the acceptance and suitability of the natural-host egg. The test should be performed two to four times per year, depending on the rearing system (number of generations reared on the factitious hosts).

Comments

[1,2] See comments for *T. brassicae*.

Initial design: F. Bigler.
Coordinator: S.A. Hassan.

Trichogramma dendrolimi Matsumura (Hymenoptera: Trichogrammatidae)

Test conditions

Temperature:	23 ± 2°C
RH:	75 ± 10%
Light regime:	16L : 8D
Rearing host:	*E. kuehniella* *S. cerealella*

Species identification

The species is specified on the label and verified by the producer.[1]

Quality control criteria

Sex ratio	⩾ 50% females; 100 adults assessed on ten release units each or 5 × 100 adults of bulk material; at least weekly or batch-wise test if batches were exposed to special treatments (e.g. storage).
Number of females[2]	As indicated on label; determined as for sex ratio.
Fecundity and longevity	⩾ 75 offspring per 7 days per female; 50% of females should live at least 7 days; $n = 30$; monthly or batch-wise test.
Natural-host parasitism	⩾ 10 parasitized hosts per 4 h per female

Description of testing methods

Fecundity and longevity	30 females (age 24 h) are confined individually in glass tubes; at least 200 factitious host eggs (< 24 h) are glued with water on a small cardboard strip; a small droplet of honey and a droplet of water are added directly to the wall of the vial. Eggs of *E. kuehniella* (< 24 h old) are UV-irradiated and provided on day 1 and removed after day 7; fresh eggs of *S. cerealella* are provided on days 1, 3 and 5. The number of living adults is recorded after day 7. Egg cards are incubated and the number of black eggs is counted not earlier than day 10. Minimum fecundity after day 7 is 75 eggs per female; mortality after day 7 is < 50%; at least monthly test or batch-wise if batches were exposed to special treatments (e.g. storage procedures, long-range shipments).
Natural-host parasitism	30 females (age 24 h) are confined individually in tubes; approx. 40 fresh eggs of natural host (*C. pomonella* or *A. orana*, as available; < 24 h old) are added for 4 h; honey and water are provided as described above; after the separation of the eggs from the females, they are incubated for 3 days; the number of black eggs is counted; mean number of black eggs ⩾ 10 per female. This test is an indirect measure of the acceptance and suitability of the natural-host egg. The test should be performed two to four times a year, depending on the rearing system (number of generations reared on the factitious hosts).

Comments

[1,2] See comments for *T. brassicae*.

Initial design: F. Bigler.
Coordinator: S.A. Hassan.

Acknowledgements

European and North American producers of natural enemies are thanked for cooperation in the design and testing of quality control guidelines. Development of quality control guidelines was financially supported by the Commission of the European Communities, Directorate General for Agriculture, Concerted Action CT93–1076 'Designing and Implementing Quality Control of Beneficial Insects: Towards More Reliable Biological Pest Control'. It does not necessarily reflect its views and in no way anticipates the commission's future policy in this area. F. Bigler and CAB International gave permission for reproduction of Fig. 19.1 and the corresponding text.

References

Bigler, F. (1994) Quality control in *Trichogramma* production. In: Wajnberg, E. and Hassan, S.A. (eds) *Biological Control with Egg Parasitoids.* CAB International, Wallingford, UK, pp. 93–111.

Calkins, C.O. and Ashley, T.R. (1989) The impact of poor quality of mass-reared Mediterranean fruit flies on the sterile insect technique used for eradication. *Journal of Applied Entomology* 108, 401–408.

Dutton, A. and Bigler, F. (1995) Flight activity assessment of the egg parasitoid *Trichogramma brassicae* (Hym.: Trichogrammatidae) in laboratory and field conditions. *Entomophaga* 40, 223–233.

Evans, O.G. (1953) The genus *Iphiseius* Berl. (Acarina – Laelaptidae). *Proceedings of the Zoological Society of London* 124, 517–526.

Greenberg, S.M. (1991) Evaluation techniques for *Trichogramma* quality. In: Bigler, F. (ed.) *Quality Control of Mass Reared Arthropods. Proceedings 5th Workshop IOBC Global Working Group, 25–29 March 1991.* Wageningen, The Netherlands. Swiss Fedeeral Research Station for Agronomy, Zurich, pp. 138–145.

Hassan, S.A. and Wen Qing Zhang (2001) Variability in quality of *Trichogramma brassicae* (Hymenoptera: Trichogrammatidae) from commercial suppliers in Germany. *Biological Control* 22, 115–121.

Luczynski, A. and Caddick, G. (2000) *Assessing Accuracy of Shipments of* P. persimilis *and* E. formosa *and Field Performance of* E. formosa. Technical Report 99-09, British Columbia Greenhouse Vegetable Research Council, 28 pp.

O'Neil, R.J., Giles, K.L., Obrycki, J.J., Mahr, D.L., Legaspi, J.C. and Katovich, K. (1998) Evaluation of the quality of four commercially available natural enemies. *Biological Control* 11, 1–8.

Posthuma-Doodeman, C.J.A.M., van Lenteren, J.C., Sebestyen, I. and Ilovai, Z. (1996) Short-range flight test for quality control of *Encarsia formosa. Proceedings Experimental and Applied Entomology,* Nederlandse Entomologische Vereniging 7, 153–158.

Prezotti, L. and Parra, J.R.P. (2002) Controle de qualidade em criacoes massais de parasitoides e predadores. In: Parra, J.R.P, Botelho, P.S.M., Correa-Ferreira, B.S. and Bento, J.M.S. (eds) *Controle Biologico no Brasil: Parasitoides e Predadores*. Manole, São Paulo, pp. 295–311.

Silva, I.M.M.S., van Meer, M.M.M., Roskam, M.M., Hoogenboom, A., Gort, G. and Stouthamer, R. (2000) Biological control potential of *Wolbachia*-infected versus uninfected wasps: laboratory and greenhouse evaluation of *Trichogramma cordubensis* and *T. deion* strains. *Biocontrol Science and Technology* 10, 223–238.

van Lenteren, J.C. (1998) *Sting Volume 18, Newsletter on Biological Control in Greenhouses. Special Issue on Quality Control.* Wageningen Agricultural University, Wageningen, 32 pp.

van Lenteren, J.C. and Tommasini, M.G. (1999) Mass production, storage, shipment and quality control of natural enemies. In: Albajes, R., Gullino, M.L., van Lenteren, J.C. and Elad, Y. (eds) *Integrated Pest and Disease Management in Greenhouse Crops.* Kluwer Academic Publishers, Dordrecht, The Netherlands, pp. 276–294.

van Schelt, J. and Ravensberg, W.J. (1990) Some aspects on the storage and application of *Trichogramma maidis* in corn. In: Wajnberg, E. and Vinson, S.B. (eds) *Proceedings of the 3rd International Symposium on* Trichogramma *and Other Egg Parasitoids, 23–27 September 1990, San Antonio, USA.* Les Colloques de l'INRA 56, Paris, pp. 239–242.

van Lenteren, J.C., Schettino, M., Isidoro, N., Romani, R. and van Schelt, J. (2002) Morphology of putative female sex pheromone glands and mating behaviour in *Aphidoletes aphidimiza. Entomologia Experimentalis et Applicata* 102, 199–209.

20 Basic Statistical Methods for Quality Control Workers

E. Wajnberg
INRA, 37, Blvd du Cap, 06600 Antibes, France

Abstract

The quality of mass-produced beneficial insects and mites needs to be checked regularly, but mass-rearing companies often do not have personnel with knowledge of the statistical techniques that have to be used to do this. Therefore, a number of simple and easy-to-perform statistical methods are described in this chapter. Quality control workers normally collect a lot of numerical data, and these data should then be handled using proper and accurate statistical methods, in order: (i) to describe the main features of the mass-produced animals; and (ii) to check whether these animals satisfy the predefined standards. This chapter provides the basic terminology and knowledge concerning the statistical methods for quality control of mass-produced beneficials. Working examples are presented throughout the chapter, to help in the understanding of the concepts of the use of statistical methods in this case. This chapter should be considered as a starting-point, leading the interested reader to more advanced statistical methods, which can be found in several textbooks that are mentioned at the end of the chapter.

Introduction

Entomologists and/or producers of beneficial insects and mites, who have to use quality control (QC) procedures, always have to handle a lot of numerical data collected during mass rearing. Once these data are collected, tools are needed: (i) to summarize the information available; and (ii) to check whether their mass-reared animals satisfy predefined criteria, such as size and fecundity. The statistical methods that are used can lead to wrong conclusions about the quality of the natural enemies if the basic terminology, assumptions and procedures are not known and accurately understood. The aim of this chapter is to provide the reader with a minimum of knowledge needed to summarize the data collected and to test whether the mass-produced organisms meet the preset quality criteria. In the first two sections, the necessary terminology is specified and a short description of the usual statistical distributions that QC workers have to handle is given. Normally, several traits are quantified simultaneously on mass-reared animals. Therefore, in the next sections, a thorough description is given on how to analyse each quantified trait separately, together with a presentation of methods to take two quantified traits simultaneously into account.

The aim of this chapter is not to provide the reader with a deep knowledge of statistical

© CAB International 2003. *Quality Control and Production of Biological Control Agents: Theory and Testing Procedures* (ed. J.C. van Lenteren)

theory and methods. However, with the help of the explanations given and the working examples dealing with QC presented throughout the chapter, the reader will acquire the necessary concepts and methods to solve most of the problems encountered with QC in mass-rearing and production of beneficial organisms. Thus, this chapter can be considered as a starting-point, leading the interested reader to understand more advanced methods that can be found in several textbooks mentioned at the end of the chapter.

Basic Terminology

Consider that a given parasitoid species is mass-produced and that the average fecundity of the females needs to be quantified. All mass-reared females form, in statistical terminology, the *population*. This population is the unit of interest and the one we want to describe. Of course, the fecundity of all the females constituting the population cannot be measured. Instead, a *random sample* has to be collected. In this case, the word 'random' is important. It means that the sample has to be a correct representation of the whole population. The animals belonging to the sample have to be collected randomly, and it would thus be a wrong method to only use the first emerging females, or the biggest, etc. After taking a correct random sample, the fecundity of each female of this sample is quantified. Each value obtained is called an *observation*, and the total number of observations is the *sample size*. Using all the observations, the sample can be described. For example, we can compute the average fecundity of the females (see below for a detailed description of the method to be used here). However, our goal is not to describe the sample, but the population from which the sample was taken. Fortunately, most of the parameters used to describe a sample can be used to compute, for each of them, what is called an *estimation* of the corresponding parameter describing the population. When the descriptive parameter is an average value, as for fecundity, then it can be shown that the best estimation of the average of the population is exactly the average value computed for the sample.

Usual Statistical Distributions

QC workers are collecting data that may have different statistical characteristics. Therefore, the nature of the data collected has to be clearly identified. QC workers usually work with two types of variables: regular quantitative traits (e.g. size, fecundity) or proportions (e.g. sex ratio, percentage emergence). Quantitative traits follow a normal (or Gaussian) distribution and proportions follow a binomial distribution. Other statistical distributions (i.e. Poisson, exponential, etc.) may also be encountered, but it is in most cases possible to transform the corresponding data into a normal distribution.

The normal distribution

As indicated by its name, this is the most commonly encountered distribution. It is used to describe the distribution of a continuous quantitative trait that is quantified for each individual of the sample, such as the size of an organism. This distribution is also regularly used for discrete variables (i.e. variables that can only be expressed as integer values, such as the number of eggs). The normal distribution is often expressed as the well-known 'bell-shape' curve, and its symmetry around the average value of the trait studied is its main feature. This distribution can be completely described with only two parameters: its average, and its variance. A method to compute these two descriptive parameters is given below. Data that are normally distributed are expressed using clearly identified units: longevity of adults (in days) or size of pupae (in mm).

The binomial distribution

Sometimes we have to describe the number of times that a particular event will occur among all the individuals of the sample. The event may or may not occur, and the measured trait is expressed as a percentage of the occurrence of the event, such as the insect being a female or male (sex ratio) or the insect having emerged or not (percentage emergence). In this case, the trait studied is estimated from measures done on several individuals simultaneously, and ranges from 0.0 to 1.0 (or from 0% to 100%). This second type of variable, which cannot be expressed in clearly identified units, is supposed to follow a so-called binomial distribution. This distribution is usually non-symmetric. This type of variable can also be described completely with only two parameters: the probability of occurrence of the event, and the number of individuals used to quantify it. A description of the method used to compute the two parameters is also given below.

Describing a Sample with a Regular Quantitative Trait

When only one quantitative trait is measured for each individual of a sample, several descriptive statistics have to be computed in order to summarize the main features of the population from which the sample has been drawn. In this case, the whole data set corresponds to all the observations of the sample and two types of parameters have to be computed: (i) position parameters (e.g. arithmetic mean, median) give the order of magnitude of the values; and (ii) dispersion parameters (e.g. range, variance, standard deviation, standard error) provide information regarding the dispersion of the values around a position parameter (usually around the arithmetic mean). The sample size is n, and the observations are $x_1, x_2, \ldots, x_n$.

Position parameters

- Arithmetic mean:

$$\bar{x} = \frac{x_1 + x_2 + \ldots + x_n}{n} = \frac{\sum_{i=1}^{n} x_i}{n}$$

- Median: this parameter is defined as the middle value when all observations are arranged from lowest to biggest. Thus, by definition, half of the observations are below the median, and half are above the median.

Dispersion parameters

- Range: maximal value – minimal value
- Variance:

$$\sigma^2 = \frac{(x_1 - \bar{x})^2 + (x_2 - \bar{x})^2 + \cdots + (x_n - \bar{x})^2}{n-1}$$

$$= \frac{\sum_{i=1}^{n} (x_i - \bar{x})^2}{n-1} = \frac{\sum_{i=1}^{n} x_i^2}{n-1} - \frac{n \times \bar{x}^2}{n-1}$$

- Standard deviation: $\text{SD} = \sqrt{\sigma^2}$
- Standard error: $\text{SE} = \frac{\text{SD}}{\sqrt{n}}$

When a sample is described by a quantitative trait, at least three values must be provided: (i) the sample size; (ii) a position parameter (e.g. the mean); and (iii) a dispersion parameter (e.g. the standard error). Finally, there is no way to describe accurately a sample without a graphical representation. Two possibilities are available here: either a histogram of the full distribution of all the observations; or a distribution summary, by plotting, for example, the interval [$\bar{x}$ − SE; $\bar{x}$ + SE] in a graph.

Working example: fecundity of *Trichogramma*

Thirty mated females of *Trichogramma brassicae* (i.e. an egg parasitoid used for biological control of the European cornborer in Europe), less than 24 h old, are isolated each in a single test-tube for 7 days with more than 200 *Ephestia kuehniella* eggs as hosts. Then females are removed and the number of black (i.e. parasitized) eggs is counted 3 days later. The recorded data were:

75	81	60	90	57	67	69	57	79	81	64	62	54	86	65
85	59	76	25	62	75	70	64	46	84	61	45	63	67	20

From these data, the following descriptive parameters can be computed: (i) n = 30; (ii) $\bar{x} = 64.97$ eggs; (iii) median = 64.5 eggs; (iv) range = 70 eggs; (v) variance = 263.48; (vi) SD = 16.23 eggs; and (vii) SE = 2.96 eggs. Figure 20.1 shows a histogram presenting the full distribution.

Describing a Sample with One Trait Expressed as a Percentage

When the trait is a percentage, several statistical parameters have to be quantified in order to summarize the main features of the population from which the sample was drawn. In this case, the percentage measured corresponds to the proportion of the observations of the sample showing a given characteristic (e.g. proportion of individuals that are females: sex ratio). Four descriptive statistics have to be computed:

- The total number of individuals for which the percentage was determined: N
- The number of individuals showing the characteristic of interest: x
- The percentage itself: $p = \frac{x}{N}$
- Its standard error: SE $= \sqrt{\frac{p \times (1-p)}{N}}$

In this case, both the percentage and its standard error have to be provided for an accurate description of the sample. Here also a graphic representation can be made and often a pie chart is enough (see below).

Working example: sex ratio of *Trichogramma*

In a population of 500 *T. brassicae*, 163 wasps were males and 337 were females. So the sex ratio (% females) was: 337/500 = 0.674 (i.e. 67.4% females), and its standard error was

$$\sqrt{\{0.674 \times (1-0.674)\} / 500} = 0.021$$

(i.e. 2.1% females). Figure 20.2 shows the pie chart corresponding to this data set.

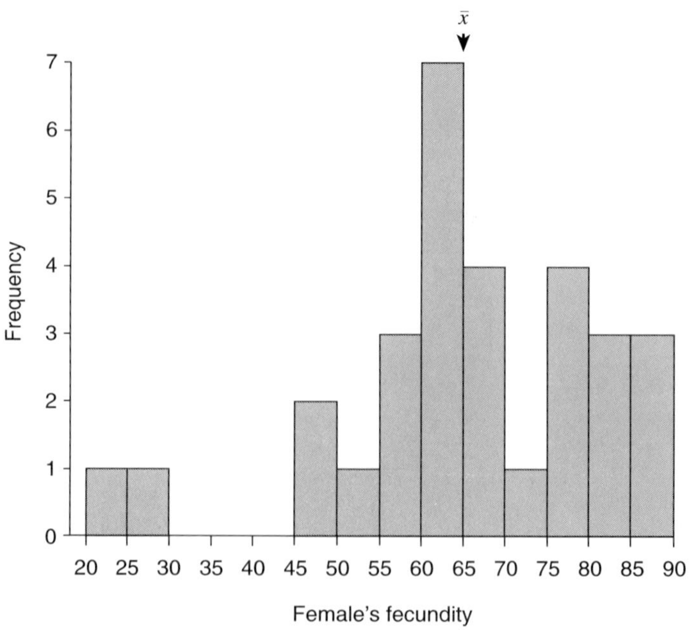

Fig. 20.1. Frequency distribution of the data collected to quantify the fecundity of mated *T. brassicae* females.

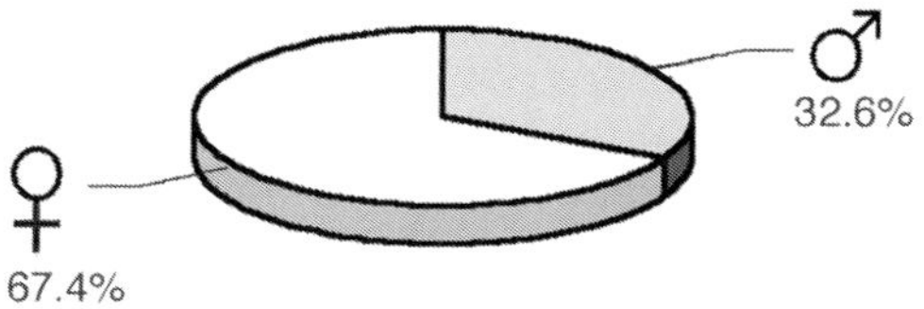

Fig. 20.2. A graphical representation of the estimated sex ratio in *T. brassicae*.

Building a Confidence Interval Around the Mean: a Way to Compare the Sample with a Predefined Standard

Once a sample is described with the parameters presented above (e.g. mean, standard error, etc.), one must check whether the corresponding mass-reared animals satisfy a predefined standard (i.e. a QC criterion). For this, a statistical procedure is available, based on the estimation of a confidence interval around the estimated value of interest. The procedure is the same both for the average value of a regular quantitative trait (e.g. fecundity) and for a trait expressed as a percentage (e.g. sex ratio). The confidence interval of the estimated parameter, at a 5% probability level, is computed using the following equation:

$$[\textit{parameter} - 1.96 \times \text{SE}; \textit{parameter} + 1.96 \times \text{SE}]$$

where *parameter* is the computed average or percentage of the sample, and SE is its standard error. So, for a regular quantitative trait (e.g. fecundity), the confidence interval of the average value, at a 5% probability level, is:

$$\left[\bar{x} - 1.96 \times \frac{\text{SD}}{\sqrt{n}}; \bar{x} + 1.96 \times \frac{\text{SD}}{\sqrt{n}}\right]$$

and, for a percentage (e.g. sex ratio):

$$\left[p - 1.96 \times \sqrt{\frac{p \times (1-p)}{N}}; p + 1.96 \times \sqrt{\frac{p \times (1-p)}{N}}\right]$$

Then, if the predefined standard is not within this confidence interval, the hypothesis that the estimated parameter and the guideline are equal is rejected with a risk of 5% – that is, with a 5% chance that the hypothesis rejected is actually true. Saying that the confidence interval is estimated at a 5% probability level means that, if 100 equivalent samples had been used to estimate 100 confidence intervals, the true (unknown) average value or percentage representing the population would occur in only 95 of the confidence intervals computed. One can, of course, make more reliable decisions at much lower probability levels, e.g. 1%. In this case, the value '1.96' should be replaced by a larger value (e.g. a value of '2.58' should be used for a decision at a probability level of 1%).

A very important point to make here is that the procedure explained above is valid only if the sample size is at least 30. This point is often violated in QC and leads to wrong conclusions, both positive and negative, about the quality of natural enemies.

Working example for a regular quantitative trait: does the fecundity of *Trichogramma* meet the quality control criterion?

The fecundity of 30 *T. brassicae* females has been quantified and an average value of 64.97 eggs per female was obtained, with an SE of 2.96 eggs per female (see the working example above). The predefined fecundity criterion is a minimum of 40 eggs per female. The confidence interval at a probability level of 5% is [64.97 – 1.96 × 2.96; 64.97 + 1.96 × 2.96] resulting in [59.17; 70.77]. The standard of 40 eggs per female is not within this confidence interval. So we conclude, with a risk of 5%, that the average fecundity of the sample differs significantly from the standard fecundity. In the population tested, the fecundity is higher. For the mass rearing, there should of course be no problem in this case.

Working example for a percentage: does the sex ratio of *Trichogramma* meet the quality control criterion?

The sex ratio (% females) estimated from a sample of 500 adults of *T. brassicae* was 0.460 (with an SE of 0.022). According to the QC guidelines, a standard of 0.50 (i.e. 50% females) is required. The estimated confidence interval at a 5% probability level is [0.460 – 1.96 × 0.022; 0.460 + 1.96 × 0.022], which results in [0.417; 0.503] or, in terms of percentages, [41.7%; 50.3%]. The calculated confidence interval of the sample includes the standard value of 0.50. So the conclusion is that the estimated sex ratio is not significantly different from the standard value, although it is somewhat lower.

Studying the Correlation Between Two Quantitative Traits

Usually, more than a single quantitative trait is quantified with data from one sample, and there are more accurate statistical methods available than the ones described above to handle several traits simultaneously. In this chapter, only methods for simultaneously handling two quantitative traits will be presented. When the two quantified traits are regular quantitative values measured simultaneously on each individual of the sample, it is possible to check if there is a significant relationship between them. For this, the covariance of the population has to be estimated first from all the observations of the sample. This is done using the following equation:

$$\text{covariance} = \frac{\sum_{i=1}^{n}(x_i - \bar{x}) \times (y_i - \bar{y})}{n-1}$$

$$= \frac{\sum_{i=1}^{n}(x_i \times y_i)}{n-1} - \frac{n \times \bar{x} \times \bar{y}}{n-1}$$

Using the obtained value, the linear correlation between the two quantified traits is then:

$$r = \frac{\text{covariance}}{\sigma_x \times \sigma_y}$$

where σ_x and σ_y are the standard deviations of the variables x and y, respectively.

A bivariate plot is used for a graphical representation, with one trait on the x axis, and the other on the y axis. Each point corresponds to an observation in the sample. Sometimes, a bivariate confidence interval can be added to the graph (i.e. ellipse), but special statistical computer programs are needed for this (see example below). As can be seen in Fig. 20.3, the correlation coefficient ranges from –1 to +1. A value of –1 indicates that there is a strictly negative linear relationship between the two traits. A value of +1 indicates a strictly positive linear relationship. A value of zero shows the absence of any relationship. Intermediate values correspond to intermediate situations.

It is important to determine whether the correlation is significant. This can be done by calculating t with the following formula:

$$t = \sqrt{\frac{r^2 \times (n-2)}{1-r^2}}$$

where n is the sample size. If the computed value is larger than 1.96, there is a significant relationship between the two traits studied, at a probability level of 5%. Again, this test is valid only if the sample size is at least 30. For QC problems, such a statistical procedure can be very useful. Indeed, the existence of a significant relationship between two quantitative traits can lead to a decision to quantify only one of them, thus saving both time and money.

Working example: are longevity and fecundity correlated in Trichogramma?

Thirty mated *T. brassicae* females are isolated in test tubes with more than 200 *E. kuehniella* eggs as hosts, for 7 days. Then both the number of parasitized (i.e. black) eggs (i.e. fecundity, expressed in eggs per female) and the number of days females remained alive (i.e. longevity, expressed in days) are quantified. The recorded data for the 30 females were as in Table 20.1.

From these data, the following descriptive statistics can be computed: (i) $n = 30$; (ii) $\bar{x} = 19.3$ days and $\bar{y} = 99.93$ eggs; (iii) $\sigma_x = 4.921$ days and $\sigma_y = 17.836$ eggs; and (iv) covariance = 51.469. The correlation coefficient is thus: 51.469/(4.921 × 17.836) = 0.586, and the corresponding t value is:

$$\sqrt{0.586^2 \times (30-2) / (1-0.586^2)} = 3.83,$$

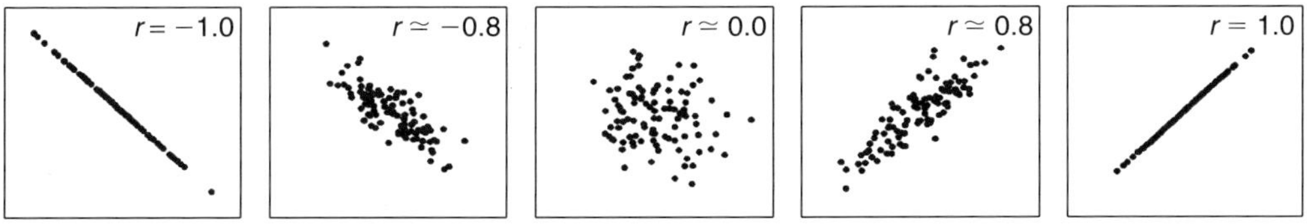

Fig. 20.3. Examples of relationships between two quantitative traits with different values of the linear correlation coefficient.

Table 20.1. Data for longevity and fecundity in *Trichogramma brassicae*.

Female number	Longevity	Fecundity	Female number	Longevity	Fecundity	Female number	Longevity	Fecundity
1	16	103	11	20	97	21	20	113
2	24	104	12	19	106	22	21	67
3	23	115	13	26	140	23	23	100
4	18	95	14	14	98	24	23	103
5	17	64	15	15	106	25	21	93
6	15	95	16	22	126	26	15	112
7	9	85	17	20	109	27	13	71
8	33	140	18	20	106	28	13	88
9	24	92	19	23	106	29	24	93
10	13	73	20	17	98	30	18	100

which is greater than 1.96. So there is a significant positive correlation between females' fecundity and longevity, with a probability level of 5%. Figure 20.4 shows the graph describing this significant relationship.

Studying the Correlation Between Two Traits Expressed as Percentages

When the two quantified traits are percentages computed simultaneously with data from the same sample (e.g. sex ratio and adult mortality), the significance of a relationship between them can also be calculated. For this, a first step consists of building a so-called contingency table. If the two traits studied are sex ratio and rate of adult mortality, then such a table would look like that in Fig. 20.5.

From this table, the t value can be computed as follows:

$$t = \sqrt{\frac{(a+b+c+d)\times(ad-bc)^2}{(a+b)\times(c+d)\times(a+c)\times(b+d)}}$$

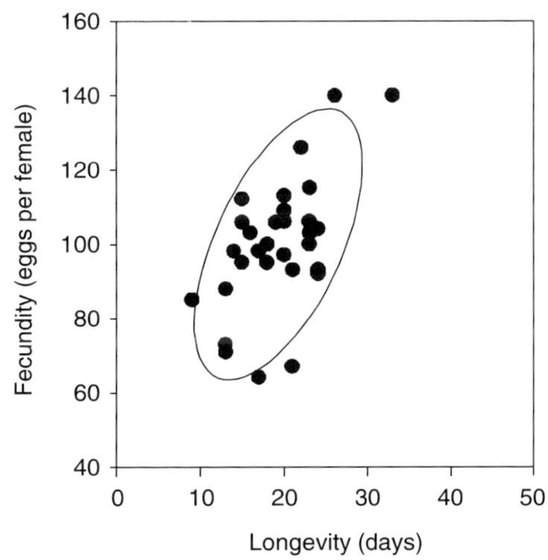

Fig. 20.4. Relationship between the longevity and the fecundity of *T. brassicae* females. The ellipse represents a 5%-risk confidence interval of the bivariate average values.

If the obtained t value is greater that 1.96, there is a significant relationship between the two traits at a 5% probability level. For the example here, this would mean a significant difference in adult mortality between males and females.

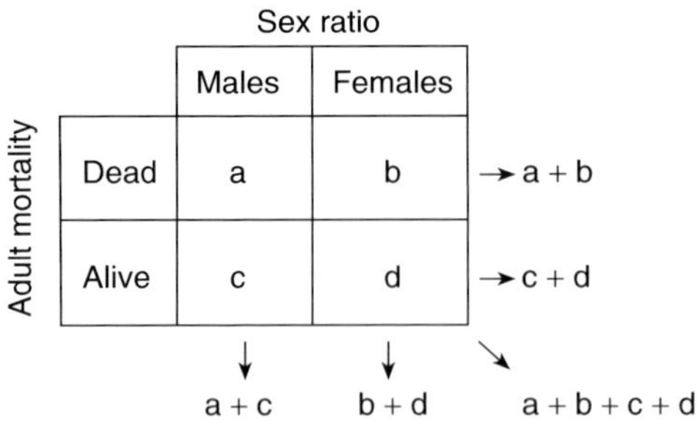

Fig. 20.5. Contingency table for relationship between sex ratio and adult mortality.

Working example

A short-distance flight test was conducted for *Encarsia formosa* (J.C. van Lenteren and co-workers, unpublished data). The flight ability was compared between parasitoids emerging from old or new pupae, and the data obtained were as in Table 20.2.

So the t value is:

$$t = \sqrt{\frac{145 \times (30 \times 23 - 53 \times 39)^2}{83 \times 62 \times 69 \times 76}} = 3.192$$

which is greater than 1.96. Thus, the flight ability of *E. formosa* significantly differs between parasitoids emerging from old and new pupae.

Studying the Relationship Between a Regular Quantitative Trait and a Percentage

Sometimes, both a quantitative trait and a percentage are estimated simultaneously with data from one sample. In this situation, since the quantitative trait is measured on all individuals separately but these individuals are used all together to estimate the percentage, a real correlation cannot be estimated between the two traits. What can be done instead is to compare the two average values of the quantitative trait computed after splitting the sample according to the binomial trait expressed as a percentage. For example, suppose that both the sex of adults and their longevity are quantified. The comparison can be done by comparing the males' and females' average longevity. In order to do this, the whole sample must first be split according to the two categories from which the binomial trait is studied. Then the two subsamples obtained are described for the quantitative trait measured with standard descriptive parameters (i.e. n, $\bar{x}$, σ_x) and a graph can be produced showing the distribution summaries of the two subsamples. Finally, the t value is computed:

$$t = \frac{|\bar{x}_1 - \bar{x}_2|}{\sqrt{\frac{\sigma_1^2}{n_1} + \frac{\sigma_2^2}{n_2}}}$$

where n_1, $\bar{x}_1$, σ_1^2 are the descriptive parameters of the first subsample, and n_2, $\bar{x}_2$, σ_2^2 are those of the second subsample. If the obtained value is greater than 1.96, there is a significant difference between the average values of the quantitative trait between the two subsamples, at a probability level of 5%. This test is valid only if the sizes of the two subsamples (n_1 and n_2) are both greater than or equal to 30.

Table 20.2. Results of short-distance flight test.

	Flight	No flight	Total
Old pupae	30	39	69
New pupae	53	23	76
Total	83	62	145

Working example: is there a difference in adult longevity between males and females in *Trichogramma*?

Both the sex ratio and adult longevity were quantified for *T. brassicae*. Thirty females and 30 males were measured. For all of them, the number of days that adults remained alive was quantified. The data obtained were:

Longevity of females in days:

16	15	9	27	19	15	20	19	12	7	17	16	16	24	26
17	16	15	10	2	10	14	15	9	19	18	21	23	17	11

Longevity of males in days:

13	11	15	7	9	15	9	11	14	7	9	12	8	16	16
17	10	6	8	6	16	10	2	13	11	6	0	9	3	5

From these data, the following descriptive statistics can be computed: (i) for the females, $n_1 = 30$, $\bar{x}_1 = 15.83$ days, $\sigma_1^2 = 31.18$; (ii) for the males, $n_2 = 30$, $\bar{x}_2 = 9.80$ days, $\sigma_2^2 = 19.41$; and the test is thus:

$$t = |15.83 - 9.80| / \sqrt{31.18/30 + 19.41/30} = 4.64,$$

which is greater than 1.96. So there is a significant difference between male and female adults longevity, at a probability level of 5%. Figure 20.6 shows the graph summarising this difference.

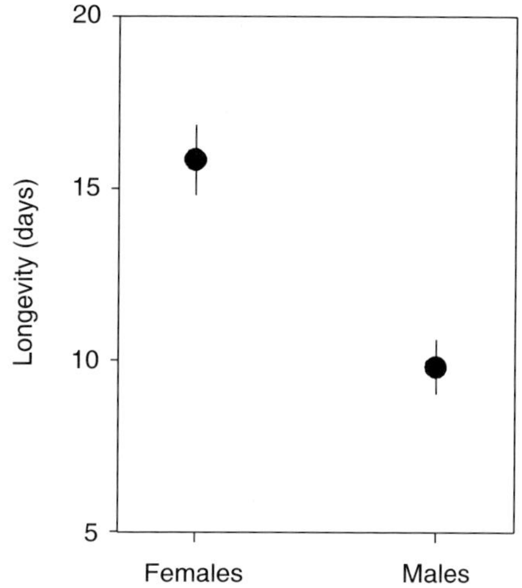

Fig. 20.6. Average (± SE) longevity of females and males of *T. brassicae*.

Conclusions

QC workers usually collect a lot of quantitative and qualitative data. These data have to be synthesized properly to assess the quality of beneficial organisms as biocontrol agents. For this, statistical tools provide substantial help, both in summarizing the main features of the samples collected and in checking whether the mass-reared animals are satisfying predefined criteria. The statistical procedures discussed in this chapter are those which only consider samples described by one or two variables. Further information about these tests can be found in statistical handbooks. There are a number of software packages that include the above-described tests, such as: (i) SAS® (SAS Institute, Inc. (1990); http://www.sas.com/products/sassystem/index.html); (ii) STATISTICA® (http://www.statsoftinc.com/); (iii) SPSS® (http://www.spss.com/); and (iv) SYSTAT® (http://www.spssscience. com/systat/). Of course, studying the quality of a mass-produced animal cannot be carried out by describing only two traits. So, in most of the cases, more than two traits are measured simultaneously. For example, in order to assess the quality of the aphid parasitoid *Aphelinus abdominalis*, at least five biological traits have to be quantified: (i) emergence rate; (ii) sex ratio; (iii) adult size; (iv) female fecundity; and (v) adult mortality. In order to handle these kinds of data sets, the procedures described here should be generalized using multivariate (also called multidimensional) methods. More accurately, multidimensional correlations can be computed and tested in order to see whether there is some redundancy between the different traits studied. Besides this, multidimensional confidence intervals can also be constructed in order to check whether the averages of the different traits quantified satisfy the predefined multidimensional standard. However, the corresponding methods, which are just a generalization of those presented here, cannot be used without some specific statistical computer software package running on microcomputers or on mainframes, and these are not described here. Finally, the only way to prove that a mass-produced biocontrol agent is of a good quality is to check, after it has been released in the field, whether its pest-control efficiency is sufficient. Therefore, there is also a need for statistical methods to determine the correlation between such a field efficiency and the simple traits that are quantified in the laboratory. As the corresponding methods are also based on multivariate approaches, special statistical computer programs are needed, but they are not presented here. A description of multivariate approaches can be found in Morrison (1990), Harris (2001), Tabachnick and Fidell (2001) and Johnson and Wichern (2002); Tabachnick and Fidell (2001) can be recommended because of its practical approach.

References

Harris, R.J. (2001) *A Primer of Multivariate Statistics*, 3rd edn. Lawrence Erlbaum, Mahwah, New Jersey, 609 pp.

Johnson, R.A. and Wichern, D.W. (2002) *Applied Multivariate Statistical Analysis*, 5th edn. Prentice Hall, Englewood Cliffs, New Jersey, 834 pp.

Morrison, D.F. (1990) *Multivariate Statistical Methods*, 3rd edn. McGraw Hill College, New York, 222 pp.

SAS Institute, Inc. (1990) *SAS/STAT User's Guide*, Release 6.07, Vol. 1. SAS Institute, Cary, North Carolina.

Tabachnick, B. and Fidell, L.S. (2001) *Using Multivariate Statistics*, 4th edn. Allyn and Bacon, Boston, Massachusetts, 966 pp.

Index

Browse Read and Buy

www.cabi.org/bookshop

ANIMAL & VETERINARY SCIENCES
BIODIVERSITY CROP PROTECTION
HUMAN HEALTH NATURAL RESOURCES
ENVIRONMENT PLANT SCIENCES
SOCIAL SCIENCES

CABI *Publishing*
A division of CAB International

Online BOOK SHOP

Subjects
Search
Reading Room
Bargains
New Titles
Forthcoming

Order & Pay Online!

★ FULL DESCRIPTION ★ BUY THIS BOOK ★ BOOK OF THE MONTH

Tel: +44 (0)1491 832111 Fax: +44 (0)1491 829292